Ecologica
Physiology of **Insects**

Ecological and Environmental Physiology of **Insects**

Jon F. Harrison
School of Life Sciences, Arizona State University
H. Arthur Woods
Division of Biological Sciences, University of Montana
Stephen P. Roberts
Department of Biology, Central Michigan University

OXFORD
UNIVERSITY PRESS

Great Clarendon Street, Oxford OX2 6DP

Oxford University Press is a department of the University of Oxford.
It furthers the University's objective of excellence in research, scholarship,
and education by publishing worldwide in

Oxford New York

Auckland Cape Town Dar es Salaam Hong Kong Karachi
Kuala Lumpur Madrid Melbourne Mexico City Nairobi
New Delhi Shanghai Taipei Toronto

With offices in
Argentina Austria Brazil Chile Czech Republic France Greece
Guatemala Hungary Italy Japan Poland Portugal Singapore
South Korea Switzerland Thailand Turkey Ukraine Vietnam

Oxford is a registered trade mark of Oxford University Press
in the UK and in certain other countries

Published in the United States
by Oxford University Press Inc., New York

First published 2012

British Library Cataloguing in Publication Data
Data available

Library of Congress Cataloging in Publication Data
Data available

Typeset in Adobe Garamond Pro by Cenveo Publisher Services
Printed and bound by
CPI Group (UK) Ltd, Croydon, CR0 4YY

ISBN 978–0–19–922594–1 (Hbk)
ISBN 978–0–19–922595–8 (Pbk)

10 9 8 7 6 5 4 3 2 1

We dedicate this book first to our wives, Julie, Creagh, and Jennifer, who nurtured our families and paid the bills while we got to bury our heads in the sands of insect environmental physiology.
Second, we dedicate this to our scientific "mothers" and "fathers" who collected the data and developed the conceptual models we've stolen and amended: V.B. Wigglesworth, Torkel Weis-Fogh, Liz Bernays, Mary Chamberlin, George Bartholomew, Peter Miller, Reg Chapman, Neil Hadley, John Phillips, Todd Gleeson, Joel Kingsolver, Ray Huey, John Lighton, Tim Bradley and Stefan Hetz come to mind, but there are many others.

Ecological and Environmental Physiology Series (EEPS)
Series Editor: Warren Burggren, University of North Texas

This authoritative series of concise, affordable volumes provides an integrated overview of the ecological and environmental physiology of key taxa including birds, mammals, reptiles, amphibians, insects, crustaceans, mollusks, and fish. Each volume provides a state-of-the-art review and synthesis of topics that are relevant to how that specific group of organisms have evolved and coped with the environmental characteristics of their habitats. The series is intended for students, researchers, consultants, and other professionals in the fields of physiology, physiological ecology, ecology, and evolutionary biology.

A Series Advisory Board assists in the commissioning of titles and authors, development of volumes, and promotion of the published works. This Board comprises more than 50 internationally recognized experts in ecological and environmental physiology, providing a combination of both depth and breadth to proposal evaluation and series oversight.

The reader is encouraged to visit the EEPS website for additional information and the latest volumes (http://www.eeps-oxford.com/). If you have ideas for new titles in this series or just wish to comment on EEPS, please do not hesitate to contact the Series Editor, Warren Burggren (University of North Texas; burggren@unt.edu).

Volume 1: Ecological and Environmental Physiology of Amphibians
Stanley S. Hillman, Philip C. Withers, Robert C. Drewes, Stanley D. Hillyard

Volume 2: Ecological and Environmental Physiology of Birds
J. Eduardo P.W. Bicudo, William A. Buttemer, Mark A. Chappell, James T. Pearson, Claus Bech

Volume 3: Ecological and Environmental Physiology of Insects
Jon F. Harrison, H. Arthur Woods, Stephen P. Roberts

Contents

Acknowledgments

We thank and acknowledge the current community of working insect physiologists and comparative and evolutionary physiologists, whose hard work is pushing the boundaries of our field so fast that any text can only be a blurry snapshot. Warren Burggren deserves a special thank you for envisioning and directing this series, which acknowledges that evolutionary history makes the environmental physiology of each taxon in some way unique. We thank our students and colleagues who made comments on portions of this manuscript, including Arianne Cease, Thomas Förster, Melanie Frazier, Sue Nicolson, Brent Sinclair, Peter Piermarini, James Waters, John VandenBrooks, Jaco Klok, Sydella Blatch, and Keaton Wilson. Melanie Frazier and Lee McCoy graciously hosted and fed us through a week of furious writing at their lovely home on Beaver Creek, and provided many helpful scientific comments and discussions. Their dog Pierce graciously allowed us into the living room and permitted us to pet him. The Whiteley Center at the Friday Harbor Laboratories was a wonderful and peaceful location for two weeks of intensive writing. Milcho Penchev, Elana Niren, Adam House, Stephanie Heinrich, and Brian Foster did the yeoman's work of helping with the reference library and preparing figures. The IOS Division of the National Science Foundation has funded much of the research described here, including most of the work from our own labs. We thank our institutions for being supportive of research and scholarly activity, especially a sabbatical from ASU to JFH.

1
Introduction

1.1 Overview of this Book

Ecological and environmental physiology is the study of how physiological systems respond, over different timescales, to variation in physical and biological environments. This introduction describes the ecological importance of insects, some of the advantages of insects as experimental models, a brief introduction to insect phylogeny and evolution, and the themes of the book. Chapter 2 is designed for the physiological ecologist without a background in entomology. It introduces some basic anatomy and physiology of insects by system, and provides references for detailed information on these topics. The heart of the book (Chapters 3–6) examines responses of insects to the classic environmental parameters that vary across the planet and that determine insect distributions. Chapter 3 examines the pervasive effects of temperature on insect physiology, ecology, and evolution, covering heat balance, thermoregulation, and mechanisms of thermotolerance. Chapter 4 examines the role of water, covering organismal balancing and mechanisms for coping with deficiency or excess. Chapter 5 examines the role of nutrient availability and its effects on growth, body size, and reproduction. Chapter 6 examines the role of oxygen, focusing on the structure and function of the respiratory system and its plasticity in response to changing supply and demand. In choosing these variables we excluded others of interest, including light, pollutants, and the biotic environment (social interactions, parasites, and disease), but we could not include a careful coverage of these in a space-constrained book. The concluding chapters of the book consider insect-specific methodological approaches in environmental physiology (Chapter 7) and propose a set of conclusions and future directions (Chapter 8).

There are multiple ways to write such a book. One approach is the comprehensive summary. However, the field is simply too large, and the coverage too uneven, to make such an approach feasible. For a few species (e.g. *Drosophila melanogaster*) there is a wealth of genetic and cellular studies, and some on environmental physiology. Many other fundamental questions have been addressed only in a few scattered species. We have therefore taken a different approach that, we hope, will both summarize important ideas and highlight gaps.

In each chapter, topics are organized around a set of fundamental questions. Each major chapter begins by 'defining the problem'. In this section, we cover what is known about pathologies that ensue when responses to environmental change are inadequate, and provide theoretical background. We hope to use this section to introduce the functional and theoretical relevance of the physiological responses described later in each chapter.

Three themes run through each chapter. The first is that different physiological responses occur over different timescales. These responses include passive buffering, short- and medium-term active responses, developmental plasticity, cross-generational plasticity (i.e. epigenetic inheritance), multigenerational evolution (within population selection), and deep evolutionary responses. The second theme is that active responses require control systems. Here we address what is known about sensors, integrators, and effectors for each type of response. The third theme is trade-offs, because all organisms have limited resources (e.g. materials, energy, space, time) the enhanced capacity for one function may reduce the capacity for others. Indeed, trade-offs have also constrained our writing of this book: we have sought to identify provocative patterns and interesting directions for future research at the expense of providing exhaustive summaries.

Such an approach will inevitably run into the problem of data scarcity in some areas and overload in others. Our solution is to point out holes in the literature and to give short shrift to many other areas. We ask our un-cited colleagues for their forgiveness.

> 'It is a capital mistake to theorize before one has data. Insensibly one begins to twist facts to suit theories, instead of theories to suit facts.'
> Sherlock Holmes in *A Scandal in Bohemia* by Arthur Conan Doyle.

We attempt to follow Sherlock Holmes's advice, but—reader beware!

1.2 The Ecological Importance of Insects

Insects often are the most important heterotrophs in terrestrial ecosystems. During outbreaks of the gypsy moth, *Lymantria dispar* (L.) (Lepidoptera: Lymantriidae), nearly 100% of available host leaves can be consumed (Kosola et al., 2001). A recently observed locust swarm in Morocco was 230 km long, at least 150 m wide, and contained an estimated 69 *billion* locusts (Ullman, 2006). While insect outbreaks occur infrequently, they are common in many ecosystems and can have long-lasting effects on communities. During such outbreaks, insects are thought to function as keystone species that reduce abundance of other dominant species, which can subsequently increase biodiversity (Carson et al., 2004). Insects also have a major role in secondary production due to their high abundance and assimilation efficiencies relative to vertebrates (Price, 1997). In a study of energy flow through old temperate fields, insects (primarily Orthoptera and Hymenoptera) accounted for 80% of total energy assimilation by heterotrophs (Wiegert and Evans, 1967). In agriculture, insecticide use is reported to increase plant production by 30% or more, suggesting that insect herbivory exerts tremendous loads (Price, 1997). These pesticides can have major effects on humans and non-target wildlife (Matson et al., 1997), indicating that we need to better understand and control insects.

In other ecological situations, insects typically have more modest effects on nutrient transfer through ecosystems. A variety of studies have suggested that the standing crop of insects is relatively low compared to plant and vertebrate biomass, and that, in most ecosystems (particularly forests), insects consume less than 10% of net primary productivity (Weisser and Siemann, 2004). However, these studies have emphasized that even when abundances are low, insects do have strong effects on plant investment in secondary metabolites and structures, and on rates of mineralization and availability of phosphorus and nitrogen (Weisser and Siemann, 2004). In grasslands, insect herbivory has been shown to increase nutrient cycling and plant productivity (Belovsky and Slade, 2000). These studies together suggest that the most important effects of insects on ecosystems are indirect; for example, selective herbivory and pollination can strongly affect plant species' composition (Weisser and Siemann, 2004). Pollination is a particularly critical contribution of insects. In addition to being essential for most wild angiosperms, insects pollinate 84% of the world's food crops, representing one third of global food production, with an estimated annual replacement value of $68–433M for the USA alone (Allsopp et al., 2008).

1.3 Insects as Models for the Study of Ecological Responses

The jagged exoskeletons, protruding horns, large compound eyes, lacey wings, feathery antennae, sharp stings, and painful bites of insects have always fascinated humans. Their extremes in size, shape, and lifestyle have even served as models for our concept of alien. In one sense, these divergent life forms provide an opportunity to understand how the physical and chemical aspects of the environment shape all life; for example similar biomechanical principles can explain wing-to-body ratios in birds and insects (Dudley, 1992). Despite the obvious differences, many profound similarities exist between humans and insects, such as the virtually identical organization of major biochemical pathways (glycolysis, Krebs cycle, transcription, and translation). We (humans and insects) shared a common ancestor around 600 million years ago, more than half of insect and vertebrate genes are orthologous, and these have a mean sequence identity of about 46% (Consortium, 2006). This fundamental similarity is one reason why it is so difficult to find insect-specific pesticides. Some of the most significant questions about how life works have first been asked of insects, with the answers later found to apply to all eukaryotes.

All organisms respond to environmental variation, and the goal of environmental physiology is to determine the proximate mechanisms and functional significance of such responses, as well as the mechanisms that allow different species to occupy different niches. Proximate questions ask about mechanisms: how do organisms sense change? What regulatory or compensatory responses occur? What role does the neuroendocrine system play in the response? Are specific genes activated? How effective is compensation? Ultimate questions ask *why* organisms show such responses, and the answers often represent amalgams of current selection and historical constraint. Do responses enhance fitness, or do they indicate stress-induced pathologies? Do organisms evolve different types of physiologies in response to spatial scale (local vs. global climate change)? What historical pathways have made some insects sensitive to change and others hardy? What fitness trade-offs are associated with different physiological phenotypes?

Insects are uniquely suited for the study of physiological responses to environmental variation because one often can ask both proximate and ultimate questions. Although the small size of insects can make physiological measurements challenging (e.g. sampling blood repeatedly from individuals), sensor miniaturization is overcoming these limitations, and most standard physiological tools can be applied to at least some insects. The genomes of more than 20 insects have been completely sequenced and many more (more than100) are in the pipeline. The genes of no other animal are currently manipulated as easily as those of *Drosophila melanogaster*. Ease and cost-effectiveness of rearing many individuals in small spaces makes insects uniquely suited for powerful experimental designs. Insects can often be easily observed in large numbers in the field and then collected and reared in the laboratory. The ethics of using insects for physiological research are less complicated than for vertebrates, and therefore many animal care and use committees actively promote insects as model organisms.

Many insect species are also exceptionally suited to answering ultimate questions. The short lifespan of insects allows students to examine development or evolution during a graduate thesis, an option not available for many vertebrate studies. The wide diversity and number of species allows comparative studies with much greater scope than is possible in most vertebrate groups (though the lack of phylogenies for many insect groups remains a challenge). Many insects can be reared and mated in the lab, allowing long-term natural and artificial selection experiments. Such laboratory selection studies can only be conducted in a handful of small vertebrate species. The ability to address ultimate questions with repeatable laboratory experiments provides researchers with exceptionally powerful tools for understanding evolutionary responses of insects.

1.4 Insect Evolutionary History and Phylogenetic Diversity

Insects evolved at least 400 million years ago, probably in the early Devonian or late Silurian, and therefore are among the earliest land animals (Grimaldi and Engel, 2005; Glenner et al., 2006). Recent molecular evidence suggests that hexapods are most closely related to branchiopods, a freshwater crustacean group that includes water fleas and fairy shrimp (Glenner et al., 2006; Mallatt and Giribet, 2006). This finding suggests that the common ancestor of branchiopods and hexapods evolved in fresh water (Fig. 1.1), and that crustaceans, rather than myriapods, (centipedes, millipedes) are the most closely related major taxa to insects (Glenner et al., 2006; Bradley et al., 2009). Thus insects can be considered terrestrial crustaceans.

Most modern insect orders and some modern families had evolved by the Cretaceous (ca. 100 million years ago, Fig. 1.2). This long evolutionary history has important implications for insect environmental physiology. In particular, major groups of insects are only distantly related. For example, polyneuropterans such as grasshoppers and cockroaches have not shared a common ancestor with Panorpida (including flies and moths) since the Carboniferous, more than 300 million years ago. This deep evolutionary history means that many physiological characteristics of insects diverge strongly along taxonomic lines.

Consideration of insect evolution can help us understand which features are derived in insects. The most primitive insects are the Archeognatha (bristletails) and the Zygentoma (silverfish) (Fig. 1.3). These groups possess basic insect characteristics (segmented, six-legged adults, external mouthparts, antennae without muscles beyond the initial segment, a chordotonal organ in the base of the antennae, compound eyes, ocelli, tracheal system, Malpighian tubules). However, they

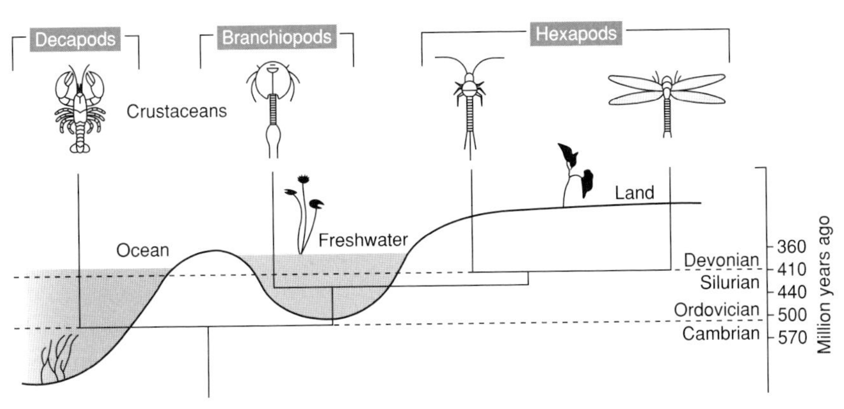

Fig. 1.1 Evolution of hexapods. From Glenner et al. (2006), used with permission from AAAS.

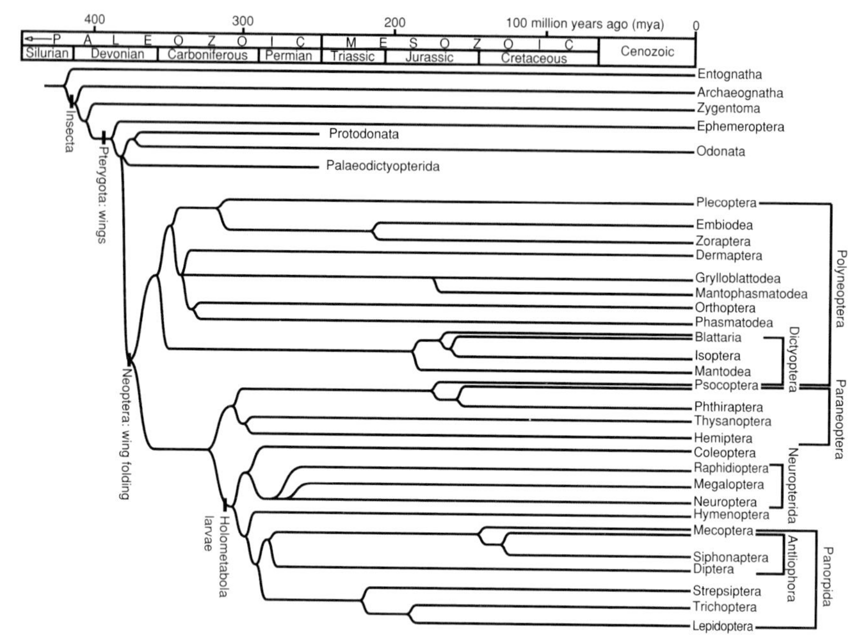

Fig. 1.2. Insect phylogeny in relation to geologic time. From Grimalidi and Engels (2005), with permission from Cambridge University Press.

also exhibit several key primitive characteristics, including lack of wings, a single mandibular articulation, and projections on the abdomen that may be remnants of ancestral legs (Grimaldi and Engel, 2005). Like other relatively primitive insects, they are hemimetabolous, without a pupal stage marking a profound change between juvenile and adult stages.

Phylogenetic and fossil data provide the best support for the hypothesis that insects evolved first on land, with aquatic forms evolving later, and independently, in many groups. One class of evidence for this hypothesis is that all of the known non-insect hexapods (Entognatha, which have internal mouthparts and include springtails, proturans, and diplurans) are terrestrial. This suggests that terrestriality evolved in ancestors to the Entognatha and insects (the Ectognatha), after fresh-water derived hexapods colonized land in the late Devonian (Glenner et al., 2006). In addition, the earliest true insects (Archaeognatha: bristletails and Zygentoma: silverfish) are completely terrestrial. The oldest insect fossils, from the Devonian, Carboniferous, and Permian, are all terrestrial, with no clear aquatic insect fossils known before the late Triassic (Grimaldi and Engel, 2005). More support for this hypothesis is that the air-filled tracheal system of insects would have been unlikely to evolve in aquatic organisms. Nearly all insects have

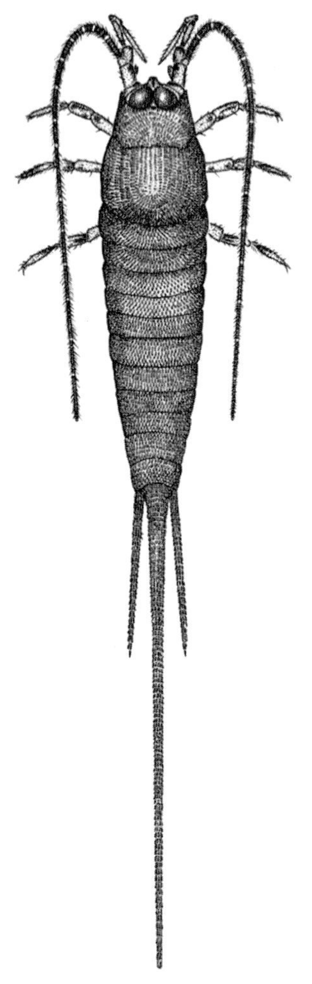

Fig. 1.3 A bristletail, *Nesomachilis* sp. [Archaeognatha: Meinertellidae], illustration by Geoff Thompson, reprinted with permission from the Queensland Museum, 1985.

a terrestrial adult stage, and those that don't—certain beetles and hemipterans—are clearly derived from terrestrial ancestors. Finally, insects are virtually excluded from marine environments, suggesting that the insect bauplan and physiology are poorly suited for deep waters. Most likely, this is because possession of a tracheal respiratory system precludes deep diving due to excessive buoyancy (Maddrell, 1998) or hydrostatic pressure effects on tracheal system function.

While the major orders of insects are at least 100 million years old, extant species are much younger, with most estimates ranging from 2 to 10 million years (Grimaldi and Engel, 2005). Extensive studies from the Pleistocene have demonstrated that beetles dramatically changed their distributions as glaciers advanced and receded, allowing them to maintain similar climate and plant niches, but at different places and times (Coope, 1979). These data strongly support the hypothesis that characteristics of species, such as preferred temperature, can be relatively invariant over millions of years, as long as dispersal allows habitat-tracking.

The most remarkable aspect of insects is their diversity. About one million species of insects have been described, which represents at least half of documented global species diversity (Gullan and Cranston, 2005). High insect diversity likely relates to their environmental physiology. First, small size allows insects to occupy small ecological niches compared to vertebrates, and to better mitigate macroenvironmental conditions by choosing microhabitats. Flight provides tremendous advantages relative to terrestrial animals in foraging and dispersal. High reproductive output provides the grist for natural selection, and the capacity to rapidly produce new, successful biotypes. The tremendous flexibility of insect morphology and physiology, especially regarding capacities to feed and reproduce on different plants, certainly plays a key role in the great biodiversity of this group.

1.5 Themes of this Book

1.5.1 Timescale of Environmental Responses

During environmental challenges, organismal responses will depend strongly on the duration of exposure. For the sake of simplicity, physiological responses can be divided into six functionally distinct stages, based on temporal duration (Fig. 1.4).

These six temporal classes clearly overlap (e.g. the distinction between macromolecular remodeling and developmental plasticity relies on sufficient time for altering a developmental pathway, a continuous rather than binary response), but these definitions provide a useful paradigm for understanding the duration of exposure on organismal response.

1.5.1.1 Passive Buffering

Changes in environment produce changes in internal state. When sunlight falls on a shaded insect, temperature rises to a new steady-state. Consumption of a base-generating food quickly raises the blood and tissue pH. The result is a passive, physical and chemical change in the state of the organism from A_0 to A_1 that stem from no active response by the animal.

However, state changes for given environmental changes are not the same for all organisms, as they depend on many different organismal characteristics. We divide them arbitrarily into structural properties and biochemical equilibria.

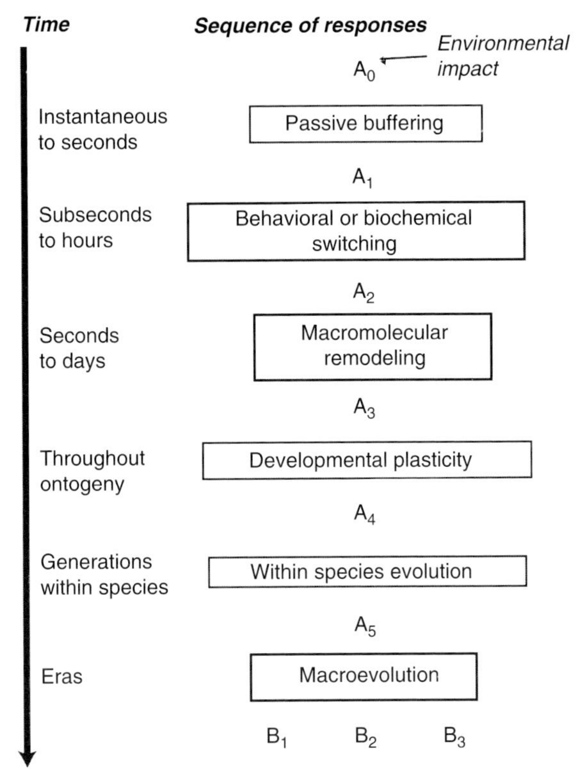

Fig. 1.4 Effect of duration of exposure on type of response to the environment.

Structural characteristics are important components of passive buffering. For example, the size of an insect determines how quickly it will heat or cool and how quickly it will dehydrate. For an insect suddenly exposed to sunlight, its exposed surface area, color, and reflectance determines its new steady-state temperature. For a phytophagous insect consuming a base-generating leaf, a smaller pH change will occur when hemolymph and tissue space are larger.

Biochemical equilibria also contribute to passive physical and chemical buffering. According to Le Chatelier's principle, in an equilibrium reaction:

$$A + B \leftrightarrow C + D \tag{1}$$

Increasing A leads to increased formation of C and D. A classic example of such an effect is buffering of acid.

$$H^+ + A^- \leftrightarrow HA \tag{2}$$

Protons (H^+) can bind to anions (A^-) such as proteins or inorganic phosphate to produce protonated forms, reducing the concentration of H^+. Higher concentrations of buffering compounds (A^-) lead to a reduced change in pH for a given acid or base load. Similar chemical equilibria can affect the change in state for many environmental impacts such as changes in nutrient or ion intake.

1.5.1.2 Behavioral or Biochemical Switching

The first active responses to environmental variation are usually neural or biochemically-mediated modification of the function of existing proteins or structures. Neural activation of locomotion is one example: in response to excessive solar heating a grasshopper walks into the shade. The activity of many enzymes, channels, and transporters can be altered by membrane electrical activity, changes in phosphorylation state, or intracellular calcium levels. For example, glycogen phosphorylase is activated by de-phosphorylation, which promotes glycolytic flux and ATP generation, a common response to environmental perturbation. Another mechanism for altering activity of proteins is by changing their location. For example, activation of ion and fluid secretion in Malpighian tubules (e.g. in response to fluid-loading) is accompanied by transfer of channels and transporters from intracellular vesicles to the apical membrane (Bradley and Satir, 1979).

Changes in protein activity are sometimes intrinsic responses of individual cells to environmental perturbation but can also be mediated by the neuroendocrine system. Modulation of protein activity is generally accomplished by water-soluble hormones/neurotransmitters with membrane-bound receptors, such as octopamine or leukokinin. Hormone-receptor binding is followed by intracellular signaling cascades of molecules such as G proteins, cyclic AMP, protein kinases, calcium, diacylglycerol, and inositol phosphates. These responses can be very rapid, occurring within fractions of seconds, and can be sustained for long periods.

1.5.1.3 Macromolecular Remodeling

The next kind of response is active alteration of the concentration or identity of functional and structural molecules (proteins, lipids, and carbohydrates), a process we term "macromolecular remodeling." Protein concentration can be increased either by changing gene transcription, post-transcriptional regulation (alternative splicing), altering translation, or by changes in protein breakdown. The control of gene transcription and post-translational modification has been extensively studied in *D. melanogaster*, and environmental factors can regulate the mapping of gene to phenotype through many mechanisms; some of which will be covered in subsequent chapters. For example, induction of heat shock proteins (Hsp) in response to high temperatures occurs via binding of transcription factors to the *hsp* promoter. Heat shock transcription factors can be converted from an inactive monomer to active trimer in response to decreased free levels of Hsp, which disappear as they bind to denatured proteins (Fujimoto and Nakai, 2010). Similarly, low cellular levels of oxygen can directly lead to reduced degradation of a transcription factor called hypoxia inducible factor (HIF) via reduced hydroxylation of one of its subunits. This stabilizes the protein and allows it to bind to response elements of many genes (Gorr et al., 2006).

Again, changes in macromolecular composition are sometimes intrinsic to individual cells but can also be mediated by neuroendocrine systems. For example, in insects, juvenile hormone and ecdysone are known to cause wide-spread changes in gene transcription (Riddiford, 2008; Loof, 2008). Such changes can be very rapid but generally occur over minutes or longer. Such active responses to the environment have also been termed acclimation (Huey and Berrigan, 1996), a term sometimes restricted to genomic (Wilson and Franklin, 2002) or reversible (Angilletta, 2009) responses.

1.5.1.4 Developmental Plasticity

The same genotypes exposed to different environments over ontogeny can result in very different phenotypes, a phenomenon known as developmental plasticity. Developmental plasticity represents long-term changes in macromolecular composition resulting from environmental effects on developmental processes. These changes in developmental trajectories may produce alternative phenotypes and are sometimes irreversible. Some changes resulting from developmental plasticity are major, such as presence or absence of wings, wing color and shape, diapause, number of generations per season, lifespan, age at reproduction, number of ovarioles, and flight muscle size (West-Eberhard, 2003). The genetic mechanisms of developmental plasticity are under intensive study. Some dramatic examples of polyphenism, such as the striking differences in physiology and morphology between honey bee queens and workers, are at least partially caused by nutritional regulation of the genome structure via differential methylation (Chittka and Chittka, 2010).

1.5.1.5 Multi-generational Responses

Beneficial, genetically-based responses of populations to environmental change (evolution by natural selection) are well-documented in insects. It has been estimated that a fertile 1-mg female *D. melanogaster* could, if all her progeny reproduced and lived forever, accumulate in her descendants the entire mass of the Earth within 25 generations! Of course, high insect mortality saves the planet and ensures that over long time periods, fruit fly biomass is approximately constant. High mortality rate combined with high reproductive output gives strong responses to natural selection.

The cleanest examples of within-species evolution come from laboratory studies. Laboratory studies provide the means to separate developmental plasticity from evolutionary responses; one can rear the products of selection in common environments, preferably for multiple generations to remove parental effects. Selection, full-sib and half-sib designs with controlled matings allows measurement of the heritability and additive genetic variance associated with a particular phenotype (Falconer, 1989). *Drosophila* have been shown to evolve substantially larger body sizes, longer lifespans, different upper thermal knock-down temperatures, and greater desiccation resistance in less than 10 generations (Gibbs, 1999; Partridge and Fowler, 1993; Rose, 1984; Gilchrist and Huey, 1999). These studies demonstrate that substantial heritable variation exists, and that with proper encouragement from the environment, major changes in important parameters can occur rapidly.

The role of within-species natural selection in explaining variation in the field can be investigated by comparing populations or families of the same species collected across an environmental gradient. In order to separate developmental plasticity and parental effects from true changes in allele frequencies (evolution), individuals must be reared in common environments. Such experiments have shown, for example, that high latitude populations of multiple *Drosophila* species have evolved larger bodies than low latitude conspecifics (Gilchrist et al., 2004).

Non-DNA-mediated, cross-generational responses to the environment may also occur at the individual level; these are often termed parental effects. Broadly defined, such non-genetic inheritance effects include parenting behavior (e.g. learning, changes in juvenile habitat), investment in the eggs, and inherited epigenetic factors (Bonduriansky and Day, 2009). The parents can alter the offspring's phenotype by transferring nutrients, RNA, hormones, or parasites through the gametes. Nutrition of the mother often affects egg size and nutrient content, with important effects on the development rate and size of offspring (Mousseau and Dingle, 1991; Holbrook and Schal, 2004). In locusts, crowded conditions cause the female oviduct to secrete an unknown substance that increases the likelihood that offspring will develop into migratory forms (Simpson and Miller, 2007). In some moths, transfer of nuptial gifts at mating endows the offspring with nutrients or defensive compounds (Dussourd et al., 1988; Smedley and Eisner, 1996). Epigenetic inheritance includes transmission from parent to offspring via sequence-independent chromosomal changes (e.g. cytosine methylation), RNA (e.g. selective silencing of offspring genes via RNAi in the eggs), and transmission of molecules that can serve as structural templates (e.g. prions and membranes, (Jablonka and Gaz, 2009). Most demonstrations of such mechanisms in insects have focused on structural features such as eye color; for example selection for eye color can result in rapid phenotypic change without associated genetic change (Sollars et al., 2002). Exposure to high temperatures has been shown to have heritable effects on histone structure and DNA methylation (Bantignies and Cavalli, 2006).

1.5.1.6 Macroevolution

Over the long term, evolution within-species drives divergence among species. Species evolve adaptations to the environment independently, and inter-specific comparisons provide a powerful tool for understanding adaptation to different environments. The last two decades have been accompanied by an explosion of new comparative tools that allow us to control for the degree of relatedness of the species examined (Garland et al., 2005). For example, such approaches have revealed that desert *Drosophila* are more resistant to desiccation than mesic *Drosophila* species, and that the differences primarily relate to differences in rates of water loss (Gibbs and Matzkin, 2001).

1.5.2 Control Systems

Like all organisms, insects use energy to convert environmental materials into integrated physiological machines that preserve and transfer information over time. Life has achieved this success by evolving an incredibly complex machine (in *Drosophila melanogaster*, over 20,000 genes and perhaps 100,000 proteins, (Oliver and Leblanc, 2004; Culliton, 2001). These proteins can be arranged into different cell types within an organism and expressed differentially in response to environmental variation. In addition, the structure and function of many proteins can be altered rapidly by changes in phosphorylation, ion-binding, or charge state. In the face of this complexity, it is difficult and dangerous to make simple physiological models. However, the essence of science is to derive simplified models that capture the essentials of pattern and process. For physiology, we believe that a good starting point for understanding how any "within-individual" physiological system responds to the environment is to use a simplified control-theory model.

As noted above, the environment directly and passively alters physiological systems. For example, grasshoppers exposed to solar radiation become hotter. Rising body temperature will have direct effects on many aspects of the insect's physiology; for example, metabolic rate will rise because the enzymes underlying it go faster. A first step in understanding environmental responses is therefore to understand the direct, passive effects of the environmental variable. In many cases this is non-trivial, and for many environmental responses even this first step has not been taken.

State changes in response to environmental variation produce active organismal responses (behavioral or biochemical modification of existing proteins, changes in concentration or identity of proteins, Fig. 1.4). Changes in state, including rises in body temperature and the subsequent rises in metabolic rate, can stimulate active responses by the insect. Active, physiological responses require *sensors*, which detect changes in state, and *integrators*, which receive

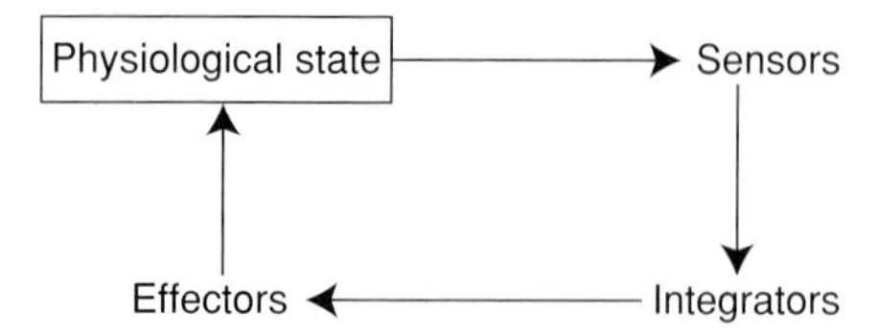

Fig. 1.5 A simple control system.

information from the sensors that they then compare to a set-point. These integrators in turn control *effectors* that influence physiological state (Fig. 1.5). For the solar-heated grasshopper, temperature-sensitive neurons change their firing rate, sending afferent information to ganglia in the CNS that function as comparators. *Integrators* compare the current "state" to an internal set-point, and produce outputs, via effectors, to respond to this change in state. In a classic negative feedback response, integrators and effectors produce some action that counters the direct effect of the environment, returning the state toward the integrator's set-point. In this example, the effector is the locomotory system, and the net effect of walking into the shade is a return of body temperature toward some set-point (behavioral thermoregulation).

In this case, and for many cases in insect physiology, we know the organismal-level response, but the specifics of individual control steps are poorly understood, and often the disparate components of control systems have been studied in disparate species, which hampers our ability to develop integrated understanding. For example, a variety of insects have been shown to possess multipolar antennal thermosensors and cuticular and central thermoreceptors (Chapter 3, (Schmitz and Wasserthal, 1993)); but these have not been studied in grasshoppers. Similarly, basic properties of the insect thermoregulatory center (analogous to the vertebrate hypothalamus) are just now being described, but so far only for *Drosophila* (Hamada et al., 2008). In contrast, centers that drive walking locomotion in Orthoptera have been located in the metathoracic ganglia and are well-studied (Burrows, 1996). Throughout the text, we cobble together these disparate studies to outline possible mechanisms for control of insect responses to the environment. However, given the evolutionary diversity of insects, readers should accept such proposed networks with a healthy dose of caution.

Modern control theory, both in engineering and physiology, involves more than simple negative feedback. Feed-forward control occurs when the system is designed to produce changes that anticipate future needs. A prime example is the observation that neuronal centers driving flight-muscle contractions are connected via interneurons to neuronal centers driving ventilation (Ramirez and Pearson, 1989). Many insect behaviors and physiological changes are triggered by external cues (e.g. change in day length) or internal clocks, and these can also provide feed-forward regulation that enhances insect responses to the environment (e.g. shifts in the timing of reproduction, (Bradshaw and Holzapfel, 2006), induction of diapause (Denlinger, 2002)). Of course, most environmental responses are mediated via multiple sensory and regulatory pathways, resulting in complex integration. In a few model systems, insect physiologists have begun to explore such complex regulation systems, but in most cases we can only keep our minds open to their probable existence.

1.5.3 Trade-offs in Physiological Systems

The final theme of the book, that we attempt to weave into each chapter, is the concept of trade-offs. All organisms are integrated systems constrained by limited resources (space, nutrients, energy, time). Sometimes trade-offs can be identified, and these cases provide some of the most interesting general insights into life histories and physiological design (Stearns, 1992; Zera and Harshman, 2001). For example, winged morphs of *Prokelisia* planthoppers invest more in flight muscle mass, and can disperse farther, at the expense of feeding and reproducing relative to brachypterous (wingless) forms (Huberty and Denno, 2006; Zera and Zhao, 2006).

Perhaps just as interesting is the observation that sometimes expected trade-offs do not occur. For example, although a variety of studies support a trade-off between fecundity and lifespan, recent studies in which reproduction was stopped by ablating the germ line found no extension of lifespan (Barnes et al., 2006). Also, in many cases, there is little correlation between critical thermal maxima and minima across species, suggesting that resistance to these stresses do not require trade-offs (Chown et al., 2004).

2
Basic Insect Functional Anatomy and Physiological Principles

This chapter is for the reader without a background in entomology or insect physiology, for example, a student in behavior or ecology, or perhaps a vertebrate environmental physiologist with developing interests in insects. It contains an abbreviated description of the lifecycles, anatomy, and physiology of insects. Omitted among these descriptions are basic sensors and the endocrine and reproductive systems. We do so because sensors and hormonal control are covered extensively within the individual chapters, and because we focus relatively little on the ecological and environmental physiology of insect reproduction. More detailed information on systems we cover (and omit) in this chapter is presented in the texts by Chapman (Chapman, 1998), Nation (Nation, 2008), and Klowden (Klowden, 2009).

2.1 Insect Lifecycles

Insect lifecycles are complex: they consist of distinct stages that unfold in sequence. The number of stages is three ("hemimetabolous") or four ("holometabolous"). Three-stage insects, which are more basal, have egg, juvenile, and adult stages. The four-stage insects—the more derived groups, of flies, bees and wasps, beetles, and moths and butterflies—have egg, juvenile (larva), pupa, and adult. Anatomical and physiological specializations of the different life stages are critical for allowing insects to cope with environmental variation and to achieve rapid growth (juvenile stages) and strong dispersal (adult stages). A key developmental innovation that allows seamless transitions between these functional stages is the pupa, during which the insect completely overhauls its physiology. These complex anatomical changes entail constraints—that is, genetic and morphological linkages between stages that may constrain how they evolve.

Although a few insects bear live young, most lay eggs. Eggs cannot move, and many of their physiological systems are incompetent, at least initially, making them vulnerable to biotic and abiotic threats. However, like embryos of other animals, insect eggs have traits that reduce their vulnerability (Hamdoun and

Epel, 2007). Most are enclosed by a tough covering, called the chorion, that protects them from mechanical damage, and by lipid coatings that protect them from desiccation (Hinton, 1981). Insect eggs are provisioned with nutrients (yolk) that buffer them against varying energetic needs, and which are critical for eggs that enter diapause. In addition, insect eggs have a variety of plastic responses to cope with environmental variation. In embryos, segment boundaries are regulated precisely despite variation in temperature that should alter production of morphogens (Houchmandzadeh et al., 2002). However, studies of the heat shock response in *Drosophila* eggs shows that early in development they do not induce heat shock proteins and are very heat-sensitive (Welte et al., 1993). Eggs also have plastic mechanisms for controlling the traffic of gases across the eggshell in response to changing metabolic demands (Woods et al., 2005; Zrubek and Woods, 2006).

Insects grow through a series of juvenile stages (instars) separated by molts during which the exoskeleton is shed. The number of molts in the growth period varies widely, from three in many flies to ten or more in some beetles, and probably is flexible in most species. Wingless insects, such as bristletails, grow without substantial external changes other than development of reproductive structures in the adults. In insects such as cockroaches, grasshoppers, dragonflies, and bugs, which have wings as adults, the wings develop gradually on the external cuticle during later instars. In other winged insects (holometabolous), the younger stages (called larvae) look very different from adults and usually occupy different ecological niches. In this group, the wings develop internally in the pupal phase. Juvenile stages must feed, assimilate, and grow, and thus responses of juvenile digestive and growth systems to environmental variation are very important. Prominent features of adult insects include reproduction and the ability to disperse, so environmental effects on these traits are likely to be of fundamental importance.

Some insects have even more complicated lifecycles. Eusocial insects, like honeybees, have queens that produce both sterile workers and fertile, daughter queens. Aphids often have parthenogenic stages. Parasitic insects can have very complex lifecycles, with endoparasitic and free-living stages, and dramatically different morphologies and lifecycles for different genders (Matthews et al., 2009; Kathirithamby, 2009). The diversity in insect life histories illustrates the astounding plasticity and evolutionary malleability of this animal group.

2.2 The Cuticle

The cuticle covers the outer surfaces of insects and also the luminal faces of tracheal tubes, foreguts, and hindguts. Cuticle is multifunctional (Vincent and Wegst, 2004): it is the primary barrier between insects and many of their external threats, such as diseases, predators, parasites, and desiccation; it plays a structural role as the exoskeleton, which requires it to assume, even in the same insect, multiple different stiffnesses and elasticities; and it provides a platform for the coloration, sensors, and bristles that insects present to the world.

Despite this functional diversity, the basic structure of cuticle is strongly conserved—spatially within individual insects, from one developmental stage to the next, and across insect species. Most cuticle is organized into three layers (Fig. 2.1): an envelope, an epicuticle, and a procuticle (Locke, 2001). Each of the three layers forms from different components in a stereotyped sequence (Fig. 2.2).

First, the envelope is laid down at the surface of the plasma membrane, usually above plaques that occur at the tips of microvilli. The envelope consists of proteins, lipids, and waxes (Schwarz and Moussian, 2007), but the details of its composition are poorly known. Next, the underlying cells lay down the epicuticle, which assembles on the inner face of the envelope from precursors secreted from the cells. The epicuticle consists of proteins but no chitin (Schwarz and Moussian, 2007), and the proteins are probably cross-linked by quinones, which confer

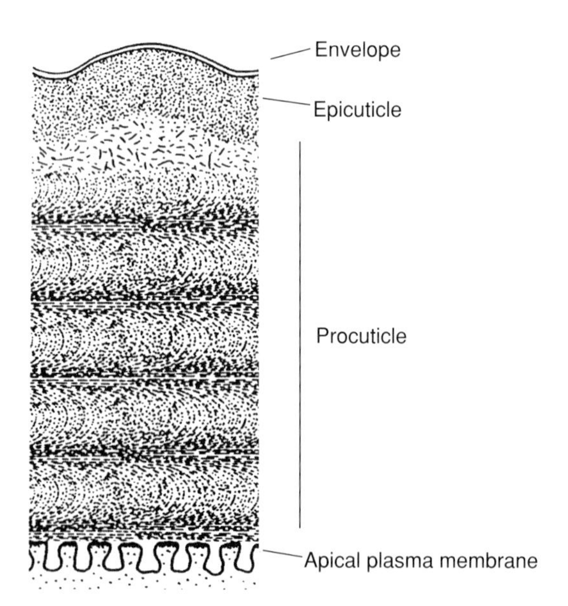

Fig. 2.1 Insect cuticle is made from three layers differing in their origin, structure, and function. The envelope is formed at the cell surface. It is membrane-like, and serves as the primary permeability barrier. It is a surface for lipid layers and template for patterns. The epicuticle assembles on the inner face of the envelope. It is structurally isotropic, except for the channels which traverse it. The procuticle assembles at the cell surface and forms the largest proportion of the cuticle. It is a fibrous composite stabilized in various ways to function as the exoskeleton and as a nutritional reserve of carbohydrate and protein. From Locke (2001), with permission from Elsevier.

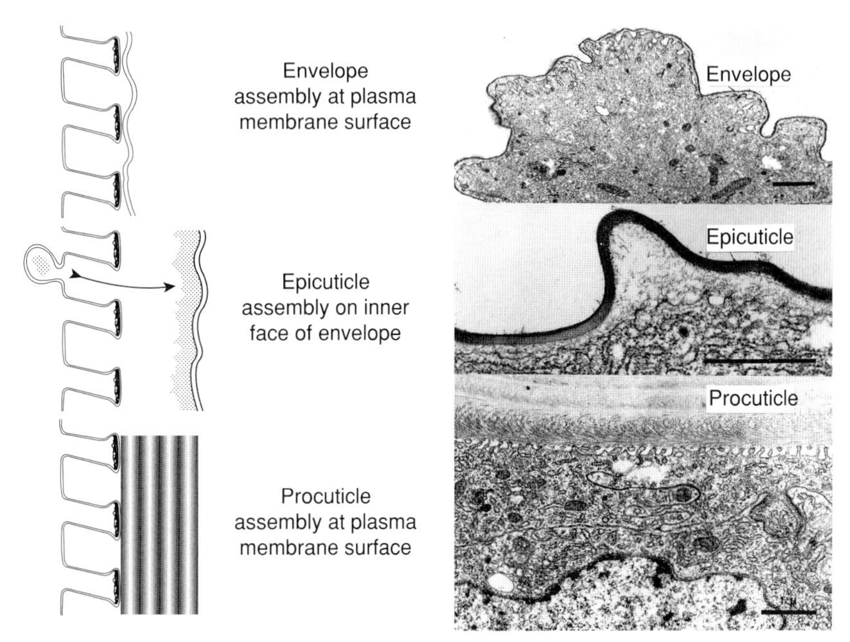

Fig. 2.2 (Left) The three layers in insect cuticle are each laid down by a different process. The envelope is laid down at the plasma membrane surface. The epicuticle is assembled on the inner face of the envelope, at first just above the cell surface, later away from the cell surface by the reassembly of components exocytosed into the cuticle space. The procuticle, usually hundreds of time wider than the other two layers, is assembled at the cell surface. It is typically formed from laminae, single layers of fibers with parallel orientation to one another within each lamina but changing in orientation by a constant amount from lamina to lamina to give the familiar helicoidal pattern. (Right) Electron micrographs of the three phases of cuticle deposition in molting larvae of *Calpodes*. From Locke (2001), used with permission from Elsevier.

stiffness (Locke, 2001). Like the envelope, the composition of the epicuticle is poorly known. Finally, the cells form procuticle, which usually is hundreds of times thicker than the other layers. The procuticle consists of stacked laminae of chitin fibers. In each lamina, the fibers are oriented parallel to one another, but succeeding lamina have fibers offset by some angle, which gives the typical helicoidal orientation visible in cross-sections of cuticle (Fig. 2.3). The procuticle can be further subdivided into an outer exocuticle and an inner endocuticle. The exocuticle usually is more electron-dense owing to its higher degree of sclerotization.

Below, we cover just two aspects of cuticle—the sequence of events during molting and metamorphosis by which insects form new cuticle and shed their old ones; and the processes contributing to the different properties of cuticle.

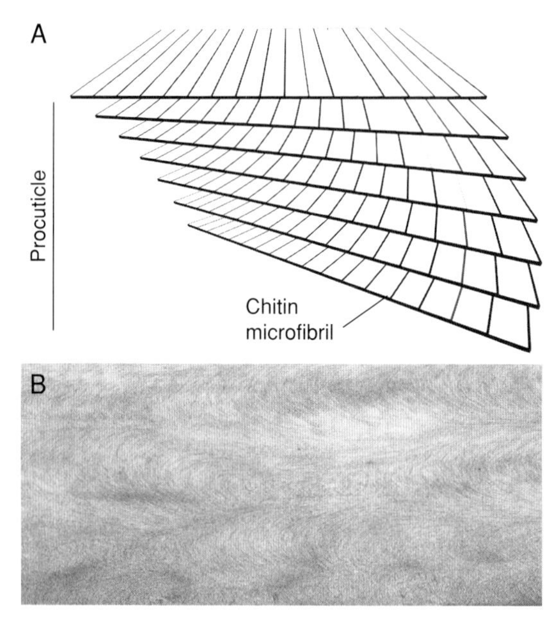

Fig. 2.3 Schematic for the organization of chitin in procuticle, as described by Bouligand (1965). (A) Chitin microfibrils are arranged into flat sheets called laminae. The orientation of microfibrils in successive sheets is offset by 7–8°. (B) In oblique section, this arrangement of fibrils gives the typical illusion of arches running through the procuticle. From Moussian (2010), used with permission from Elsevier.

2.2.1 Molting

Having a cuticular exoskeleton entails a series of functional challenges and opportunities that are unfamiliar to the average endoskeletal reader. For one, unlike bones, exoskeletons do not grow once they are laid down. Therefore, juvenile insects face the problem of outgrowing their current exoskeletons, which forces them to stop feeding (and growing) and to go through a quiescent period (the molt) during which they form a new, larger exoskeleton. They then shed the old exoskeleton (ecdysis) and, with their new exoskeleton, start feeding and growing again. At the beginning of the new juvenile stage (instar), the new exoskeleton may be loose and pliable, so that the growing tissues gradually expand into it.

The sequence of events that occurs during molting, and the cellular and molecular events that underlie them, are increasingly well known (Fig. 2.4). The overall sequence is driven by titers of edysteriods, the insect steroid hormones, circulating in the hemolymph (Charles, 2010). In turn, the ecydsteroid pattern is driven by the prothoracic glands, which secrete the ecdysone or 3-dehydroecdysone. These are converted by target tissues into a much more active form of the hormone, 20-hydroxy-ecdysone.

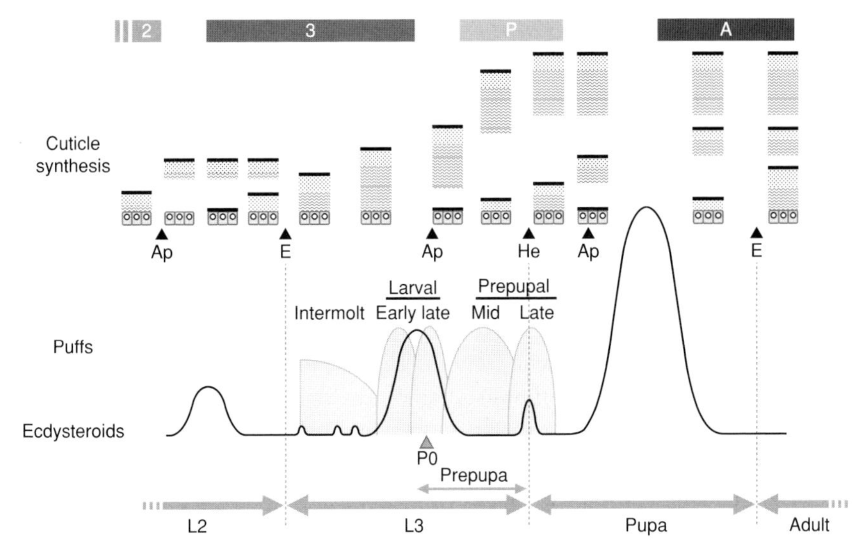

Fig. 2.4 Schematic of cuticle production and shedding across stages in *Drosophila melanogaster.* The horizontal bars (at top) represent periods of cuticle synthesis (2 is L2 larva, 3 L3 larva, P pupa, A adult), and horizontal arrows at bottom represent developmental stages. The sequence of cuticle synthesis is shown in the upper middle section. Epidermal cells are shown in gray with small white "nuclei" inside them. Secretion of the cuticle envelope is not shown. Pictured first, the epidermal cells secrete the epicuticle (black bars), then the exocuticle (wavy lines) and, after ecdysis (E), the endocuticle (dotted lines). The timing of major events is shown by black arrows: apolysis (Ap), ecdysis (E), and head eversion (He). Below this panel is another one showing concentrations of ecdysteroids over time (black line) and associated puffs in salivary glands (gray). From Charles (2010), used with permission from Elsevier.

During molting cycles, titers of edysteroids rise and fall in stereotyped patterns, and different patterns of genetic and cellular activity are associated with different phases of these peaks. We follow the description given by Charles (Charles, 2010). When ecdysteroids begin to rise, the old cuticle detaches from the underlying epidermis, an event known as apolysis. Epidermal cells then secrete molting gel into the space; the gel contains inactive forms of enzymes that will later be activated to digest the old cuticle. Once ecdysteroids are near peak levels, the epidermal cells secrete the envelope and epicuticle (Locke, 2001). Subsequently, once the ecdysteriod levels begin to fall, the epidermal cells secrete the bulk of the cuticle—the procuticle—and they continue to do so for some time after ecdysteroids reach basal levels. The two layers of procuticle (exocuticle and endocuticle), correspond to procuticular layers produced before and after ecdysis, respectively.

2.2.2 Functional Properties of Cuticle

Insect cuticle also must fulfill multiple biomechanical roles (reviewed by (Vincent and Wegst, 2004)). Obviously cuticle provides mechanical support to the body. In particular, the inner sides of the exoskeleton have bumps and ridges to which internal muscles attach. Besides providing support, cuticle also sustains bending and twisting forces, and can store elastic energy and transmit muscle movements in important ways. For example, flying sphinx moths have, in their thoraces, a set of orthogonal flight muscles. On one axis the flight muscles run vertically and connect the top and bottom of the thorax, and on the other they run horizontally front to back. To flap their wings, moths contract the two sets in an offsetting pattern, which alternately compresses the thorax dorso-ventrally and longitudinally. These deformations in turn drive the wings up and down (summarized by (Chapman, 1998)). Other examples of exoskeletal movements that depend strongly on the biomechanical properties of cuticle include fleas jumping (Sutton and Burrows, 2011) and explosive mandible closing in trap-jaw ants (Patek et al., 2006). However, it can be just as important for insects to have flexible sections of cuticle—in joints, around hinges, and between body sections (Vincent and Wegst, 2004). For example, cuticle is much thinner and more flexible between abdominal segments.

The properties of cuticle stem from its molecular architecture and are influenced by a number of factors, including composition and degree of cross-linking (Andersen, 2010). The main proposed mechanism for cuticle hardening invokes sclerotization—a tightly controlled process in which cuticular proteins are extensively cross-linked. In the first steps of sclerotization, tyrosine is converted to dopa and then dopamine. Subsequently, dopamine is transformed into two acyldopamines, NADA and NBAD. When sclerotization starts, NADA and NBAD are transported into the cuticular matrix, where the enzyme laccase drives their oxidation into reactive *ortho*-quinones which, via other enzymes, can be isomerized into *para*-quinonemethides. The quinones then react with one or two nucleophilic amino acids to form mono- or di-adducts, respectively. In general, NADA is more likely to form di-adducts (cross-links) than is NBAD; therefore, the properties of any section of cuticle depend on the relative levels of the two precursors and on the relative levels of laccase and the isomerizing enzymes (Andersen, 2010).

Besides sclerotization, another process that can influence cuticle hardness is dehydration; dehydration has even been proposed as an alternative to sclerotization (Vincent and Hillerton, 1979; Hillerton and Vincent, 1979). During dehydration, protein chains come into close contact and can establish strong non-covalent interactions (sclerotization involves covalent bonds). The appeal of the dehydration hypothesis is that insect cuticle is known to lose water and, moreover, direct evidence for cuticular cross-linking has been difficult to obtain.

Regarding this controversy, Andersen (Andersen, 2010) writes "None of the arguments are sufficient to prove that the changes in cuticular mechanical properties occurring during sclerotization are due either to dehydration or to crosslink formation, and it is likely that dehydration and crosslinking take place concomitantly, both contributing to the changing properties." Other ingredients that contribute to cuticle hardness and resistance to abrasion, especially in some insect mandibles, are zinc and manganese (Quicke et al., 1998).

Some cuticular structures, especially those involved in sound production (tympanic membranes), jumping, and flight, may be subjected to rapid cycles of stress and release. Weis-Fogh discovered that many of these structures contain the protein resilin (Weis-Fogh, 1960; Bennet-Clark, 2007). Resilin has remarkable biomechanical properties, among them almost perfect elasticity: one can stretch resilin for weeks without any significant plastic deformation. In insects, resilin is found in many joints and tendons, where it resists breakdown from repeated stress and can store elastic energy. In some cases, resilin is combined with stiffer chitinous cuticle into an energy-storing mechanism. For example, froghoppers (Hemiptera: Cercopoidea) are among the most impressive known jumping insects. Jumping froghoppers have complex bow-shaped internal skeletal elements called pleural arches, which store much of the energy needed for long jumps (Burrows et al., 2008). The pleural arches combine the elasticity of resilin with stiffer cuticle, which allows them to store energy during slight deformation by slow muscle contraction.

2.3 Insect Muscles

2.3.1 Muscle Cells

Insects have only striated muscle, although it differs from vertebrate striated muscles in certain ways. As in vertebrate striated muscle, insect muscle cells are fused into multinucleate muscle fibers, which are covered in a membrane called the sarcolemma. However, some visceral muscle cells of the gut are uninucleate. Many invaginations of the sarcolemma, called T-tubules, penetrate deep into the muscle fiber and are important for depolarizing the fiber to cause contraction. Inside the muscle fiber are many mitochondria, the sarcoplasmic reticulum (SR), and myofibrils, which are end-to-end series of sarcomeres. Insect muscle fibers tend to have high mitochondria and high tracheolar densities because they are highly aerobic, although anaerobic catabolism is found in a few insects muscles such as caterpillar intersegmental muscles, grasshopper hindleg muscles, and muscles of chironomid larvae (Gäde, 1985). The SR is an extensive membrane-bound network running longitudinally throughout the muscle fiber and is the transient repository of intracellular Ca^{2+} that is released to allow contraction when the fiber is depolarized.

The sarcomeres, which shorten during contraction to generate force, contain actin and myosin myofilaments and other proteins. The thin filaments are composed of actin, tropomyosin, and the troponin complex proteins. The thick filaments are composed of primarily of the motor protein myosin and to a lesser degree paramyosin, which makes up the structural core of the thick filament. Titin is a massive sarcomeric protein with spring-like properties that connects the ends of the sarcomere (the Z-lines) to its middle (the M-line) and contributes to the elasticity of relaxed muscle (Ziegler, 1994). Projectin, kettin, and flightin (an insect-specific protein) contribute to stiffness of flight muscle (Bullard et al., 2006) and play a role in the stretch-activation of insect flight muscle, a phenomenon discussed below.

The basic contractile mechanism of myofilament sliding and the proteins that regulate cross-bridge cycling and contraction are similar to those of striated muscle in other organisms, and will not be reviewed here in great detail. These regulatory proteins include troponin (subunits I, T, and C), tropomyosin, calmodulin, and Ca^{2+}-dependent Ca^{2+} binding protein. The process of excitation-contraction coupling in insect muscle is also similar to that in striated muscle of other organisms, although the asynchronous flight muscles possessed by certain insect orders have unique mechanisms of neural control and activation, which we discuss below.

2.3.2 Neuromuscular Control

Most insect muscles are classified as synchronous because they contract synchronously with nervous signals from their innervating motor neurons. In these muscles the ratio of contractions to incoming nerve signals is 1:1. Synchronous insect muscles can contract at frequencies up to several hundred hertz, although most operate well below 100 hz. In a typical vertebrate muscle, individual muscle fibers are innervated by a single neuron at one or few neuromuscular junctions, which allows the contraction of such relatively large muscles to be regulated according to the number of fibers recruited. However, the small muscles of insects often possess too few fibers to be regulated via progressive fiber recruitment. The contraction of synchronous insect muscle is instead regulated by means of individual muscle fibers that are multiply innervated by each of several motor neurons, including both fast and slow excitatory neurons as well as inhibitory neurons (Usherwood, 1975). The designations fast and slow refer to the speed of contraction generated by the neurons rather than the speed of signal transduction through the neurons. Fast neurons cause single, large depolarizations that yield a rapid, singular contraction, while slow neurons initiate smaller, more variable depolarizations and likewise variable contraction intensities (Burrows, 1996). Furthermore, repeated firing of slow neurons causes a summation of depolarization that further permits a small muscle consisting of few fibers to yield graded contractions.

Insect excitatory motor neurons release L-glutamate at the neuromuscular junction, which initiates depolarization of the sarcolemma and the ultimate release of Ca^{2+} by the SR. Insect inhibitory motor neurons release the neurotransmitter γ-aminobutyric acid (GABA), which suppresses muscle depolarization by causing an influx of Cl^- that hyperpolarizes the sarcolemma. Some insect muscles are innervated by neurons that release octopamine or other neuromodulators that affect muscle function in a variety of ways such as enhancing the responsiveness to L-glutamate and increasing the availability of lipid fuel substrates via release of adipokinetic hormone.

Many myotropic neuropeptides modulate the contractile properties of insect muscle. Most insect neuropeptides have only been studied *in vitro* and about half of them exhibit myostimulatory or myoinhibitory properties (Altstein and Nassel, 2010). Proctolin stimulates contraction as a co-transmitter of L-glutamate, and FMRFamides enhance the contraction of gut visceral muscles and the abdominal extensor and ovipositor muscles in grasshoppers (Lange and Cheung, 1999). Pyrokinins and myosuppresins also help regulate the rhythmic contractions of gut and oviposition muscles, while crustacean cardioacceleratory peptides (CCAP) stimulate contraction of the heart.

2.3.3 Flight Muscle

Because of the aerodynamic costs of supporting a flying insect's weight, the flight muscles are the largest (comprising up to 50% of total body mass) and most energy-demanding of insect muscles. For example, the flight muscles of a flying honeybee consume approximately 400 ml O_2 g^{-1} hr^{-1}, which is about 3 times greater than in the flight muscles of hovering hummingbirds and 30 times higher than in elite human athletes at maximum performance. Insect flight muscles are completely aerobic and devoid of enzymes of anaerobic respiration pathways; not surprisingly, mitochondria and tracheolar spaces together can make up over half their volume. Vigoreaux (Vigoreaux, 2006) provides a comprehensive review of the biology of insect flight muscle.

Insect flight is powered and controlled by three categories of flight muscles. The first are the direct flight muscles that insert at the base of the wing and help depress the wing directly during contraction as well as move the wing through several degrees of freedom through their contraction and relaxation. The second, indirect flight muscles, do not attach to the wings but are instead inserted and attached to the inner surfaces of the thorax (Fig. 2.5). Contraction of the indirect flight muscles causes the wings to move via deformation of the thorax. Specifically, the dorso-ventral muscles extend vertically from the tergum to the sternum and depress the tergum during contraction, which causes the wings to elevate. The dorso-longitudinal muscles run fore and aft just under the tergum and their contraction causes the tergum to elevate and the wings to depress. The elevation of the wings via the dorso-ventral muscles is ubiquitous in insects, and in more evolutionarily-derived insect orders the indirect dorso-longitudinal muscles predominate over direct muscles in wing depression. Finally, accessory muscles attach to the thorax and are used to alter the shape and mechanical properties of the thorax for the purposes of modulating wing kinematics and power during steering and maneuvers.

Synchronous flight muscle is an ancestral trait in the pterygote insect orders, and species with synchronous flight muscles have wingbeat frequencies typically ranging from 5 to 100 Hz (Dudley, 2000). The frequency of synchronous flight muscle contractions is limited to a few hundred Hz due to the duration of motorneuron repolarization and ensuing refractory period. However, the insect orders with synchronous flight muscles are relatively large and long-winged, which imparts wingbeat frequencies well below the theoretical maximum contraction frequency of synchronous muscle.

Insects from several advanced orders (Thysanoptera, Hemiptera, Homoptera, Psocoptera, Coleoptera, Hymenoptera, Strepsiptera, and Diptera) possess an evolutionarily derived type of flight muscle called asynchronous flight muscle (Pringle, 1949). This type of flight muscle allows for (but does not dictate) extremely fast contraction frequencies (up to 1000 Hz) in these insects, which tend to have small wings relative to body mass (i.e. are heavily wing-loaded). The signature neuro-physiological feature of asynchronous flight muscle is a ratio of contractions to incoming nerve signals that is far greater than 1:1 (typically ranging from 5:1 to 25:1). Impulses from motor neurons initiate and generally sustain oscillatory contractions of asynchronous flight muscles, but the wingbeat frequency is determined primarily by the mechanical and resonance properties of the wings and thorax instead of the rate of firing by motor neurons as in synchronous flight muscle. This is clearly demonstrated by trimming the wings of asynchronous fliers (which increases wingbeat frequency) and synchronous fliers (which has no effect on wingbeat frequency). Due to its high contractions frequencies, asynchronous flight muscle typically operates with lower strain than synchronous flight muscle (Dudley, 2000). Asynchronous muscles also have a higher elastic stiffness than synchronous muscles due to high densities of structural proteins like projectin and flightin (Bullard et al., 2006).

Anatomically, asynchronous muscles differ from synchronous muscles in a few key ways. The sarcomeres in asynchronous muscle tend to have a larger diameter than those of synchronous muscle and are shorter due to a greater degree of overlap between thin and thick filaments, resulting in relatively short, low-force contractions. The SR of asynchronous muscle is greatly reduced, reflecting a lower dependence on Ca^{2+} cycling through the SR. This affords asynchronous flight muscle a considerable energy saving per contraction compared to synchronous muscle (although mass-specific rates of metabolism and power production in asynchronous fliers are typically greater due to higher contraction frequencies).

When motor neurons are firing above a threshold frequency, asynchronous flight muscles are in an electrically activated state during which sarcoplasmic Ca^{2+} levels are constantly elevated and contractions occur at a frequency determined primarily by resonance properties of the thorax and wings and secondarily by small accessory muscles (which also regulate the contraction amplitude of asynchronous flight muscles). In this active state, asynchronous flight muscles become very sensitive to stretching, with wing depressor muscles activated (after a slight delay) upon stretching via contraction of antagonistic wing elevator flight muscles, and vice-versa. In typical striated muscle, elevated Ca^{2+} causes tropomyosin to physically realign to expose actin's myosin binding sites. In asynchronous flight muscles, the binding of Ca^{2+} to troponin C is necessary, but not sufficient, for tropomyosin displacement and the exposure of myosin binding sites. The muscle must also be stretched for these critical events to occur that allow cross-bridge formation, albeit with a slight delay characteristic of stretch activation (Perz-Edwards et al., 2011). Sustained oscillatory contractions of asynchronous fliers

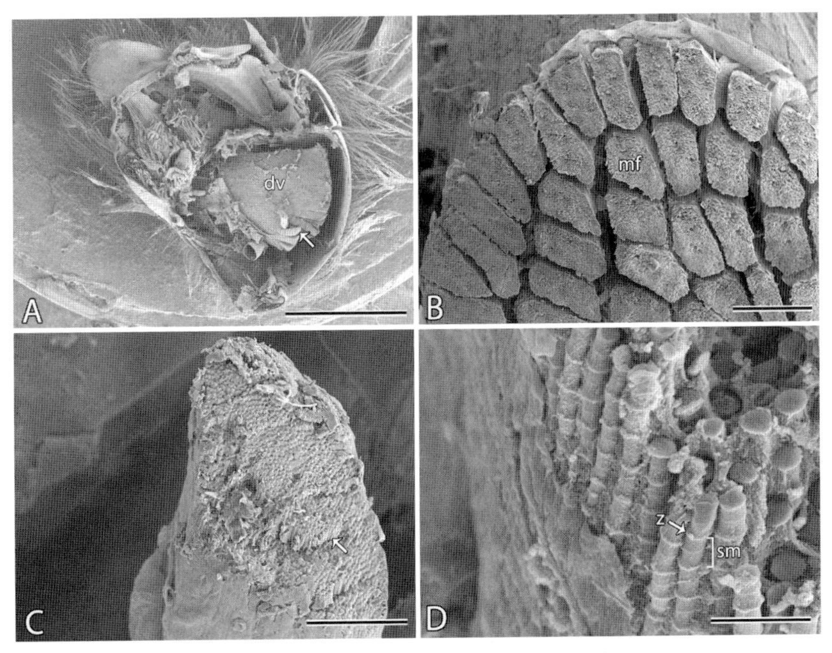

Fig. 2.5 SEM micrographs of *Bombus impatiens* (Hymenoptera: Apidae) flight muscle. (A) Mid-sagittal section revealing dorso-ventral indirect flight muscle (dv). The arrow is pointing to an individual muscle fiber. Bar = 2 mm. (B) Individual muscle fibers (mf) of a flight muscle cross-section. Bar = 200 µm. (C) A single flight muscle fiber cross-section revealing myofibrils (arrow). Bar = 50 µm. (D) Myofibrils, with Z-lines (z) and individual sarcomeres (sm) shown. Bar = 5 µm. Images courtesy of Danielle Kopke.

are also facilitated by a generally high degree of elasticity in the flight apparatus that results in significant storage of wing inertial energy in the first half of a stroke (and its release in the second half).

Asynchronous flight muscle is an evolutionarily-derived trait found in many exopterygote and endopterygote orders (Fig. 2.6). Although asynchronous flight muscle exists only in recently-evolved orders, there is debate about whether the trait is polyphyletic or monophyletic in origin. Several orders within the

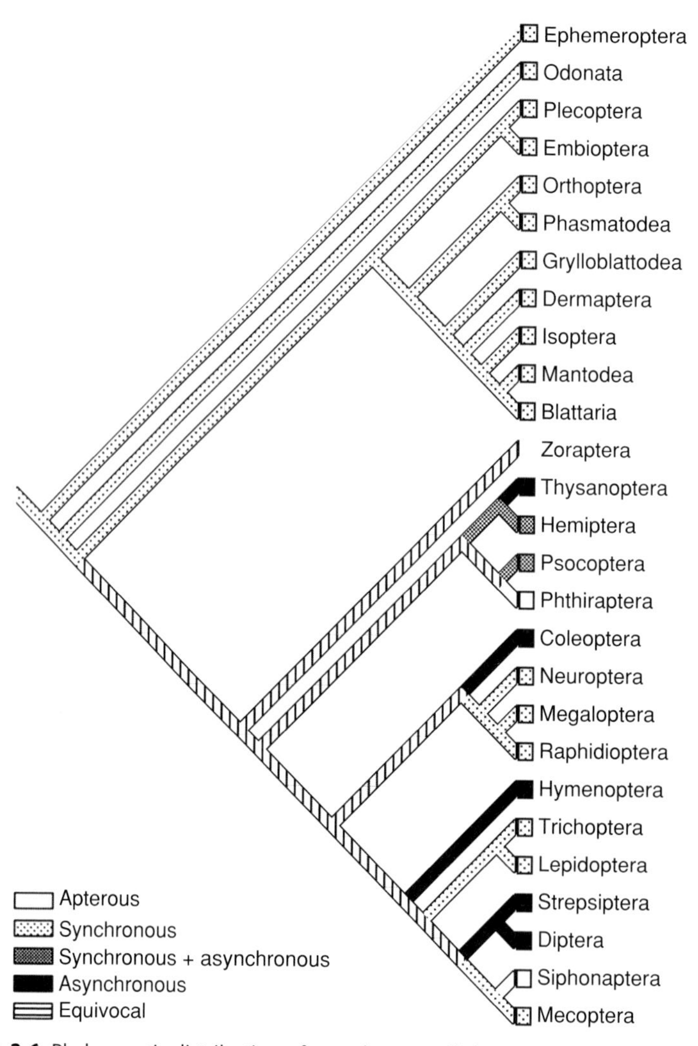

Fig. 2.6 Phylogenetic distribution of asynchronous flight muscle. From Dudley (2002), with permission from Princeton University Press.

"asynchronous clade" include some species with synchronous flight muscles, which would indicate several evolutionary reversions to synchronous flight muscle under a monophyletic hypothesis. Furthermore, flight muscles have not been physiologically characterized in each of the orders, making the evolutionary patterns of asynchronous flight muscles difficult to resolve (Dudley, 2000).

2.3.4 Flight Muscle Fuels

Flight muscles require the most energy of all insect tissues because of their high contraction frequencies and the comparatively low chemo-mechanical efficiency (5–10%) of hovering flight (Roberts et al., 2004; Lehmann and Heymann, 2006). Even so, insect flight muscles are completely aerobic and powered by a variety of substrates depending on several factors including flight duration, ecological context (short flights vs. migration), and taxonomy. High-energy phosphates, particularly arginine phosphate, phosphorylate ADP during the first moments of flight, after which ATP must be replenished by glycolysis, the citric acid cycle, and the electron transport chain via the breakdown of glycogen in the muscle cells and other fuels circulating in the hemolymph.

The disaccharide trehalose is the primary circulating carbohydrate used to fuel insect flight muscles, and many insects, including most dipterans and hymenopterans, rely almost exclusively on trehalose for this purpose. After initial stores of circulating trehalose are depleted during a flight bout, additional fuels are mobilized from the fat body via stimulation by adipokinetic hormone (AKH). Dipterans and hymenopterans convert fat body glycogen to trehalose to replenish circulatory stores, while migratory orthopterans convert fat body triacylglycerols to diacylglycerols that are transported to and metabolized by the flight muscles. Lipid substrates are also commonly used to power flight muscles in homopterans, coleopterans, lepidopterans, and odonates. Because of the insolubility of lipids in hemolymph, they are transported from the fat body to the flight muscles by the carrier protein lipophorin.

Glycolysis in the flight muscles of some insects involves a glycerol phosphate shuttle that recycles NAD^+, which is necessary because NAD^+ exists at only catalytic levels in insect flight muscle and because glycolytic NADH is produced in the sarcoplasm at a rate far greater than it can be transported into the mitochondria for oxidation at the electron transport chain. In vertebrate muscle and non-flight muscles of insects, reduction of pyruvate via NADH oxidation (i.e. lactate fermentation) is the mechanism by which excess NAD^+ levels are restored during oxygen-limiting conditions. In the glycerol phosphate shuttle, sarcoplasmic NADH is oxidized to NAD^+ by the enzyme glycerol phosphate dehydrogenase during the reduction of its substrate dihydroxyacetone phosphate to glycerol-3-phosphate. Glycerol-3-phosphate easily passes into mitochondria, where its conversion back to dihydroxyacetone phosphate reduces FAD to $FADH_2$ (an electron carrier that is oxidized at the electron transport chain). Dihydroxyacetone phosphate returns to the sarcoplasm where it can again be reduced by NADH. In summary, the glycerol phosphate shuttle relies on glycerol-3-phosphate to act as a highly membrane-permeable electron carrier between glycolysis in the sarcoplasm and the electron transport chain in the mitochondria.

Some dipterans and beetles use derivatives of the amino acid proline to "prime" specific steps in the citric acid cycle during the initiation of a flight bout. However, tsetse flies and certain beetles use proline as the primary flight fuel. In this pathway, hemolymph proline enters the muscle mitochondria and is oxidized to glutamate, which then becomes α-ketoglutarate via the transamination of pyruvate into alanine. The α-ketoglutarate enters the citric acid cycle to be broken down for ATP synthesis, while the alanine is transported to the fat body where it is converted back to proline via the addition of a 2-carbon molecule derived from stored lipids. Hence, the proline pathway is essentially a way of delivering lipids, 2 carbons at a time, to the citric acid cycle as α-ketoglutarate.

2.3.5 Development and Senescence of Muscles

The properties of muscles change significantly throughout the lifespan of an insect. We will limit our brief discussion here to temporal changes in the muscles of adult insects, a stereotypic trajectory involving the maturation (and functional improvement) of muscle immediately following eclosion, the attainment of peak muscle performance for some period, and the senescence of muscle capacity that occurs with advanced age. Facets of this trajectory are particularly well-described in *Libellula* spp. dragonflies and honeybees.

In preparation to establish and defend mating territories, male *Libellula* dragonflies double the mass of their flight muscles soon after emergence and concurrently improve calcium sensitivity in the muscles via regulated troponin T isoform expression. These changes greatly improve flight muscle thermogenic and power-producing capacity, but at a slight cost to efficiency (Marden et al., 1998). In the 1–2 weeks between eclosion and the onset of foraging behavior, honey bees increase levels of flight muscle cytochrome, glycogen, and troponin T 10A, and elevate V_{max} in hexokinase, phosphofructokinase, pyruvate kinase, citrate synthase, and cytochrome oxidase (Herold and Borei, 1963; Harrison, 1986; Schippers et al., 2006; Schippers et al., 2010). These modifications of the cellular respiration pathway combine to help honey bees increase their maximum flight metabolic rate, kinematic performance and aerodynamic power several fold during this period (Harrison and Fewell, 2002; Vance et al., 2009).

Senescence of insect flight muscle is probably due to the effects of accrued oxidative and mechanical stress on their genetic, biochemical, and structural components (Williams et al., 2008; Fernandez-Winckler and da Cruz-Landim, 2008; Zheng et al., 2005; Miller et al., 2008, reviewed by Martin and Grotewiel,

2006). Not surprisingly, several studies indicate that flight muscle senescence may be a function of cumulative muscle use (or metabolism) instead of chronological age. For example, honeybees with high daily flight activity forage fewer total days, deplete glycogen reserves earlier in their lives, and die sooner, than bees with lower rates of daily flight activity (Neukirch, 1982; Schmid-Hempel and Wolf, 1988). When lifetime flight activity of house flies is restricted, their longevity increases, while protein carbonylation and DNA oxidation in flight muscle mitochondria are attenuated (Yan and Sohal, 2000; Agarwal and Sohal, 1994).

The onset, rate, and duration of muscle maturation, peak performance, and senescence can be strongly influenced by life history strategies among insects, and can be strongly affected by environment. For example, when honeybees are prevented from becoming foragers (i.e. remain in-hive nurse bees their whole lives), they suspend their flight muscles in an "immature" physiological state, never reaching the kinematic capacity of forager bees (Vance et al., 2009). In what is called the oogenesis-flight syndrome, many insects regulate energy allocation and growth of flight muscles vs. reproductive structures, with investment in the former during migratory and non-reproductive periods and investment in the latter during non-migratory and reproductive periods (Zera and Denno, 1997; Marden, 2000; Johnson, 1969; Lorenz, 2007). In univoltine species exhibiting the syndrome, the switching of such allocation typically occurs only once, while in long-lived species the cycling of energy allocation between flight muscles and reproductive tissues occurs seasonally or more frequently. Pine engraver beetles (*Ips* spp) display perhaps the most exquisite control of flight muscle plasticity, with cycles of muscle histolysis and regeneration occurring in as little as ten days depending upon breeding opportunities in their galleries (Robertson, 1998).

2.4 Insect Respiratory Systems

Like vertebrates, insects are mostly aerobic—they use oxygen as the final electron acceptor and produce carbon dioxide from organic substrates. Unlike vertebrates, insects exchange gases between tissues and the environment primarily using tracheal systems (Fig. 2.7). For a vertebrate physiologist, the tracheal system can seem bizarre: it is a set of air-filled tubes that ramify throughout the body and deliver oxygen essentially to every mitochondrion. At one end, the system connects to the outside of the insect via a system of usually occludable portholes, termed spiracles, which occur in pairs, usually one pair to a segment. At the other end of the tracheal system, sprouting from the finest tracheal tubes, are the tracheoles, whose thin walls and high surface-to-volume ratios allow them to be the site of most gas exchange. The organization of the tracheal system varies dramatically among insects, with spiracle numbers ranging from 0 to 22 and with

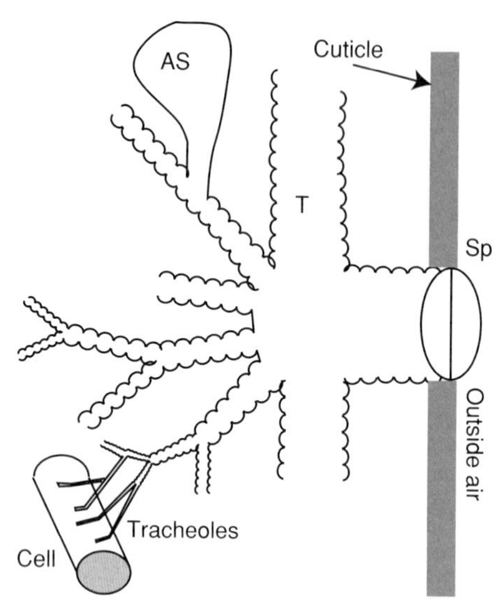

Fig. 2.7 An overview of the tracheal system. Spiracles (Sp) in the integument connect to tracheae (T), which branch repeatedly within the insect, eventually leading to the tracheoles, which are located near most cells. In some insects, soft, collapsible air sacs (AS) occur. Redrawn based on Harrison (2009), with permission from Elsevier.

tracheal branching patterns varying widely across species, between body regions, and during the developmental stages of holometabolous insects.

Spiracles can be highly variable both within and between insects (Schmitz and Wasserthal, 1999). Some Apterygote and larval insects lack valves in their spiracles and therefore have trachea that are always open to the atmosphere. However, most spiracles have valve-like structures that allow them to close (Fig. 2.8). Often the outside of the spiracle is fringed with hairs or covered with a filter to help resist entry of dust, water, or parasites. The valve that closes the spiracle takes various forms but can appear as two lips of cuticle or as a bar of cuticle that pinches the trachea shut (Fig. 2.8). Often the valve is recessed behind the atrium, an enclosed space behind the external opening. The movement of the valve depends on the action of muscles that insert on it, as well as elastic ligaments. Some thoracic spiracles have only a closer muscle, passive elastic ligaments open the valve when the closer muscle is relaxed (Fig. 2.8). The most common situation (true for all abdominal spiracles studied to date) is for the spiracle to have both opener and closer muscles.

The tracheal wall contains an epithelial cell layer that secretes a basement membrane, which forms the outermost layer of the tracheal wall. Along the inner wall, the tracheal cells secrete the intima, which contains protein and chitin fibers. The intima forms spiral folds known as taenidia. In general, it is thought

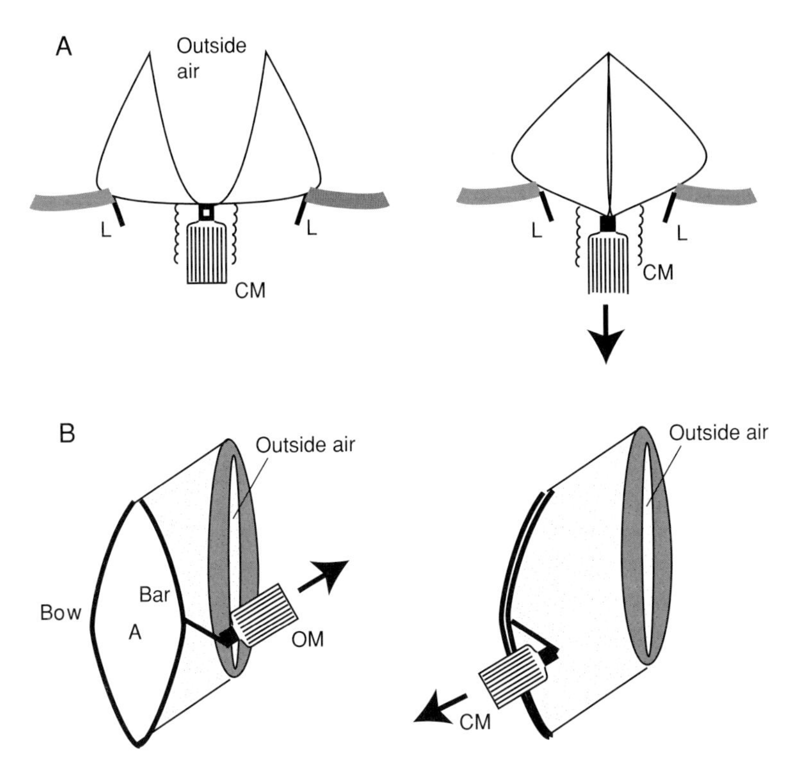

Fig. 2.8 Spiracular structure and function. A) Schematic side view of the locust metathoracic spiracle, a one-muscle spiracle with an external valve. When the closer muscle (CM) contracts, the lips of the valve come together. When the closer muscle relaxes, ligaments (L) attached to the cuticle passively pull the valve lips apart. B) Schematic view from the inside of the animal of the abdominal spiracle of a butterfly, a two-muscle spiracle with an internal valve. When the opener muscle (OM) contracts, it pulls on a lever which pulls the bar away from the bow, opening the atrium (A). When the closer muscle (CM) contracts, it pulls the bar against the bow, collapsing the walls of the atrium together. Redrawn from Chapman (1998), with permission from Elsevier.

that taenidia prevent tracheae from collapsing inward, much like the rings of a vacuum hose, yet allow them to flex. However, in the tracheae of some insects, the taenidia are widely spaced and the intima is quite flexible, so that the trachea are compressible (Wasserthal, 1996; Westneat et al., 2003). Because tracheal walls are relatively thick, oxygen is not believed to pass out of the larger trachea in appreciable amounts. Thus, the primary function of the trachea is to transport gases between the tracheoles and the spiracles. However, because carbon dioxide can diffuse more easily through tissue than can oxygen, it is possible that a significant portion of the carbon dioxide emitted passes through the tracheal walls (Schmitz and Perry, 1999).

Tracheoles, the small tubes that form the terminal endings of the tracheal system, are 1 to 0.1 microns in diameter (Wigglesworth, 1983). Tracheoles are formed within single tracheolar cells. These tracheolar cells have many branches, some of which contain air-filled channels (tracheoles) connecting to the air-filled lumen of the trachea. Tracheoles are particularly dense in metabolically active tissues such as flight muscle. Most tracheoles occur outside of the cells of the insect's body, but sometimes in histological sections they appear to press into cells, particularly in flight muscle. These tracheoles are believed to enter the flight muscle via infoldings of the muscle plasma membrane, and so they are actually extracellular. The high tracheolar densities and penetration of flight muscle cells by tracheoles allows flying insects to achieve mass-specific oxygen consumption rates that are among the highest in the animal kingdom.

In many insects, portions of the tracheal system are enlarged to form thin-walled air sacs. These air sacs lack taenidia and therefore collapse and expand with variation in hemolymph pressure (Harrison, 2009). Air sacs are common in adult flying insects of many orders including Diptera, Hymenoptera, Lepidoptera, Odonata, and Orthoptera. These air sacs serve as bellows and enhance the movement of air through the tracheal system. Tracheal lungs, or aeriferous tracheae, are similar to air sacs in that they are thin-walled expansions of the tracheal system with few taenidia that float freely in the hemolymph (Mill, 1998). Their primary function is believed to be oxygen delivery to tissues that lack tracheoles, such as hematocytes, and some ovaries and Malpighian tubules. In this case, oxygen diffuses through the thin wall of the tracheal lung, through the hemolymph, and to the tissues. Tracheal lungs have been reported for Lepidopteran larvae (Locke, 1998), which apparently lack hemocyanin in the hemolymph; it would be interesting to look for such structures in the insects that utilize respiratory pigments in the blood.

The development of the tracheal system has been well-studied in *Drosophila melanogaster* (Ghabrial et al., 2003; Weaver and Krasnow, 2008). In flies, the complete larval tracheal system develops, in the embryo, from only 80 cells per body segment without further mitosis or cell death. Primary branching of the tracheae is induced by secretion of a protein homologous to vertebrate fibroblast growth factor (called Branchless in *Drosophila*) secreted by local tissues. The primary branches are formed of 3–20 cells organized into tubes. Some cells become terminal cells and ramify to become tracheoles.

In general, the size of the tracheal system increases with age in order to support the increased gas exchange needs of the larger insect. During molts, the cuticular lining of the trachea is drawn out of the spiracle with the old integument (Wigglesworth, 1981). Both tracheoles and tracheae are fluid-filled in newly hatched insects, and fluid fills the space between the old and new tracheae at each molt (Wigglesworth, 1983). This fluid is replaced with gas shortly after hatching or molting. Major changes in tracheal structure, including changes in spiracle number and tracheal system organization, can occur at each molt and during the pupal period for endopterygote insects. In *Drosophila*, the adult tracheal system is formed from both imaginal disc cells and larval tracheal cells that proliferate and alter their morphology (Weaver and Krasnow, 2008).

Changes in tracheal system structure are not limited to molting periods, because the tracheoles can change structure within an instar. In the event of injury or oxygen-deprivation, local tracheoles grow and increase their number of branches (Wigglesworth, 1983; Jarecki et al., 1999). If no undamaged tracheoles are nearby, damaged tissues produce cytoplasmic threads that extend toward and attach to healthy tracheoles. These threads then contract, dragging the tracheole and its respective trachea to the region of oxygen-deficient tissue (Wigglesworth, 1983).

Insect gas exchange occurs in a series of steps. Oxygen molecules first enter the insect via the spiracle, then proceed down the branching tracheae to the tracheoles. The terminal tips of the tracheoles are sometimes fluid-filled, so at this point gas transport may occur in a liquid medium rather than air (Wigglesworth, 1983). Oxygen then must move across the tracheolar walls, through the hemolymph, across the plasma membranes of the cells, and finally through the cytoplasm to the mitochondria. Carbon dioxide generally follows the same path in reverse.

A key question, and one that has a long historical record over the past 100 years, is how oxygen moves from the spiracles to the tracheoles. One possibility is diffusion. In a now-famous paper, Krogh (Krogh, 1919) calculated that, for most insects at rest, diffusion alone would more than supply adequate oxygen. Nevertheless, insects often are not at rest, and ensuing studies throughout the twentieth century showed that convection in tracheae plays a primary role in moving oxygen—in both active *and* resting insects. At this point, we know that insects have evolved a spectacular set of mechanisms that move air by bulk flow: abdominal and thoracic compressions, compressible internal air sacs, hemolymph driven tracheal compression, and ram ventilation during flight. By coupling convection to control of spiracle opening, insects can achieve unidirectional air flow through the main tracheal branches. Even small insects (e.g. *Drosophila*) appear to ventilate their tracheal systems under some circumstances (Lehmann, 2002).

The mechanisms by which insects achieve convective gas exchange are complex and varied. In many insects, well-coordinated actions of muscles and spiracles produce regulated convective air flow through the tracheae and spiracles (Miller, 1966; Weis-Fogh, 1967). Most commonly, convection is driven by contractions of respiratory muscles attached to the body wall, which produce increases or decreases in body volume, causing compressible portions of the tracheal system to inflate or deflate.

One common mechanism by which insects move air through tracheae is abdominal pumping. Expiratory muscles connect the ventral and dorsal

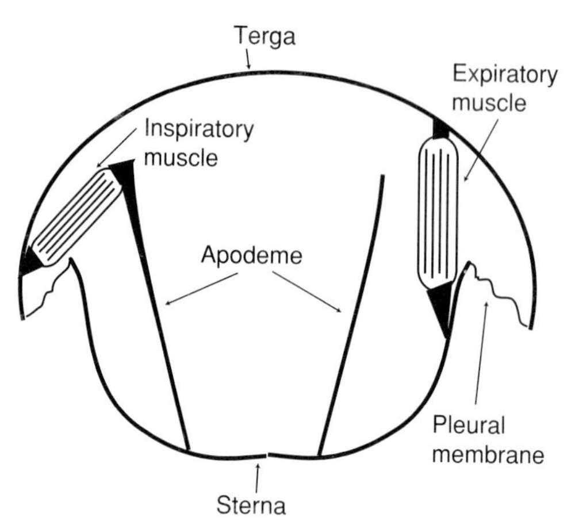

Fig. 2.9 Intersegmental muscles that drive abdominal pumping in grasshoppers. Contraction of expiratory muscles compress the terga (top) and sterna (bottom) of the abdominal cavity together. Contraction of inspiratory muscles lift the terga upwards, expanding the abdominal volume. Reprinted from Harrison (2009) with permission from Elsevier.

cuticular plates. In response to neuronal central pattern generators, these contract, pulling the dorsal terga and ventral sterna together (Fig. 2.9), and the tip of the abdomen inward as the cuticular plates slide over the flexible pleural and intersegmental membranes (Hustert, 1975; Burrows, 1996). Inspiration may be passive, and result from cuticular elasticity, or contractions of inspiratory muscles attached to tall sternal apodemes and the lower edge of the terga can lift the terga relative to the sterna, and expand abdominal volume and the air sacs (Fig. 2.9).

In some adult cockroaches, grasshoppers, bees, beetles, and dragonflies that have been examined, abdominal pumping is coordinated with spiracular opening in a manner that produces directed, unidirectional flow through the body (Weis-Fogh, 1967; Miller, 1974). During inspiration, air flows in through open thoracic spiracles. During expiration, the thoracic spiracles close and air flows out via open abdominal spiracles.

Muscles that have the primary purpose of driving hemolymph circulation, such as the heart and ventral diaphragm, also play a role in creating convective ventilation (Wasserthal, 1996). Pumping of accessory muscles at the base of the wings, antennae, and legs, pushes hemolymph and air into these appendages (Hustert, 1999; Pass, 2000). In a variety of adult Lepidoptera, Diptera, and Hymenoptera, the heart occasionally reverses pumping direction, shifting hemolymph from the thorax to the abdomen (Miller, 1997). As the hemolymph accumulates in one body compartment, air sacs are compressed (Wasserthal, 1996).

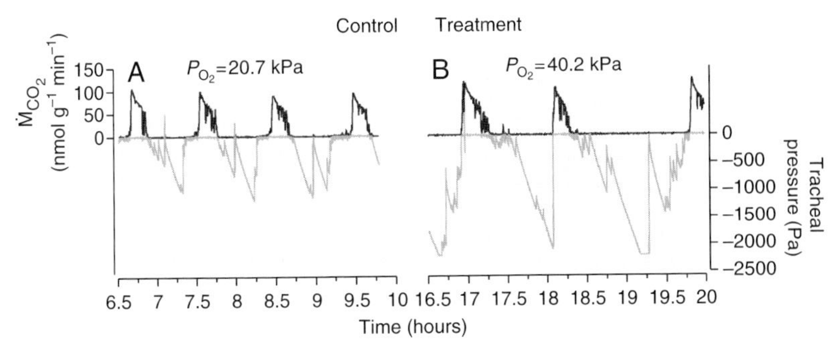

Fig. 2.10 Discontinuous gas exchange in a lepidopteran pupa, *Samia cynthia*. Upper traces (dark gray): CO_2 release; lower traces (light gray): intratracheal P_{O2}, kPa. A. During normoxia, note drop in tracheal P_{O2} during spiracular closed periods when no CO_2 is emitted. B. In hyperoxia, tracheal P_{O2} can decrease further, and the closed period is longer. From Terblanche et al. (2008), with permission from The Company of Biologists.

Of particular interest to insect respiratory physiologists is a pattern of intermittent ventilation, called discontinuous gas exchange cycles (DGCs), shown by some insects under some circumstances (usually when their metabolic rates are low). During DGCs, the spiracles can be completely sealed (no gas exchange) for extended periods of time, accompanied by decreasing tracheal P_{O2} (Fig. 2.10). The duration of the closed period is affected by tracheal P_{O2}, as hyperoxia extends the closed period, at least in lepidopteran pupae (Fig. 2.10). The closed period is sometimes followed by a flutter phase, which can be characterized by microopenings of the spiracles that allow air to move in by bulk flow down a total pressure gradient, which has the effect of letting oxygen in while preventing carbon dioxide release (Hetz and Bradley, 2005). However, fluttering is not a long-term solution, as carbon dioxide builds up in the tracheal system and in the body fluids. Full spiracular opening is believed to be generally caused by rise in internal P_{CO2} to a threshold (Hetz and Bradley, 2005; Chown et al., 2006).

Elaborate processes tend to attract adaptive explanations, and DGCs are no exception. The proposed hypotheses include that DGCs: lower respiratory water losses (the "hygric hypothesis"); promote oxygen uptake and carbon dioxide release in microhabitats that are hypoxic and hypercapnic (the "chthonic hypothesis"); prevent toxically high levels of oxygen in the tissues (the "oxygen hypothesis"; (Hetz and Bradley, 2005)); and are an emergent property of separate control systems for oxygen and carbon dioxide (the "emergent hypothesis"). No clear winner has yet emerged (Chown et al., 2006).

Both hemoglobin and hemocyanin play roles in insect gas exchange, with their distribution being at least partly related to phylogenetic history (Fig. 2.11). Hemocyanin occurs in the hemolymph of most primitive insect orders including

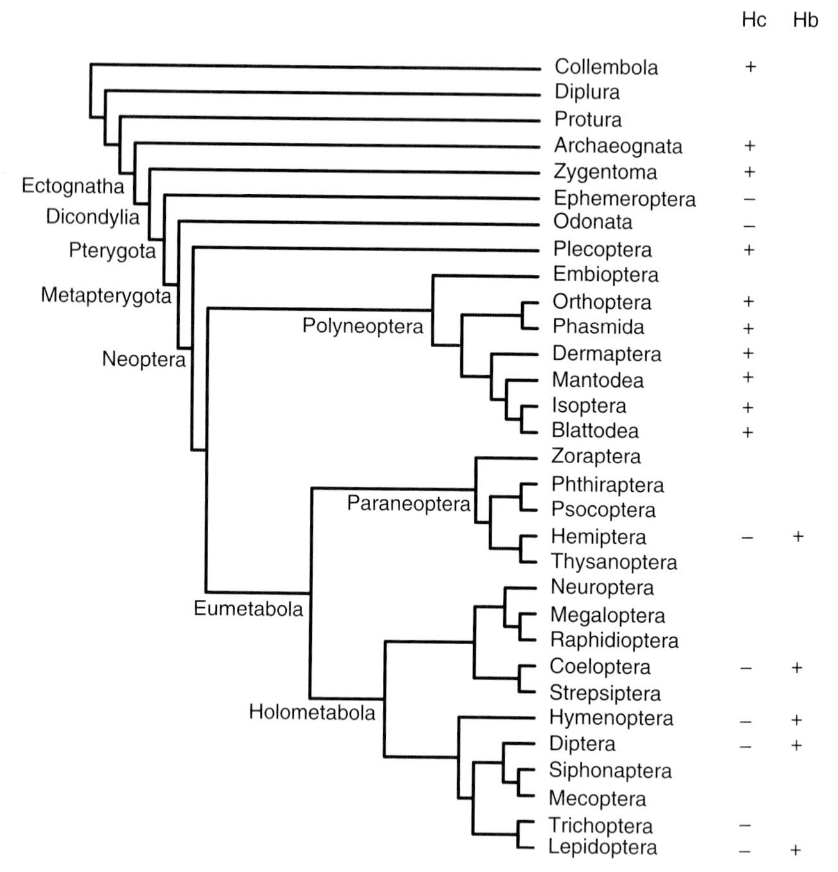

Fig. 2.11 Distribution of hemocyanin (Hc) and hemoglobin (Hb) in insects. Hemocyanin has been found in the hemolymph in many primitive insect orders. Hemoglobin has been found intracellular throughout the insects. From Burmester and Hankeln (2007), used with permission from Elsevier.

Collembola, Zygentoma, Plecoptera, Orthoptera, Isoptera, and Blattoidea, but not in the higher holometabolous insects, providing strong support for the crustacean ancestry of insects (Pick et al., 2009). Interestingly, hemocyanin appears to have been secondarily lost in some taxa: Protura, Diplura, Ephemeroptera, and Odonata (Pick et al., 2009). The functional importance of hemocyanin in insects still is unknown, but at least in one stone fly, hemocyanin occurs at high concentrations in the hemolymph and has an appropriate oxygen affinity to be playing an important role in oxygen delivery for this species (Hagner-Holler et al., 2004). It has also recently been shown that hemocyanin is expressed in embryos and first instars of the ovoviviparous cockroach, *Blapticadupia*, suggesting a role in embryonic oxygen delivery (Pick et al., 2010). This promises to be an exciting area for future research.

In aquatic larval and pupal chironomids, hemoglobin occurs only in the hemolymph (Burmester and Hankeln, 2007). These hemoglobins have high affinity for oxygen, which helps them extract and store oxygen from hypoxic environments. Some other aquatic insects have specialized tissues that contain high concentrations of intracellular hemoglobin, which also extract and store oxygen. One example is larvae of the bot fly, *Gasterophilus intestinalis*, which parasitizes horse intestines and obtains oxygen from intermittently available bubbles swallowed by its host. Backswimmers (Hemiptera) have specialized organs that contain hemoglobin and are penetrated by tracheoles. When backswimmers dive, oxygen is transferred from the air bubble they carry to the hemoglobin, stabilizing their buoyancy and prolonging their dive duration (Matthews and Seymour, 2006).

Intracellular hemoglobins have also recently been detected in tracheoles and a variety of other tissues in diverse insects, including the fruit fly, *Drosophila melanogaster*, the honeybee, *Apis mellifera*, and the mosquitos, *Aedesa egypti* and *Anopheles gambiae*. The concentration of hemoglobin in these tissues is unknown. These hemoglobins are similar to crustacean hemoglobins, suggesting that intracellular hemoglobins are common and primitive in insects. The functions of these hemoglobins are as yet poorly understood, but may include oxygen storage for use during hypoxia or burst performance, enhancement of oxygen transport within tissues, detoxification of reactive oxygen species, and oxygen sensing and signaling (Gleixner et al., 2008).

2.5 Insect Circulatory Systems

Unlike the closed circulatory system of vertebrates, in which vessels control where blood goes, insects have open circulatory systems—relatively open spaces (hemocoel) in which hemolymph, the insect blood, bathes the tissues. Insect hemolymph has some of the same functions as vertebrate blood. It carries nutrients and wastes, distributes endocrine signals, moves immune cells, and disperses heat (Klowden, 2009; Nation, 2008). However, unlike vertebrate blood, hemolymph in most insects does not carry significant amounts of oxygen; oxygen is physically dissolved in hemolymph but not concentrated by respiratory pigments. This difference from vertebrates reflects the fact that insects use an alternative system, the tracheal system, for primary gas transport. However, some insects have hemocyanin in their hemolymph at concentrations sufficient to transport significant oxygen (Pick et al., 2009; Burmester and Hankeln, 2007; Hagner-Holler et al., 2004), and physiological studies of these insects are urgently needed to determine the importance of blood-borne gas exchange in these species. The circulatory system can also participate in gas exchange by ventilating the tracheal system (Wasserthal, 1996; Miller, 1997).

Although the insect circulatory system is open, it nevertheless can generate complicated patterns of flow (Glenn et al., 2010) (Figs 2.12 and 2.13). The main pump is the dorsal vessel, a tube that runs just under the dorsal cuticle for most of the length of the insect. In the abdomen, the dorsal vessel usually is pulsatile and is called the heart (Fig. 2.14). The heart is supported by pairs of alary muscles at each segment, and is surrounded by pericardial cells and fat body cells. The pericardial cells, heart, alary muscles, and dorsal diaphragm are linked together with connective tissue (Curtis et al., 1999), and thus the pericardial cells and dorsal diaphragm may play a structural role in limiting heart expansion as performed by the pericardial sac in mammals.

A diversity of feedbacks, usually based on factors associated with metabolism, affect the heartbeat, including carbonate ions, norepinephrine, serotonin, pyruvate, lactate, and glutamate (Slama, 2010). In the thorax and head, the dorsal vessel usually is not pulsatile (though it can have pulsatile elements; (Wasserthal, 1999)). In some insects, and usually only in later developmental stages, the heart can undergo periodic reversal, in which contraction starts at the anterior end and moves posteriorly (retrograde) (Fig. 2.12). Retrograde circulation probably serves

Heartbeat	Mechanism	Frequency	Circulation	Conical chamber	Apertures
Anterograde I	Synchronic	~ 4 Hz	Thoracic	Contracted	Closed
Anterograde II	Synchronic	~ 4 Hz	Abdominal	Dilated	Open
Retrograde	Peristaltic	1–2 Hz	Abdominal	Dilated	Open

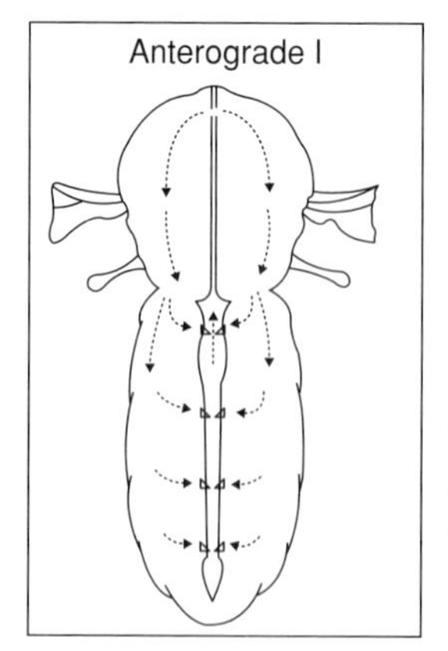

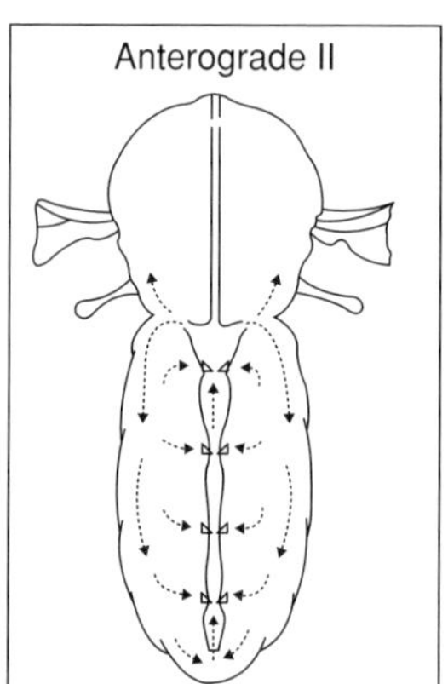

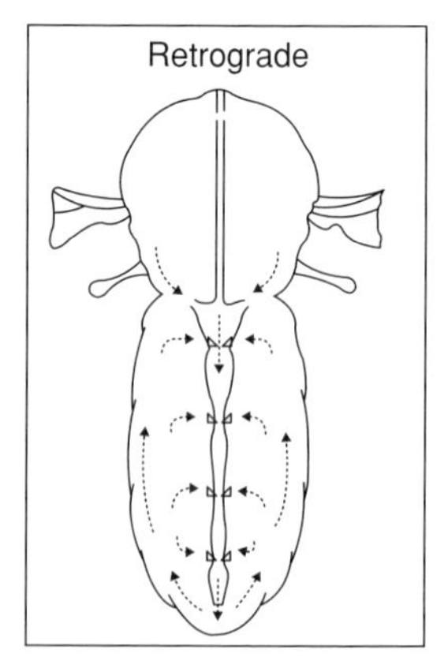

Fig. 2.12 Three different patterns of circulation in adult *D. melanogaster*: anterograde I, anterograde II, and retrograde. A single fly can switch between all three patterns. From Sláma K (2010), with permission from the *European Journal of Entomology*.

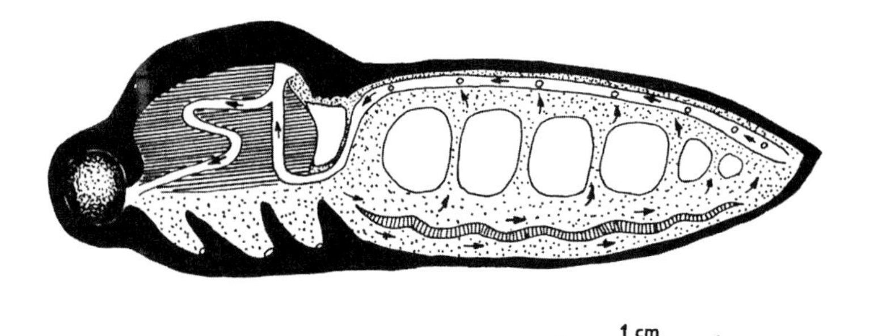

Fig. 2.13 Schematic diagram of patterns of hemolymph flow in an adult *Manduca sexta* during anterograde pumping. The pulsatile heart in the abdomen sucks in hemolymph through pairs of small holes (ostia) and pumps it anteriorly through the aorta. The aorta loops through the thorax and ends in the posterior head. Hemolymph returns posteriorly through a more open hemoceol, which in many adult insects has large air sacs (white regions). Horizontal cross-hatching = thoracic muscles; vertical cross-hatching = ventral diagram; solid black = insulating scales. From Heinrich (1971a), with permission from the The Company of Biologists. REQ permission June 30, 2011.

to mix or flush hemolymph in the posterior parts of the abdomen (Slama, 2010). The nature and control of heartbeat reversal has been controversial. Good reviews of insect circulation can be found in (Jones, 1977; Wasserthal, 1996; Miller, 1985; Miller, 1997).

2.5.1 Homologies Between Insect and Vertebrate Hearts

Although insect hearts differ profoundly from vertebrate hearts, a series of studies over the past 15 years has suggested that the two taxa share fundamental similarities with respect to embryological origins and genetic bases of heart formation (Bodmer, 1995; Bodmer et al., 2005). In particular, during gastrulation, the mesoderm is subdivided into visceral, cardiac, and somatic mesoderm. A key gene in this process is the homeobox transcription factor Tinman (*tin*), and its actions appear to be similar in humans and flies. Mutant flies lacking *tin* do not develop any heart precursor cells, suggesting that the gene is a key determinant of cardiac cell lineages (Bodmer, 1993; Azpiazu and Frasch, 1993). The downstream regulatory targets of Tin are increasingly well characterized (Akasaka et al., 2006), and homologs of *tin* have been found in all vertebrates examined, probably with similar functional roles. Sláma suggests that insects contain pacemaker regions in the terminal heart ampoule that are analogous to the atrioventricular pacemaker in the human heart (Slama, 2006). As a consequence of these similarities, *Drosophila* has become a model insect system for studying the genetics, development, and pathology of human hearts (Bier and Bodmer, 2004; Ocorr et al., 2007).

2.5.2 Heartbeat: Patterns and Controversies

The past decade has seen substantial progress, and controversy, in understanding the control of insect heartbeat. Intrinsic, myogenic functions can clearly be important in determining cardiac function (Rizki, 1978; Slama, 2006; Slama, 2010; Slama and Lukas, 2011). In addition, insect hearts receive neural innervation that modulates cardiac function (Ai and Kuwasawa, 1995; Dulcis et al., 2001; Dulcis and Levine, 2005; Tublitz and Truman, 1985; Tublitz, 1989; Veenstra, 1989; Nichols et al., 1999). Different sections of the heart are innervated by glutamatergic neurons, cutting or stimulating these neurons can alter patterns of heartbeat (speed or direction), and application of neuropeptides (proctolin, corazonin, CCAP) can dramatically speed up the heart rate. In addition to the neural control of heart function, at least in *Drosophila*, heart rate is also modulated by a variety of circulating neurohormones. Injections of CCAP, serotonin, dopamine, acetylcholine, octopamine, and norepinephrine raise heart rates, while Dromyosupressin, DPKQDFMRFamide, and PDNFMRFamide slow heart rate (Johnson et al., 1997; Nichols et al., 1999), with some variation in control mechanisms depending on developmental stage (Zornik et al., 1999). In many insects, cardiac reversal can be induced by a variety of external factors, such as changing light levels or altered olfactory and visual stimuli (Kuwasawa et al., 1999; Dulcis et al., 2001), indicating some sort of cardiac reflex.

2.5.3 Accessory Pulsatile Organs

In insects, hemolymph movement is driven by the pulsatile, abdominal heart. However, because insects have open circulatory systems, and because their hemocoels are small and support low-velocity flows (giving very low Reynolds numbers, such that flow is dominated by viscous effects), they may not be able to circulate hemolymph readily through narrow appendages and possibly also through the head. For local circulation, therefore, many insects use accessory pulsatile organs (Pass, 2000) (Fig. 2.15). Normally these local pumps are disconnected from, and operate independently of, the heart.

The locust *Schistocerca gregaria*, for example, has a pulsatile organ located in the trochanter of the middle thoracic leg (Hustert, 1999). It moves hemolymph toward the tip of the leg along a ventral sinus; at the tip of the leg, the hemolymph exits into a dorsal sinus and flows back to the main hemocoel. When locusts actively ventilate their tracheae, pumping of the leg-heart is synchronized with abdominal movements; at rest, when ventilation is not active, the pumping of the leg-heart is myogenic (not synchronized). Besides appendages, even the heads of some insects may need accessory circulation from local pumps. Wasserthal (1999) described a cephalic pulsatile organ in the blowfly *Calliphora vicina*. The organ is associated with the terminus of the aorta in the posterior head; its

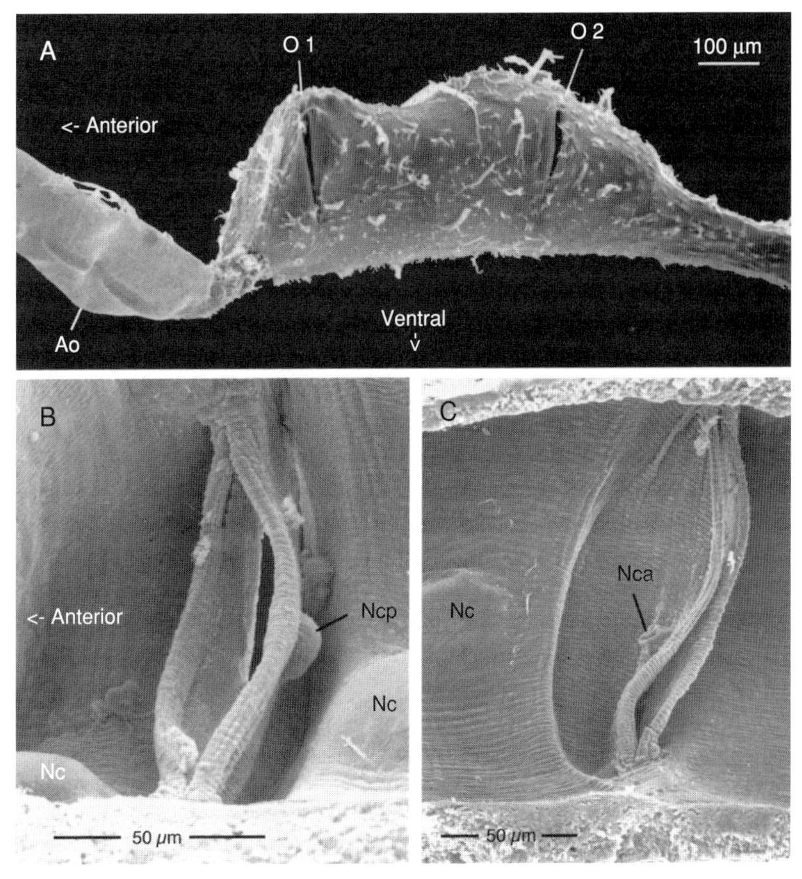

Fig. 2.14 SEMs of heart and ostia in the blow fly *Calliphora vicina*. (a) Enlarged heart chamber with two pairs of incurrent ostia (O1 and O2) and posterior part of aorta (Ao). (b) First ostium with two ostial lips (cells) protruding into the heart lumen. (c) Second ostium with both ostial lips oriented backward. Nc = nucleus of heart wall cell; Nca = nucleus of anterior ostial cell; Ncp = nucleus of posterior ostial cell; Same orientation in a–c. From Wasserthal (1999), with permission from Elsevier.

incurrent flow comes from the opening of the aorta, and its excurrent flow is directed onto the rear of the brain (Fig. 2.14). Functionally, the organ sucks hemolymph into the head and distributes it within the head space.

2.5.4 Using Hemolymph to Move Heat

Hemolymph moves biological currencies of many different types. For the purposes of this book, one of the most important is heat. A common view is that

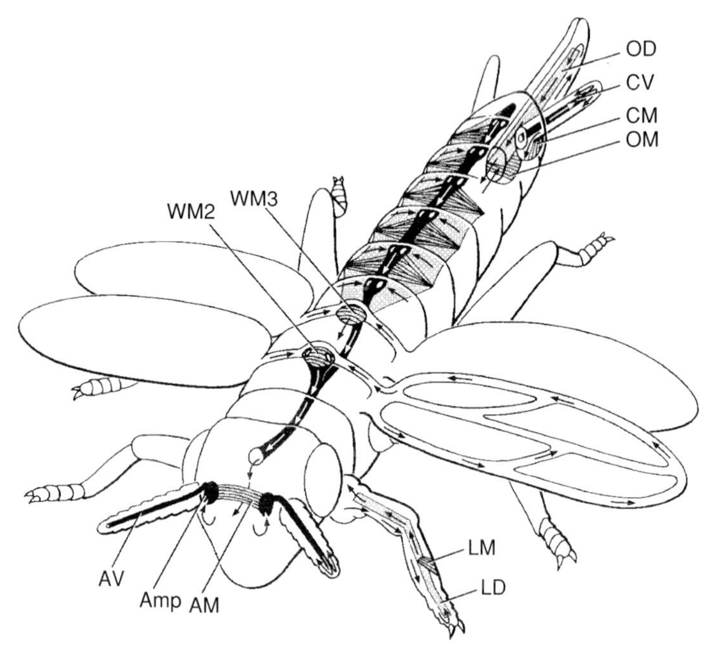

Fig. 2.15 Diagram of an idealized insect with the maximum possible set of circulatory organs. Vessels are in solid black, diaphragms and pumping muscles are in gray; arrows indicate directions of hemolymph flow. Central body cavity: dorsal vessel composed of anterior aorta and posterior heart region (with paired ostia, dorsal diaphragm, and alary muscles); ventral diaphragm concealed. Antennae: ampullae (Amp) with ostia connected to antennal vessels (AV); pumping muscle (AM) associated with ampullae. Legs: diaphragm (LD) pulsatile owing to associated pumping muscle (LM). Wings: dorsal vessel muscle plate with ostia pair in mesothorax (WM2); separate pumping muscle in metathorax (WM3). Cerci: cercal vessel (CV) with basal suction pump (CM). Ovipositor: each valvula with non-pulsatile diaphragm (OD) and basal forcing pump (OM). Among insect species, the functional morphology of the accessory pulsatile organ in a given body appendage may be heterogeneous. From Pass (2000). Reprinted, with permission, from the *Annual Review of Entomology*, Volume 20 © 2000 by Annual Reviews www.annualreviews.org.

insects are necessarily ectothermic poikilotherms—because they have small body sizes and lack metabolisms specialized for homeothermy. Nevertheless, it has become clear over the past 35 years that many insects, especially flying adults, have sophisticated mechanisms for moving heat around in their bodies by controlling flows of hemolymph. Indeed, by finely controlling patterns of circulation, some flying adults can thermoregulate precisely around set points in excess of 40°C (Heinrich, 1993).

For an insect to establish body temperatures higher than ambient requires a source of heat. One obvious source of heat is radiation from the sun, and many insects bask to heat up, sometimes to temperatures well above ambient. The best known basking group is butterflies, which use at least four kinds of behavioral thermoregulation (Clench, 1966). In butterflies, heat gains and losses vary in interesting and controversial ways with wing position, wing color, and thoracic color and hairiness (Wasserthal, 1975; Heinrich, 1993). An alternative source of heat, studied best in flying moths, is the flight motor itself. The thorax of flying moths is packed to overflowing with muscle fibers and associated machinery (mitochondria, sarcoplasmic reticulum). For these muscles to generate enough power for flight requires that they sustain very high rates of mass-specific metabolism. However, of the energy flux through these muscles, less than 10% ends up generating lift; the rest is "lost" to inefficiency—it becomes heat. This heat can be highly useful (reviewed in (Heinrich, 1993)): when the thorax is hot, the muscles beat faster and more readily generate lift sufficient for flight. Moths therefore often go through a "shivering" warm up, in which they contract thoracic muscles isometrically until the thorax is hot enough for takeoff.

Heat from the flight motor, and high thoracic temperatures, do not, by themselves, imply *thermoregulation*. Regulation requires feedback—a set of sensors and control mechanisms by which moths actively alter body temperature toward some set point. In an extraordinary set of papers starting in 1970, Heinrich and others (reviewed in (Heinrich, 1993)) established that flying moths, especially sphinx moths, show exquisite thermoregulation of body temperature before and during flight. They do it by modulating where hemolymph goes. Hemolymph moving forward from the heart is directed through the thoracic aorta, which loops through the thoracic muscles. There the hemolymph picks up heat. This suggests that moths could control temperature in the thorax by retaining heated blood within the thorax or by shunting it to other, cooler locations, such as the abdomen. Indeed, in *M. sexta*, heart pulsations appear to be suppressed during pre-flight shivering (Heinrich and Bartholomew, 1971), which reduces the flow of heat from thorax to abdomen. Flying moths face the opposite problem of dumping heat; indeed, many flying moths are restricted to flying in cooler ambient temperatures, probably to avoid overheating (Heinrich, 1993). Moths dump heat by shunting greater flows of hemolymph from the thorax, where it picks up heat, to the relatively cool abdomen, which has relatively less insulation and acts as a radiator (Heinrich, 1970b; Heinrich and Bartholomew, 1971; Heinrich, 1993). Heart rate also increases dramatically at high thoracic temperatures (Heinrich, 1971). Some small moths specialized for flying in cold weather have sophisticated counter-current heat exchangers in their thorax and abdomen (summarized in (Heinrich, 1993)).

A poorly resolved question about thermoregulation in moths is the location and nature of the thermosensors. In the silk moth *Hyalophora cecropia* it appears to reside in the thoracic ganglia (Hanegan and Heath, 1970; Hanegan, 1973). In *M. sexta*, experimentally heating the thorax to more than 40°C led to rapid, high-amplitude heartbeats, whereas heating just the abdomen did not (Heinrich, 1970a; Heinrich and Bartholomew, 1971). However, heating the thorax in moths with transected nerve cords did not raise heart rate. These results imply a thoracic temperature sensor for *M. sexta* too.

2.5.5 Coordination of Circulation and Ventilation

Besides heat, there is another currency that hemolymph distributes: pressure. Changes in internal pressure affect both flow of hemolymph and ventilation of the tracheal system. In some cases, especially in adult insects, this linkage is so tight that circulation and ventilation should be considered a single phenomenon, one that reflects the joint actions of intersegmental muscles, hemolymph, and compliant elements of the tracheal system (Wasserthal, 1996; Miller, 1997). In the relatively open hemocoels of larvae, hemolymph pressure rapidly equalizes everywhere. Contraction of the intersegmental muscles can therefore collapse compliant parts of the tracheal system (Slama, 1988)—flexible tracheal tubes and air sacs—but not readily move hemolymph in bulk (Wasserthal, 1996). In adults, by contrast, the hemocoel often surrounds many enlargements of the tracheal system (Wasserthal terms these enlargements the "aerocoel"), which partition hemolymph into functionally distinct compartments. Moreover, there can be substantial physical constriction between abdominal and thoracic spaces, which further isolates hemolymph compartments. In adults, therefore, pressure differences can arise between different hemolymph spaces, which can drive flow. In addition, heartbeat reversal can cause a slow shift of hemolymph from thorax to abdomen, with concurrent expansion of thoracic air sacs and compression of abdominal sacs (Wasserthal, 1996). A new technique recently applied to the hemolymph-ventilation coupling is x-ray videography. In conjunction with respirometry, this approach has provided clear views of the coordination between collapsing tracheal elements and expiration of carbon dioxide ((Socha et al., 2008; Greenlee et al., 2009); Chapter 6).

In our view, the circulatory system of insects is the most understudied of insect physiological systems, and it seems likely that further research will bring many important advances in ecological and environmental physiology. One fundamental question concerns the control of flow through the various tissues. In vertebrates, flow to various tissues is tightly regulated by arterial and capillary sphincters, with flow matched to metabolic demand. How is hemolymph flow driven through insect tissues, and regulated according to need? Insect hemolymph has much higher levels of carbohydrates (primarily trehalose) and amino acids than vertebrates. Weis-Fogh pointed out that this is likely related to the need to maintain substrate delivery sufficient to match nutrient needs despite relatively slow blood flow, and suggested that nutrient limitation, rather than oxygen delivery, may be more critical to metabolically active tissues such as flight muscle (Weis-Fogh, 1964). However, as yet, there has been little quantitative investigation of substrate delivery via the blood, though it has been shown that flight metabolism in bees can be correlated with blood sugar levels (Gmeinbauer and Crailsheim, 1993; Crailsheim, 1988).

2.6 Metabolic Systems in Active and Resting Insects, and in Diapause

Metabolism in insects is much like it is in all eukaryotes. In general, the essential amino acids are similar to those in vertebrates (Harrison, 1995). The need for vitamins has been little-studied in insects, but the data available suggest that insects and vertebrates require similar compounds (Dadd, 1961; Fraenkel and Blewett, 1947; Fraenkel and Blewett, 1943). Although some normal vitamins, like folate, are reportedly synthesized by insects, they probably are produced instead by microbes living inside them (Blatch et al., 2010). Glycogen is the primary carbohydrate reserve, primarily stored in fat body and muscle.

However, there are striking biochemical and metabolic differences between insects and vertebrates that scientists trained primarily in vertebrate biochemistry and physiology should be aware of (Table 2.1). First, the fat body, a tissue distributed loosely throughout the abdomen, serves the functions of vertebrate liver and adipose tissues (Keeley, 1985, Arrese and Soulages, 2010). The fat body is a major organ of intermediary metabolism and nutrient processing and storage. It has multiple cell types specialized for different functions (Arrese and Soulages, 2010). Lipids (mostly triacylglyerols), carbohydrates (mostly glycogen), and proteins (e.g. hexamerins) are stored here. Storage/transport/binding proteins (e.g. hexamerins, vitellogenin) are synthesized and released into the hemolymph, as are lipophorins, the major blood lipoproteins. Trehalose, the major blood sugar, is synthesized here and released into the hemolymph.

The fat body is critical in energy metabolism of insects. During feeding stages, nutrients are stored to support demands in metamorphosis, adult reproduction, or diapause. Egg development is supported by transfer of nutrients from the fat body (e.g. as vitellogenin). The lipid stores of the fat body are also important for pheromone and steroid hormone production.

Accumulation of triglyceride in lipid droplets within the adipocytes, and their release during lipolysis, depends on adipocyte-associated proteins called Lsd proteins, which bear sequence homology to PAT proteins that perform similar functions in vertebrates (Arrese and Soulages, 2010). Lsd1 seems to function similarly to perilipin in vertebrates. Phosphorylation of Lsd-1 in response to adipokinetic hormone (AKH) leads to lipolysis in *Manduca sexta*, and Lsd1 expression is low in feeding larvae but high in flying adults (Arrese et al., 2006; Patel et al., 2005;

Table 2.1 Some striking differences in the biochemistry and physiology of metabolism in insects relative to vertebrates

Trait	References
Fat body serves combined functions of liver (e.g. detoxification, lipoprotein synthesis) and adipose tissue	(Arrese and Soulages, 2010; Locke, 1998b)
Use of a disaccharide (trehalose) as the primary blood sugar	(Matthews and Downer, 1974; Moreau and et.al., 1984; Steele and Hall, 1985; Hayes and et.al., 1986; Becker et al., 1996; Ali and Steele, 1997; Fields et al., 1998; Blatt and Roces, 2001; Chen et al., 2002)
Use of diacylglycerols as the primary lipid transported in hemolymph	(Haruhito and Haruo, 1982; Chino, 1981; Downer and Chino, 1985; Soulages et al., 1996)
Use of storage proteins (hexamerins)	(Burmester et al., 1998; Hahn and Wheeler, 2003; Pan and Telfer, 1996)
High levels (often > 100 mmol l^{-1}) of organic acids, raising hemolymph osmotic pressures to about 400 mosmol l^{-1}	(Chen and Wagner, 1992; Leonhard and Crailsheim, 1999; Hanrahan et al., 1984)
Inability to synthesize cholesterol	(Clayton, 1964; Behmer and Elias, 1999)
Use of glutamate as the primary excitatory neuromuscular neurotransmitter	(Burrows, 1996)
Octopamine is the best-known "flight or fight" or stress hormone/neuromodulator	(Roeder, 2005; Adamo, 2008)
No known sex steroid hormone	
Hindgut utilizes primarily amino acids (proline) secreted by the Malpighian tubules as metabolic fuel	(Chamberlin and Phillips, 1982b, Chamberlin and Phillips, 1982a)
Generally limited use of anaerobic metabolism	(Gäde, 1985)

Arrese et al., 2008). In contrast, Lsd2 seems to promote lipid accumulation and is expressed at all developmental stages (Teixeira et al., 2003).

Glycogen is used to support the metabolic expenses of locomotion and to synthesize both chitin and the alcohol sugars for cold tolerance. Glycogenolysis is regulated by a phosphorylase under the control of various hormones. Glucose products of glycogenolysis are converted by trehalase to trehalose, which is exported to the hemolymph. Trehalose concentrations vary widely in hemolymph (5–50 mmol l^{-1}) and are reduced by starvation (Thompson, 2003). In addition to its role in energy metabolism, trehalose and other sugar alcohols are important in tolerating freezing (Storey, 1997), anoxia (Chen et al., 2002), and heat stress (Watanabe et al., 2002), because they stabilize protein structure.

The fat body can directly sense hemolymph nutrient levels and can respond accordingly. In mosquitoes, hemolymph amino acid levels, which are high after a blood meal, are transported into the fat body, stimulating the vitellogenesis needed for egg production via the target of rapamycin (TOR) pathway and a GATA transcription factor (Attardo et al., 2005; Hansen et al., 2005; Park et al., 2006).

As in vertebrates, multiple hormones regulate nutrient storage and release from the fat body. Adipokinetic hormone (AKH, released from the corpora cardiaca) stimulates lipolysis and glycogenolysis associated with flight, but the hormonal control of lipolysis during starvation is unknown (Gäde and Auerswald, 2003; Arrese and Soulages, 2010). Octopamine, the invertebrate counterpart to epinephrine, also stimulates lipolysis and glycolysis (Roeder, 2005). AKH and octopamine act via intracellular calcium, cAMP, and protein kinases to activate catabolic enzymes (Arrese and Soulages, 2010).

In insects, transport of lipids through the blood (hemolymph) differs in interesting ways from transport in vertebrates. Lipophorins synthesized by the fat body collect diacylglycerols (rather than triacylglycerols as in vertebrates) from the midgut, and transfer the diacylglycerols to the fat body without internalization of the lipoprotein into the cell as occurs in mammals (Law and Wells, 1989). The lipophorins are re-usable shuttles that also carry diacylglycerols from the fat body to other tissues (Soulages et al., 1996).

One aspect of insect biochemistry that differs from vertebrates is their lack of capacity to synthesize cholesterol; thus they are completely dependent on dietary intake of cholesterol-related compounds to synthesize their cell membranes and steroid hormones (Law and Wells, 1989). Herbivorous insects convert phytosterols to cholesterol, but only certain phytosterols can be utilized (Behmer and Elias, 2000; Behmer and Nes, 2003).

In their blood, insects generally have much higher levels of small organic compounds, primarily amino acids and organic acids, than do vertebrates. Proline is particularly common and important in insect hemolymph. Proline is synthesized in the fat body from alanine and acetyl-coA and then released into the hemolymph under the control of adipokinetic hormone (Bursell, 1981; Auerswald and Gäde, 2002). The tsetse fly and the Colorado potato beetle preferentially fuel flight with proline, recycling alanine back to the fat body so that proline essentially serves as a transporter of acetyl units from the fat body triglycerides to the flight muscle (Gäde and Auerswald, 2002). The organic acids of the hemolymph also contribute significantly to hemolymph buffering capacity (Harrison et al., 1990).

Diapause is an important component of the energetics of many insect species during periods when food is unavailable or conditions unacceptably harsh, such as during winter or prolonged dry spells. Diapause can stretch over many months, sometimes years. It is generally triggered by photoperiod, at least in temperate insects (Nelson et al., 2010). Hahn and Denlinger recently reviewed the energetics of insect diapause (Hahn and Denlinger, 2011). Prior to diapauses, insects accumulate metabolic reserves, or face high mortality during diapause (Hahn and Denlinger, 2007). Triacylglycerol, synthesized in the fat body, is usually the predominant stored fuel, though there can also be substantial accumulation of storage proteins, hemolymph amino acids, and carbohydrates, especially the alcohols glycerol and sorbitol (Hahn and Denlinger, 2011).

In diapause, metabolic suppression is an important means of conserving energy and nutrients. Low body temperatures can help (Irwin and Lee, 2003). Metabolic suppression during diapause results from selective suppression of growth and reproduction, with atrophy of some tissues such as the digestive and reproductive tract (Hahn and Denlinger, 2011). Interestingly, as in mammalian hibernators, some insects alternate deep metabolic depression with bouts of high metabolism (Denlinger et al., 1972). The bouts are associated with increased rates of protein synthesis and possibly repair (Joplin et al., 1990).

Diapause seems to be regulated by several hormones, including juvenile hormone and insulin (Hahn and Denlinger, 2011). In *Drosophila*, disruption of insulin signaling induces a diapause-like state (suspended reproduction and accumulation of energy reserves), which can be reversed by application of juvenile hormone mimics (Tatar et al., 2001; Tatar and Yin, 2001). Insulin signaling also seems to promote reproductive diapause and nutrient accumulation in *Culex* mosquitos (Sim and Denlinger, 2008). These data suggest a common model in which prediapause insects depress insulin levels, leading to accumulation of the transcription factor FOXO, which then promotes nutrient accumulation and reproductive suppression (Hahn and Denlinger, 2011). Termination of diapause may be associated with elevation of juvenile hormone and/or ecdysteroids (Denlinger et al., 2005).

2.7 Insect Feeding and Digestive Systems

2.7.1 Morphology and Basic Physiology

The incredible diversity in insect feeding strategies corresponds to similar diversity in the morphology and physiology of the feeding and digestive systems. Here we focus on similarities and differences between these systems in insects and vertebrates, with whom some readers will be more familiar. More detailed physiology background is available in the recent insect physiology texts from Klowden (Klowden, 2009) and Nation (Nation, 2008).

The mouthparts of insects are complex and highly varied. In the head and cranial region of the thorax, there is close association between the mouthparts, salivary glands, digestive tract, and the subesophageal ganglion which is a primary regulator of digestive activity (Fig. 2.16).

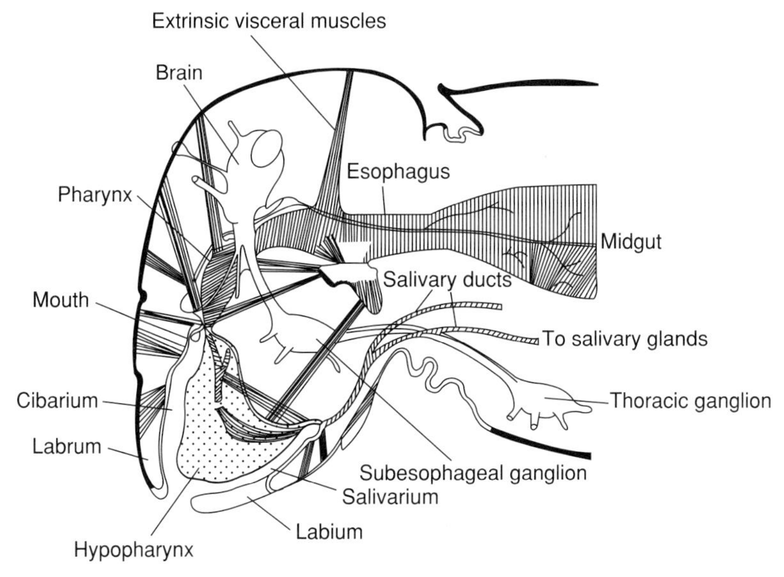

Fig. 2.16 Section of the head of a grasshopper, showing the pharynx, esophagus, midgut, anterior portion of the central nervous system (brain, subesophageal ganglion, thoracic ganglion) and some of the mouthparts. Reprinted from Snodgrass (1935).

In most insects, the digestive tract is divided morphologically and functionally into foregut (often called a crop), midgut, and hindgut (Fig. 2.17). Salivary glands empty into the buccal cavity, and typically produce a watery, enzyme rich fluid to lubricate the food and initiate digestion. The foregut often stores recently ingested food, and mixes ingested food with secretions. The proventriculus serves as a valve to control food passage into the midgut, as a filter to separate solids and liquids, or as a grinder to break up solid material, depending on the insect and its food source. The midgut is where most digestion and absorption occurs. The hindgut combines the functions of the distal regions of the vertebrate nephron and the colon, being a tight epithelium that modifies the ion and water content of the feces.

The generalized digestive structure varies enormously across insects. At the intersection of the foregut and midgut many insects, especially Orthoptera, have blind-ended pouches (caeca) that promote nutrient breakdown and may be major sites of nutrient absorption (Dow, 1986). In insects such as the cricket, it has been demonstrated that most digestion occurs in the caeca (Woodring et al., 2007a). Several types of insects have loops in the digestive tract that facilitate transport of materials between sections. For example, in aphids, the wall of the crop contacts the wall of the hindgut, and aquaporins at this site likely permit water transport down an osmotic gradient between these gut compartments

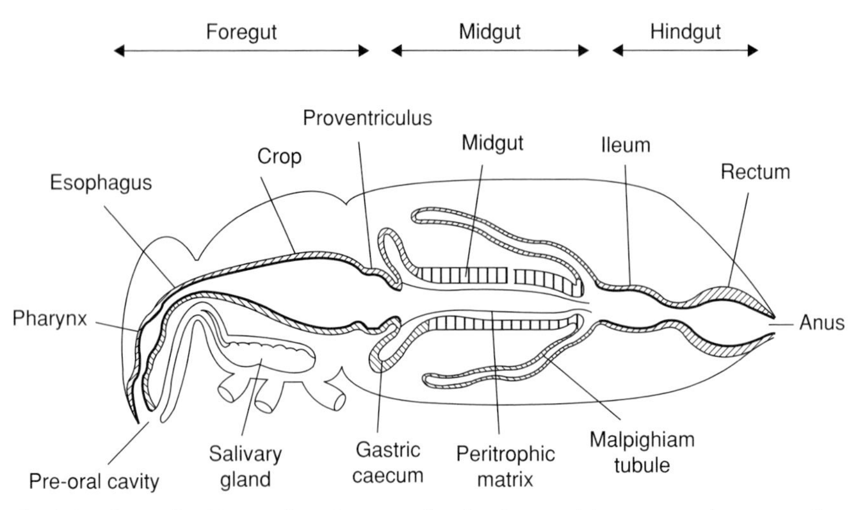

Fig. 2.17 Generalized insect digestive tract, showing foregut (pharynx, esophagus, crop), midgut (with cecae, Malpighian tubules), and hindgut (ileum, rectum). From Dow (1986), used with permission from Elsevier.

(Shakesby et al., 2009). Similarly, the cryptonephridial complex found in some Coleoptera and Lepidoptera connect the posterior part of the Malpighian tubules to the hindgut; active K^+ transport into the tubule lumen creates an osmotic gradient that draws water out of the rectal lumen into the Malpighian tubules, reducing fecal water loss (Phillips, 1970).

Another fascinating variation is the discontinuous guts of some larval true bugs, ant lions, bees, wasps, and ants. In these species, the midgut and hindgut are not connected during some of the larval stages, and food accumulates in the midgut until later developmental stages when defecation occurs. This morphology may function to prevent self-soiling of the larvae (Woodring et al., 2007b). Readers interested in the diversity of these systems are referred to: (Dadd, 1970; Applebaum, 1985; Labandeira, 1997; Chapman, 1998).

Most insects line their guts with multiple layers of aperitrophic matrix rather than the secreted mucin commonly found in vertebrates (Hegedus et al., 2009). The peritrophic matrix may be secreted at the junction of the foregut and midgut (Dermaptera, Isoptera, larval Diptera, and some Lepidoptera), or along most of the midgut (most orders). The peritrophic matrix consists of a lattice of chitin fibrils structured by binding proteins. Proteins (peritrophins) interact with chitin to form anionic glycosaminoglycans (Hegedus et al., 2009). Glycans (glycoproteins called mucins) fill the interstitial spaces, allowing the peritrophic matrix to resist digestion by proteases and to create a molecular sieve. Macromolecules in the food are unable to pass through the peritrophic matrix, and are partially digested there by proteases, lipases, and amylase. The products of these steps (e.g. dipeptides, disaccharides) pass through the peritrophic matrix, and then are further digested by exopeptidases in the midgut lumen but outside the peritrophic membrane and by enzymes associated with the microvilli of the midgut (e.g. trehalase, aminopetidase). Water may be secreted by the posterior midgut and absorbed by the caeca or anterior midgut, creating a flow of fluid that runs counter to the anterior-to-posterior movement of the food bolus, enhancing digestion and absorption (Dow, 1986).

Some insects do not have peritrophic membranes, such as adult ants (Hymenoptera), most adult moths and butterflies (Lepidoptera), Psocoptera, adult fleas (Siphonaptera), Bruchidae (Coleoptera), and Phthiraptera (Terra, 2001). Hymenoptera and Lepidoptera feed almost exclusively on low-molecular weight substances such as sugar, which does not require luminal digestion, perhaps explaining the lack of peritrophic membrane in these groups (Terra, 2001). Heteroptera (sap-sucking insects) also lack a peritrophic membrane, but have evolved a perimicrovillar membrane that serves a similar function (Woodring et al., 2007b).

2.7.2 pH and Redox of the Digestive Tract Lumen

The pH of the gut lumen is, in most cases, actively regulated at values different from those of the hemolymph (Cooper and Vulcano, 1997; Dow and Harvey, 1988). For most insects, the pH of the gut regions varies, with a general trend toward acidic crops, neutral to alkaline midguts, and neutral to acidic hindguts. This general pattern shows many variations, some of which are related to phylogeny (Clark, 1999). Larvae of many endopterygote orders have highly alkaline regions within their midguts (particularly Lepidoptera, Coleoptera, and Diptera). All members of the Mecopterida except the derived dipteran lineage Muscomorpha have alkaline midguts. This suggests that alkaline midguts are ancestral within this group and that the very acidic midguts of the muscoid and tephritid flies are derived (Clark, 1999). Among the higher termites, most subfamilies exhibit extreme alkalinity in the hindgut (Brune and Kuhl, 1996). Lower termites and the Macrotermitinae have a simpler gut morphology and more neutral digestive lumens (Bignell and Anderson, 1980). Exopterygote insects usually have more neutral midguts, suggesting that this is the ancestral condition for Insecta, but phylogenetic studies are lacking.

Redox conditions vary with diet in some insects, but can be regulated in some species and gut compartments (Johnson and Felton, 2000; Johnson and Felton, 1996a). For example, the midguts of wool- and fur-feeding insects are highly reducing environments, and this appears to be critical for cleavage of the disulphide bonds of keratin (Applebaum, 1985). Some plant allelochemicals (e.g. juglone) can produce strong effects on midgut redox conditions (Johnson and

Felton, 1996a). Manipulation of redox levels with a thiol reducing agent reduced gut proteinase activities in the corn earworm (*Helicoverpa zea*), but did not affect growth or digestive efficiencies, suggesting that insects can compensate for moderate gut redox disruption (Johnson and Felton, 1996b).

2.7.3 Midgut Function: Digestion

The midgut epithelia is a single-cell layer that contains multiple cell types, including endocrine cells. Best described are columnar cells or principal cells, which have abundant apical microvilli and basolateral mitochondria and seem specialized for absorption (Billingsley and Lehane, 1996). These cells also contain extensive endoplasmic reticulum and are believed to also be major sites of digestive enzyme production. These cells have limited lifespan, and are regenerated by stem cells present in groups called nidi.

Goblet cells occur in the midgut of lepidopteron and ephemeropteran larvae. These have an invaginated shape, with an inner cavity that communicates to the midgut lumen via an apical pore formed of inter-digitated microvilli (Billingsley and Lehane, 1996). Goblet cells are believed to be responsible for using active transport to create the large electrochemical and pH gradient across the epithelia of these insects that drives reabsorption of most nutrients and ions.

Digestive enzymes are produced by the columnar cells of the midgut epithelia, and are secreted in response to the presence of food, stretch of the gut, and neuroendocrine regulation associated with the enteric nervous system and local endocrine cells (Terra et al., 1996; Terra, 1990). Digestive enzyme synthesis can be stimulated by exposure to soluble food at the transcriptional and translational level (Lehane et al., 1995). Protein digestion to oligopeptides begins within the peritrophic membrane, and is carried out by endopeptidases, which hydrolyze peptide bonds within the polypeptide chain (Terra et al., 1996; Terra, 1990). In many insect orders (e.g. Diptera, Lepidoptera, Hymenoptera, and some Coleoptera) these enzymes, including elastases, trypsins, and chymotrypsins, are members of the serine protease superfamily and are highly active under conditions of alkaline pH. Other insect groups (Hemiptera and some Coleoptera) have acidic midguts and utilize cysteine proteases such as cathepsins (Terra et al., 1996; Terra, 1990). The oligopeptides produced by the endopeptidases are then digested by exopeptidases (carboxypeptidases and aminopeptidases) into amino acids and dipeptides. Some of these exopeptidases are associated with the microvilli, and some are secreted by the columnar cells (Terra et al., 1996).

The expression of different types of protein-digesting enzymes seems to vary spatially in a manner consistent with the progress of protein digestion and absorption through the midgut. For example, in the human body louse, exopeptidases are secreted in the anterior midgut compartment where blood is stored after feeding, while exopeptidases are secreted in the posterior regions of the midgut

(Waniek, 2009). Similarly, in the sugar cane weevil, *Sphenophorus levis*, cathepepsin begins protein digestion in the anterior midgut, while aminopeptidases are concentrated in the posterior midgut where they complete protein digestion (Soares-Costa et al., 2011).

Soluble carbohydrates are digested by amylase, lysozyme, glucosidase, galactosidase, trehalase, and fructosidase (Terra et al., 1996). Lipids are digested by lipases and esterases secreted by the columnar cells of the midgut, and by the salivary glands in some cases (Terra et al., 1996). The role of cellulases and hemicellulases such as are found in mammalian ruminants had not generally been considered to be important in most herbivorous insects, excepting specialized wood and detritus feeders such as termites (see Chapter 5). However, increasingly, it is becoming clear that many insect herbivores do possess cellulase and xylanase activity in their guts, and this can be due both to intrinsic cellulases and microbial forms (Shi et al., 2011). Thus the physiological importance of cellulose/xylan digestion in insect herbivores may require re-evaluation (Chapter 5). Types of digestive enzymes produced by insects varies with phylogeny and diet (Chapter 5). For example, *Oncopeltus fasciatus*, a true bug (Heteroptera) feeds on seeds containing large amounts of lipid, some protein, and very little carbohydrate. This bug secretes large amounts of lipase and proteases, but very little amylase (Woodring et al., 2007b).

2.7.4 Midgut Function: Transport and Absorption

Strong alkalization of up to pH 12 is a characteristic feature of midgut regions of many larvae of endopterygote insects, including members of the orders Coleoptera, Diptera, Trichoptera, and Lepidoptera (Clark, 1999). Insect midgut ion transport has been studied most intensively with lepidopteran larvae, particularly *Manduca sexta*. Data for *Manduca* strongly suggests that the strong alkalization of the midgut lumen is driven by an apical proton V-ATPase in the goblet cells (Wieczorek et al., 2009). Proton transport is linked to apical exchange of 2 H^+ for one K^+, thus causing K^+ secretion into the lumen (Wieczorek et al., 2009).

In dipterans (mosquitoes, *Drosophila melanogaster*), the proton-transporting V-ATPase is located on the basolateral membrane of cells of the anterior midgut (Shanbhag and Tripathi, 2005; Onken et al., 2006). Inhibitors of the V-ATPase strongly reduce alkalization of the midgut (Onken et al., 2008). Serotonin is a strong stimulator of luminal alkalization (Onken et al., 2008). However, unlike in lepidopterans, there is no evidence for a role for apical $K^+/2H^+$ antiporters or Cl^-/HCO_3^- exchangers, so the mechanisms by which basolateral proton transport are linked to luminal alkalization in mosquitos are unclear (Onken and Moffett, 2009).

Amino acid absorption by the midgut can occur through multiple mechanisms. When luminal concentrations of amino acids are high, most transport can be accomplished by uniporters, driven by the diffusion gradient for that amino acid (Wolfersberger, 2000). However, at least in *Manduca* and *Bombyx mori*, uptake of low concentrations of amino acids from the midgut lumen is accomplished by a symporter that transports K^+ with the amino acid; this uptake is driven by the K^+ and lumen positive electrical gradients, and so is secondarily dependent on the V-ATPase and $K^+/2H^+$ transport system in the goblet cells (Wolfersberger, 2000).

Fatty acids are absorbed by diffusion, primarily in the midgut (Turunen and Crailsheim, 1996). Fatty acid binding proteins occur within the midgut epithelia, and likely contribute to facilitating diffusive uptake of fatty acids (Turunen and Crailsheim, 1996). Sterols diffuse into the midgut epithelia, where they are esterified to polar molecules. The primary lipid released by the midgut into the hemolymph is diacylglyerol rather than triacylglycerol as in vertebrates. Fat-binding proteins called lipophorins transport the diacylglyerols and sterols to target tissues (Soulages et al., 1996; Downer and Chino, 1985; Chino et al., 1981).

Monosaccharide absorption has been little-studied in insects, however there is evidence that, at least in some insects, the system may be quite similar to that of vertebrates. In aphids, cation-linked secondary active transport of monosaccharides occurs on the apical membrane; inhibitor studies suggest these are similar to vertebrate SGLT1 transporters (Caccia et al., 2007). Basolateral transport of glucose occurs by facilitated diffusion using a GLUT-like transporter (Caccia et al., 2007). Similarly, in the hemipteran, *Dysdercus peruvianus*, there is evidence for facilitated hexose transporters analogous to the GLUT transporters of mammals, as well as a secondary active hexose uptake system that may utilize K^+ rather than Na^+ gradients for energizing uptake (Bifano et al., 2010). Especially in nectar-feeders, passive diffusional uptake by uniporters seems likely, as there is a strong glucose gradient from the midgut lumen into the midgut cells, with absorbed glucose being converted to the disaccharide trehalose (Turunen and Crailsheim, 1996). In most insects, trehalose is released from the midgut epithelium into the hemolymph, and is the dominant blood sugar.

2.8 Insect Renal Systems

The system controlling water and salt balance in most insects consists of two interacting parts—the Malpighian tubules (MTs) and the hindgut. The hindgut is further divided into ileum and rectum, which may be functionally divergent. Figure 2.17 shows an idealized insect gut, with the blind-ended Malpighian tubules emptying into the junction between mid- and hindgut.

The functional strategy embodied by the tubule-hindgut system is to throw out the bathwater—baby and all—then to recover constituents worth saving, including water, ions, and other small molecules (Beyenbach, 2003). Insects *secrete* primary urine rather than producing it by (glomerular) filtration, as vertebrates do, because insect blood pressures typically are too low to drive filtration. The mammalian kidney can turn over the entire extracellular fluid volume about every 2 hours. Because they rely on secretion, insects might be expected to process extracellular fluids more slowly. Not so—a recently fed *Aedes aegypti* mosquito can process its entire extracellular fluid in about 7 minutes (Beyenbach 2003).

The Malpighian tubules drive fluid into the lumen by secreting ions, primarily Na^+ and K^+. In pursuit of electrical neutrality, Cl^- follows passively. The energy for ion secretion is provided by V-type H^+ATPases and possibly also Na^+/K^+ ATPases. V-type H^+ ATPases inhabit the apical membrane at high densities and usually are associated with mitochondria. The protons they pump then move back into the epithelium in exchange for Na^+ and K^+, via one or more kinds of antiporters. The salts—NaCl and KCl—secreted into the lumen draw water and other blood constituents into the tubule by osmosis (Beyenbach, 2003). Recent reviews of transport mechanisms in MTs are available from (Wieczorek et al., 2009; Harvey, 2009; Beyenbach et al., 2010). In the hindgut, insects have mechanisms for selectively reabsorbing water, ions, and nutrients; this absorption too is energized by V-type H^+ ATPases and Na^+/K^+ ATPases.

2.9 Insect Nervous System

2.9.1 Neurons

As is the case in all organisms, the nervous system of insects permits rapid chemo-electrical communication for activating and coordinating events at target cells and tissues that are innervated. The evolution of neurons and nervous systems far pre-dates insects, and so basic insect neuronal anatomy and function is similar to most other animals. Therefore, we are omitting a comprehensive review of basic neuronal anatomy and function due to space constraints, and refer readers to excellent treatments of the subject in (Nation, 2008) and (Klowden, 2009). In summary, insect nervous systems are comprised of sensory (or afferent) neurons, motor (or efferent) neurons, interneurons, secretory neurons, and glial cells. Insect neurons may be either monopolar, bipolar, or multipolar.

Sensory neurons carry signals from sensory structures to the central nervous system (CNS). Insects sensory neurons are usually bipolar, with somata (i.e. neuronal cell bodies) that tend to be located near the receptor (unlike the case in vertebrates), short dendritic processes projecting into the receptor, and long axons extending to the CNS. Motor neurons innervate and regulate the contraction of muscle (see Section 2.3.2), and the somata of motor neurons and interneurons are situated peripherally within neuronal aggregates called ganglia in the

CNS. The interior of a ganglion, called the central neuropil, is comprised of axons and dendrites of many neurons and synapses between them. Because of the overall bilateral symmetry of insect musculature, motor neurons are typically paired on each side of a ganglion, with the axon of a neuron most often extending towards muscle on the same side as the soma residing in the ganglion. However, dorsal unpaired medial (DUM) motor neurons are not paired, but project an axon out of each side of a ganglion to a gland or muscle.

Interneurons are the most common neurons of the insect nervous system, and synapse with all other neuron types in complex networks coordinating sensory receptors, integration/control centers, and effectors. Local interneurons are typically paired and contained entirely within a ganglion, while intersegmental interneurons (also paired) project axons through inter-ganglionic connectives and create synapses in other ganglia. There are also local and intersegmental DUM interneurons. Interneurons can be spiking (active via classic depolarization waves transmitted along the axon) or non-spiking (lacking axons and active within ganglia via graded changes in membrane polarity).

Secretory neurons are specialized neurons that secrete chemical messengers (neuromodulators and neurohormones) into the hemolymph or directly onto target cells. The effects of neurosecretory substances on target cells persist longer than events controlled by neurotransmitters such as synaptic transmission and muscle contraction. Neurohormones delivered to the hemolymph are released through a specialized structure on the secretory neuron called a neurohemal organ, while neuromodulators are released at synaptic contacts.

2.9.2 Central Nervous System

The brain, ventral ganglia, and ventral nerve cord comprise the insect central nervous system. The brain is actually composed of three fused ganglionic masses: the protocerebrum, the deutocerebrum, and the tritocerebrum. These bilateral masses lie atop the esophagus and are collectively termed the supraesophogeal ganglion. Posterior to the supraesophogeal ganglion and ventral to the esophagus is the subesophogeal ganglion. The insect brain can be made up of over 1 million neurons, compared to approximately 100 billion in the human brain and 302 in the brain of the nematode worm *Caenorhabditis elegans*. The insects with the largest bodies have the biggest brains, but in ants, brain volume scales with body mass with a coefficient of less than 1 (Wehner et al., 2007; Seid et al., 2011), just as in vertebrates.

The protocerebrum is the most anterior region of the supraesophogeal ganglion and features the neurosecretory centers of the brain—the pars lateralis and the pars intercerebralis—as well as the optic lobes, the corpora pedunculata (or mushroom bodies), and the central body complex. The optic lobes process input from the eyes, while the mushroom bodies, which receive input from the olfactory lobes of the deutocerebrum, are the center of sensory integration and learning. Many specific functions of the central body complex remain unknown, but it plays an important role in higher locomotion control.

The deutocerebrum, lying just behind the protocerebrum, is made of several sensory neuropils that receive input from chemosensory neurons of the antennae and mechanosensory neurons on the antenna and regions of the head cuticle. The antennal lobes are neuropils of the deutocerebrum that receive chemosensory input, while mechanosensory input is directed to the antennal mechanosensory and motor center, which also projects efferent neurons to muscles and glands in the head. The tritocerebrum, situated beneath the deutocerebrum, receives sensory input from the mouthparts and innervates muscles of the mouthparts and the visceral nervous system (discussed below). Short connectives bridge the tritocerebrum to the subesophageal ganglia, the first ganglia of the ventral nerve cord, which innervates muscles of the mouthparts and neck and receives sensory input from these same regions. The subesophageal ganglia also help coordinate motor patterns controlling walking, flying, and breathing.

The ventral nerve cord, with interspersed ventral ganglia, projects posterior from the head along the main axis of the body. There are three thoracic ganglia, one per segment, that generate motor patterns for flying and walking. They may be completely or partly fused into a metathoracic ganglion, an evolutionarily derived condition prominent in hymenopterans and dipterans. The number of abdominal ganglia is highly variable, ranging from eight in thysanurans (one per segment) to none. In many orders, the forward-most abdominal ganglia have fused with thoracic ganglia into one metathoracic ganglion. The posterior abdominal ganglia are often fused to form a terminal abdominal ganglion that controls hindgut and reproductive organs.

2.9.3 Peripheral and Visceral Nervous Systems

The peripheral nervous system comprises most of the affector and effector neurons branching from the CNS. These branches are usually bundles of intermixed sensory and motor neurons, although some carry only one type of neuron. Such is the case with the ocellar nerve, which is comprised only of sensory nerves sending signals from the ocelli to the protocerebrum. Overall, the peripheral nervous system innervates and delivers sensory information from the muscles, cuticle, and reproductive organs. The visceral nervous system controls and receives feedback from organs and systems that maintain the internal environment of the insect. These include the gut, heart, endocrine glands, and certain mouthparts, and their control is necessary for regulating various aspects of ingestion, digestion, absorption, and excretion. The neuropils of the visceral nervous system include the tritocerebrum (and several smaller, peripheral ganglia that it innervates) and many nerves of the terminal abdominal ganglion.

2.9.4 Glial Cells and the Blood–Brain Barrier

The composition of insect hemolymph is different than the intracellular composition of neurons, and any direct contact between the two would impair the neuron's ability to generate action potentials. Glial cells adjacent to neurons secrete sheaths around the neurons that act as an ionic and electrical insulator for the neuron. These sheaths are absent at the synapse where neurotransmission occurs. Parker and Auld (2006) comprehensively review the function of insect glial cells, which help direct neuron growth and degradation, provide structural support and nutrition for neurons, and help remove excess neurotransmitter at synapses. The sheaths surrounding insect neurons are not perfectly analogous to the myelin sheaths surrounding vertebrate neurons, because the sheaths around insect neurons are continuous along their length, while the Schwann cells that wrap around vertebrate neurons are spaced, leaving gaps between them along a neuron without insulation that permit saltatory, or jumping, propagation of action potentials. Hence, saltatory propagation does not occur in insects, which is probably of little consequence due to their small size (and reduced propagation distances) relative to vertebrates.

In addition to secreting a simple sheath around individual neurons, multiple layers of glial cells create a complex, layered network of channels and membranes around ganglia and connectives. The outermost layer, the neural lamella, is a thick sheath of connective tissue secreted by an underlying layer of glial cells that form the perineurium. These layers form the insect blood–brain barrier that protects the CNS from toxins in the hemolymph and ionic challenges due to the hemolymph's chemical composition (Treherne and Schofield, 1981). The glial cells of the perineurium are overlapping and connected to each other with septate and tight junctions to minimize paracellular movement of ions and other substances between the extracellular fluids of the CNS neurons and the hemolymph. The regulated movement of Na^+ and K^+ across the blood–brain barrier, occurring transcellularly through NA^+/K^+ pumps and ion channels, is particularly important for maintaining CNS function, especially when hemolymph ion concentrations change due to feeding or exercise. A recent study of the cockroach blood–brain barrier by (Kocmarek and O'Donnell, 2011) indicates that mechanisms for regulating extracellular K^+ flux differ between the connectives and ganglia, resulting in K^+ following a cyclic path via net K^+ influx at the connectives and balancing K^+ efflux at the ganglia. However, the function of this cyclic path is unknown.

2.9.5 Insect Neurotransmitters, Neuromodulators, and Neurohormones

Like neurons, neurotransmitters far pre-date insects, and not surprisingly insects share many of the same neurotransmitters with other animals. Acetylcholine is the primary excitatory neurotransmitter and is released by sensory neurons and interneurons. The main excitatory neurotransmitter of the insect neuromuscular junction is L-glutamate, while GABA is an important inhibitory neurotransmitter at the neuromuscular junction and at synapses within the neuropil. GABA can act to either hyperpolarize post-synaptic membranes (thus requiring a stronger excitatory signal to reach threshold depolarization), or inhibit the vesicular release of an excitatory neurotransmitter from the pre-synaptic membrane.

Nitric oxide (NO) is a gaseous cellular messenger that acts in several physiological systems. Typical neurotransmitters are stored in vesicles and used at very precise times and locations at the synapse, but NO is made on-demand in a calcium-dependent reaction catalyzed by cytoplasmic NO-synthase (NOS) and diffuses freely across cell membranes into target cells. Hence, an NO-producing neuron may affect nearby target cells with which it has no synaptic connections. NO can bind to transcription factors or soluble guanylate cyclase (sGC), which upon binding to NO synthesizes the second messenger cGMP, a molecule that can have a wide variety of effects on the target cell. In insects, NO signaling occurs in a variety of neurological functions including neuronal motility and regeneration, sensory processing, learning and memory, and motor pattern modulation (reviewed by (Heinrich and Ganter, 2007)).

The insect nervous system secretes a diverse array of peptide neurohormones to control a variety of physiological and biochemical processes. Many are discussed in several sections throughout this chapter, but space limits prevent a thorough, stand-alone description of their functional diversity, and readers are directed to recent reviews by (Gade et al., 2006; Nassel and Homberg, 2006; Mercier et al., 2007; and Mykles et al., 2010). Prothoracicotropic hormone (PTTH) is a large neuropeptide produced by neurosecretory cells of the corpus cardiacum or corpus allatum. PTTH is released by these glands via environmental cues and acts on the prothoracic glands, which respond by releasing the ecdysteroid molting hormone into the hemolymph to cause molting. Adipokinetic hormones (AKHs) are a family of related neuropeptides that are synthesized and released by neurosecretory cells of the corpus cardiacum. The primary function AKH is the regulation of lipid motility to fuel flight (see Section 2.3.4).

The biogenic amines octopamine and tyramine are important, tyrosine-derived compounds that control many aspects of insect behavior and metabolism (Roeder, 2005). Both the hormones and their receptors are homologous to the adrenergic counterparts in vertebrates (Roeder, 2005). Octopamine-producing DUM and VUM (ventral unpaired median) cells occur throughout the insect body including the brain. As in vertebrates, these hormones act via G proteins and are released from a few, widely-arborizing neurons. The physiological roles of tyramine are poorly known, but octopamine is considered the insect stress, or "flight-or-fight," hormone (Roeder, 2005). In addition to its role in stimulating the fat body to release fuels into the hemolymph (discussed below), octopamine functions along with cardioacceleratory peptides to stimulate heart muscle function (Prier et al., 1994) and increase the power output of flight muscle (Malamud et al., 1988). Octopamine also stimulates ventilation (Bellah, 1984). It can also elicit a variety of behaviors (Sombati and Hoyle, 1984) and modulate the input of virtually all sensory systems (Roeder, 2005). Octopamine also appears to suppress the immune system, as corticosterone does in vertebrates (Adamo, 2008).

2.10 Energizing Nutrient and Water Transport across Insect Epithelia

In insects, as in all animals, nutrient uptake and water regulation both depend on convincing nutrients and water to go where they otherwise would not. Two related principles underlie nutrient and ion transport. First, nutrients generally are more concentrated in insect tissues than they are in the foods they eat. Therefore, insects need some way to force nutrients up their concentration gradients. Second, water is not subject to primary transport itself; it has to be led by osmosis, which also requires moving ions. Thus, nutrient and water transport both depend on ion transport. In turn, ion transport itself is energized by proton ATPases.

2.10.1 Ion Transport from ATPases

Na^+ and K^+ are not moved by primary active transport (transported directly by ATPases). Rather, primary active transport moves protons, by V-type H^+ ATPases with Na^+ and K^+ moved themselves by H^+/cation antiporters, which tap the large electrical potentials produced by primary proton transport (reviewed by (Harvey, 2009); (Wieczorek et al., 2009)). In midguts, the electrochemical gradients established for Na^+ or K^+ can also be used to drive amino acid uptake via cation-amino-acid symport (see (Harvey, 2009)). Below we discuss the roles of both V type H^+ ATPases and P-type Na^+/K^+ ATPases.

The discovery of the role in ion transport played by V-type H^+ ATPase emerged from work on the midguts of *Manduca sexta*, which are highly electrogenic and support rapid nutrient absorption. From the apical membranes of goblet cells in *Manduca* midgut, (Schweikl et al., 1989) purified ATPases and showed that they were the vacuolar type (V-type ATPases)—a surprising finding because V-type ATPases had previously been associated with intracellular organelles. V-type ATPases transport protons rather than metal cations (Wieczorek et al., 1989). The cations themselves (in *Manduca*, K^+) are transported by a K^+/H^+antiporter (Wieczorek et al., 1989; Wieczorek et al., 1991), which uses the large lumen-positive electrical gradient, established by V-type ATPases, to drive K^+ into the lumen. The key stoichiometry of this process is the 2 H^+:1 K^+ exchange required by the antiporter (Azuma et al., 1995); because the exchange is not electroneutral, it allows the antiporter to tap the energy in the electrical gradient.

V-type ATPases also energize ion, and thus water, transport in the Malpighian tubules of mosquitoes (Filippova et al., 1998; Beyenbach et al., 2000; Wieczorek et al., 2009) and possibly also in hindguts (Smith et al., 2008). In Malpighian tubules, V-type ATPases pump protons across the apical face into the tubule lumen. The protons are then exchanged, in a 2:1 ratio (protons:cations), for sodium or potassium, which builds up in the lumen and draws in water by osmosis to form primary urine.

Although V-type H^+ ATPases have received most of the attention, there is another transporter, Na^+/K^+ATPase, that may be able to establish ion gradients for moving water. For example, in larvae and adults of *Ae. aegypti*, both V-type H^+ ATPases and P-type Na^+/K^+ ATPases are distributed in all important osmoregulatory organs (Patrick et al., 2006). In distal parts of the Malpighian tubules (the strongly secreting part), V-type H^+ ATPases were present in principal cells (both in the cytoplasm and on the apical membrane), and Na^+/K^+ ATPases were present in stellate cells but not the principal cells. In more proximal sections of the Malpighian tubules, Na^+/K^+ ATPases were on basal membranes and V-type H^+ ATPases on apical membranes. Interestingly, larval rectal tissues contained only the Na^+/K^+ ATPase, suggesting that it energizes uptake of ions from urine; however, later work (Smith et al., 2008) found rectal V-type H^+ATPases. Other species show similar overall patterns of localization of the two ATPases (Okech et al., 2008; Smith et al., 2008), establishing "beyond any reasonable doubt that both ATPases are abundant in insect plasma membranes, usually in the same cells" and probably both kinds provide primary energy for other ion- and water-moving processes.

3
Temperature

3.1 Defining the Problem

The effects of temperature on the activity, distribution, and abundance of insects (as least terrestrial species) likely exceed those of any other abiotic factor. Because most insects are poikilothermic ectotherms, their life histories unfold only if their body temperatures fall within a range that permits the necessary biochemical, developmental, physiological, and behavioral functions. As a result, insects have evolved an impressive array of physiological and behavioral adaptations affecting every route of heat flux between an insect's body and its environment: convection, conduction, radiation, and evaporation. Moreover, for a small fraction of insect species, metabolic heat supplements environmental heat and helps them achieve and maintain necessary minimum body temperatures and, in some cases, challenges them with extreme high temperatures. Despite the efficacy and diversity of thermoregulatory mechanisms, insects often experience extreme high and low body temperatures that, in the absence of specific mechanisms for thermotolerance, can injure or kill them. As with all aspects of insect environmental physiology, the duration of exposure to thermal variation has strong effects on patterns and mechanisms of response. Thermal biology reveals some of the best examples of trade-offs, illustrating advantages of specializations that do much to explain the diversity of insects. This chapter addresses the following questions:

3.1 What are the consequences of variation in insect body temperature?
3.2 What are the sensory and control systems that allow insects to respond to thermal variation?
3.3 What are the mechanisms of heat balance, and how do insects regulate body temperature?
3.4 What are the mechanisms of thermotolerance that protect insects from extreme body temperatures?

3.1.1 Non-pathological Consequences of Thermal Variation

The physiology, development, and behavior of insects are intimately linked to temperature. Over a broad range of body temperatures, enzyme-catalyzed chemical reactions become faster with increasing temperature as greater fractions of enzymes and their substrates achieve velocities yielding Gibbs free energy values above their activation energies. Hence, within a finite range, essentially all enzyme-dependent reactions and physiological processes are positively affected by temperature. Of course at high enough temperatures, when proteins and membranes start to fall apart (see below), performance no longer increases with temperature, but instead plateaus and then declines precipitously (Figs 3.1 and 3.2). In fact, temperature–performance relationships have characteristic shapes and are known as thermal performance curves (Huey and Kingsolver, 1989; Angilletta, 2009).

The effects of temperature on biochemistry drive generally positive relationships between body temperature, performance, and fitness over a broad range of temperatures, and these benefits are reflected in diverse patterns of behavioral and physiological thermoregulation (Chown and Nicolson, 2004; Chown and Terblanche, 2007; Heinrich, 1993). The increases in growth rate with higher temperatures can enable insects to achieve more generations per season and higher population growth rates (Frazier et al., 2006), and to reproduce in ephemeral environments. Examples abound showing that metabolism and growth increase, and longevity decreases (Fig. 3.3), with increasing, non-pathological

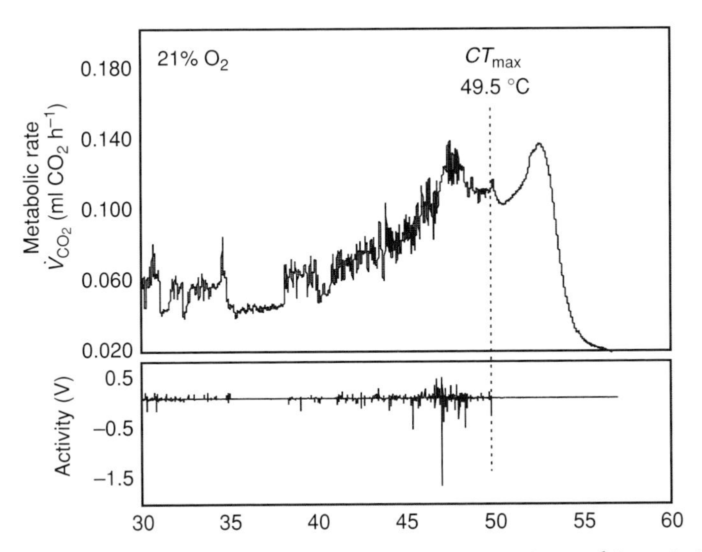

Fig. 3.1 Metabolic rate and activity level during warming at 0.25°C min^{-1} for an individual beetle, *Gonocephalum simplex* (Tenebrionidae). Activity is in arbitrary units (V), where both negative and positive values indicate activity. The critical thermal maximum (CT_{max}) is the temperature at which activity ceases, and the large release of CO_2 at 52°C occurs as the insect expires. From Klok et al. (2004), with permission from The Company of Biologists.

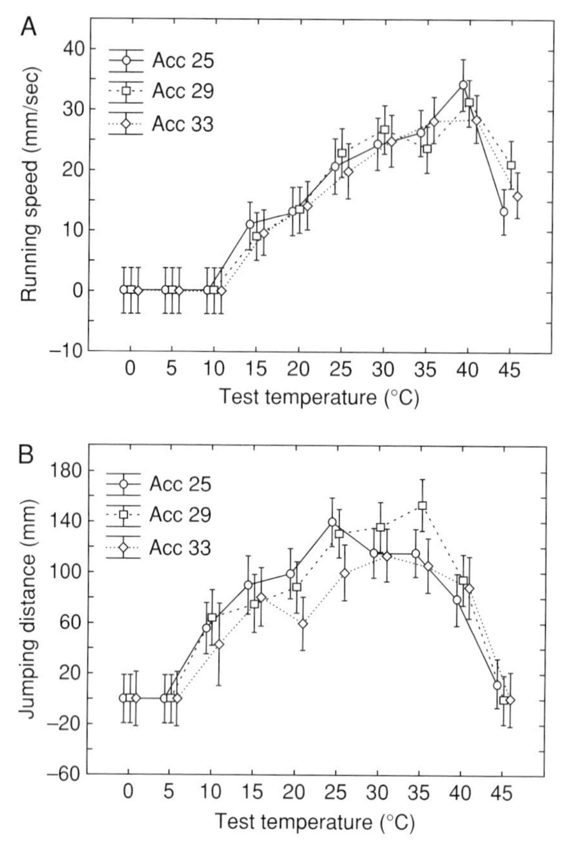

Fig. 3.2 Thermal performance curves for (A) running speed and (B) jumping distance in the cricket *Acheta domesticus* following acclimation at 25, 29, and 33°C (thermal performance curves not statistically different among the acclimation groups). Modified from Lachenicht et al. (2010), with permission from Elsevier.

temperature (Fischer and Fiedler, 2001; Gould et al., 2005, Legaspi and Legaspi, 2005; Matadha et al., 2004; Gomez et al., 2009; Wang and Tsai, 2001).

Although insects have diverse strategies for achieving high body temperatures (Section 3.3), and for biochemical compensation for cool temperatures (Section 3.4), broad-scale analyses suggest that, in general, warmer is better for insects (Fig. 3.4), insofar that insects from warmer climates (and higher temperatures of optimal performance) are able to achieve higher rates of growth (Frazier et al., 2006). The effect is large, with a 1°C drop in optimal temperature giving an approximately 10% decrease in growth rate.

This positive relationship between temperature, growth, and reproduction may be an important component of higher diversity and abundance of insects at

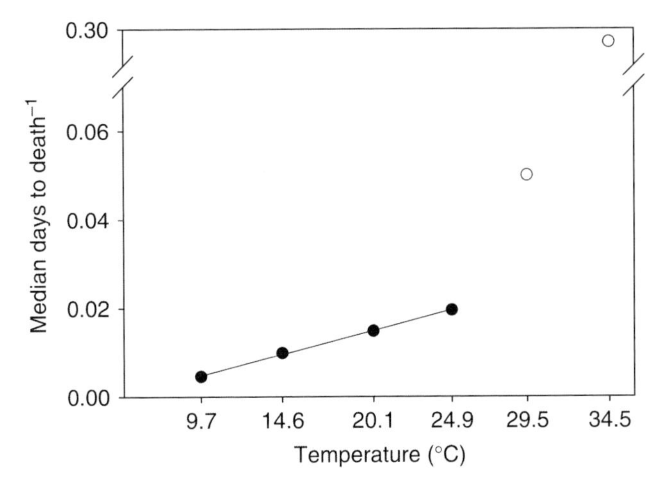

Fig. 3.3 Live hot, die young. Rate of senescence (i.e. the inverse of longevity) vs. temperature for *Copitarsia corruda* (Lepidoptera: Noctuidae). The linear relationship from 9.7 and 24.9°C (closed symbols) reflects thermally benign conditions, while the values at 29.5 and 34.5°C (open symbols) deviate from the linear relationship and reflect stressful temperatures. From Gomez et al. (2009), with permission from Cambridge University Press.

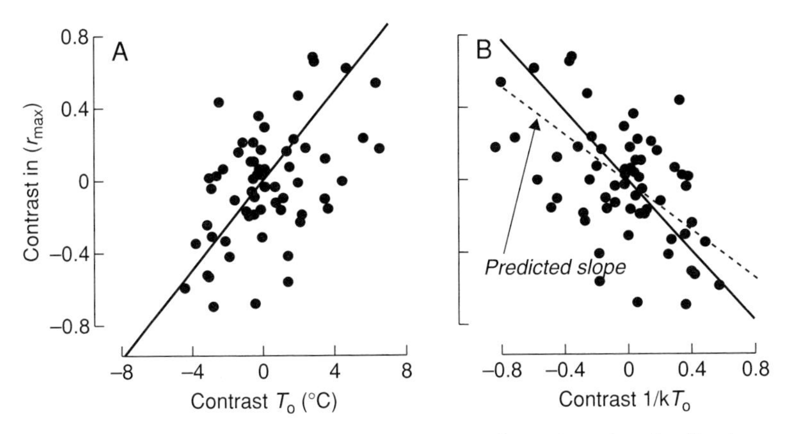

Fig. 3.4 Warmer is better. Left: Maximum intrinsic growth rate (r_{max}, female offspring per female day) vs. optimal temperature (T_o) for 65 insect species, corrected for phylogeny. Right: The slope of the temperature effect is significantly steeper than predicted by Boltzman's constant. From Frazier et al. (2006), with permission from The University of Chicago Press.

lower latitudes (Huey and Kingsolver, 1989; Savage et al., 2004; Frazier et al., 2006).

3.1.2 Pathologies of Temperature Extremes

3.1.2.1 Macromolecular and Cellular Damage

When cells get too hot or cold, cellular macromolecules and their functional aggregates tend to dissociate, form non-functional aggregates, or both. At extreme high temperatures, proteins assume non-native configurations that impair function and increase the likelihood of irreversible binding to other unfolded proteins. Additionally, membranes become hyperfluid, which can alter a variety of functions, including solute transport and synaptic transmission. Even when macromolecular functions are not disturbed, high temperatures can elevate metabolism and oxygen demands to the point that growth and development are impaired in insect eggs or small insects that rely on oxygen diffusion (Frazier et al., 2001; Woods and Hill, 2004).

In the cold, enzyme reactions and other processes are slower, and membranes become hyperviscous and may transition to the gel phase. Although these effects may be reversible, they are often pathological even in the absence of freezing—a phenomenon termed non-freezing cold injury—and persist even after re-warming. Cold-induced apoptosis and loss of ion homeostasis also contribute to non-freezing cold injury (Yi et al., 2007; Kostal et al., 2007; MacMillan and Sinclair, 2011b). If ice formation begins it usually proceeds rapidly and, in freeze intolerant species, causes lethal damage to cells and tissues. There are several sources of damage. First, extracellular ice (which has a lower vapor pressure than biological fluids) forms before intracellular ice, causing an osmotic gradient between the unfrozen extracellular fraction and the cytoplasm. The transmembrane osmotic gradient causes water to leave the cell, dehydrating it and increasing the concentration of its interior solutes. Cellular dehydration increases the viscosity of the cytoplasm, disrupts enzyme function, and impairs pH and ion regulation within the cell. As extracellular ice accumulates, the unfrozen fraction of the extracellular space also becomes smaller, and cells must reside within narrow, unfrozen channels, where they may be compressed or shrunken to a volume below that necessary for survival (Mazur, 1984). The cell membrane may be damaged by ice crystal formation due to cell shrinkage and mechanical injury, such as perforation.

3.1.2.2 Behavioral Impairment

The best known thermal pathologies of insects are various behavioral problems that occur during and after exposure to extreme temperatures. Full or partial loss of motor control occurs from transient exposure to extreme temperatures, and the condition is called variously: thermotorpor, heat coma, heat stupor, heat knock-down, cold stupor, chill-coma, etc. Among these terms, torpor represents general immobility, stupor refers more specifically to reversible immobility caused by extreme temperatures, and coma describes permanent immobility (Vannier, 1994; Chown and Nicolson, 2004). Knock-down has been studied primarily in *Drosophila* and refers to the reversible inability of adult flies to cling to surfaces in a vertical, baffled tube (Gilchrist and Huey, 1999). Thermal stress during sub-adult stages can also impair long-term mobility in later life stages. For example, adult *D. melanogaster* previously heat-shocked at specific times during the pupal stage have problems walking and flying, even without gross morphological injury to legs or wings (Roberts et al., 2003).

What causes temperature-induced loss of mobility? It appears to reflect the summed effects of thermal stress on enzymes, membranes, and cytoskeletal properties of neurons, but few data are available. Temperature's effects on pattern generators for ventilation and locomotion have been studied and reviewed by Robertson and colleagues, who described thermal impairment of central pattern generation, action potentials, and synaptic transmission in *Drosophila* and *Locusta* (Karunanithi et al., 1999; Xu and Robertson, 1994; Gray and Robertson, 1998). Less is known about subcellular mechanisms underlying these effects. Heat stress impairs regulation of cytosolic Ca^{2+} (Klose et al., 2008; Klose et al., 2009), and increasing free cytosolic Ca^{2+} can be toxic via upregulated catabolic enzymes that accelerate breakdown of the cytoskeleton, DNA, and organelles. Hyperthermia likewise impairs K^+ homeostasis, and elevated extracellular K^+ promotes neural failure in the ventilatory central pattern generator of locusts (Rodgers et al., 2007). Cytoskeleton-dependent neural function is also sensitive to heat shock, as locusts treated with a cytoskeletal stabilizer (concanavalin A) have less heat-induced disruption of motor pattern generation compared to untreated locusts (Garlick and Robertson, 2007; Klose et al., 2004).

Cellular pathologies from cold are even less well understood, but substantial progress has been made in recent years. Intracellular ice damages cell membranes and organelles, while subsequent thawing leads to osmostically-driven water fluxes across membrane wounds. Extracellular ice, as described above, promotes hyperosmotic conditions in the unfrozen extracellular fraction (Lee Jr, 1991), which draws water out of the cell, causing intracellular damage from elevated ion concentrations and excessive cell shrinkage.

Cold injury in the absence of ice formation stems primarily from impaired membrane function (Fig. 3.5). Macmillan and Sinclair provide an excellent, detailed review of the mechanisms of chill-coma (MacMillan and Sinclair, 2011a). When held at –5°C, non-diapause adults of the bug *Pyrrhocoris apterus* do not freeze but show massive redistribution of ions and water between hemolymph and tissues, with Na^+ and water going into the tissues and tissue K^+ going into the hemolymph (Kostal et al., 2004; Kostal et al., 2007). In this case, chilling injury occurred because membrane-bound Na^+/K^+-ATPases were unable to counteract

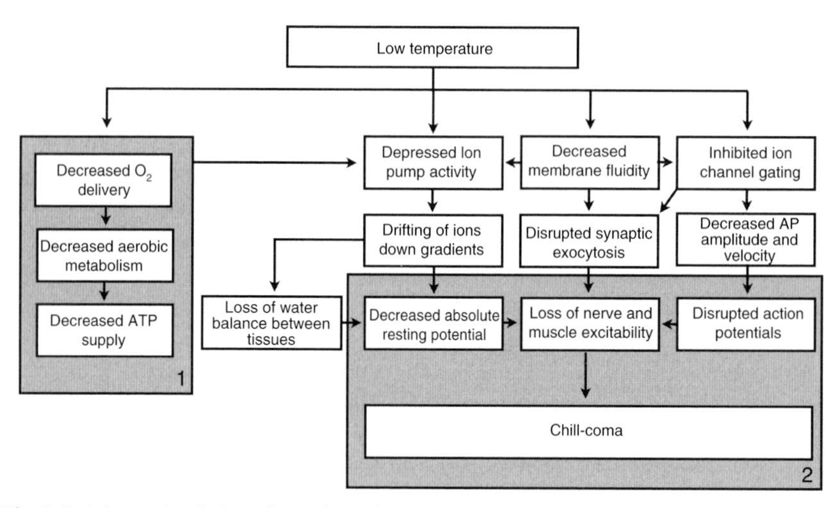

Fig. 3.5 Schematic of physiological mechanisms of neuromuscular transmission failure thought to underlie insect chill-coma. (1) Mechanisms predicted by oxygen and capacity limited thermal tolerance (thought to occur on a whole-animal level) although current evidence suggests these are unlikely to occur in insects (see Macmillan and Sinclair 2011). (2) Proposed mechanisms of chill-coma currently supported by empirical data. AP: action potential. From MacMillan and Sinclair (2011a), with permission from Elsevier.

passive movements of Na^+ and K^+, leading to partial depolarization of cell membranes. Similar effects have been found in neurons of insects in chill-coma (Heitler et al., 1977; Hosler et al., 2000). Prolonged cold shock also depletes metabolic fuels in *Drosophila*, probably due to the high ATP requirements of restoring gradients and proteins during a period when feeding and digestion is impaired (Overgaard et al., 2007). Chilling injury and coma in the cricket *Gryllus pennsylvanicus* results from the loss of ion homeostasis in organ systems, specifically disrupted muscle ion equilibrium potentials brought on by tremendous hemolymph water and ion loss to the gut following cold injury to the gut epithelium (MacMillan and Sinclair, 2011b).

3.1.2.3 Developmental Instability

Thermal variation can induce developmental plasticity that improves the "match" between performance and prevailing thermal conditions (see Section 3.4), but thermal stress experienced early in life can also disrupt development and yield abnormal, functionally-impaired adults. Developmental delay or arrest is probably the best-documented developmental pathology caused by extreme temperatures. All insect species (and often specific sub-adult stages within a species) have upper and lower developmental threshold temperatures, which together define the temperatures beyond which no development occurs.

Research, mostly on *Drosophila*, has established that extreme temperatures during development can be teratogenic, defined as disrupting development to the extent that later stages exhibit pathologies. Although many factors can be teratogens (Faruki et al., 2007), temperature is probably the most ecologically relevant. The efficacy of a teratogen depends on its identity, the duration and intensity of exposure to it, and the developmental stage at which exposure to it occurs. For example, heat shock applied to wild-type *D. melanogaster* can induce different cuticular abnormalities depending on the developmental stage at which it is imposed. Abdominal segmentation is disrupted by embryonic heat exposure (Maas, 1949; Welte et al., 1995), while wing defects follow larval or pupal heat exposure (Goldschmidt, 1935; Milkman, 1962; Mitchell and Lipps, 1978; Mitchell and Petersen, 1982). Far less is known about how thermal stress disrupts development of internal structures and physiology. Daily exposure to high temperatures during larval and pupal development of *Drosophila* attenuates development of the *corpus pedunculata* (mushroom bodies), a paired structure in the insect brain that controls olfactory memory (Fig. 3.6). These flies perceived odors

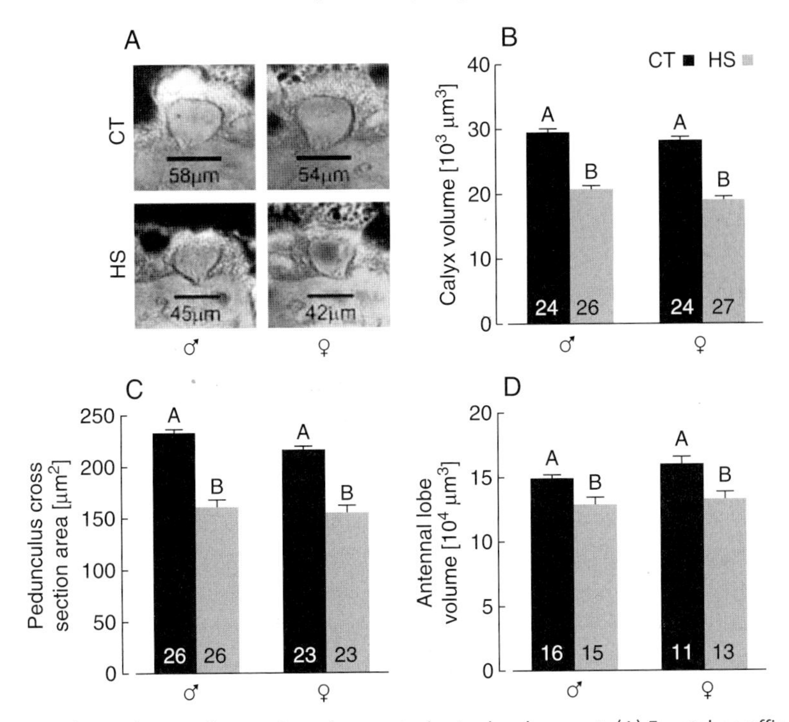

Fig. 3.6 Thermal stress disrupts *D. melanogaster* brain development. (A) Frontal paraffin sections of mushroom body calyces at their broadest point. Mushroom bodies (A, B, and C) and olfactory lobes (D) are smaller in heat-stressed (HS) flies than in control (CT) flies. HS: 23°C rearing temperature, plus daily exposure to 39.5°C for 35 min throughout larval and pupal development. CT: constant 23°C rearing temperature. From Wang et al. (2007).

and avoided shocks just as well as untreated flies, but were less able to *associate* shocks with particular odors (Wang et al., 2007).

Abnormalities produced by teratogens often resemble defined phenotypes of known genetic mutations, and so were termed "phenocopies" (Goldschmidt, 1938). These injuries represent stress-induced disruption of otherwise normal development, and do not appear in non-stressed offspring of phenocopied individuals. While many phenocopies of well-characterized mutations have been identified, there are only a few descriptions of how environmental stress targets specific genes or gene products to disrupt development. Regarding thermal teratogenesis, embryonic heat shock disrupts abdominal segmentation in *D. melanogaster* by delaying turnover of the protein encoded by the gene *fushi tarazu*, which regulates *bithorax*-complex genes that encode body segment patterning (Welte et al., 1995).

High temperatures can also be teratogenic in the field. For example, natural phenocopies occur in orchard populations of *D. melanogaster* along gradients of temperature stress (Roberts and Feder, 1999). Temperatures of necrotic fruit on the ground are extremely variable and can reach 45°C when sunlit, causing dramatic mortality in the larvae they contain (Roberts and Feder, 1999; Feder et al., 1997; Warren et al., 2006). In wild populations, up to 15% of individuals surviving peak summer temperatures had substantial wing and abdominal abnormalities, and flies eclosing from sunlit fruit exhibited deformities two to three times more frequently than did flies from nearby fruits in deeply shaded areas (Roberts and Feder, 1999). Furthermore, larval crowding in necrotic fruit compounds the risk of heat-induced teratogenesis (Warren et al., 2006).

Cold temperatures can also be teratogenic. Honeybee pupae cooled to 20°C during the pink-eye stage yield adults with deformed wings and non-functional stingers (Chacon-Almeida et al., 1999). Wing deformities arise in nearly one-half of fall armyworm (*Spodoptera frugiperda*) adults following pupal rearing at constant 15°C or in a fluctuating 10.8–17.5°C regime (Simmons, 1993). Cold exposure of pharate adults in the flesh fly *Sarcophaga crassipalpis* disrupts eclosion rhythm and eclosion behavior, suggesting cold-injury to neuronal function (Yocum et al., 1994). Development of the reproductive system is also susceptible to cold, as egg viability and egg laying rate are reduced in *Musca domestica* that experience cold-shock as pharate adults (Coulson and Bale, 1992).

3.2 Sensing Temperature and Control

Many insects try to maintain body temperatures within specific ranges, and they do so using behavior and physiology described elsewhere in this chapter. Although insects sense and respond to temperature gradients, how they do so is not well known, especially at the molecular level. Thermoregulatory control requires

(A) sensors that detect temperature variation, (B) integrators or control centers that compare body temperature to a setpoint temperature, and (C) effectors serving thermoregulatory mechanisms/behaviors that drive body temperatures towards the setpoint temperature when the integrator detects a mismatch. Thermoregulatory control elements are well-described in mammals, but this is not true for insects, especially regarding the sensors and integrators.

3.2.1 External Thermosensors

Insects detect temperature using internal and external thermosensory neurons. The external neurons innervate thermosensory sensilla that are reasonably conserved across insects and have been described previously by others (Altner and Loftus, 1985; Steinbrecht, 1984; Nation, 2008), and so will not be reviewed in detail here. Typically, each sensilla contains a thermosensory neuron, another sensitive to moist air, and a third sensitive to dry air. This triad is the most common architecture, although thermoreceptive sensilla make up only a small fraction of all receptors on the surfaces where they occur (antennae, wings, legs, etc.). Even so, thermo/hygrosensory sensilla show considerable variation in the number of sensory neurons, their dendritic anatomies, and the stimuli they perceive (cold, warm, humid, and dry). Most recent research on insect thermosensors focuses on the functional diversity of infrared (IR) receptors and the molecular genetics of thermosensitive neurons, which we describe in the following sections.

Certain beetles in the families Buprestidae and Acanthocnemidae, and at least one bug in the family Aradidae, possess specialized IR receptors. These insects are attracted to forest fires, where they lay eggs on fire-damaged logs and debris. They probably use their IR receptors both to detect forest fires from a distance and to avoid harm when close. Buprestids in the genus *Melanophila* have approximately 70 IR sensilla in pits on the ventral thorax (Fig. 3.7). Individual sensilla transduce IR radiation into a physical event via thermal expansion of the cuticle and underlying fluid, which deforms the dentrite of the single mechanoreceptive sensory neuron innervating each IR sensilla (Schmitz and Bleckmann, 1997; Vondran et al., 1995).

The buprestid *Merimna atrata* has abdominal IR organs, each consisting of a specialized IR-absorbing area innervated by one thermosensitive, multipolar neuron. The primary dendritic branches of the neuron ramify into several hundred closely packed terminals enveloped in a mass of mitochondria-rich glial cells. This IR sensor is functionally and structurally similar to that of pit vipers in that IR radiation is absorbed by a special surface and the increase in temperature is measured by a heat-sensitive neuron (Schmitz and Trenner, 2003; Schmitz et al., 2000; Schmitz et al., 2001). The Australian beetle *Acanthocnemus nigricans* (Acanthocnemidae) has a pair of infrared (IR) receptor organs on the first thoracic segment. Each organ consists of an IR-absorbing disc above an air-filled cavity.

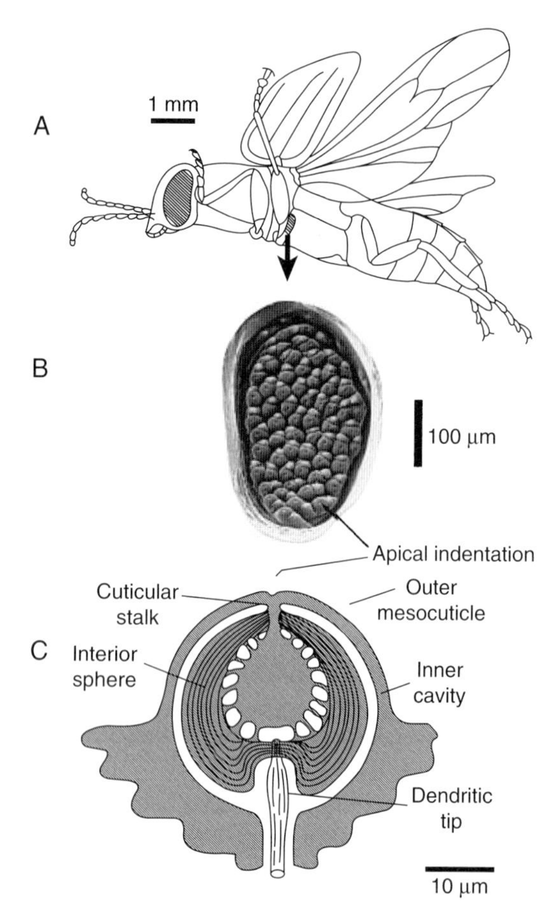

Fig. 3.7 Infrared (IR) pit organs of the fire-seeking beetle *Melanophila acuminata*. (A) During flight the mesothoracic legs are lifted to expose the IR pit organs next to their coxae. (B) The IR pit organ composed of 62 domed IR sensilla, each with an apical indentation and accompanied by a wax gland. (C) Schematic drawing of the cuticular components (shaded areas) and the dendritic tip of the sensory cell of an IR sensillum. A small stalk arising from the apical indentation connects a large endocuticular sphere—situated within an inner cavity—to the outermost thin mesocuticle. A sensory neuron innervates the sphere, which expands due to absorption of IR radiation and causes a deformation of the dendritic tip. From Schmitz and Bleckmann (1998), with permission from Springer Science + Business Media.

Inside the disc, roughly 30 heat-sensitive, multipolar neurons contact the outer cuticle. As in *M. atrata*, IR absorption by the *A. nigricans* disc causes temperature to rise, which is detected by the receptors. The system is highly sensitive in part because its air-filled cavity reduces its thermal mass (Kreiss et al., 2005; Schmitz et al., 2002).

The aradid bug *Aradus albicornis* has IR-sensitive, dome-shaped sensilla on the ventral prothorax. Each sensillum has a fluid-filled compartment enclosed in a round cuticular shell that is innervated by the dendrite of a mechanoreceptive neuron coupled to the inner fluid compartment (Schmitz et al.; 2008, Schmitz et al., 2010). These photomechanic sensilla are structurally and functionally similar to the IR sensilla of *Melanophila* beetles. The similarities between the IR detectors in *M. atrata* and pit vipers and between those of *A. albicornis* and *Melanophila* beetles are two remarkable examples of convergent evolution.

3.2.2 Neuromolecular Thermosensation

Research on insect thermosensation has focused on *D. melanogaster* and revealed that, like mammals, insects possess internal thermosensors (Sayeed and Benzer, 1996; Zars, 2001) and rely heavily on members of the transient receptor potential (TRP) superfamily of ion channels to sense temperature (Fig. 3.8). In *D. melanogaster*, three TRP channels (*painless*, Pyrexia, and dTRPA1) mediate heat avoidance (Lee et al., 2005; Rosenzweig et al., 2005; Sokabe and Tominaga, 2009; Tracey et al., 2003; Viswanath et al., 2003) and two (TRPL and TRP) mediate cold avoidance (Rosenzweig et al., 2008). Knock-down of these TRP channels by mutation or RNA interference impairs thermotaxis towards optimal growth or survival temperatures (dTRPA1, Pyrexia, TRPL, TRP) and nociceptive behaviors (*painless*). *Drosophila* thermotaxis is also influenced by molecules

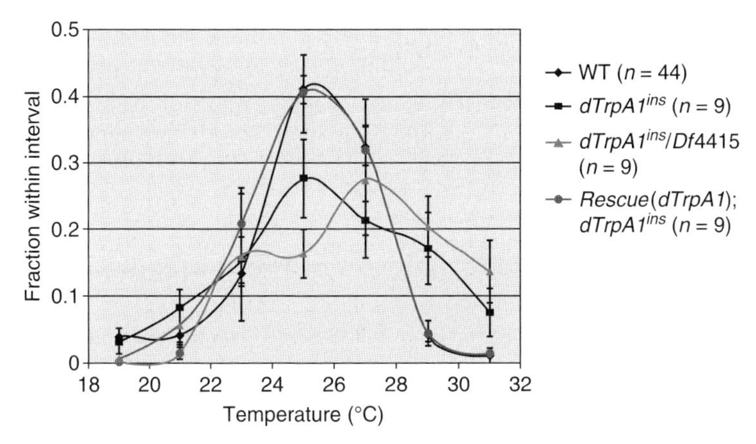

Fig. 3.8 *dTrpA1* is required for warmth avoidance in *D. melanogaster*. Spatial distribution of wild-type flies, *dTrpA1* loss-of-function mutants and a *dTrpA1* rescue mutant along a thermal gradient. Wild-type and *rescue(dTrpA1); dTrpA1^{ins}* flies prefer the optimal growth temperature (25°C), but *dTrpA1^{ins}* and *dTrpA1^{ins}/Df4415* flies do not avoid warmer temperatures. Adapted from Hamada et al. (2008a), with permission from Nature Publishing Group.

other than TRPs; for example, thermotaxis is disrupted by mutations of certain genes encoding proteins of histamine signaling pathways and in wild-type flies treated with pharmacological antagonists of histamine receptors (Hong et al., 2006).

Little is known about the diversity of thermosensitive TRP channels across insect species, although recent research is beginning to address this issue. The hymenopterans *Apis mellifera* and *Nasonia vitripennis* lack dTRPA1, but possess a hymenopteran-specific TRPA (HsTRPA) that, when expressed in cultured human embryonic kidney cells, is activated by high temperatures (Matsuura et al., 2009). The malaria mosquito *Anopheles gambiae* TRPA1 (agTRPA1) is warmth-activated when expressed in *Xenopus* oocytes (Wang et al., 2009), and closely-related, putatively thermosensitive TRPA1s are present in the flour beetle *Tribolium castaneum*, the human body louse *Pediculus humanus corporis*, and the mosquitoes *Culex pipiens* and *Aedes aegypti* (Hamada et al., 2008). These results suggest that TRP channels may be generally used by insects to sense heat and select habitats.

The role and location of TRP channels in internal, thermoregulatory neural control circuits, and the nature of these circuits in general, are just beginning to be understood. Two pairs of anterior cell (AC) neurons in the *D. melanogaster* brain express dTRPA1 and are important for thermosensation (Hamada et al., 2008). Rescued dTRPA1 expression in AC neurons restored normal warmth-sensing in *dTRPA1* knockout flies, and warmth-responsive G-CaMP fluorescence of AC neurons was similarly restored by transgenic dTRPA1 expression in knock-out flies. Likewise, RNA interference selectively targeting *dTRPA1* in the AC neurons impaired heat sensing by adult flies. Combined, these results show that dTRPA1 expression in the AC neurons is both necessary and sufficient for normal thermotaxis in *D. melanogaster*. The function and anatomy of the insect thermoregulatory integrator, and its coordination with sensors and effectors, await description.

3.3 Heat Balance and Mechanisms of Thermoregulation

For the majority of insects, exchange of heat energy with the environment determines their rate of living. Net heat exchange with the environment (Fig. 3.9) reflects net heat transfer by conduction (Q_{cond}), convection (Q_{conv}), radiation (Q_{rad}), evaporation (Q_{evap}), and metabolic heat production (MHP):

$$\Delta\mathrm{H} = Q_{cond} + Q_{\mathrm{conv}} + Q_{\mathrm{rad}} + Q_{\mathrm{evap}} + \mathrm{MHP} \qquad (1)$$

At a steady body temperature, ΔH is zero. MHP is always positive and Q_{evap} is typically negative (except when water vapor condenses onto the cuticle), though

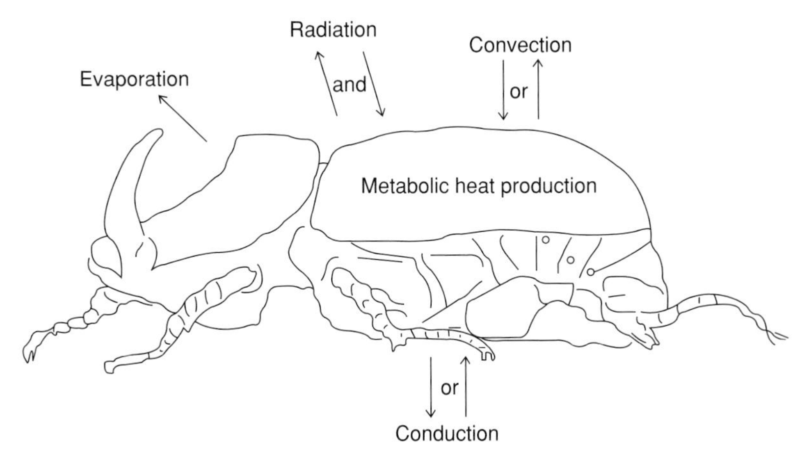

Fig. 3.9 Routes of heat gain and loss in an insect. Metabolic heat production may be sufficiently high in some insects to elevate body temperature (T_b) well above air temperature (T_a). All insects lose heat by evaporation, but a few actively cool their bodies by evaporation. All insects gain long-wave and short-wave radiation, and lose long-wave radiation. Short-wave radiant heat gain (sun-basking) allows many insects to elevate T_b above T_a. If $T_b > T_a$, then the insect loses convective heat, and if $T_b < T_a$, then the insect gains convective heat. If $T_b >$ substrate temperature (T_{sub}), then the insect loses conductive heat, and if $T_b < T_{sub}$, then the insect gains conductive heat.

both can be negligible. If the insect is in direct sun, Q_{rad} will almost always be positive. Thus, for many small insects with negligible MHP, radiant solar energy input is balanced by losses via conduction, convection, and evaporation. With little solar irradiance, insects with limited MPH will have body temperatures very similar to environmental temperatures due to high conductive and convective heat exchange with their environment.

3.3.1 Conduction

Conduction is the direct transfer of heat between bodies in physical contact. The transfer of heat occurs down a thermal gradient (from higher to lower temperature), and the rate of transfer (Q_{cond} in J sec^{-1}) depends on the temperature differential ($T_2 - T_1$, or ΔT, in °C), the area of physical contact (A in m^2), the conductivities of participating materials (k in J sec^{-1}°C^{-1} m^{-1}), and the length of the thermal gradient (x in m). These variables interact via Fourier's law of heat transfer such that:

$$Q_{cond} = -kA\Delta T/x \tag{2}$$

The negative sign indicates that the heat flux occurs *down* the thermal gradient. This equation illustrates that heat transfer is enhanced by greater surface area and thinner cuticles. The equation is often simplified when there is little idea of k, A, and x by collapsing these variables into a single term called thermal conductance (C_{cond}, defined as $-kA/x$ with the units J $sec^{-1}{}^{\circ}C^{-1}$) where:

$$Q_{cond} = C_{cond} \Delta T \quad (3)$$

Thermal conductance reflects the propensity of heat to move down a gradient, more specifically the amount of heat that passes per unit time per temperature differential through a barrier of particular k, area, and thickness. Another useful parameter is thermal resistance, which is the inverse of thermal conductance (hence with units of °C sec J^{-1}). Also related is a parameter called the heat transfer coefficient (with the units J $sec^{-1}{}^{\circ}C^{-1}$ m^{-2}) that indicates the quantity of heat passing per unit time per temperature differential through unit area of a barrier of particular thickness.

The relative importance of conduction compared to other sources of heat flux is rarely considered because conduction is often coupled with convection (see below). Conduction is certainly the primary determinant of body temperature of insects living in water, soil, and semi-solid substrates, and is likewise an important heat pathway for many superterranean insects. For example, some grasshoppers and beetles press their bodies onto warm substrates to elevate body temperatures, or alternatively "stilt" above hot substrates to prevent overheating (Henwood, 1975; Anderson et al., 1979; Whitman, 1987). Stilting allows Saharan *Cataglyphis* ants to remain active during midday, when air temperature 4 mm above the ground may be 10°C lower than ground surface temperature (Gehring and Wehner, 1995). Ant brood temperatures are probably determined primarily by conduction, as they sit in still air in direct contact with dirt. Workers regulate brood temperature by moving the brood within the nest in response to daily and seasonal changes in nest temperature gradients (Penick and Tschinkel, 2008). Adult bumblebees use conduction to incubate their brood by pressing their abdomens (warmed by metabolic heat produced from flight muscle contractions and transferred to the abdomen by convective blood flow) onto their brood cells, warming them up to 20°C above the temperature of surrounding air (Heinrich, 1972; Gardner et al 2006). This behavior accelerates brood development and ultimately allows bumblebees to occupy regions with abbreviated growing seasons (Heinrich, 1993).

3.3.2 Convection

Convection is the transfer of heat between a body and the fluid (air in the case of a terrestrial insect) around it. Convective heat exchange occurs across a fluid boundary layer (δ) covering the insect and separating it from the free-stream air (or water), where

$$Q_{cond} = -kA\Delta T/\delta \quad (4)$$

Here, k refers to the conductivity of air. As with conduction, more surface area speeds heat transfer. The depth of the boundary layer air over the cuticle depends on wind speed, with δ decreasing exponentially with wind (Chappell, 1982). Certain large, strongly endothermic species that fly in cold air, such as bumblebees and winter- and night-flying noctuid moths, possess a layer of long, dense pubescence over the cuticle that increases δ and reduces convective heat loss (Heinrich, 1993). As k, A, and δ are often unknown, they are often collapsed into a single term called convective conductance (C_{conv}, defined as $-kA/\delta$ with the units J sec^{-1}°C^{-1}) where

$$Q_{conv} = C_{conv}\Delta T \quad (5)$$

The C_{conv} can be calculated from the cooling rate of dead, dried insects suspended in air. The dead insect is implanted with a thermocouple, heated radiatively, and C_{conv} is calculated from the cooling curve and the thermal capacitance of dried tissue (May, 1979; Heinrich, 1993). Ideally, this is done at various wind speeds, and then convective heat loss in the field can be estimated if local wind speeds over the insect are known. In some cases (e.g. flying insects), convective heat loss can be calculated from Equation 1 and measurements of body temperature, net radiative exchange, total EWL, and metabolic heat production (Roberts and Harrison, 1999). In cases where the insect is on a substrate and the contact architecture is complex, conductive and convective conductances are sometimes summed into a total thermal conductance (Dzialowski, 2005; Pincebourde and Casas, 2006).

Flying honeybees, bumblebees, and some moths use regional heterothermy to maintain high thoracic temperatures. Restricting high temperatures to small volumes can reduce combined conductive and convective heat losses. In flight, these insects use metabolic heat to generate high flight muscle temperatures (above 35°C), even in cool air. Simultaneously, temperatures of the head and abdomen are often much lower, sometimes dramatically so in cool air. Heat is efficiently retained because the temperature gradient driving convective heat loss from the abdomen and head (with their relatively large surface areas) remains low, and the insulation on the thorax increases the boundary layer thickness and thus minimizes convective heat loss from that region (Heinrich, 1993).

Bumblebees and sphinx moths use one additional mechanism, countercurrent heat exchange, to retain heat in the thorax. This is not so for the honeybee because its aorta is highly convoluted within the petiole, and it effectively and obligatorily exchanges heat between the warm hemolymph leaving the thorax, and cooler aortic hemolymph traveling forward into the thorax from the abdomen. In both bumblebees and sphinx moths, the degree of petiolar countercurrent heat exchange can be varied with air temperature, such that the abdomen is cool when the air is cool (reducing convective heat loss to air and elevating thorax temperature) and the abdomen is warm when air temperature is warm (increasing heat loss to air and reducing thorax temperature) (Heinrich, 1993).

3.3.3 Radiation

All objects warmer than 0 °K emit electromagnetic energy (radiant heat) at wavelengths that depend on the surface temperature and emissive properties of the object. The predominant wavelength (λ_m, in μm) of peak spectral emittance from a surface is a function of surface temperature (T, in K) according to Wein's law:

$$\lambda_m = 2897 / T \tag{6}$$

Insects in nature receive direct or reflected solar (i.e. short-wave) radiation as well as extremely varying levels of incoming long-wave radiant energy emitted by other objects in their environment. They absorb incoming radiation in fractions that depend on the absorptive and reflective properties of their cuticle. Insects also emit radiation. Thus, radiative heat exchange depends on the net gains and losses:

$$Q_{\text{rad}} = Q_{\text{rad-in}} - Q_{\text{rad-out}} \tag{7}$$

An insect in nature receives radiant energy from two primary sources, the sun (surface temperature = ~5800 K; λ_m = ~0.5 μm) and objects in the environment, which are much cooler and yield the Earth emittance spectrum (surface temperatures = ~288 K; λ_m = ~10 μm). Hence the radiant load on an insect is the sum of short-wave (i.e. solar; 0.3–4 μm) radiation and long-wave (i.e. Earth emittance; 4–80 μm) radiation. The radiation received by an insect is highly variable, with radiation coming as direct, reflected, or scattered light, depending on season, time of day, surroundings, cloudiness, and so on (Dzialowski, 2005; Pincebourde and Casas, 2006).

The degree to which an insect absorbs irradiance depends on the cuticular absorptivity (a; the fraction of incident radiation at given wavelength that is absorbed), reflectivity (r, the fraction of incident radiation at a given wavelength that is reflected), and transmissivity (t; the fraction of incident radiation at a given wavelength that is transmitted) such that $a + r + t = 1$. For long-wave irradiance, insect cuticle (like most animal surfaces) is essentially a black body, with a approaching 1. However, the a and r of animal coats and surfaces can be quite

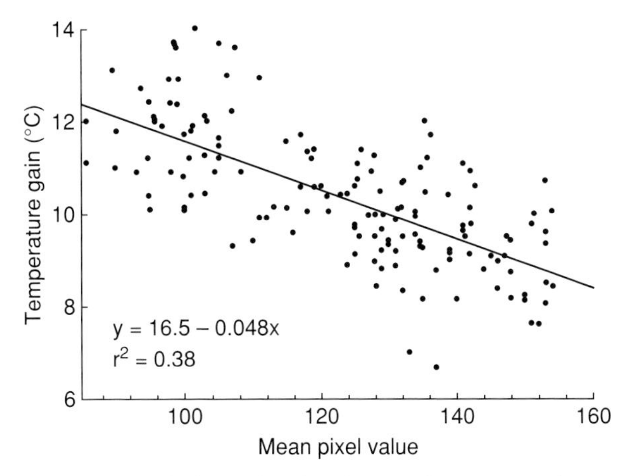

Fig. 3.10 Grasshoppers with dark cuticles absorb more short-wave radiation and heat faster than those with lighter cuticles. Heating rate vs. cuticular melanization in *Melanoplus sanguinipes* exposed to irradiance (approximately 900 W m^{-2}) from a xenon lamp with spectral qualities similar to the solar spectrum. Pixel values representing melanization range from 0 (black) to 255 (white). From (Fielding and DeFoliart, 2005), with permission from the Orthopterists Society.

variable within short-wave spectra, with significant thermal consequences. The short-wave reflectance and absorptive properties of insect cuticle varies within and across species (Fig. 3.10), with darker colors sometimes associated with more radiative absorbance (Fielding and DeFoliart, 2005; Majerus, 1998, Willmer, 1982).

Insects also emit radiation according to the Stefan–Boltzman equation:

$$Q_{\text{rad-out}} = \varepsilon_s \sigma A T_s^4 \tag{8}$$

where ε_s is the body surface emissivity (assumed in insects and other organisms to be nearly 1), σ is the Stefan-Boltzmann constant (5.67 x10^{-8} W m^{-2} K^{-4}), A is radiative surface area (m^2) and T is surface temperature (°K). Equation 8 shows that radiative heat loss depends strongly on body temperature and on surface area. Net radiative exchange is the difference between absorbed and emitted radiation (Equation 7). Due to the complexity of the radiative environment and the strong influence of insect surfaces and body orientation on radiative absorption, the most practical approach to assessing radiative exchange is to use the operative temperature models of insects (see Chapter 7).

Behavioral regulation of radiant heat flux (especially short-wave heat gain) is common in insects. The most general strategy is shade seeking vs. solar basking, examples of which have been described for butterflies, caterpillars, dragonflies, grasshoppers, beetles, flies, and cicadas (Heinrich, 1993). Many of these insects move to sunlit areas to warm themselves and may assume postures that maximize heat gain. When their body temperatures become too warm, they seek shade and orient to reduce solar heat gain. This behavior and its thermal consequences in nature are especially well-described for the flightless lubber grasshopper *Taeniopoda eques* (Whitman, 1986; Whitman, 1988), which maintains body temperatures of 32–36°C throughout most of the day. Good thermoregulation is necessary for these grasshoppers to meet their degree-day requirements for development prior to the onset of low autumn temperatures (Whitman, 1986; Whitman, 1988).

3.3.4 Evaporation

The latent heat of evaporation (LHE) for water is approximately 2450 J g^{-1} at 25 °C (44 kJ mol^{-1}), which means that water evaporation from the surface of an insect (or any object) has a major cooling effect. Total evaporative heat loss (Q_{evap}) from an insect is LHE times the sum of transcuticular water loss (EWL_{cut}), respiratory water loss (EWL_{resp}), and water vapor loss from the mouth or anus (EWL_{ext}):

$$Q_{evap} = \mathrm{LHE}(\mathrm{EWL}_{cut} + \mathrm{EWL}_{resp} + \mathrm{EWL}_{ext}) \quad (9)$$

Evaporative water loss across the cuticle (EWL_{cut}) depends on the vapor density differential between the insect (usually assumed to be saturated at the surface temperature) and the surrounding air ($\rho_{vsur} - \rho_{vair}$, or $\Delta\rho$), the surface area (A) from which evaporation is occurring, the water vapor diffusion coefficient (D), and the path length (e.g. across the cuticle + boundary layer of air) for evaporative water loss (x) according to the equation:

$$\mathrm{EWL}_{cut} = DA\Delta\rho/x \quad (10)$$

The terms *DA/x* together constitute the conductance of the cuticle for water, and so can be combined into a cuticular water conductance that determines the rate of water loss at a given vapor pressure differential. For most insects, cuticular water loss depends strongly on the type and quantity of cuticular lipids (see Chapter 4). Because the saturating vapor pressure of water increases exponentially with temperature, the gradient for water loss to dry air increases exponentially with surface temperature, so higher temperatures tend to increase losses.

Respiratory water loss occurs because the tracheal air is generally assumed to be near saturated with water, as tracheal surfaces must be highly permeable to oxygen and carbon dioxide. Water escapes by diffusion and convection from open spiracles down the vapor pressure gradient (see diffusion and convection equations in Chapter 6). Respiratory water loss is increased by: 1) increasing number, degree, and duration of spiracular opening; 2) increased vapor pressure gradients for water (reduced humidity or elevated temperature); 3) increased temperature or reduced air pressure (both increase the diffusion coefficient for water); and 4) increased convective ventilation of the tracheal system.

In insects, evaporation is an uncommon thermoregulatory strategy simply because they tend to contain little water and have high surface area-to-volume ratios. However, this can be an important strategy for an insect exposed to temperatures near the maximum tolerable temperature, because this is the only mechanism that depresses body temperature below operative temperature. Insects that consume water-rich diets, such as honeybees and cicadas, utilize evaporation to stay active in temperatures well-above those that strongly inhibit behavior in most other insects. When the heads of honeybees are heated to 45°C, they regurgitate a droplet of fluid and extend their proboscis to disperse the droplet and promote evaporation (Heinrich, 1980a). When flying in air above 43°C, this behavior increases evaporative water loss seven-fold and cools the head and thorax to significantly below air temperature (Louw and Hadley, 1985; Roberts and Harrison, 1999; Woods et al., 2005).

Males of the desert cicada *Diceroprocta apache* "sing" at midday, when air temperatures can approach 50°C, from trees in the summer in the southwest deserts of the US. These xylem feeders seek shade at these times, but also rely on evaporative cooling to prevent overheating while staying active in hot weather. When air temperatures exceed 41°C, *D. apache* greatly increases evaporation (4–7 times baseline levels) across the dorsal surfaces of the thorax and abdomen (Fig. 3.11), which are covered in dense (up to 55,000 cm^{-2}) aggregates of cuticular pores (Toolson, 1987; Toolson and Hadley, 1987; Hadley et al., 1989).

3.3.5 Metabolic Heat Production

All insects produce endogenous heat as a byproduct of ATP hydrolysis and other exothermic chemical and mechanical processes. For the vast majority of insects, metabolic heat has little or no effect on body temperature, since their metabolic rates are low and their high surface area-to-volume ratios lead to relatively fast heat exchange with the environment. However in some insects, particularly large flying species, heat production can be high enough that body temperatures, particularly of the thorax, are elevated above that of the surrounding air. In some cases, the temperature excess is constant over a range of air temperatures, indicating that the insect has little physiological control over either metabolic heat production or avenues of loss. In some strongly endothermic insects, including large bees and sphinx moths, variation in metabolic heat production contributes to thoracic temperature control during flight, brood incubation, and social defense (Heinrich, 1993). The ability to maintain and regulate elevated body temperatures via active control of metabolic heat production greatly broadens the

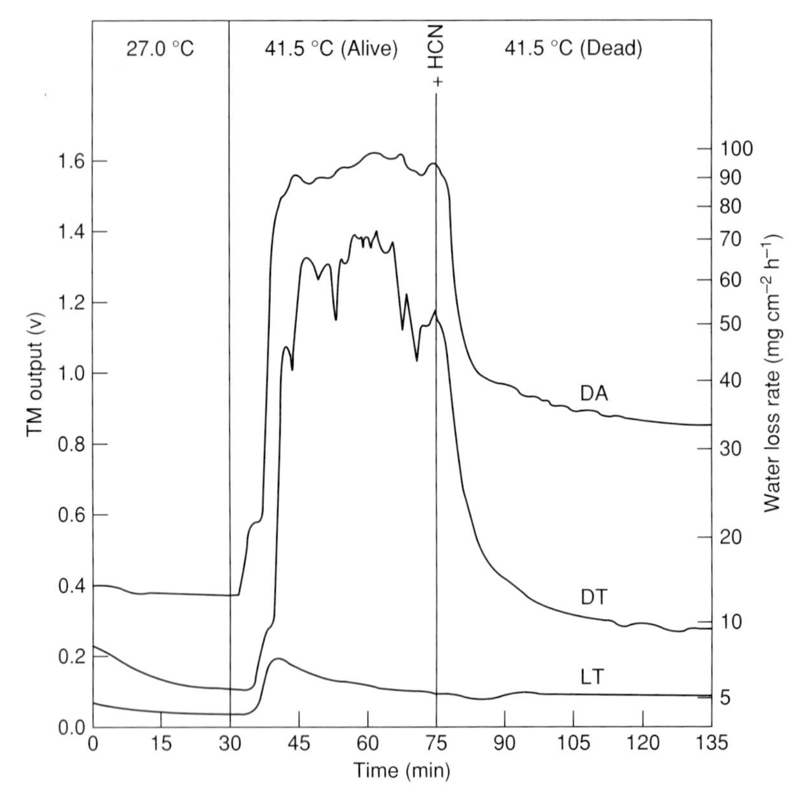

Fig. 3.11 Sweating in the desert cicada *Diceroprocta apache*. Water loss vs. time and body temperature for an individual *D. apache*. At body temperatures of 27 and then 41.5°C, cuticular transpiration was measured from the dorsal thorax (DT), lateral thorax (LT), and dorsal abdomen (DA) using a ventilated capsule (~1 mm^2 covered area). High densities of cuticular pores cover the DT and DA, but not the LT. The water loss rate quickly drops after the cicada is killed, even though its body temperature is maintained at 41.5°C, illustrating that the cicada actively controls evaporative cooling. Modified from Hadley et al (1989), with permission from The Company of Biologists. .

thermal niche of these species, which are often specialists in cold or thermally variable environments.

The insects that regulate metabolic heat production do so by varying the frequency or intensity of flight muscle contractions, often independently of obvious wing or thoracic movements (e.g. bumblebees and large beetles during pre-flight warm-up). (Newsholme et al., 1972) proposed that bumblebees and other bees decouple metabolic heat from flight muscle contraction by futile cycling of fructose-6-phosphate (F6P) via synchronous phosphofructokinase (PFK) and fructose-1,6-bisphosphatase (FbPase) activity (these enzymes operate on similar substrates in opposite directions). Numerous arguments and observations support the theoretical, qualitative possibility of such warming, including an inverse relationship between F6P cycling and air temperature, cessation of F6P cycling at flight initiation, and FbPase inhibition by cytoplasmic Ca^{2+} released during stretch and neuronal flight muscle activation (Clark et al., 1973; Storey, 1978; Greive and Surholt, 1990). However, whether the mechanism provides enough heat for significant endothermy remains under question because (a) no experiments have demonstrated significant thoracic warming via metabolic heat when cross-bridge cycling is irrefutably absent and (b) calculations of *in vivo* PFK/FbPase cycling in bumblebees (based on measurements of PFK and FbPase activities in several species) yield rates of heat production that are much less than that required to elevate thorax temperature to that necessary for flight on a cool day (Staples et al., 2004).

Pre-flight warm-up of flight muscles via contraction has been demonstrated conclusively in moths, bees, wasps, katydids, scarab beetles, syrphid flies, and cicadas, and the rate of warm-up depends on a variety of environmental, morphological, and taxonomic variables (Heinrich, 1993). For some species, pre-flight warm-up accompanies wing "shivering," while other species hold their wings motionless and warm up silently or with an audible buzzing sound. Honeybees decouple thermogenic contractions from wing movements by rapid excitation of the dorso-longitudinal muscles causing unequal shortening of these muscles relative to the dorso-ventral muscles; this effect forces the structures joined to the mesophragma against a mechanical stop (Esch et al., 1991). Warm-up permits flight across wide ranges of air temperature, and also aids singing in katydids and cicadas and brood incubation in bees (Kleinhenz et al., 2003).

Perhaps the most gruesome use of metabolic heat is thermo-execution of hornets (*Vespa* spp.) by Japanese honeybees (*Apis cerana japonica*). Hornets are predators whose prey includes honeybee larvae, and a few hornets can kill all adult honeybees in a colony (up to 30,000 individuals) in a matter of hours prior to gorging on their larvae. However, *A. c. japonica* has evolved an effective defense that that depends on metabolic heat. Hundreds of honeybees engage an attacking hornet (whose mass can reach two grams, or 20 times that of a bee), clinging to and engulfing it so that the hornet, rendered incapable of flight, is covered with a thick layer of bees. The honeybees' stingers cannot penetrate the hornet's cuticle, but instead the bees warm themselves to body temperatures approaching 47°C (Fig. 3.12), which are not lethal to the honeybees but exceed the hornet's upper lethal temperature by 2°C, killing the hornet within a few minutes (Ono et al., 1987; Ono et al., 1995; Ken et al., 2005).

Variation in metabolic heat production as a means of regulating body temperature *during* flight is a topic that has received considerable attention in the past decade. Some sphinx moths and bees maintain constant flight metabolic rates across a range of air temperatures and thus regulate temperature during

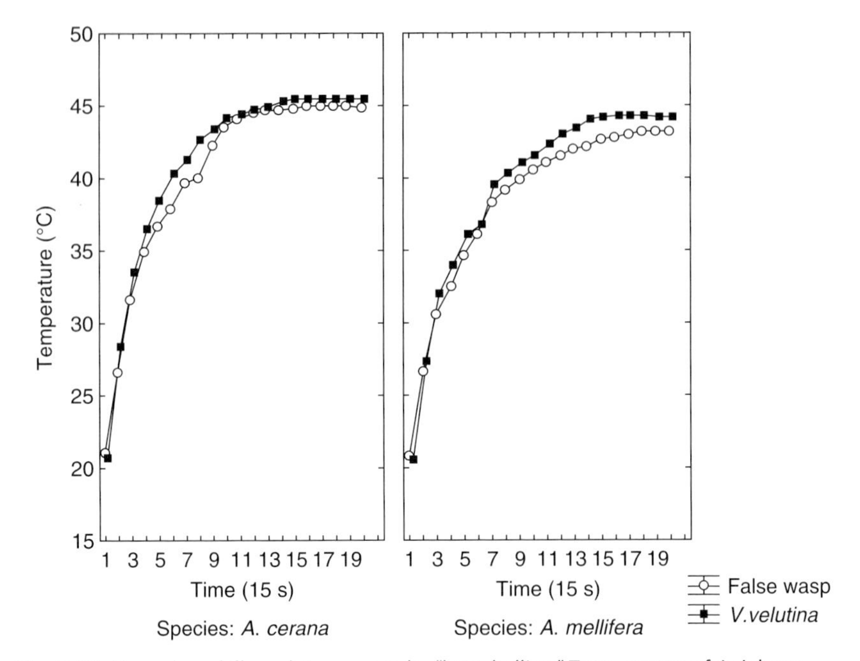

Fig. 3.12 Honeybees kill predatory wasps by "heat-balling." Temperature of *Apis* bees conglomerating on (but not stinging) a *Vespa* wasp or wasp model over 5 minutes. *Apis cerana* conglomerates reach 47°C, which is lethal to the enclosed wasp. *Apis mellifera* conglomerates reach only 45°C, which is insufficient to kill the enclosed wasp. From Ken et al (2005), with permission from Springer Science + Business Media.

flight using other mechanisms, primarily heat dumping to the abdomen (Heinrich, 1970; Heinrich, 1971; Heinrich, 1975; Heinrich, 1976). However, nearly a dozen bee species from five genera lower their wingbeat frequency *while hovering* as air temperatures rise (Fig. 3.13), and while doing so maintain stable or moderately increasing muscle temperatures (Unwin and Corbet, 1984; Spangler and Buchmann, 1991; Spangler, 1992; Harrison et al., 1996; Roberts et al., 1998; Borrell and Medeiros, 2004; Roberts, 2005). The endothermic dragonflies *Anax junius* and *Epitheca cynosura* also decrease wingbeat frequency during flight at high air temperatures, and in *A. junius* and *Vespa crabro* drones there is a corresponding decrease in flight speed at higher air temperatures (May, 1995a; Spiewok and Schmolz, 2006). In these species, reductions in wingbeat frequency, flight performance, and thoracic temperature excess during flight in warm air are likely coupled to reduced flight metabolic rates and therefore lower heat loads.

Several studies support the hypothesis that these species indeed adjust flight kinematics (i.e. lowering wingbeat frequency) in order to maintain minimum

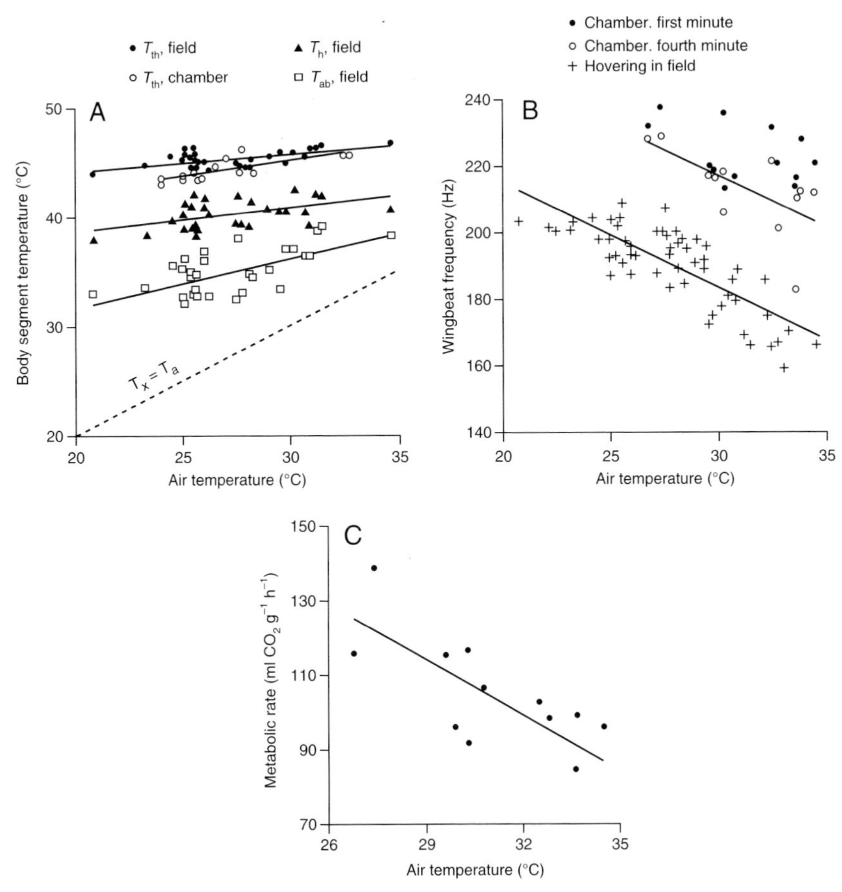

Fig. 3.13 Metabolic depression via lowered wingbeat frequency helps the highly endothermic desert bee *Centris pallida* prevent overheating during flight. (A) Thorax, abdomen, and head temperatures (T_{th}, T_{ab} and T_h, respectively) vs. air temperature (T_a) for flying *C. pallida*. T_{th} is regulated very precisely between 43 and 46°C. (B) Wingbeat frequency vs. T_a during flight in chamber and in the field. Wingbeat frequency is lower at high T_as for *C. pallida* flying in the field and after 4 minutes of flight in a chamber. (C) Metabolic rate vs. T_a for *C. pallida* flying in a chamber. Adapted from Roberts et al. (1998).

muscle temperatures required for flight in cool air and to prevent overheating of flight muscles during activity in hot air. An inverse relationship between flight metabolic rate and air temperature has been shown for honeybees, *Centris pallida* (Fig. 3.13) and *Vespa crabro* (Harrison et al., 1996; Roberts et al., 1998; Roberts and Harrison, 1999; Schmolz et al., 1999), and heat budget calculations support this relationship for *A. junius* (May, 1995b). However, some studies measuring honeybee metabolic rate as a function of air temperature have not found an inverse relationship (Heinrich, 1980b; Woods et al., 2005), and it is clear that variables such as insolation, perception of the environment, physiological state, parasite load, and the range of air temperatures considered can affect the ability or propensity of flying honeybees to vary metabolism for thermoregulation (Harrison et al., 2001; Harrison and Fewell, 2002; Woods et al., 2005). Definitive evidence for this thermoregulatory mechanism awaits simultaneous measurement of metabolic rate and flight kinematics as functions of air temperature or solar irradiance *within* individual insects.

3.4 Thermotolerance Mechanisms

3.4.1 Instantaneous and Short-term Responses

For mobile insects, behavior is usually the main mechanism of thermoregulation. Behavioral changes alter body temperature by altering avenues of thermal balance, as described in the preceding sections of this chapter. In addition, physiological responses (e.g. sweating in cicadas in response to heat, metabolic heat generation in honeybees in response to cold) can also occur rapidly. If neither behavior nor physiological mechanisms can maintain body temperatures within non-stressful limits, then insects must rely on acute thermotolerance mechanisms (i.e. rapid cold hardening or the heat-shock response).

Despite different names and induction temperatures, rapid cold hardening and the heat shock response share certain features (Denlinger, 2010; Feder, 1999; Feder and Hofmann, 1999; Richter et al., 2010; Sorensen et al., 2003). Briefly, both responses are inducible, cellular phenomena in which moderate thermal stress causes rapid changes in gene expression and metabolism that protect cells from subsequent intense thermal stress that would otherwise be lethal. Most insects are capable of some degree of rapid cold hardening, and the primary feature of the response is the synthesis of heat shock proteins (HSPs) and modest amounts of small organic molecules, including sugars, polyhydric alcohols, pyruvate, and certain amino acids (Michaud and Denlinger, 2007). In combination (and in sufficient concentrations), these substances mitigate protein denaturation, protect membranes from temperature-induced phase transitions, colligatively depress freeze temperatures, and offset cellular dehydration stress caused by extracellular ice formation (see freeze avoidance and supercooling in Section 3.4.2 below). However, the role of the polyols, sugars, and amino acids accumulated during rapid cold hardening is equivocal because they occur in concentrations so low as to act in non-colligative roles. In *S. crassipalpis*, rapid cold hardening inhibits the activation of pro-caspases (Fig. 3.14), which mitigates cold-induced apoptosis (Yi and Lee, 2011). Little is known about the molecular activation and control of rapid cold hardening, although recent research

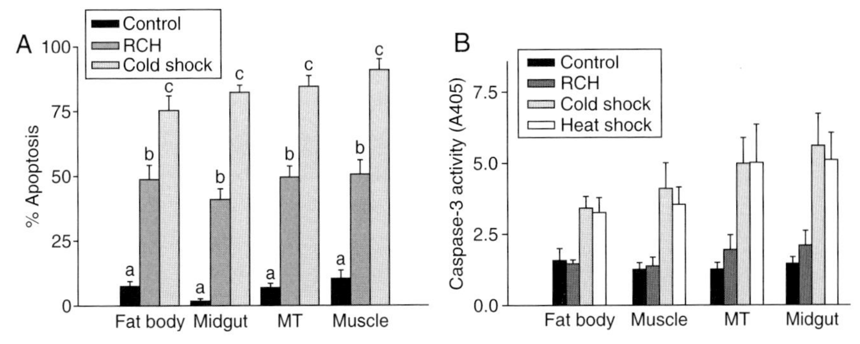

Fig. 3.14 Rapid cold hardening (RCH) mitigates cold-induced apoptosis in the flesh fly *Sarcophaga crassipalpis* by inhibiting pro-caspase activation. (A) Apoptosis in various tissues in untreated, cold-shocked (directly exposed to -8°C for 2 h) and RCH-treated (exposed to 0°C for 2h prior to a 2h exposure to -8°C) flies. (B) Caspase-3 activity in tissues from untreated, cold-shocked, RCH-treated and heat-shocked *S. crassipalpis*. Cold and hot temperature extremes induce apoptosis, and RCH inhibits the cold-induced upregulation of pro-caspases in apoptosis pathways. Adapted from Yi and Lee (2011), with permission from Springer Science + Business Media.

implicates roles for calcium signaling and p38 mitogen-activated protein kinase (Fujiwara and Denlinger, 2007b; Fujiwara and Denlinger, 2007a; Teets et al., 2008).

Hsps are a principal feature of a universal cellular stress response that protects cells from high temperatures and other protein-unfolding stresses. In response to stress, the genes encoding Hsps are rapidly turned on, and their protein protects cells by stabilizing and refolding stressed proteins or by targeting them for degradation (Feder and Hofmann, 1999; Sorensen et al., 2003). Synthesis of Hsps, while much more profound during acute heat stress, also occurs in response to acute and long-term cold. The role of Hsps in cold tolerance is poorly understood. In response to cold, many insects upregulate the 70 kD inducible form, Hsp70. Knock-down of *hsp70* and *hsp23* mRNA via RNAi reduces cold tolerance in pre-diapause larvae of *S. crassipalpis* (Rinehart et al., 2007), whereas RNAi knock-down of *hsp70* mRNA suppresses temperature-induced Hsp70 expression and recovery from chill-coma in the linden bug, *Pyrrhocoris apterus* (Fig. 3.15) (Kostal and Tollarova-Borovanska, 2009). In *D. melanogaster*, upregulation of Hsp70 is not part of the rapid cold hardening response (Kelty and Lee, 1999; Kelty and Lee Jr, 2001), although Hsp70 expression increases during recovery from cold shock (Sinclair et al., 2007).

In *D. melanogaster*, *Frost* is implicated in recovery from cold shock, but not cold hardening (Goto, 2001; Qin et al., 2005; Morgan and Mackay, 2006; Sinclair et al., 2007; Colinet et al., 2010). Flies in which *Frost* is knocked out

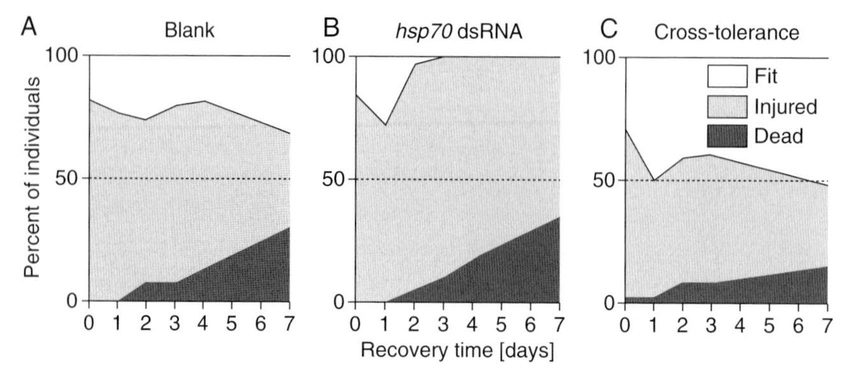

Fig. 3.15 Expression of the heat-shock protein Hsp70 mitigates cold-induced injury and death in the *Pyrrhocoris apterus* (Heteroptera). Survival and recovery of (A) blank-injected and (B) *hsp70* double-stranded RNA-injected *P. apterus* following exposure to –5°C for 5 days. Injection of *hsp70* double-stranded RNA inhibits *hsp70* mRNA concentration and Hsp70 expression. (C) Survival and recovery of *P. apterus* pretreated with a mild, Hsp70-inducing heat shock (41°C for 1h) prior exposure to –5°C for 5 days. Adapted from Kostal and Tollarova-Borovanska (2009).

recover more slowly from chill-coma (Colinet et al., 2010). The function of the *Frost* protein is unknown, although its sequence indicates that it is a mucin-like protein (Goto, 2001). Mucins act as membrane-bound barriers that protect cells from pathogens and help regulate ionic concentrations and hydration, and perhaps *Frost* is important to reestablishing ion homeostasis following cold-driven membrane damage (see Section 3.1.2 above).

Although Hsps are key agents of heat tolerance, they do not underlie all aspects of this trait. For example, in *Drosophila*, acquired heat tolerance and induced Hsp70 expression following a moderate heat shock are not well synchronized, and some heat tolerance traits occur independently of Hsp expression (Jensen et al., 2010). These observations indicate that heat tolerance is complex and likely involves other genes besides *hsps*. In *Drosophila*, several genes besides *hsps* are upregulated by heat shock factor (Hsf), the transcription factor common to all *hsps* (Jensen et al., 2008). Still other non-*hsp* genes are upregulated independently of Hsf. Most of the genes in that study have unknown functions, although one of them is putatively associated with apoptosis, proteolysis, and peptidolysis. The protein encoded by the gene contains a Bcl-2-associated athanogene (BAG) domain, which is also present in certain Hsp70 co-chaperones, where it regulates chaperone activity through binding to the ATPase domain of Hsp70 (Doong et al., 2002). Further characterization of heat-inducible, non-*hsp* genes and heat-responsive regulation independent of the Hsf system will help broaden our understanding of heat tolerance and the cellular stress response in general.

3.4.2 Plasticity: Long-term Thermal Modification of Protein Expression and Cell Biochemistry

Prolonged exposure to moderately higher or lower temperatures can trigger longer-term, typically reversible, changes in the thermosensitivity of many traits. Seasonal changes in temperature often trigger changes that allow insects to expand their thermal performance breadth. Such plasticity is often called 'thermal acclimation', although this term is used differently depending on the duration of exposure to thermal extremes, the temporal kinetics of the response, and whether the change improves fitness. For example, (Angilletta, 2009) considers thermal acclimation as 'any phenotypic response to environmental temperature that alters performance and plausibly enhances fitness,' a definition that does not distinguish between responses to acute and chronic exposures to thermal extremes and thus considers the heat shock response and rapid cold hardening as forms of acclimation. See Chapter 1 and also (Chown and Nicolson, 2004) and (Loeschcke and Sorensen, 2005) for further discussion on this topic.

This section covers plausibly beneficial cellular and physiological responses to temperature that occur on longer timescales (several hours to weeks). (Angilletta, 2009) also argues that it is most useful to apply the "beneficial acclimation hypothesis" (i.e. that environmental changes induce phenotypic changes that benefit performance in the new environment) to adult organisms or those within a specific developmental stage. This is because temperature-dependent plasticity over development is usually not demonstrated as beneficial. Rather, most organisms have some optimal developmental temperature range, outside of which they underperform (Woods and Harrison, 2002). Even so, some studies of temperature-induced developmental plasticity in insects support the beneficial acclimation hypothesis. For example, thermal-plasticity of *Drosophila* wing development may also be beneficial. For *Drosophila* and many small insects, body mass is negatively related, but wing-loading (the body mass-to-wing area ratio) is positively related, to developmental temperature. The wings of cold-reared *Drosophila* are larger because of increased cell size, but not number (Azevedo et al., 2002). (Frazier et al., 2008) tested *Drosophila* takeoff flight performance within a range of cool temperatures (14, 16, and 18°C) among flies reared over a wide range of temperatures (15, 23, and 28°C). Nearly all flies took off at the warmer flight temperature (18°C), but only flies reared at the coldest temperature (15°C) were able to take off in the coldest flight temperature (14°C; see Fig. 3.16). This experiment suggests that cold-induced developmental plasticity of wing size and shape increases flight performance in cold air.

If the (Frazier et al., 2008) study describes a case of beneficial plasticity, then flies reared at high temperatures should fly better, in warm air, than cold-reared flies (Huey et al., 1999;, Kingsolver and Huey, 1998; Kingsolver and Hedrick, 2008). Although this prediction has not been tested, cold-reared flies in the

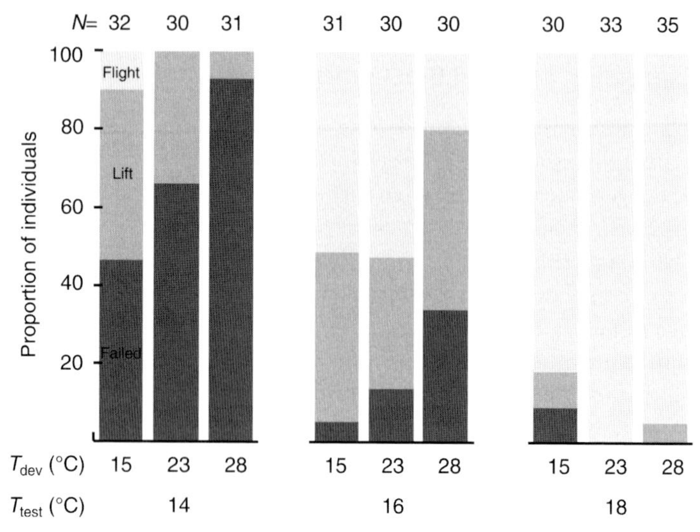

Fig. 3.16 Possible beneficial acclimation in *Drosophila melanogaster*. Cold-rearing during larval and pupal stages improves adult flight performance in cold air. Fail = unable to fly; Lift = takeoff possible, but no sustained flight; Flight = sustained flight. Cold-reared flies are aided by larger wings (and greater lift capacity), despite lower wingbeat frequencies (data not shown). Adapted from Frazier et al (2008).

(Frazier et al., 2008) study, by virtue of their much larger wings, also had lower wingbeat frequencies, a factor known to limit forward flight speed and maneuverability. This limitation predicts that warm-reared flies, with their higher wingbeat frequencies and smaller bodies and wings, would fly faster and be more maneuverable in warm air.

3.4.2.1 Cuticular Melanization

When developing in the cold, several insect species exhibit increased levels of cuticular melanization. Plasticity in surface coloration may represent an adaptive thermoregulatory trait if darker morphs from cooler developmental environments are better able to absorb short-wave (solar) radiation and attain higher body temperatures. This effect will depend on body size. Insects like *Drosophila*, which exhibits modest plasticity in melanization, are small enough so that they are completely isothermic with surrounding air (i.e. convective and conductive heat exchange predominates regardless of insolation and radiative absorption). Larger insects, however, may attain higher body temperatures if they are darker. Temperature-sensitive plasticity of melanization allows darker morphs of *Melanoplus* grasshoppers (Fielding and DeFoliart, 2005; Fielding and Defoliart, 2005), adult *Pieris* butterflies (Kingsolver and Wiernasz, 1991; Stoehr and Goux,

2008), and *Papilio* and *Battus* caterpillars (Hazel, 2002; Nice and Fordyce, 2006) to maintain higher body temperatures than lighter morphs, which improves growth and various performance indices in cooler conditions. However, this strategy comes with costs, and in Section 3.5 we discuss trade-offs among thermoregulation, thermotolerance, and other traits.

3.4.2.2 Insect Membrane Acclimation

Passive thermal effects on the phase characteristics of plasma membranes, particularly the inverse relationship between viscosity and temperature, are a significant problem for all organisms. Insects and other ectotherms have evolved an acclimation response to temperature change, termed "homeoviscous adaptation," during which membrane composition is altered to preserve membrane fluidity (Cossins and Raynard, 1987; Hazel, 1995). In this response, which typically takes several days, cold reduces phospholipid fatty acid saturation, favors ethanolamines over cholines in phospholipid head groups, and lowers levels of membrane cholesterol (cholesterol adds rigidity to fluid-phase membranes). Thus, in cold-acclimated animals, low cholesterol content combined with higher proportion of phospholipid fatty acid unsaturation and ethanolamine headgroups should yield a more disordered membrane. Such a membrane is less likely to transition, at low temperatures, from the liquid-crystal to gel phase.

Most studies of homeoviscous adaptation have focused on vertebrate ectotherms (primarily fish). Nonetheless, a few studies indicate that it is important for insects too. Thermal acclimation of membrane lipid composition—reflecting probable homeoviscous adaptation—has been described for Diptera, Lepidoptera, and Heteroptera (Bennett et al., 1997; Kostal and Simek, 1998; Kostal et al., 2003; Michaud and Denlinger, 2006; Ohtsu et al., 1998; Overgaard et al., 2006; Overgaard et al., 2008; Slachta et al., 2002; Tomcala et al., 2006). In *Drosophila*, development at 15°C alters membrane composition, strongly improves long-term adult survival at 0°C, but has little effect on short-term adult survival at temperatures between -8 and -12°C (Overgaard et al., 2008). These results and those of (MacMillan et al., 2009) suggest that homeoviscous adaptation contributes little to classic rapid cold hardening, although even short (2–8 h) cold exposures can induce changes in phospholipid fatty acid chain composition (Overgaard et al., 2005; Overgaard et al., 2006; Michaud and Denlinger, 2006). *Drosophila* raised on a high-cholesterol diet have elevated membrane cholesterol levels and cold tolerance (Shreve et al., 2007). Interestingly, no insect studies have yet addressed the role of homeoviscous adaptation in tolerating warm temperatures.

3.4.2.3 Freeze Avoidance and Supercooling

Many insects do not tolerate extracellular freezing. To survive low subzero temperatures by avoiding freezing, they lower the freezing temperature of their tissues, typically on a seasonal basis. The mechanisms underlying seasonal freeze avoidance and cold tolerance are thoroughly reviewed by (Denlinger, 2010; Doucet et al., 2009; Clark and Worland, 2008). The freezing points of insects are always lower than 0°C (the equilibrium freezing point of pure water), both because they are small and because their fluids contain organic and inorganic solutes (the latter being termed a "colligative" effect). Because internal ice is so dangerous, freeze avoidance is a common strategy in insects (although, as stated earlier, significant damage may also arise from non-freezing cold injury). To reduce the likelihood of winter freezing, many insects, over the course of days to weeks, actively depress freeze temperatures to -20°C or lower. They do so by 1) accumulating high concentrations of solutes, typically polyhydroxyl alcohols, to colligatively lower their equilibrium freezing temperature and 2) improving supercooling, the ability to chill to temperatures below the equilibrium freezing temperature without freezing.

During seasonal cold acclimation, polyols (e.g. glycerol, sorbitol, and mannitol) and other low molecular weight molecules, such as trehalose and certain amino acids, are synthesized in the fat body and released into the hemolymph. These molecules aid cold tolerance by colligative depression of the equilibrium freezing temperature and non-colligative effects such as protein and membrane stabilization. Insects minimize ice nucleators by gut voidance and masking or eliminating other surfaces in their tissues that can orient water molecules into a crystal lattice and trigger freezing. Cuticular waxes reduce inoculation from external ice (Crosthwaite et al., 2011; Olsen et al., 1998). A good example of this overall strategy is the overwintering pre-pupae of the buprestid *Agrilus planipennis*, which lower their supercooling temperature to –30°C and avoid freezing by elevating hemolymph glycerol to over 3M (Fig. 3.17), synthesizing antifreeze agents, and suppressing inoculative freezing via cuticular lipids (Crosthwaite et al., 2011). Accumulating cryoprotectants takes days to weeks and is often cued by changes in day lengths or temperatures.

Another factor that promotes supercooling is antifreeze proteins (AFPs), discovered initially in Antarctic fish (Devries, 1971). AFPs do not depress freezing temperatures colligatively but instead inhibit ice crystal proliferation by adsorbing onto the surface of newly formed ice crystals. In the absence of AFPs, ice crystals grow by adding water molecules to prism surfaces such that the crystal growth faces have a broad, low radius of curvature and low surface free energy. When antifreeze proteins are present, they adsorb onto ice crystals, especially at preferred growth sites, and prevent growth in the normal, low-radius configuration. Consequently, ice growth is restricted to spaces in between adsorbed AFPs, which forces crystal fronts to assume a high radius of curvature and significantly reduces the probability of additional water molecules joining the lattice. Thus, AFPs lower the freezing temperature of hemolymph but do not affect the melting/freezing equilibrium temperature, a phenomenon known as thermal hysteresis and initially described in *Tenebrio* cryptonephridial fluid (Ramsay, 1964).

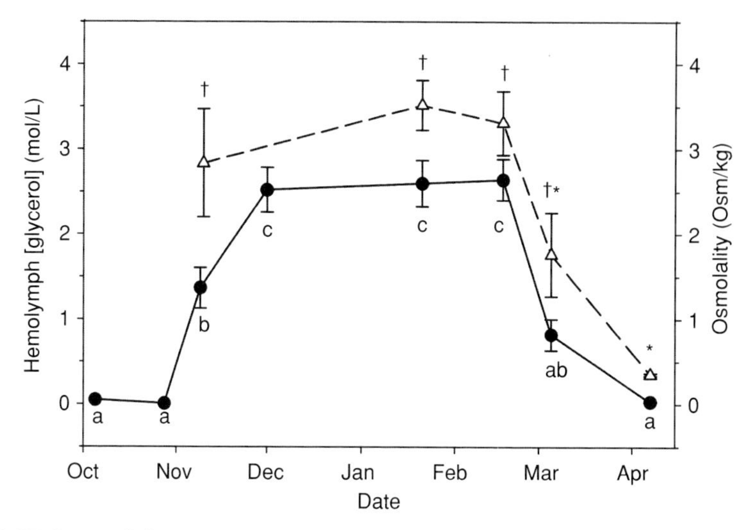

Fig. 3.17 Seasonal changes in hemolymph glycerol concentration (filled circles) and total hemolymph osmolality (triangles) in prepupae of the emerald ash borer *Agrilus planipennis* (Coleoptera: Bupresidae). High concentrations of hemolymph glycerol help *A. planipennis* depress its supercooling point to -30°C during the winter months. Adapted from Crosthwaite et al. (2011), with permission from Elsevier.

Xylomannan-based glycolipids are a newly identified class of thermal hysteresis antifreeze in insects (Walters et al., 2009, 2011).

Many insects synthesize antifreeze proteins, and sequence comparisons across species reveal little similarity, indicating that antifreeze proteins have evolved independently several times in insects (Graham and Davies, 2005). Furthermore, antifreeze proteins from individual species exist as multiple isoforms (Andorfer and Duman, 2000; Doucet et al., 2002; Tyshenko et al., 2005), some of which are differentially expressed across season, tissue, and development (Andorfer and Duman, 2000; Qin et al., 2007). Little is known about the regulation of insect antifreeze protein expression, although (Qin et al., 2007) suggest JH-mediated control as opposed to direct photoperiod or temperature cues.

3.4.2.4 Freeze Tolerance and Ice Nucleation

Relatively few insects are freeze tolerant, although the trait has evolved independently at least six times among the following orders: Blattaria, Orthoptera, Coleoptera, Hymenoptera, Diptera, and Lepidoptera (Sinclair et al., 2003). In some species freeze tolerance is restricted to specific life stages and/or is seasonally regulated, while other species are freeze tolerant year-round. Freeze tolerant insects control, to some degree, the location and temperature of freezing by concentrating antifreezes, aquaporins, ice-nucleating proteins, and other ice-nucleating agents in various body compartments. Some species rely on inoculative freezing around food particles or bacteria in the gut. Elevating membrane aquaporins, extracellular cryoprotectants, and ice nucleators allows freeze tolerant insects to rapidly deploy water and small solutes to extracellular spaces (Philip and Lee, 2010). There, ice formation is restricted to temperatures comparatively close to equilibrium freeze temperatures, and unfrozen fractions have liberal concentrations of cryoprotectants. Freeze tolerant species often produce antifreezes in cells adjacent to ice, probably to prevent recrystallization (Wharton et al., 2009). Some insects survive intracellular ice formation, but its regulation and tolerance mechanisms are largely unknown (Sinclair and Renault, 2010). Freeze tolerant insects seasonally upregulate concentrations of ice-nucleating proteins, although the control of ice-nucleating proteins is one of the least understood phenomena in insect cold tolerance. Though not freeze tolerant, the beetle *Ceruchus piceus* removes a hemolymph lipoprotein with ice nucleating activity as it enters winter diapause (apparently it doesn't need the lipoprotein during winter diapause), which helps the beetle to lower its supercooling point below -20°C without producing antifreeze proteins or polyols (Neven et al., 1986).

3.4.2.5 Diapause

Diapause is a stage-specific, environmentally-driven period of dormancy. We include a brief discussion here because diapause can be triggered by temperature, is a well-regulated physiological response (as opposed to quiescence resulting from direct, passive thermal affects), commonly involves inducible thermotolerance, and allows many insect species to withstand unfavorable environmental conditions. Bale and Hayward (2010) provide a more extensive review of diapause and thermal tolerance. During pupal diapause, *S. crassipalpis* acclimates to seasonal cold by upregulating levels of glycerol, glucose, alanine, pyruvate, and leucine (Michaud and Denlinger, 2007). At the onset of diapause in this species, glucose (a glycerol precursor) is liberated from glycogen reserves by a temperature sensitive glycogen phosphorylase (Chen and Denlinger, 1990), and this enzyme may also play a role in seasonal cold acclimation. While the endocrine mechanisms controlling diapause are receiving growing attention (Denlinger, 2002; Sim and Denlinger, 2008; Williams et al., 2006), it remains unclear how temperature sensing is functionally linked to hormonal control of diapause.

Upregulation of various Hsps during diapause, whether initiated by temperature or other factors, is a common feature across insects. Recent genomic and proteomic studies show that diapause is accompanied by the upregulation of many thermotolerance genes and proteins, primarily those involved with cryoprotection and protein damage prevention (Li et al., 2007; Ragland et al., 2010; Rinehart et al., 2007; Wolschin and Gadau, 2009). Hsp upregulation during diapause can, by an unknown mechanism, occur independently of the temperature-sensitive Hsf control system of *hsp* transcription (Clark and Worland, 2008;

Hayward et al., 2005). In addition, activation of p38 mitogen-activated protein kinase plays a role in the initiation of diapause in some species (Fujiwara et al., 2006a; Fujiwara et al., 2006b).

3.4.3 Non-genetic Inheritance of Thermotolerance Traits

Insects can inherit parental traits that are not determined by DNA sequence but instead are conferred by elements of the ancestral environment, broadly defined to include ambient conditions, the organismal soma, the intracellular cytoplasm (excluding cytoplasmic DNA), and the proteins and methyl-groups associated with the DNA molecule (Bonduriansky and Day, 2009). In insect thermobiology, there are few detailed studies of the inheritance of non-genetic, environmentally-induced traits (often called parental affects or transgenerational phenotypic plasticity). In *Drosophila*, flies from warm-reared mothers have slightly higher heat knock-down temperatures than flies from cold-reared mothers (Crill et al., 1996), while cold-acclimation of mothers improves the eclosion rates of their eggs (Rako and Hoffmann, 2006). In the butterfly *Bicyclus anynana*, hatchling caterpillars from cold-reared mothers have a higher protein:lipid ratio than offspring from warm-reared mothers (Geister et al., 2009). Together, these studies suggest that parental epigenetic effects are likely to be important in understanding insect responses to temperature.

3.5 Evolution of Thermoregulation and Thermotolerance

Most of the thermoregulatory and thermotolerance traits discussed earlier in this chapter have verified or plausible fitness benefits. Space constraints prevent us from thoroughly reviewing patterns and examples of evolutionary responses to natural and laboratory temperature gradients. Readers are directed to treatments of the subject in (Angilletta, 2009; Chown and Nicolson, 2004; Denlinger, 2010; Evgen'ev et al., 2007; Heinrich, 1993; Sorensen and Loeschcke, 2007; Strachan et al., 2011). Instead, our goal in this section will be to illustrate some examples of thermal adaptations in the context of investment or capacity trade-offs, which is one of the major themes of this book.

3.5.1 Melanization, Immunity, and Predation

Development in cool temperatures induces adult melanization, or darker color, in some butterfly species, with adaptive thermoregulatory benefits (Kingsolver, 1995a; Kingsolver, 1995b; Stoehr and Goux, 2008). However, two other plastic traits in these species also require melanin. The first is a basic insect immune response in which internal parasites or pathogenic organisms are enveloped by hemocytes, and then encapsulated in melanin. The strength of the immune response can vary according to age, nutritional status, mating history, and parasite/pathogen load. The second is pupal melanization, which is induced by dark background coloration and is probably an adaptation for crypsis.

Stoehr (2010) conducted rearing experiments to test for trade-offs among these responses in *Pieris rapae*. Rearing in yellow containers resulted in lightly pigmented pupae while red containers induced darker pupae. Wing melanization was the same in adults eclosing from pupae in either colored container, indicating no trade-off in pupal vs. wing melanization. In a second experiment, individuals were reared in warm, long photoperiod conditions (which induced light-colored wings) or cool, short photoperiod conditions (which induced dark wings). In each temperature/photoperiod treatment group, half of newly eclosed adults were implanted with a short nylon thread in their abdomens, which induced an immunological encapsulation response. Light butterflies from the warm, long photoperiod treatment had a higher encapsulation response than dark butterflies from the cool, short photoperiod treatment (Fig. 3.18). In this case, thermoregulatory investments in wing melanization during development depressed the capacity of the adult immunological response. A similar thermally-induced immunological trade-off may exist for the butterfly *B.anynana*, in which rearing temperature is positively correlated with adult fat content but negatively correlated with haemocyte number and phenoloxidase activity (Karl et al., 2011).

3.5.2 Short-term Inducible Thermotolerance vs. Long-term Survival and Fitness

Rapid cold hardening and the heat-shock response are energetically costly. Both require significant upregulation of transcription, translation, and mobilization of metabolites as described earlier in this chapter. These costs predict a trade-off between costs of surviving thermal stress and current and future energetic demands. However, the long term fitness consequences associated with short-term stress responses may not result exclusively from an energy trade-off, but could also have a non-energetic explanation, such as incomplete recovery from long-term injuries caused by short-term thermal stress. In other words, maybe both chronic injury due to thermal extremes *and* the cost of inducing acute thermotolerance mechanisms have deleterious fitness consequences.

Several studies have investigated this issue by testing the effects of acute thermotolerance induction, sometimes coupled with transgenic increases in stress proteins such as Hsp70, on investment in energy-dependent, fitness-related traits. Exposure of *Drosophila* females to a mild, Hsp70-inducing heat stress reduces egg viability for several days (Fig. 3.19), but only in flies that overexpress Hsp70 through transgenic increases in the number of *hsp70* copies (Silbermann and Tatar, 2000). In the same *Drosophila* strain with extra copies of *hsp70*, larval growth and survival is reduced compared to control strains following mild heat

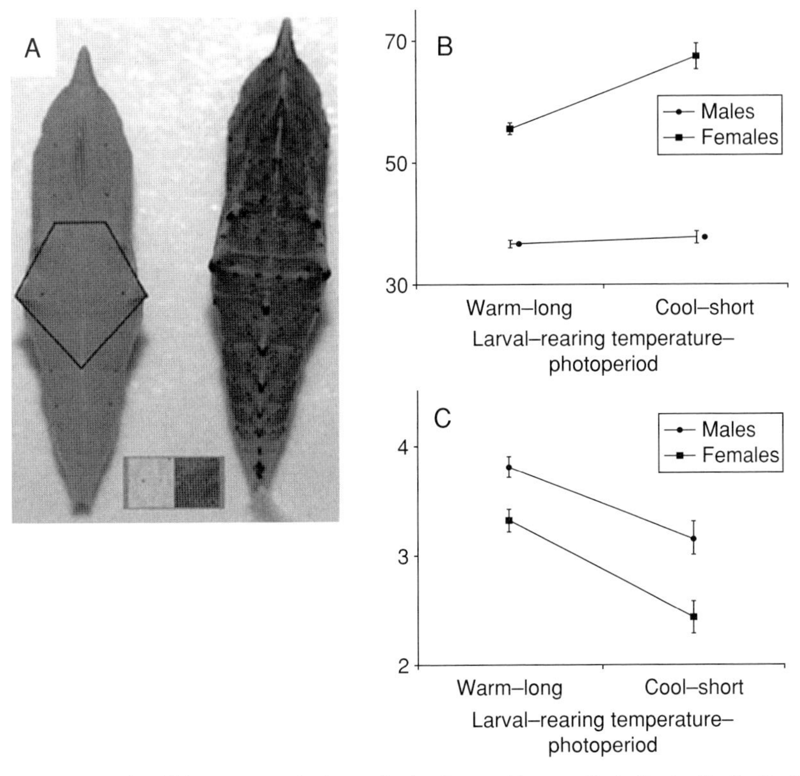

Fig. 3.18 A trade-off between cuticular melanization and immunity in the butterfly *Pieris rapae*. (A) Left: an unmelanized pupa from a larva reared on a yellow background. Right: a melanized pupa from a larva reared on a red background. Such plasticity aids in crypsis. (B) Effects of temperature/photoperiod regime during larval rearing on adult wing melanization. Rearing in cool, short-daylight conditions induces the development of darker wings, which aid in adult thermoregulation. (C) Effects of temperature/photoperiod regime during larval rearing on adult melanin encapsulation of a foreign body. Cool-reared, dark-winged butterflies have a lower encapsulation immune response than warm-reared, light-winged butterflies. Warm-Long is 25°C, 24:0 light:dark photoperiod; Cool-Short is 19°C, 8:16 light:dark photoperiod. Units are square millimeters (B) and the re-reflected square root of capsule darkness, in arbitrary units (C); higher numbers indicate greater melanization. Adapted from Stoehr (2010), with permission from Springer Science + Business Media.

stress (Krebs and Feder, 1997a; Roberts and Feder, 2000). Isofemale *Drosophila* lines from a single natural population can vary significantly in their ability to express Hsp70, and this variation positively correlates with larval heat tolerance, but negatively correlates with survival in thermally benign conditions (Krebs and Feder, 1997b).

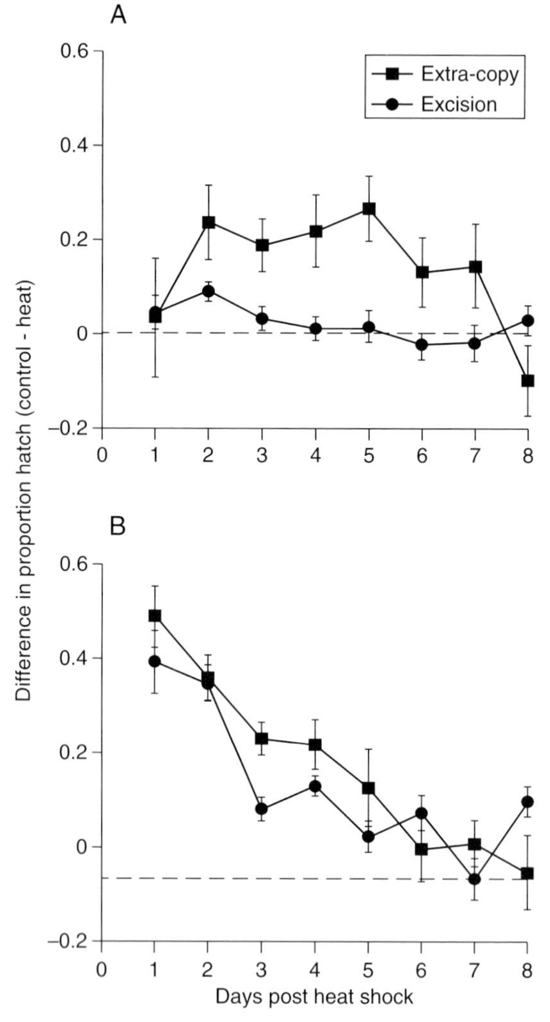

Fig. 3.19 A trade-off between Hsp70 expression and egg viability in *D. melanogaster*. Differences among proportion hatch of eggs laid by control and heated transgenic females over 8 days following a Hsp70-inducing exposure at 36°C for 30 min (A) or 60 min (B). The "excision strain" possesses the wild-type complement of inducible *hsp70* genes (10 copies). The "extra-copy strain" possesses the wild-type complement plus 12 additional copies of *hsp70* (22 copies in total), and has higher levels of inducible Hsp70 expression than the excision strain. Viability of eggs laid in the days following female exposure an Hsp70-inducing heat treatment is markedly lower in the extra-copy strain than in the excision strain. From Silbermann and Tatar (2000), with permission from John Wiley and Sons.

Induction and maintenance of cold-tolerance mechanisms can also sap investment from other traits. For example, the survival of *Drosophila* adults following a single 10-hour, -0.5 C exposure is much lower than in flies given the same duration of cold (10 hours), but in 2-hour blocks over a 5-day period, which demonstrates the short-term benefits of recovery/repair mechanisms following repeated exposures (Marshall and Sinclair, 2010). However, following daily 2-hour, -0.5 C exposures for up to 5 days, long-term reproduction is highest in flies exposed for only 2 days and lowest in flies exposed for 5 days (Marshall and Sinclair, 2010), illustrating the costs of surviving acute temperature stress versus future reproductive potential.

Relative to the number and diversity of thermotolerance and thermoregulation mechanisms that have been studied, there has been little research focusing on costs of these mechanisms and energy allocation strategies concerning them and other traits. While our coverage of trade-offs has not been comprehensive, we hope that the few examples discussed here motivate researchers in the field to explore thermoregulation and thermotolerance traits from an energetic perspective and test for allocation and trade-off strategies so as to more clearly place these traits into insect natural history strategies.

4 Water

"If there is magic on this planet, it is contained in water."
Loren Eiseley, *The Immense Journey* (1959)

4.1 Defining the Problem

Most insects are terrestrial and, because they are small and have high surface area-to-volume ratios, often face problems of conserving water. Likewise, aquatic insects living in hypersaline waters face the problem of dehydration. Compared to vertebrates, insects may be less sensitive to dehydration because they have open circulatory systems and need relatively low pressures to drive circulation. However, dehydration increases ion concentrations, which affects membrane potentials and can denature proteins. Loss of hemolymph volume can decrease circulatory transport of nutrients and hormones. For insects with hydrostatic skeletons such as caterpillars, dehydration can impede movement.

Other terrestrial insects—blood, nectar, or sap feeders—and freshwater insects face the opposite problem of too much water (Benoit and Denlinger, 2010). Blood dilution may cause water to move by osmosis into cells, impairing their function or destroying them. Blood dilution can also affect membrane potentials by changing ion concentrations, and can increase the volume and weight of the insect.

Problems of insect water balance occupy two conceptual levels. The first level, which traces to seminal studies in the 1930s, focuses on whole-animal water balance and analyzes how insects control, or attempt to control, avenues of gain and loss. This approach has identified basic mechanisms of water balance and how they vary as functions of ecological circumstance; new discoveries continue to provide major advances, and some of these—particularly in areas of hygrosensory physiology and genetics—we will cover. However, because two monographs on insects and water cover earlier discoveries (Edney, 1977; Hadley, 1994) and others have reviewed more recent work (Gibbs, 2002; Chown, 2002; Chown et al., 2004; Beyenbach and Piermarini, 2009), we provide minimal coverage of these issues here. The second, complementary level, with more emphasis in this chapter, examines cellular, molecular, and endocrine aspects of how insects control where water goes *inside* their bodies. This area has seen rapid progress, with recent discoveries providing fundamentally new ways of viewing insect water problems.

From our perspective, the key is to connect this molecular progress back to the environments in which insects live. That is, the water and ion problems that an insect faces depend on the physical attributes of its environment, especially whether it is terrestrial or aquatic, and on the resources available to it. Of those resources, the ions available, and the relative amounts of energy available for moving those ions around, are critically important. This is because water moves toward concentrated ions by osmosis. Making these connections will leverage modern molecular progress to provide new insights into whole-organism water balance.

4.2 Whole-Insect Water Balance

Changes in an insect's water content reflect differences between rates of water gain and loss. Water is gained by drinking, from ingested food, and by aerobic metabolism (metabolic water). In addition, some insects can absorb water vapor orally or rectally. Water is lost by excretion, via the cuticle, and via the spiracles. Analyzing the relative magnitudes of these routes, and their modification by ecological and evolutionary circumstances, has been a core problem in insect physiological ecology over the past 75 years (Edney, 1977; Bernays, 1990; Hadley, 1994; for a recent summary of water balance in bees, see (Nicolson, 2009)). Recent progress in the molecular genetics of *Drosophila* is also providing insight into sensors for both humidity and liquid water (Benton, 2008).

4.2.1 Drinking and Ingestion

Insects drink free water according to their hydration status. For example, locusts (*Locusta migratoria*) took larger meals (dry mass) after drinking and, conversely, took larger drinks after feeding (Raubenheimer and Gade, 1994). In a subsequent study (Raubenheimer and Gäde, 1996), locusts given food and water ingested both food and water and grew substantially. However, in the absence of water, locusts had very low rates of food consumption and lost mass, perhaps as a means of avoiding additional water loss in the feces. Likewise, in the absence of food, locusts drank very little water. These data together were interpreted as indicating that locusts make feeding and drinking decisions by assessing their hemolymph, which integrates both fluid and feeding history (Simpson and Raubenheimer, 1993).

Many herbivorous insects obtain most or all of their water from the plants they consume. Because insects contain much higher concentrations of protein (and often free carbohydrates and phosphorus) than do plants, they often must

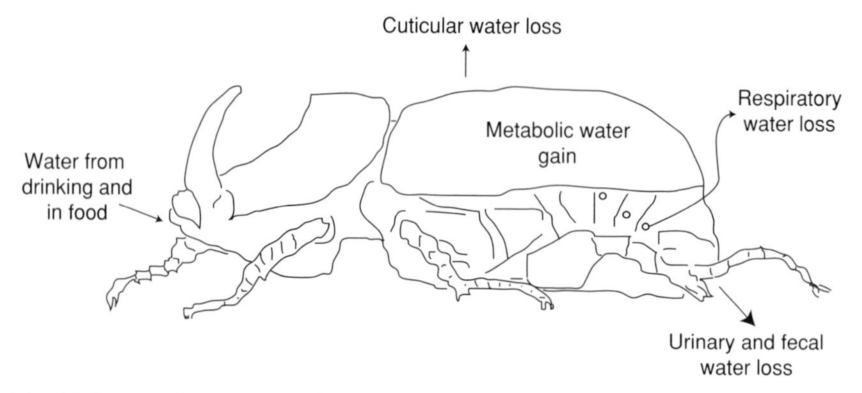

Fig. 4.1 Routes of water gain and loss in insects. Insects gain water by eating and drinking, and as a byproduct of aerobic respiration. They lose water across the cuticle, from the spiracles ("respiratory water loss"), and in the feces. Because the primary urine, formed by the Malpighian tubules, is secreted into the gut upstream of the rectum, urinary and fecal losses are usually not distinguishable. The relative magnitudes of the various routes vary enormously among taxonomic groups.

consume large amounts of plant material just to obtain adequate macronutrients (Sterner and Elser, 2002). This leads to excessive water intake. Insects may therefore prefer dry relative to moist plants (Lewis and Bernays, 1985).

A special form of drinking, notable for its use in some extremely arid environments, is fog-drinking, which tenebrionid beetles do in southwest Africa. In the Namib Desert, precipitation is sparse but nighttime fog relatively common. Many tenebrionids drink fog that has condensed on plant surfaces (Seely, 1979; Seely et al., 1983). Some beetles dig fog-collecting trenches perpendicular to the wind (Seely and Hamilton, 1976), and others (two species of *Onymacris*) have specialized elytral surfaces for condensing fog directly (Hamilton and Seely, 1976). The elytra have arrays of hydrophilic bumps that effectively condense fog and hydrophobic troughs that carry the condensed droplets toward the head, where they become available to the beetle for drinking (Parker and Lawrence, 2001). Fog-collecting elytra have inspired engineered materials with similar surfaces (Zhai et al., 2006), which could be useful for water harvesting surfaces, and microfluidic devices (e.g. 'lab-on-a-chip').

Some insects can take up water from humid air. The desert cockroach, *Arenivaga investigata*, can cause water to condense onto hydrophilic bladders on its mouthparts (O'Donnell, 1977). Some beetle larvae (*Tenebrio*; (Machin et al., 1982)) and millipedes (Wright and Westh, 2006) take up water from humidified air via the rectum. *Tenebrio* use high-osmotic pressure rectal fluids to absorb water vapor (Machin et al., 1982). The record-holder for water vapor absorption from the least humid air is the firebrat, *Thermobia domestica* (Noble-Nesbitt, 1970), which can absorb vapor in relative humidities as low as 43%. The vapor-absorbing epithelium occurs in specialized rectal sacs that have the highest densities of mitochondria of any known transporting epithelium (Noble-Nesbitt, 2010).

Other insects specialize on high-water foods, for example, blood-sucking bugs and mosquitoes, phloem and xylem feeders, and nectar-foraging bees. In species taking intermittent but large loads (e.g. blood), the problem is to activate diuresis (urine production) rapidly during feeding and then to shut it down once water balance is reestablished. Diuresis usually occurs via massive stimulation of fluid secretion by Malpighian tubules (Maddrell et al., 1969; Maddrell et al., 1971; Pietrantonio et al., 2001' Veenstra, 1988), and much of the interest in this problem now focuses on understanding upstream signaling and control. In species taking chronically water-rich meals—such as homopteran xylem feeders—other morphological and physiological solutions are apparent: some xylem-feeding Homoptera have a derived arrangement of the midgut and Malpighian tubules called the "filter chamber" (Gouranton, 1968; Cheung and Marshall, 1973) (see Fig. 4.2). In the filter chamber, the anterior and posterior portions of the midgut are closely apposed, and xylem ingested into the initial midgut is rapidly filtered so that solutes are retained and water passes into the terminal midgut and then into the hindgut. Work on filter chambers of the homopteran *Cicadella viridis* has demonstrated that water movement is facilitated by an aquaporin named AQP*cic* (see Section 4.6.1) (Beuron et al., 1995; Le Cahérec et al., 1996,; 1997).

4.2.2 Metabolic Water

Aerobic catabolism of organic substrates produces water. For glucose oxidation, the equation is:

$$C_6H_{12}O_6 + 6O_2 \rightarrow 6CO_2 + 6H_2O. \quad (1)$$

Different substrates produce different amounts of water: per gram of substrate, carbohydrate produces 0.56 g of water, lipid 1.07 g of water, and protein 0.40 – 0.50 g of water (Schmidt-Nielsen, 1997). Metabolic water is sometimes mistaken as a "free" source of water, that is as a waste product from metabolic processes that happens to contribute to water balance. However, the metabolism that produces it depends on other water-losing processes, including oxygen acquisition (which leads to respiratory water loss) (Nicolson and Louw, 1982; Willmer, 1988; Lehmann et al., 2000; Woods and Smith, 2010) and disposal of nitrogenous wastes (i.e. in urine). In other words, even if substantial metabolic water is produced, the net water gain from metabolism can be small or nonexistent.

However, during intense metabolic activity such as flight, insects may produce large quantities of metabolic water, which can impose a water load. This situation has been studied best in large bees (Bertsch, 1984; Nicolson and Louw, 1982;

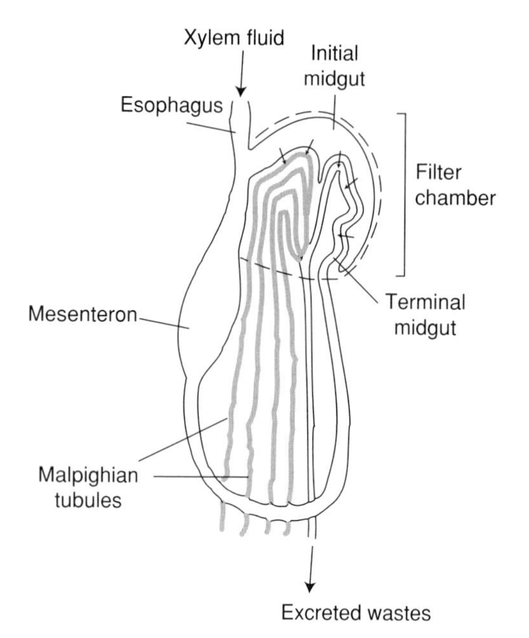

Fig. 4.2 Schematic representation of the filter chamber of some xylem-feeding Homoptera. When the insect imbibes xylem, it first goes into the initial midgut. From there, much of the water is shunted (small arrows), by osmosis, into the terminal midgut and proximal Malpighian tubules (shaded tubes). This water movement is facilitated by aquaporins in the filter chamber epithelia (see Section 4.4). The solutes left behind in the initial midgut move into the mesenteron and around the middle loop of the midgut, where they are digested and absorbed. This derived arrangement helps xylem feeders obtain sufficient nutrients from very dilute foods. Redrawn from Le Cahérec et al. (1997), with permission from Springer Science + Business Media.

Louw and Nicolson, 1983; Willmer and Stone, 1997; Roberts et al., 1998; Nichol, 2000; Nicolson, 2009), although some analysis has been done for species in other orders (Showler and Moran, 2003; Lehmann et al., 2000). In flying *Drosophila*, metabolic water can replace up to 70% of evaporative water losses (Lehmann et al., 2000); in resting *Drosophila*, this value declines to less than 10%. This is a circumstance of flight-related changes in both metabolic water gain and (mostly respiratory) water loss, where metabolic water nearly offsets massive water loss during flight, but is only a small fraction of modest water loss during rest. In bees, nectar ingestion imposes large water loads, and oxidation of sugars during flight further exacerbates the problem. Bertsch calculated, for temperate *Bombus lucorum* foraging on nectars containing 50% sugar or less (most nectars), that metabolic water produced during foraging sends individuals into water surplus (Bertsch, 1984). Other studies on a South African carpenter bee,

Xylocopa capitata (Nicolson and Louw, 1982; Louw and Nicolson, 1983), suggest likewise that metabolic water production can pose a homeostatic challenge; to rid itself of the load, *X. capitata* urinates frequently during flight. Similarly, honeybees are in net water surplus when flying at cooler air temperatures but net water deficit when flying in warm temperatures due to evaporative cooling and thermoregulatory flight metabolic depression (Roberts and Harrison, 1999).

4.2.3 The Malpighian Tubules and Hindgut: Water Movement and Excretion

In most insects, the key organs that carry out water and ion regulation are the Malpighian tubules and the hindgut, which together are functionally equivalent to the vertebrate kidney. The Malpighian tubules are blind-ended and produce, at their distal end, the primary urine. The primary urine flows along the tubule until it is dumped into the alimentary canal, at the junction between the mid- and hindgut (see Fig. 2.1). The fluid then can be highly modified by water and selective ion re-uptake across the rectum (physiological details reviewed by (Beyenbach and Piermarini, 2009)). We now know substantial detail about how these tissues move ions and water (see Sections 2.8 and 4.6.3), and about the endocrine systems that control them, though the major species in which progress has been made tend to be those that face unusual water problems (blood feeders, mosquito larvae).

Ecological studies suggest that renal modulation of water excretion occurs on a short-term basis in response to diverse challenges. For example, Woods and Bernays studied wild larvae of *Manduca sexta* feeding on a natural host plant (*Datura wrightii*) in southern Arizona (USA). Larvae experienced water stress from two sources—increasing daytime temperatures coupled to low relative humidities (giving high rates of evaporative water loss), and attacks from tachinid flies, which prompted larvae to regurgitate and defecate large quantities of liquid. Both factors led larvae subsequently to produce much drier fecal pellets, probably as a mechanism of water homeostasis (Woods and Bernays, 2000).

Renal physiology can also play a role in resource portioning between nectar-foraging insects and the flowers they visit. Willmer (1988) studied foraging adults of two species of *Xylocopa*, *X. sulcatipes* and *X. pubescens*, both of which occur in oases in Israel and forage on the same species of flowering plant, *Calotropis procera*. Willmer builds the case that *X.sulcatipes* is finely co-adapted with patterns of nectar production by *C. procera*. *X. sulcatipes* is thermally suited to foraging during the times of day when the flowers produce the most nectar. Moreover, the quantity of nectar the flowers produce is just about right to offset the modest respiratory losses that the bee incurs moving to and from flowers. By contrast, *X. pubescens*, which is a larger-bodied bee, is in water surplus during foraging, from relatively large inputs of metabolic water (see section above). The water it gains from *C. procera* sends it into water surplus, which necessitates frequent urination. Willmer argues that the longer co-evolutionary history between *X. sulcatipes* and the flower have resulted in fine balancing of water status of both nectar and insect (Willmer, 1988).

4.2.4 Cuticular Water Loss

In general, the cuticle is thought to be the major avenue of water loss from insects (Hadley, 1994), although this dogma has become controversial (Chown, 2002). The chitinous and protein components of the exoskeleton of insects provide little protection from water loss. Insects depend instead on cuticular lipids secreted by epithelial glands for waterproofing (Ramsay, 1935; Beament, 1945; Wigglesworth, 1945, 1957). These lipids can be quite effective, especially for insects with undamaged cuticles. Abrasion or handling can dramatically increase cuticular permeability, an effect that forms the basis of some schemes for insect control. For example, "inert dusts," which traditionally have involved sand or kaolin and more recently have been silica dusts, can be added to stored grains. They have little or no effect on humans but prevent insect infestations by abrading or adsorbing the insect's cuticular waxes, leading to desiccation and death (Golob, 1997).

Lipids on the insect surface appear often to be solid at low or moderate temperatures; higher temperatures melt them, which allows much faster water loss. This basic observation has led to the lipid-melting hypothesis, which states that relative waterproofing of insect cuticle depends on the amounts and kinds of lipids in the surface layer (Gibbs, 2002a). The studies of Gibbs and colleagues provide a detailed view of how lipid properties affect waterproofing, with the main conclusion being that solid-phase lipids provide good waterproofing whereas liquid-phase (melted) lipids do not. This finding suggests that insects living at high temperature or in particularly arid environments should have cuticular lipids that melt at higher temperatures. In general, they appear to do so, both inter- and intraspecifically (Gibbs, 2002a: Rourke, 2000).

Differences in lipid melting points, in turn, depend on both chain length and degree of branching and saturation. Longer chains melt at higher temperatures, and so do those with fewer branches and more saturation (fewer double bonds) (Gibbs, 2002a) (see Fig. 4.3). In general, effects of chain length appear small whereas those of branching and saturation are large.

These findings suggest a mechanistic model in which species or populations exposed to hotter, drier conditions should possess greater total quantities of cuticular lipids that are both longer and more saturated. However, a study by Gibbs et al. (2003) of 30 species of mesic and arid *Drosophila* species found no consistent patterns to support this prediction (see Section 4.3). Thus, although the lipid-melting hypothesis has been supported by multiple lines of evidence,

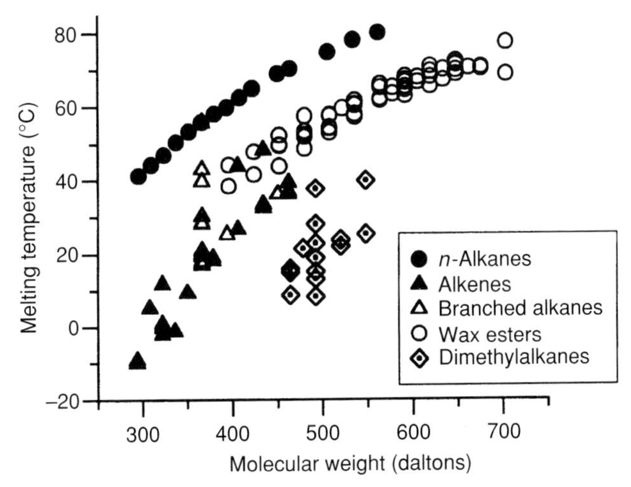

Fig. 4.3 Effects of structural differences on melting points of pure synthetic hydrocarbons and wax esters. From Gibbs (2002a), with permission from Elsevier.

major studies suggest that at the evolutionary level, our understanding of lipids and waterproofing remains incomplete (Gibbs et al., 2003).

Significant evidence suggests that individual insects alter their cuticular permeabilities over short time periods in response to hydration status or ambient humidity. Grasshoppers show evidence of reduced cuticular transpiration at low humidities (Roberts et al., 1994), and some blood-feeding insects show markedly higher (three-fold) cuticular permeability when their abdomens are distended by taking a meal (Benoit and Denlinger, 2010). A long-running set of experiments, using a variety of techniques applied to the American cockroach, *Periplaneta americana*, suggests both short- and long-term control (Noble-Nesbitt et al., 1995). Using tritiated water, Noble-Nesbitt et al. (1995) showed an abrupt decrease in water efflux from confined cockroaches when the airstream was switched from humid (RH 68%) to dry. Earlier work focused on long-term changes in cuticular permeability, and whether those changes were under endocrine control. Noble-Nesbitt and Al-Shakur showed, using decapitated cockroaches, that injecting brain homogenates from fully hydrated cockroaches into desiccated (decapitated) cockroaches led to significant and large increases in rate of water loss (measured gravimetrically (Noble-Nesbitt and Al-Shakur, 1988)). These results suggested some neurohemal factor that stimulates cuticular permeability, although the results were confounded by contributions of both cuticular and respiratory water losses to gravimetrically measured mass loss. Machin et al. (1992) repeated Nesbitt and Al-Shakur's (1988) experiments using a ventilated capsule technique to measure just cuticular water loss from the pronotum (excluded respiratory losses). Although cuticular water losses differed in magnitude from those found by Noble-Nesbitt's group, the effects of brain homogenates on permeability were essentially the same, suggesting that there is an as yet unidentified mechanism for modulating cuticular permeability.

4.2.5 Respiratory Water Loss

Gas exchange is a double-edged sword—insects must open their tracheal systems to obtain oxygen and emit carbon dioxide, but in doing so they also incur water losses ("respiratory water losses"). The relative importance of respiratory water loss has been contentious (Chown, 2002). Careful, intraspecific studies measuring both cuticular and respiratory losses have largely concluded that, for resting insects, respiratory transpiration constitutes a far lower proportion of total water loss (usually $< 20\%$) than does cuticular transpiration (Hadley, 1994; Chown, 2002), suggesting that there should be little selective pressure to reduce respiratory water loss *per se*. Even so, adaptations that reduce respiratory water loss could still be important in avoiding desiccation for insects that live in very arid environments or go long periods without drinking (Chown et al., 2006; Williams et al., 2010).

Discontinuous gas exchange cycles (DGC) (see Fig. 4.4) have long been thought to reduce respiratory water loss (Buck and Keister, 1955; Buck, 1962; Levy and Schneiderman, 1958; Kestler, 1985), an idea now known as the "hygric hypothesis." It is based on the observation that intra-tracheal pressures decline during the closed phase, leading to bulk flow of air into the tracheal system during the flutter phase, which delivers oxygen without water loss (Chown et al., 2006). Although recent studies support earlier findings that tracheal pressures can become sub-atmospheric (Hetz and Bradley, 2005; Terblanche et al., 2008), experimental tests of the role of DGC in reducing respiratory water loss have been conflicting.

A number of studies have demonstrated that insects are not more likely to use DGCs when they are desiccated. In fact, desiccation reduces the occurrence of DGC in grasshoppers and beetles, perhaps because of general effects of stress or changes in hemolymph buffering of CO_2 (Hadley and Quinlan, 1993; Chappell and Rogowitz, 2000; Rourke, 2000). Exposure to high humidities did not affect DGC patterns in diapausing moth pupae (Terblanche et al., 2008).

Multiple studies in ants have found little evidence for a role of DGC in reducing respiratory water loss. First, some CO_2 emission occurs during the flutter phase, arguing against the simple inward bulk flow mechanism proposed to prevent water loss (Lighton, 1996). Second, conversion of DGC to continuous gas exchange by exposure to hypoxia did not affect respiratory water loss in individual ants of the species *Camponotus vicinus* (Lighton and Turner, 2008). Third, comparison of *Pogonomyrmex* ants using different patterns of gas exchange found no association between gas exchange pattern and the ratio of carbon dioxide to respiratory water emission (Gibbs and Johnson, 2004).

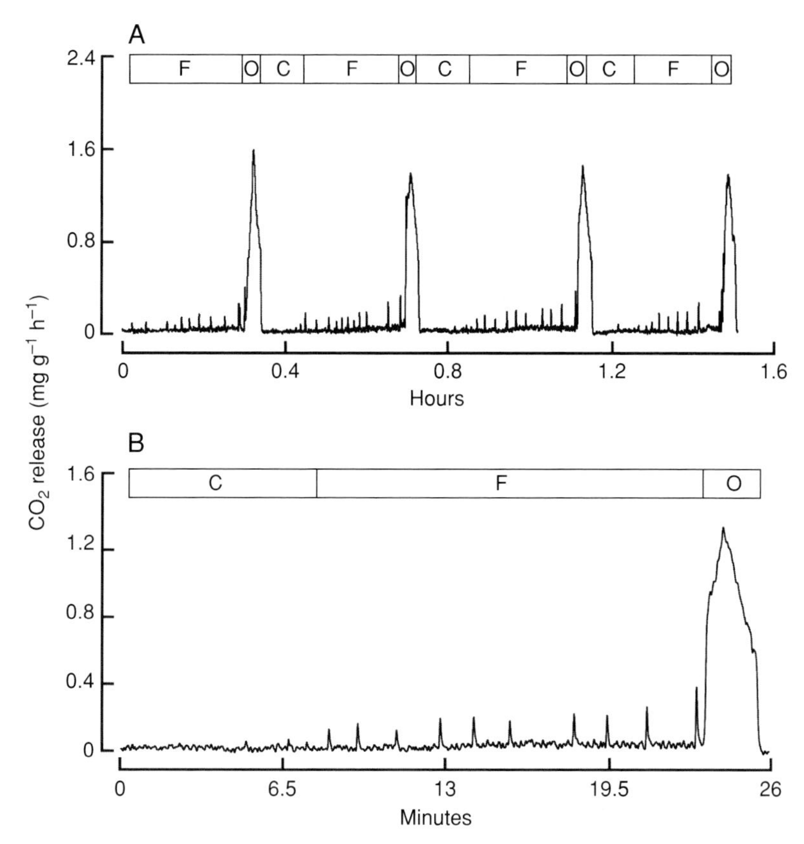

Fig. 4.4 (A) Discontinuous gas exchange cycles (DGC) in a decapitated ant, *Camponotus* sp., from Uganda, probably via a single spiracle. (B) Expanded view of DGC. In upper bars, C = closed phase, F = flutter phase, and O = open phase. Redrawn from Lipp et al. (2005), with permission from The Company of Biologists.

However, some recent data do suggest that DGC reduces respiratory water loss. Acute exposure to low humidity reduced CO_2 burst volume in the open phase, consistent with spiracular behavior that might reduce respiratory water loss rates (Terblanche et al., 2008). Chronic exposure of *Nauphoeta cinerea* to low humidities reduced the duration of the open phase of DGC and rates of body water loss, which would reduce respiratory water loss (Schimpf et al., 2009). However, because respiratory and cuticular water losses were not separated in this study, and there is evidence for modulation of cuticular water permeability in roaches (see above), these results are not definitive. Most convincingly, a recent study of larval *Erynnis propertius* (Lepidoptera) found that when individuals used DGC they had lower respiratory water loss rates than when the same individuals exchanged gases continuously (see Fig. 4.5) (Williams et al., 2010).

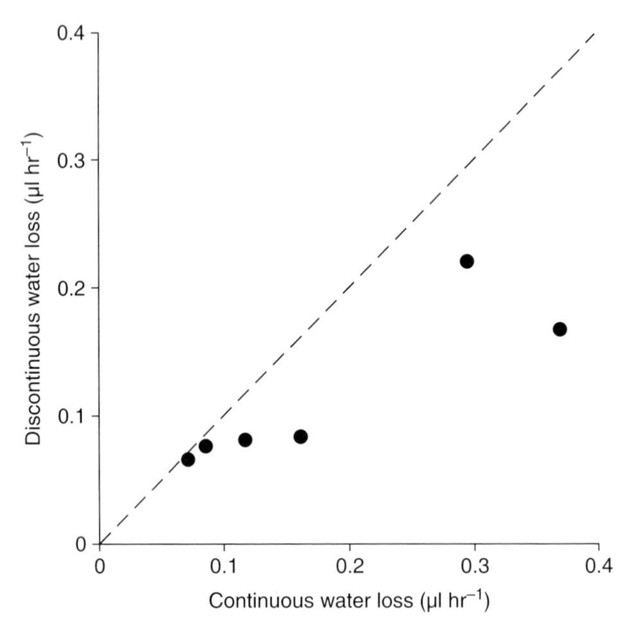

Fig. 4.5 Water loss during continuous versus discontinuous gas exchange in individual larvae of *Erynnis propertius* (Hesperiidae). Larvae switch naturally between discontinuous and continuous gas exchange. The line indicates equivalent water losses in both modes. Redrawn from Williams et al. (2010), with permission from The Royal Society and the author.

In addition, from a comparative perspective, several studies have demonstrated that arid-adapted insects are more likely to exhibit DGC (Duncan and Byrne, 2000; Chown and Davis, 2003) (Marais et al., 2005). Also, insects from warm, dry habitats tend to have longer-duration DGC (White et al., 2007), consistent with a role in reducing respiratory water loss, especially if long cycles have extended closed or flutter phases.

In summary, although data exist that support the hygric hypothesis for the evolution and maintenance of the DGC, the contradictions in the compiled data are striking, and further experiments will be required to determine whether those contradictions stem from the diversity of insects or from the experimental methods used. To understand respiratory adaptations to desiccating environments, it may be useful to focus away from DGCs *per se* and instead examine the details of gas exchange mechanisms. For example, Duncan and Byrne showed that beetles from xeric habitats use relatively greater amounts of convection in tracheal gas transport (Duncan and Byrne, 2005). Kestler has pointed out that using convection or increasing gradients for O_2 and CO_2 (e.g. by reducing tracheal conductance) can reduce the respiratory water loss at any given metabolic rate, regardless of whether DGC is utilized (Kestler, 1985). Given the diversity of gas exchange patterns among insects and the many possible routes for modifying respiratory water loss, it may not be surprising that DGC patterns are not clearly related to respiratory water loss. Indeed, a number of other hypotheses for the origin and function of DGCs have been proposed (Chown et al., 2006), and some of them are discussed in Chapter 6 on oxygen.

4.2.6 Sensing Humidity and Water: Hygrosensors

One obvious way for insects to control rates of water loss is by positioning themselves in suitable humidities. Doing so requires some means of evaluating humidity, and hygrosensitive sensilla have been identified and characterized in a variety of groups (Yokohari and Tateda, 1976; Yokohari, 1978; Itoh et al., 1984; Iwasaki et al., 1995; Sayeed and Benzer, 1996). The molecular details, however, are only now coming into focus (Benton, 2008), spurred on by *Drosophila*. Using electrophysiology, Yao et al. identified one type of coeloconic sensilla, occurring on the third antennal segment of *Drosophila*, that responded strongly to changes in humidity (Yao et al., 2005). Using molecular genetics and behavioral assays, Liu et al. later demonstrated that hygroreception is mediated by receptors encoded in two genes, *nanchung* (*nan*) and *water witch* (*wtrw*), both members of the transient receptor potential (TRP) channel family (Liu et al., 2007). Deficiency mutants in either gene impaired the ability of flies to distinguish moist from dry air. Moreover, electrophysiological recordings showed that *nan* is involved in sensing dry air and *wtrw* in sensing moist air. Why flies have two separate systems is unknown, but their dual nature may give flies the ability to distinguish fine differences in humidity (Liu et al. 2007). The evolutionary diversification of TRP genes among insects has recently been explored (Matsuura et al., 2009).

A related problem for insects is to taste water. They do so using water-sensitive neurons in sensilla, which in *Drosophila* project onto specific regions of the subesophageal ganglion (Inoshita and Tanimura, 2006). Cameron et al. (2010) subsequently identified the molecular basis of water sensing in *Drosophila*: it depends on an osmosensitive ion channel, pickpocket 28 (ppk28), a member of degenerin/epithelial sodium channel family (Cameron, 2010). Other than in *Drosophila*, little is known about water taste in insects, and even in *Drosophila* little work has linked the molecular findings to physiological or ecological events.

4.3 Evolution of Desiccation Resistance in *Drosophila*

Our understanding of how insects adapt to arid conditions has emerged primarily from comparative studies (Zachariassen et al., 1987; Hadley, 1994;

Addo-Bediako et al., 2001). A complementary, experimental approach is to select on some aspect of water relations and measure evolutionary responses, using laboratory populations of insects (Bradley et al., 1999; Gibbs, 1999). In an effort that used laboratory natural selection to select for desiccation resistance in adult *D. melanogaster* for over 20 years (>200 generations), several lines have been analyzed for changes in the various aspects of water balance discussed in Section 4.2 (Folk and Bradley, 2005). Together with a parallel set of studies on natural evolution of water relations in desert versus mesic *Drosophila* (Parsons, 1970; Hoffmann et al., 2001, 2005; Gibbs and Matzkin, 2001; Gibbs et al., 2003), these studies provide a remarkably detailed view of the evolution of insect water relations, and they highlight similarities and differences in laboratory versus natural responses to desiccation stress (Gibbs, 2002b).

In principle, populations of flies (or any insects) subjected to regular desiccation could evolve resistance in any of three ways (Gibbs and Matzkin, 2001): 1) by storing more water in their bodies, either as actual bulk water or potential metabolic water; 2) by decreasing rates of water loss; or 3) by tolerating more extreme levels of dehydration. Tolerating dehydration seems to be the least important of these, in both laboratory selected lines (Gibbs et al., 1997; but see (Archer et al., 2007)) and wild populations (Gibbs and Matzkin, 2001). Why this should be so is unclear but may represent strong constraints on acceptable lower water contents or upper solute concentrations. Of the other mechanisms, both 1 and 2 have been demonstrated in laboratory lines, but only rates of water loss (2) appear to have evolved among natural *Drosophila* populations (but see Parsons 1970).

4.3.1 Water Storage

Females from lines selected for their capacity to survive desiccation contained about 35% more water (dry masses were almost identical) than females from control lines (see Fig. 4.6) (Gibbs et al., 1997; Folk et al., 2001), and the extra water occurred primarily in hemolymph rather than tissues (Folk and Bradley, 2003). Hemolymph from these flies was not more dilute; it also contained more sodium, chloride, and trehalose (and possibly free amino acids), indicating correlated evolution in iono- and osmo-regulation (Folk and Bradley, 2005). Flies from the desiccation lines also had 80% more carbohydrate (primarily glycogen) but much less lipid than controls (Folk et al., 2001; Chippindale et al., 1998). Glycogen appears to serve as a sponge—it binds three to five times its dry mass in water and, in flies, may hold more than 60% of total bulk water (Archer et al., 2007). Glycogen-bound water becomes available as the glycogen is used in metabolism (Folk and Bradley, 2004). Metabolism of glycogen also produces metabolic water. This water, however, appeared to contribute little to overall desiccation resistance (Gibbs et al., 1997; Archer et al., 2007). Moreover, desiccated

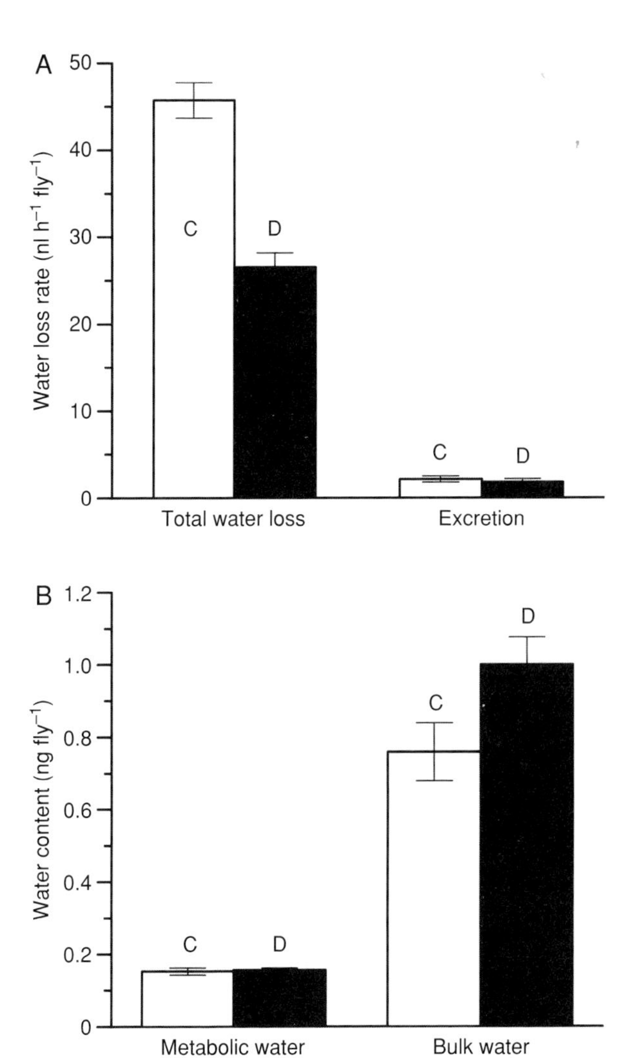

Fig. 4.6 Physiological comparison of desiccation and control lines of *Drosophila melanogaster*. Replicate lines of flies were subjected, for over 100 generations, to either desiccating or control conditions. (Upper panel) Water loss rates for control (C) and desiccation-selected (D) female flies. (Lower panel) Water reserves for control (C) and desiccation-selected (D) females. From Gibbs et al. (1997), with permission from The Company of Biologists.

flies evolved from storing lipids to storing primarily carbohydrates, even though lipid metabolism produces roughly twice the metabolic water per unit dry mass.

In contrast to the observed increases in body water content during laboratory natural selection, a comparative study of 29 species of *Drosophila* from arid and mesic habitats found no evidence that arid flies contained more body water (Gibbs and Matzkin, 2001). The authors proposed that this reflected either that 1) desert flies have easy access to free water (e.g. from necrotic patches on cactuses) and therefore do not regularly experience severe desiccation; or 2) increasing body water content negatively affects other, unmeasured performance traits that are less important for flies in laboratory culture. Of these, the most obvious is flight ability: larger flies require more energy output to stay aloft and may be less maneuverable (Lehmann and Dickinson, 2001). These pressures may be less severe in laboratory populations, which could explain why the laboratory lines, but not wild populations, were able to evolve larger size. Direct tests of the flight hypothesis have not been done.

4.3.2 Rates of Water Loss

Females from lines selected to survive desiccation also lost water at lower rates than did females from control lines (see Fig. 4.6) (Gibbs et al. 1997; Archer et al. 2008). Gibbs et al.'s comparative studies of xeric and mesic *Drosophila* (Gibbs, 2002b; Gibbs et al., 2003) came to similar conclusions came to a similar conclusion—arid flies consistently had lower rates of water loss than did mesic congeners (see Fig. 4.7). Both results are consistent with broader comparative observations on other insects (Eckstrand and Richardson, 1980; Eckstrand and Richardson, 1981; Hadley, 1994; Addo-Bediako et al., 2001).

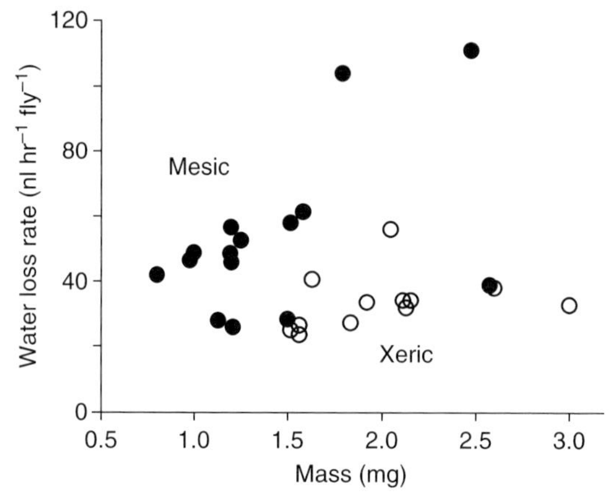

Fig. 4.7 Water-loss rates of female *Drosophila* from arid and mesic habitats. Redrawn from (Gibbs, 2002b), with permission from Elsevier.

What is the mechanism by which losses are reduced? Gibbs et al. (1997, 2003) showed in both laboratory and wild *Drosophila* that excretory losses were small (<10% of total), indicating that the main differences must be cuticular or respiratory. For the laboratory lines, Gibbs et al. (1997) argued that different rates of water loss arose from differences in cuticular permeability, although they did not show it experimentally. They showed instead that lines differed modestly in cuticular lipid composition, and invoked the argument that, in insects, respiratory losses are small enough, compared to cuticular losses (Quinlan and Hadley, 1993; Williams and Bradley, 1998), that differences in the cuticular pathways are more likely anyway to account for total differences. By contrast, in their comparative study on 29 species of *Drosophila*—which, like the laboratory study, showed that cuticular or respiratory losses must differ among species—Gibbs et al. (2003) concluded that most of the differences reflected changes in *respiratory* parameters, although again they were unable to experimentally distinguish the two sources. This conclusion was supported by the additional findings that 1) arid and mesic flies showed no consistent differences in the composition and properties of cuticular lipids; but 2) arid species had lower metabolic rates and were less active (Hoffmann and Parsons, 1993). They suggested that differences in activity could fully explain differences in metabolic rate.

4.3.3 Modes of Evolution

Collectively, the studies on selected and wild species described above have made *Drosophila* the best-studied insect system for the evolution of water balance. The studies emphasize that ostensibly common selection pressures, when applied in different contexts (field versus lab), can generate different evolutionary outcomes (Gibbs, 2002b). In *Drosophila*, reduced rates of water loss evolved in desiccation-selected laboratory lines and in arid-adapted species (although it remains unclear whether a common set of changes in cuticular and respiratory components underlie the gross changes). By contrast, evolution of increased water storage occurred in desiccation lines but not arid species. These different evolutionary responses suggest that lower-level traits can be modified in multiple different ways to achieve the same phenotype.

4.4 Hormonal Control of Water Balance

The basic physiological roles of Malpighian tubules and hindguts (described in Chapter 2) in water balance have been known for at least 70 years. New work has focused on broader aspects of information processing and coordination—how do insects measure their water content and how do they coordinate secretion and absorption by Malpighian tubules and hindgut?

The increasingly well-established answer is neurohormones, the biology of which has been summarized in a series of recent reviews (Coast, 2001; Coast et al., 2002; Nassel, 2002; Gäde, 2004; Coast, 2009). Neurohormone work has focused on identifying the structure and function of factors that stimulate water secretion or absorption, although the general lesson is that even neurohormones related to water balance encode a multiplicity of other actions (Nässel, 2002). Here we adopt Gäde's (2004) terminology—"diuretic" factors are those that increase total water loss from an insect, either by stimulating Malpighian tubule fluid secretion or by inhibiting hindgut reabsorption; "antidiuretic" factors are those that decrease total water loss.

4.4.1 Diuretic Factors

Diuretic factors are the better understood of the two groups, and five classes are known from insects.

1. The biogenic amine 5-hydroxytryptamine (5-HT, serotonin), has been known for 40 years to stimulate *Rhodnius* MT secretion *in vitro* (Maddrell et al., 1969, 1971). Since these early studies, *Rhodnius* has become a model insect for studying 5-HT's coordination of feeding-related physiological events (Orchard, 2006). 5-HT is also known to play roles in at least three other insect orders. 5-HT receptors from *Drosophila* are especially well studied (Witz et al., 1990; Saudou et al., 1992; Kerr et al., 2004) and have also been cloned and characterized from other taxa (Pietrantonio, 2001; Dacks et al., 2006; Jagge and Pietrantonio, 2008; Schlenstedt et al., 2006).
2. The corticotropin-releasing factor-related diuretic hormones (CRF-related DHs), which are related to vertebrate CRFs. Insect CRFs have been described from at least six insect orders and likely occur throughout the Insecta (Coast et al. 2002). They act via the second messenger cAMP, which stimulates Na^+ and K^+ secretion by the MTs. CRF receptors have been identified and characterized in *Manduca*, *Acheta*, and *Drosophila* (Johnson et al., 2004; Reagan, 1996).
3. The leucokinins (also called myokinins or just kinins), small peptides (6–15 amino acids), originally identified in the cockroach *Leucophaea maderae* (Holman et al., 1986) and now also known in a number of other insect orders. Kinins have diuretic activity in MTs (Hayes et al., 1989), apparently by inositol-triphosphate mediated increases in intracellular Ca^{2+} (Pollock et al., 2003). In *Rhodnius,* they also alter the properties of midgut epithelium and increase midgut contractions (Brugge et al., 2009). Kinin receptors, all G-protein coupled, are also increasingly well studied (Pietrantonio et al., 2005; Taneja-Bageshwar et al., 2008; Radford et al., 2004). In *Anopheles*, the leucokinin receptor is expressed in stellate cells of

the MT (Radford et al. 2004). Leucokinin analogs that resist hydrolysis are increasingly being developed for insect control (Nachman et al., 2009).

4. The calcitonin-like (CT-like) peptides, first identified in the cockroach *Diploptera punctata* (Furuya et al., 2000) and subsequently in other taxa (Coast et al., 2002; Te Brugge et al., 2005, 2008). CT-like peptides likely use a cAMP second-messenger system, though possibly via a different effector than the system activated by CRF-related peptides; alternatively, they may activate a separate second messenger pathway (Coast et al., 2002). A CT-like peptide contributes to post-prandial diuresis in *Rhodnius* (Te Brugge et al., 2005), though not nearly as strongly as does serontonin, and it stimulates contractions of the dorsal vessel and hindgut (Te Brugge et al., 2008). In *Rhodnius*, it appears to work together with serotonin to give finer control over post-feeding physiology (Te Brugge et al., 2008).
5. The CAPA peptides (Predel and Wegener, 2006), first identified in *Manduca sexta* (Huesmann et al., 1995). Their effects differ by species—stimulating fluid secretion by Malpighian tubules in several flies (*Drosophila*, *Musca*, *Stomoxys*; (Davies et al., 1995; Nachman and Coast, 2007)) but inhibiting secretion in *Rhodnius* (Quinlan et al., 1997) and *Tenebrio* (Wiehart et al., 2002). Secretion is stimulated by elevated levels of cGMP and involves NO and Ca^{2+} (Davies et al., 1995; Rosay et al., 1997).

Finally, the ion transport-like peptide from *Schistocerca gregaria* may promote diuresis in locusts by blocking the action of an antidiuretic peptide, the ion transport peptide (Audsley et al., 2006). Any particular insect species may have members of more than one class, and the members may act synergistically (Maddrell et al., 1993; Coast, 1995; Furuya et al., 2000; Coast et al., 2005; Holman et al., 1999; O'Donnell and Spring, 2000). For example, in *Locusta migratoria*, CRF-related DH is extremely potent in the presence of kinin but not in its absence (Coast et al., 1999).

All of the diuretic peptides listed above stimulate secretion by Malpighian tubules—and this mechanism appears to be the primary organ-level process underlying whole-body diuresis. In theory, whole-body diuresis could also be achieved by inhibiting fluid uptake in the rectum, but few studies have examined this possibility. One exception, (Coast et al., 1999), found no effect of two locust diuretic peptides on rectal ion and water transport, nor any effect of a known stimulant of ileal fluid re-absorption (ion-transport peptide) on fluid secretion by isolated Malpighian tubules. The authors concluded that the data so far indicate that the two organs, Malpighian tubules and hindgut, are regulated separately.

4.4.2 Antidiuretic Factors

The molecules described above—the diuretic peptides—stimulate water loss by increasing MT fluid secretion. A separate group—the antidiuretic peptides—inhibits water loss. Antidiuretic factors fall into two functional classes, those that inhibit fluid secretion by Malpighian tubules and those that stimulate hindgut fluid absorption (Gäde, 2004).

4.4.2.1 Inhibition of Fluid Secretion

Peptides that depress MT fluid secretion are now known from multiple species. The most widely investigated are antidiuretic factors (ADFs) from the mealworm *T. molitor* (Tenmo-ADFa and ADFb) (Eigenheer et al., 2002, 2003), which act via cGMP. Tenmo-ADFa decreases MT secretion in *T. molitor* (Wiehart et al., 2002) and also in mosquitoes (Massaro et al., 2004). Quinlan et al. (1997) showed that a CAPA peptide (CAP_{2b}) from *Manduca sexta* had potent antidiuretic effects on isolated MTs from the bug *Rhodnius prolixus*, also via a cyclic GMP (cGMP) second-messenger cascade. The *Rhodnius* peptide has subsequently been localized to neurohaemal sites of release and shown to increase 3–4 hours after feeding (Paluzzi and Orchard, 2006; Paluzzi et al., 2008). Specifically, it appears to decrease the tubule secretion stimulated after feeding by 5-HT.

4.4.2.2 Stimulation of Fluid Reabsorption

Antidiuretic peptides acting on the hindgut were known first from studies of locusts, *Schistocerca gregaria* and *Locusta migratoria* (Audsley et al., 1992) and are now known from multiple other species (Dircksen, 2009). Audsley et al. (1992) showed that ion-transport peptide (ITP) from locust *corpus cardiacum* stimulated several-fold increases in ileal Cl^-, Na^+, and K^+ transport and fluid re-absorption. Exogenous cAMP stimulated the same effects, suggesting it as a second messenger. However, the situation is complex, as ITP, but not cAMP, abolished ileal acid secretion, suggesting that ITP also activates a different second-messenger pathway. Further, there is some doubt about ITP's status *in vivo*. Audsley et al. investigated circulating levels of ITP and ITP-L in *Schistocerca* (Audsley et al., 2006). ITP and ITP-L are identical except at the C-terminus (and arise from alternative splicing), suggesting that the latter molecule acts as an ITP antagonist (Phillips et al., 1998), perhaps competing for the same binding sites on the still-unknown ITP receptor (Dircksen, 2009). Establishing the *in vivo* roles of ITP and ITP-L (and their interaction) promises to illuminate a key linkage between the neuroendocrinology of water balance and water problems posed by the environment. Finally, another peptide, chloride transport stimulating hormone (CTSH), is known to affect uptake of both chloride and fluid from the rectum (Coast et al., 2002; Phillips et al., 1996), but it has yet to be characterized.

Another class of antidiuretic peptides, the neuroparsins, also discovered in *Locusta* and *Schisctocerca*, may have antidiuretic effects on rectal tissues (Gäde, 2004). Newer work has also emphasized roles for neuroparsins in the regulation of insect reproduction and in locust phase transitions (Badisco et al., 2007).

The antagonism between diuretic and antidiuretic factors probably allows finer control over water balance in dynamic ecological circumstances. Also, diuretic and antidiuretic factors may act together to increase rates of fluid recycling within an insect (Phillips, 1982; Nicolson, 1991), which could rapidly clear unwanted or toxic molecules from the hemolymph after feeding. Antagonism between diuretic and antidiuretic hormones extends into second-messenger cascades. For example, the stimulatory effect of one of *T. molitor*'s diuretic hormones, Tenmo-DH_{37}, was reversed by an antidiuretic peptide, Tenmo-ADF (Wiehart et al., 2002). The two peptides also use alternative second-messenger systems—Tenmo-DH acts via cAMP whereas Tenmo-ADF acts via cGMP—and MT stimulation by exogenous cAMP could be neutralized by addition of exogenous cGMP (Wiehart et al. 2002).

4.4.3 Systems Integration

The molecular details of endocrine control are beginning to be understood (Beyenbach, 2003b; Gäde, 2004), including elucidation of hormone receptors and their intracellular actions (Johnson et al., 2004; Pietrantonio et al., 2005; Wiehart et al., 2002). Many outstanding questions remain, however, including how insects assess their water balance, how different systems interact at cellular and molecular levels, and how the diuretic and antidiuretic systems play out in wild insects exposed to the normal diversity of environmental change. For the rest of this section we focus on one of these questions: how do insects know whether they are in water balance? In principle, an insect attempting to estimate its water content could do so in one of at least three ways: 1) measure total hemolymph volume using stretch receptors in the body wall; 2) measure hemolymph osmotic pressure with some kind of osmosensor; and 3) take a cue from feeding and drinking events. Evidence exists for all three.

1 *Volumetric measurement*. Maddrell discovered that blood meals taken by the blood-sucking bug *Rhodnius prolixus* stretch the abdomen dorso-ventrally and activate stretch receptors in the tergosternal muscles, which then signal the mesothoracic ganglionic mass to release diuretic hormone (Maddrell, 1964). Coast (2001, 2004) demonstrated a similar mechanism in the housefly *Musca domestica*. Flies injected with saline showed subsequent diuresis proportional to the injected volume. The signal appears to be volumetric, rather than osmotic, because injection of either isosmotic saline or distilled water stimulated diuresis.

2 *Osmotic measurement*. Studies of cell volume regulation in mammals have found diverse mechanisms for osmosensing (Lang et al., 1998), and they likely involve mechanosensitive cation channels—probably stretch-inactivated potassium channels (Morris and Sigurdson, 1989)—that transduce cell swelling or shrinkage into cell-relevant signals. In insects, no osmosensor has been demonstrated unambiguously, although several studies hint at similar mechanisms. For example, Kim et al. showed in *Drosophila* that nanchung protein, the product of a gene (*nan*) in the transient receptor potential (TRP) ion channel superfamily, was mechanosensitive (Kim et al., 2003). The channel is similar to known osmosensors in *C. elegans* (O'Neill and Heller, 2005). Moreover, when *nan*cDNA was transfected into hamster cells, they became osmosensitive (Kim et al., 2003).

Other work on *Drosophila* suggests a possible intracellular osmosensor in the Malpighian tubules. Using the Ramsay assay, Blumenthal showed that tyramine stimulated MT urine secretion at low nanomolar concentrations (Blumenthal, 2003). Subsequently, he showed that saline osmolality had a strong effect on tyramine-stimulated MTs (Blumenthal, 2005). In particular, MTs stimulated by 10 nM tyramine showed a 10-fold change in transepithelial potential—from high at 200 mOsm to very low at 320 mOsm, with a linear decline in between. Blumenthal (2005) suggested that the effect is explained best by an effect of osmolality on production of inositol triphosphate or on intracellular calcium dynamics.

3 *Feeding and drinking*. It is possible that no specific physiological changes are needed to stimulate diuresis: insects could secrete appropriate hormones based on their perception of what they are eating or drinking (feed-forward regulation). Such responses may anticipate water disturbances—that is DH release by blood-feeding insects may occur simply from the act of feeding, before the meal is processed. Several examples of feeding-related diuresis may include elements of perception (Maddrell et al., 1993; Patel et al., 1995; Petzel et al., 1987).

4.5 Controlling Where Water Goes: Aquaporins

Although cells and tissues consist mostly of water, cell membranes are lipid bilayers, which alone are virtually impermeable to water. Low permeability can be a problem for tissues that support substantial water fluxes—such as the postprandial fluxes of water across MT epithelia in blood-sucking mosquitoes and bugs. In addition, water movement, or lack thereof, is important in determining how tissues and cells respond to freezing and desiccation (Ring and Danks, 1994; Kikawada et al., 2008; Philip et al., 2008). How do insect tissues support high water fluxes? More generally, how do insects control where internal water goes?

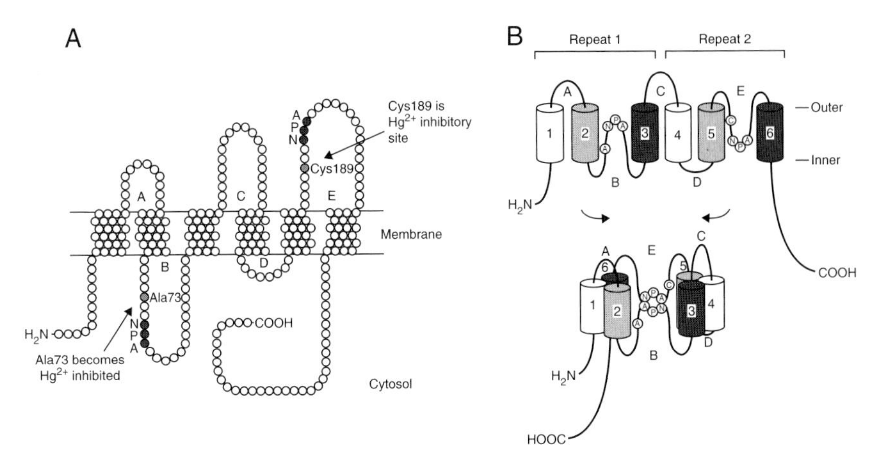

Fig. 4.8 (A) Primary structure of human AQP1, including the transmembrane domains and connecting loops (A–E). The highly conserved motif asparagine-proline-alanine (NPA) plays a key role in allowing water but not protons or other ions through the pore. (B) The hourglass model for the structure of human AQP1. The transmembrane domains consist of two tandem repeats, which fold to form the water-transporting pore. From Spring et al. (2009), with permission from The Company of Biologists.

4.5.1 Aquaporin Structure and Function—Role in Fluid Feeders

One emerging answer is aquaporins (AQPs), which are transmembrane channels that permit water movement across lipid bilayers (Fig. 4.8).

AQP-mediated water movement is passive, efficient, and highly selective. Indeed, some known vertebrate AQPs are so selective that they allow rapid water movement while excluding protons, a remarkable feat apparently accomplished by finely choreographed manipulation of hydrogen bonding within the aquaporin channel (de Groot and Grubmuller, 2005; Fu and Lu, 2007; Spring et al., 2009). Aquaporins occur widely throughout unicellular and multicellular organisms (Froger et al., 1998; Borgnia et al., 1999).

Insect AQPs have been identified and functionally expressed in only seven species (Spring et al., 2009), although there is good evidence that they occur in many others. Campbell et al. review invertebrate AQPs broadly (Campbell et al., 2008), and Spring et al. provide more focused reviews on insect AQPs (Spring et al., 2007).

Relationships of insect AQPs to their mammalian counterparts are uncertain, although they share the highest homology with AQP4 (Campbell et al., 2008). Campbell et al. (2008) constructed phylogenetic trees of insect AQPs using amino acid sequences from insects for which whole-genome sequences are complete or under way (*D. melanogaster, Aedes aegyptii, Apis mellifera, Nasonia vitripennis, and Tribolium castaneum*). Their results (Fig. 4.9) reveal two patterns. The first is that insect AQPs are divisible into three main groups: the DRIP subfamily (named for the first-known member of the group, *Drosophila* integral protein (Dow et al., 1995), the PRIP subfamily (named for the first member identified from the firefly *Pyrocoelia rufa* (Lee et al., 2001), and the BIB subfamily (which contains the *Drosophila* big brain gene). DRIP and PRIP are more closely related to each other than either is to BIB. The second pattern is that members of all three subfamilies (DRIP, PRIP, BIB) occur in each of the three major insect orders (Diptera, Coleoptera, Hymenoptera) for which sequences are available, leading Campbell et al. (2008) to propose "that all insects will contain AQPs belonging to these three sub-families."

Functional work on insect aquaporins has focused on sap-feeding insects, including the xylem-feeding leafhopper *Cicadella viridis* (Beuron et al., 1995; Le Cahérec et al., 1996; Le Caherec et al., 1997), and the phloem-feeding pea aphid, *Acyrthosiphon pisum* (Shakesby et al., 2009). Xylem feeders, like *C. viridis*, ingest enormous volumes of dilute sap, from which obtaining adequate nutrition and avoiding excess water are serious problems. By contrast, phloem feeders face the problem of ingesting sugar-laden food (phloem) containing very high osmolalities, 4–5 fold higher than insect hemolymph. In both cases, the insects use aquaporins to shunt water, but in opposite directions. *C. viridis* have "filter chambers," which are elaborate structures composed of distal gut, midgut, and Malpighian tubules together in a chamber (Le Caherec et al., 1997) that shunts water (from the dilute xylem it ingests) from the midgut to the terminal midgut and Malpighian tubules (Fig. 4.2). *Acyrthosiphon pisum*, by contrast, has a simpler gut in which a distal loop is apposed to the midgut, and water is shunted across their junction *from* the hindgut to the midgut as a re-circulator (Shakesby 2009), thereby helping to relieve the enormous osmotic load imposed by ingested phloem.

In both sap feeders, the membranes supporting water flux contain AQPs. In *C. viridis*, the aquaporin, now called AQP*cic* (Hubert et al., 1989; Beuron et al., 1995), forms homotetramers in native membranes. Proteoliposome suspensions containing AQP*cic*, when injected into *Xenopus* oocytes, gave several-fold higher membrane water permeabilities (Le Caherec et al., 1997). Immunological assays, using antibodies raised against AQP*cic*, localized the aquaporin to microvilli of the filter chamber epithelium in *Cicadella* and in six other Homoptera (Le Cahérec et al. 1997). Shakesby et al. (2009) did similar functional characterization of an aquaporin (ApAPQ1) from *A. pisum*. They also localized it to the stomach and showed, using RNAi, that aphids with transiently depressed levels of ApAPQ1 expression had higher hemolymph osmotic pressures. Such an effect is consistent with the postulated function of ApAPQ1 in recycling water from distal to proximal sections of the gut.

Given their role in urine formation, Malpighian tubules should exhibit interesting AQP biology (Spring et al., 2009). And indeed they do, at least in the few

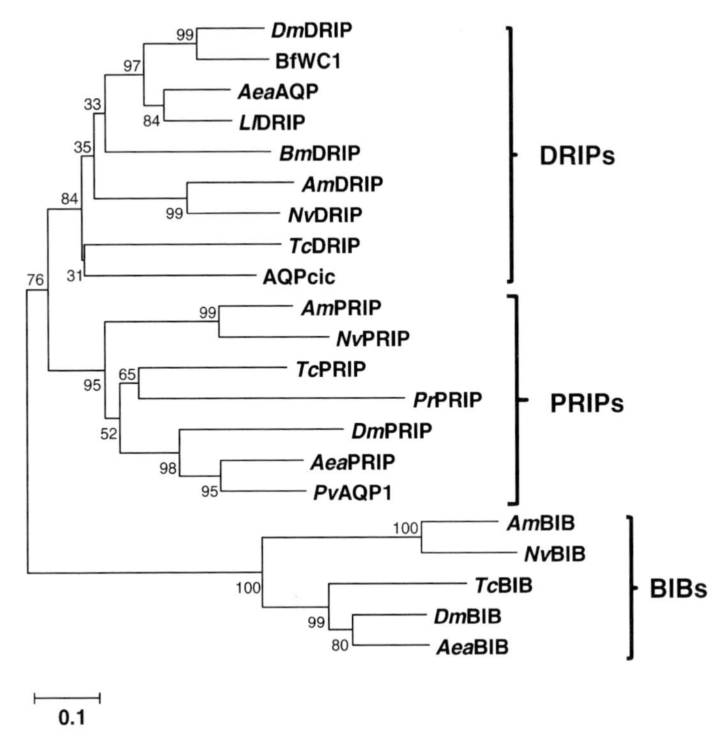

Fig. 4.9 Unrooted phylogenetic tree of insect aquaporins (AQPs), estimated from amino acid sequences. **Diptera**: Aea = *Aedes aegypti*, Dm = *Drosophila melanogaster*, Hi = *Haematobia irritans*, Pv = *Polypedilum vanderplanki*, Ll = *Lutzomyia longipalpis*; **Lepidoptera**: Bm = *Bombyx mori*; **Coleoptera**: Tc = *Tribolium castaneum*; **Hymenoptera**: Am = *Apis mellifera*, Nv = *Nasonia vitripennis*; **Homoptera**: cic = *Cicadella viridis*. For another phylogeny including non-insect taxa, see Drake et al. (2010). From Campbell et al. (2008), with permission from Springer Science + Business Media.

insects examined—*Drosophila* (Kaufmann et al., 2005), *Rhodnius*, and *Aedes*. In *Rhodnius*, which secretes copious urine after blood meals, AQP transcripts have been found throughout the Malpighian tubules (Echevarria et al., 2001; Martini et al., 2004), though native proteins *in situ* have not been demonstrated. In *Drosophila*, the DRIP aquaporin is localized to stellate cells in the Malpighian tubules and shows water-transporting capability when expressed in *Xenopus* and yeast cells, indicating that it probably facilitates water transport in tubules *in vivo* (Kaufmann et al., 2005)

Results from *Aedes aegypti* have been somewhat counterintuitive. Like *R. prolixus*, *A. aegypti* also takes large blood meals and produces copious post-feeding urine, and aquaporin cDNAs have been isolated from an *Aedes* MT library (Pietrantonio et al., 2000). However, the transcript was associated with MT-associated tracheolar cells rather than the MT epithelium. A later study (Duchesne et al., 2003) detected the protein (*Aea*AQP) in tracheolar cells and showed that it increased water permeability of *Xenopus* oocytes. Why is expression restricted to tracheolar cells when it is the MT epithelium that supports water flux? Tracheolar tips are liquid-filled, and the liquid level changes as a function of local metabolic rate—liquid is removed during high metabolic output and replaced during quiescence, likely as a means of modulating tracheolar oxygen conductance (see Chapter 6). Pietrantonio et al. (2000) suggested that *Aedes* AQPs facilitate these respiration-related changes in fluid levels (Fig. 4.10).

In a more recent study of *Ae. aegypti*, Drake et al. (2010) screened the genome for aquaporins and identified six, which they named AaAQP1–6. AaAQP1 is the same as that studied by Pietrantonio et al. (2001) and Duchesne et al. (2003). Based on studies using RT-PCR, Drake et al. found that AaAQP1 was expressed in all organs and body parts examined, with the highest expression in the Malpighian tubules (location within MT not determined, (Drake et al., 2010)). Using RNA interference to knock-down expression of AaAQP1 (and other AQPs), they found that mosquitoes with reduced levels of AQP excreted injected solutions more slowly, supporting a role for AQPs in mediating diuresis under more natural conditions (blood feeding).

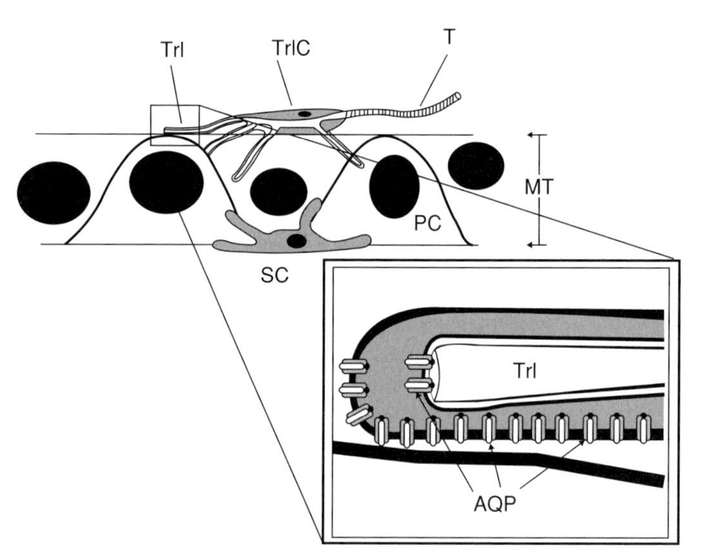

Fig. 4.10 Proposed location of *Aedes aegypti* aquaporin at the end of tracheoles associated with the Malpighian tubules. Redrawn from Pietrantonio et al. (2000), with permission from Elsevier.

Finally, some progress has been made in understanding hormonal control of AQP expression and function. A key diuretic hormone is 5-hydroxytryptamine (5-HT, serotonin). After *in vitro* treatment with 5-HT or cAMP, aquaporin transcript levels rise in *Rhodnius* MTs (Martini et al., 2004). In *Aedes*, Pietrantonio et al. (2001) cloned a 5-HT receptor from a MT cDNA library. They also showed that transcripts and the expressed protein of the 5-HT receptor were localized in tracheolar cells associated with MTs. Co-localization of the 5-HT receptor and the *Aedes* aquaporin in tracheolar cells suggests that 5-HT may directly regulate expression and function of AQPs (Pietrantonio et al., 2001, Duchesne et al., 2003).

4.5.2 Role in Desiccation and Freeze Tolerance

For insects, desiccation and freezing pose strangely similar physiological problems (Ring and Danks, 1994), both stemming from separation of water from the solutes dissolved in it (see Chapter 3 for more on freezing). During freezing, which must occur only in extracellular spaces if insects are to survive, solutes are excluded from growing ice crystals and concentrated in the remaining fluids. Their high osmotic pressures then draw water out of cells, which can cause cell and tissue death. Desiccation causes analogous problems—but because the water disappears entirely from the organism rather than freezing into unavailable lattices.

Studies on plants have shown that desiccation stress can lead to the upregulation of AQPs (Barrieu et al., 1999; Smith-Espinoza et al., 2003). These studies have prompted the search for similar desiccation-induced changes in AQPs in insects, and several interesting stories are emerging (Kikawada et al., 2008; Philip et al., 2008). Kikawada et al. (2008) studied effects of desiccation on the sleeping chironomid, *P. vanderplanki*, which occurs in rock pools in semi-arid areas of Africa. During dry periods, *P. vanderplanki* enters into a state of anhydrobiosis, the onset of which requires careful control over rates of water loss. The authors cloned two AQPs, *PvAqp1* and *PvAqp2*, which had different spatial and temporal patterns of expression during desiccation. *PvAqp1* occurred throughout larval tissues and was upregulated during desiccation, whereas *PvAqp2* was restricted to the fat body and was downregulated during desiccation. These patterns were interpreted to mean that *PvAqp2* regulates normal water homeostasis in the fat body while *PvAqp1* aids in rapid water removal during entry into anhydrobiosis. Goto et al. recently characterized a homologous AQP from the related Antarctic chironomid, *Belgica antarctica* (Goto et al., 2011).

Philip et al. (2008) studied three AQPs in the goldenrod gall fly, *Eurosta solidaginis*. Larvae overwinter in plant galls, which are exposed to low temperatures and become very desiccated, physical problems encountered by many dormant insects (Danks, 2000). Larvae become freeze-tolerant during autumn, partially by accumulating of glycerol and sorbitol (Morrissey and Baust, 1976), which allows them to survive winter temperatures down to -50°C (Storey and Storey, 1988; Lee Jr, 1991). In addition, Ramløv and Lee showed that rates of water loss for overwintering larvae are extremely low (Ramløv and Lee Jr, 2000). Philip et al. (2008) hypothesized that AQPs were involved in acquisition of both freeze- and desiccation resistance, and they showed that antibodies to mammalian AQP2, 3, and 4 recognized proteins in *Eurosta*. During experimental desiccation of larvae, AQP2 and 4 were downregulated and AQP3 was upregulated. The upregulated form is of special interest, because mammalian AQP3 is permeable to water and glycerol (although the authors did not demonstrate that *Eurosta* AQP3 itself is permeable to glycerol). The latter is tremendously important during acquisition of freeze- and desiccation tolerance in *Eurosta* and other insects. The authors proposed that "up-regulation of AQP3 may be part of a coordinated set of physiological changes that increase the permeability of water and glycerol across the cell membrane as larvae prepare for the osmotic stress associated with host plant senescence and extracellular ice formation." In a recent study based on the genome of *Aedes aegypti*, Drake et al. found no evidence for aquaglyceroporins, indicating that glycerol-transporting AQPs may not be widely distributed among all insect taxa (Drake et al., 2010).

4.6 Insects Living in Water

Compared to terrestrial environments, in which the problem is almost always water shortage, aquatic environments provide perhaps the most varied and interesting challenges. This is because the osmotic and ionic constituents of water can differ from those in insect hemolymph in two directions, depending on whether the water is fresh or saline. Insect hemolymph usually is 250–400 mOsm and contains in the order of 150 mM Na^+ and 100 mM Cl^-. In freshwater, therefore, insects sustain enormous water influx by osmosis and lose ions to their environments; their osmoregulatory systems have to dump water and gain ions. In saltwater (and some insects live in water saltier than the sea; > 1000 mOsm), by contrast, they sustain enormous water losses to, and ion loads from, the environment, just the opposite of what freshwater insects experience. Evolutionary transitions from terrestrial to aquatic living, or between fresh and saltwater, therefore can require massive physiological reorganization, primarily of Malpighian tubule and hindgut function. Remarkably, some insects show extreme euryhalinity—they can cope with rapid transitions between fresh and saltwater.

Because the problems of life in water are profound, and the associated physiological changes large, aquatic insects have played a primary role in our understanding of water and ion balance. Here we focus on aquatic insects generally and on mosquito larvae specifically. Of the world's insects that affect human health, mosquitoes rank high—because a small subset of them vector important human diseases. As a consequence, there have been significant, sustained efforts over the past century to understand mosquito ecology and physiology. These efforts inevitably hinge on understanding relationships among mosquito larvae, water, and ions. From both medical and basic perspectives, therefore, understanding larval water- and ion-balance physiology is important. Other stages—eggs, pupae, and adults—also face interesting ion and water challenges, but we do not cover them here (Bradley, 1987).

4.6.1 Habitats and Physiologies

Most mosquito larvae (> 95%) occur in freshwater (O'Meara, 1976), which Bradley (1987) defines as < 300 mOsm (though most larvae live in substantially fresher water than this implies). Because freshwater larvae have higher blood osmotic pressures and salt concentrations than does their medium, they face the twin problems of significant uptake of water by osmosis and during feeding, and significant loss of salts to the environment. Interestingly, from a whole-body perspective, mosquito larvae solve these problems much like freshwater fish do: they produce copious quantities of dilute urine, and they actively scavenge external ions, taking them up against their gradients, which can require significant inputs of energy (Bradley 1987).

By contrast, a much smaller proportion of species are saline, inhabiting water that is more than 1000 mOsm. No species is truly marine (lives directly in the open ocean) (Bradley 1987). This ecological lacuna (see also Section 4.6) cannot have arisen from osmotic problems, as many insects inhabit hypersaline waters disconnected from the world's ocean; instead it probably reflects the ecological outcome of competition with other taxa better suited to marine life in some other way. Nonetheless, for saline larvae, osmotic problems are formidable and largely reversed from those confronted by freshwater larvae. Larvae lose water by osmosis across the cuticle and from defecation; and they sustain salt loads from their environments: life in saltwater is life in a liquid desert. To cope, saline larvae have evolved a diversity of physiological strategies (see below).

Between 300 and 1000 mOsm are brackish habitats (Bradley 1987 classifies these as a subset of the more general "saline habitats"), which also are colonized by a variety of mosquitoes. In general, the physiological problems (desiccation and salt loads) for larvae in brackish water are similar to those confronted by larvae in more saline waters. However, brackish water often occurs where freshwater and seawater meet, and so conditions can be especially variable. Brackish-water mosquitoes probably require the best-developed, and most rapid, mechanisms of physiological plasticity. Patterns of osmoregulation in brackish-water species appear to be quite different: below 400 mOsm they osmoregulate and above this level they osmoconform (Garrett and Bradley, 1984).

Many other groups of insect have one or more stages with aquatic lifestyles, which face problems much like those of mosquito larvae. These groups include caddisflies, mayflies, dragonflies, damselflies, many beetles and true flies, and a few moths. Often the juvenile stage (larva or nymph) is aquatic and the adult terrestrial, which may reflect specialization of the growing stage on aquatic food resources and the adult stage on dispersal and reproduction, which probably occurs more readily in air. However, a few species complete most or all of their lifecycle underwater (e.g., (Kolsch et al., 2010)). For example, brine flies (family Ephrydridae) live in terminal lakes, which have water inlets but no outlets. Although not speciose, some species can reach incredibly high densities in hypersaline terminal lakes in the Great Basin of the US. Salinity and pH in these lakes can be extreme—upwards of 1500 mOsm and pH 10 (compared to 1000 mOsm and pH 7.8 for most oceanic sites)—and yet the flies complete almost their entire lifecycles underwater.

Below we discuss the particular physiologies that have evolved in fresh-, saline-, and brackish-water insects, primarily mosquitoes. Rather than organizing the discussion by these functional types, however, we focus on the evolutionary diversity of iono- and water-regulating processes. A subsequent section focuses on recent progress in elucidating the power supplies for moving water across epithelia in Malpighian tubules and hindgut. The burgeoning—even dazzling—diversity of water- and ion-regulating mechanisms now stands as a key example of evolutionary tinkering giving different solutions to common physiological problems.

4.6.2 Evolution and Plasticity of Mosquitoes in Response to Water and Ion Challenges

Much of the work on mosquitoes has focused on a single genus, *Aedes*, reflecting the role of *Ae. aegypti* as a vector of dengue and yellow fevers. *Ae. aegypti* has freshwater larvae, and thus much is known about its freshwater iono- and osmo-regulation. However, the genus also contains salt-tolerant species, providing close evolutionary relatives for examining diversification of pattern and mechanism.

Freshwater larvae of *Aedes* drink very little, produce large quantities of dilute urine, and actively take up ions from surrounding water (Bradley 1987). Like other insects, mosquitoes form primary urine using their Malpighian tubules (Beyenbach, 2003a; Beyenbach et al., 2010), dump that urine into the junction between midgut and hindgut, and then modify it in the rectum before elimination. The primary urine, produced by the tubules, is isosmotic with the hemolymph and rich in ions, and is therefore not particularly useful for water and ion regulation. The rectum provides the additional process, strong ion absorption, that renders the urine hypoosmotic.

Even rectal ion absorption, however, is insufficient for freshwater larvae to remain in ion balance; therefore they also absorb ions across anal papillae (AP) (Wigglesworth, 1933, 1938), paddle-like extensions from the last larval segment that have highly permeable cuticles and contain high densities of mitochondria that support transport processes (Fig. 4.11).

AP can show plasticity in response to ambient levels of ions: larvae exposed to dilute water have larger, more mitochondrial-dense papillae (Edwards, 1983; Edwards and Harrison, 1983) and exhibit rapid changes in the kinetic parameters of sodium and chloride transport (Donini et al., 2007). In *Ae. aegypti*, regulatory processes for ions versus pH interact to affect plasticity of AP (Clark et al., 2007). AP are permeable to acid-base equivalents and therefore represent a portal by which whole-organism pH can be disturbed. Adequate levels of salt allow larvae to alter the size of their AP—larvae in acidic water had smaller AP, presumably to reduce influx of acid equivalents. By contrast, in water with low salt content, larvae showed no plasticity in response to pH, presumably because the physiological need to maintain the high area for ion uptake overrode the cost of disturbing pH homeostasis.

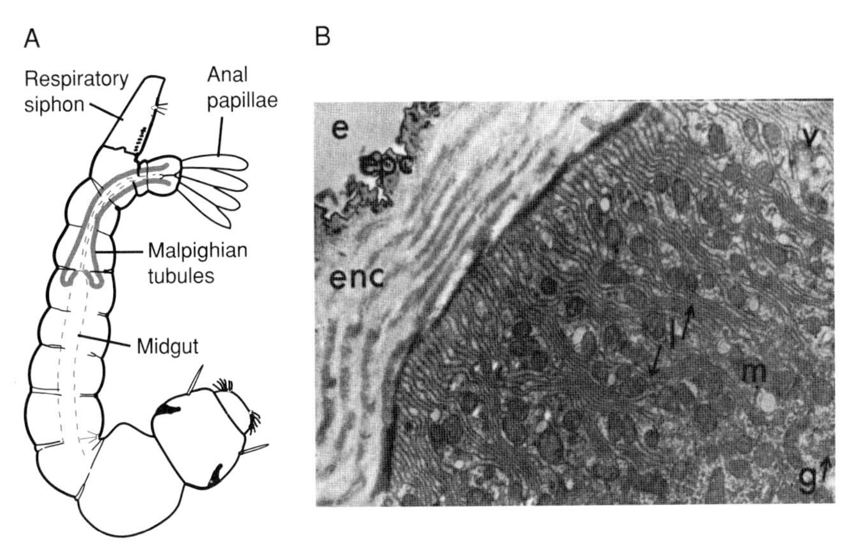

Fig. 4.11 A. Anal papillae of larva of the mosquito *Aedes aegypti*. Redrawn from Wigglesworth (1933), with permission from The Company of Biologists. B. Electron micrograph of a transverse section through the apical region of the anal papilla epithelium of *Aedes togoi* (×25,000). The apical plasma membrane is elaborated into a series of parallel lamellae, associated with abundant mitochondria. Abbreviations used: e, external medium; epc, epicuticle; enc, endocuticle; g, Golgi body; l, lamellae; m, mitochondrion; v, vesicle. From Meredith and Phillips (1973), with permission from NRC Research Press.

As discussed above, larvae living in brackish or saltwater face the opposing problems of losing water and coping with large salt loads, both of which can devastate tissues if not controlled. To maintain water and ion balances, larvae could, in principle, minimize fluxes in the problematic direction (water out, ions in) or maximize fluxes that offset these problems (by drinking water and by secreting or excreting ions at high rates). What we know about these possibilities comes from comparative studies of salt-loving species, mostly those related to *Ae. aegypti* or in nearby taxonomic groups (species of *Culex*; *Aedes* and *Culex* are *culicine* mosquitoes). A parallel literature has emerged on more distantly related halophiles in the genus *Anopheles* (anopheline mosquitoes), a large group that contains the primary malaria-transmitting species. Collectively, these studies demonstrate diverse evolutionary solutions to the problems of salt and water, and, in some species, impressive levels of plasticity.

Salt-dwelling larvae in general use some or all of the following strategies: they drink water at high rates (> 100% of the body weight per day, which causes additional salt problems, discussed below), they actively secrete salt via specialized rectal tissues, they minimize osmotic gradients by raising the osmotic pressure of their hemolymph, or they reduce cuticular permeability, sometimes by several orders of magnitude, so that less water escapes down a given osmotic gradient.

These strategies play out in different ways in different groups. In *Culex*, larvae in saltwater are osmoconformers (their hemolymph osmotic pressure is high, more or less matching external osmotic pressure). This strategy has the advantage of minimizing water loss to the environment, which in turn requires larvae to drink less, thereby reducing the requirement for ion secretion. Nevertheless, during rapid transitions to higher salinity, saline-tolerant larvae of *Culex tarsalis* maintain body volume by drinking (Patrick and Bradley, 2000a), which does incur salt loads. *C. tarsalis*, however, keeps hemolymph levels of NaCl from rising very high by rapidly eliminating it, possibly via the anal papillae (Patrick et al., 2001).

The *Culex* strategy of osmoconforming is worth additional comment. Raising hemolymph osmotic pressure above typical insect values (~350 mOsm) is potentially dangerous: allowing blood levels of sodium and chloride to rise disrupts other cellular processes. Therefore, in high salinity, larvae accumulate compounds like serine, proline, and trehalose (Patrick and Bradley, 2000a; Garrett and Bradley, 1987), compatible organic solutes that act as osmolytes without disturbing other cell functions (Fig. 4.12).

High hemolymph osmotic pressure requires similarly high cell osmotic pressures, so that cells do not shrink. In *Culex tarsalis*, different signals are used to induce proline versus trehalose: accumulation of proline was cued by ambient levels of sodium chloride, whereas accumulation of trehalose was cued by ambient osmolarity regardless of the identity of the osmolyte (Patrick and Bradley, 2000b).

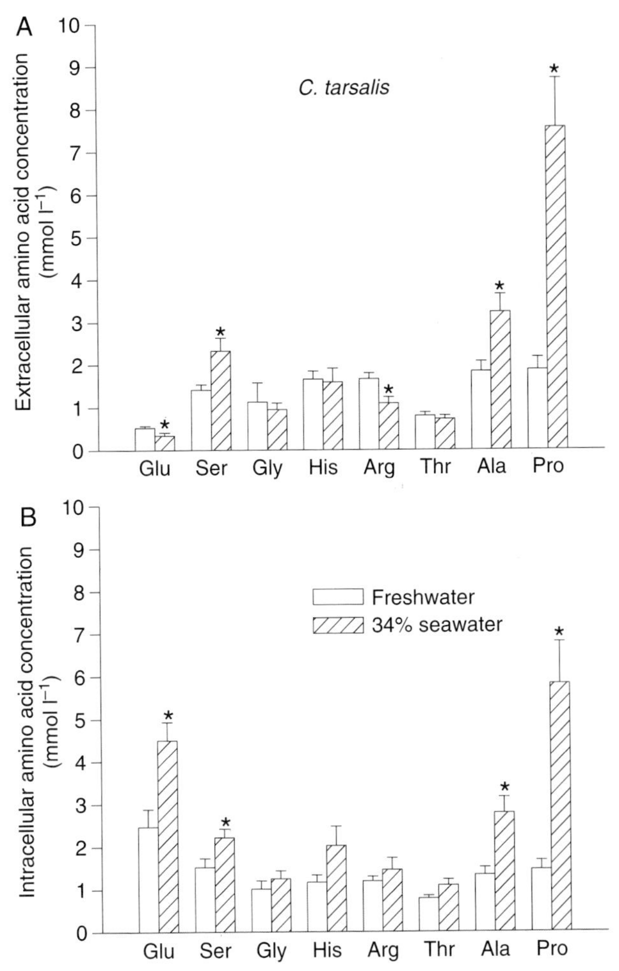

Fig. 4.12 Larvae of *Culex tarsalis* accumulate different levels amino acids when held in freshwater or 34% seawater. (A) Extracellular; (B) intracellular. Open bars are for larvae held in freshwater and hatched bars are for larvae held in 34% seawater. Redrawn from Patrick and Bradley (2000a), with permission from The Company of Biologists.

By contrast, salt-loving *Ochlerotatus* (for example, *Oc. taeniorhynchus*; also a culicine mosquito), are osmoregulators; rather than accumulating osmolytes, they maintain hemolymph osmotic pressures within a narrow range across a wide range of external osmotic pressures. This strategy requires them to drink large quantities of water and therefore to deal with enormous salt loads. In principle, anal papillae (AP) would come to the rescue. APs could secrete rather than absorb sodium and chloride, as they appear to do in *Culex tarsalis*. In fact, larvae have evolved instead to secrete salt in the rectum. Meredith and Phillips (1973) inferred active salt secretion by the rectum from ultrastructural studies indicating high capacity for ion transport; similar tissues were not observed in freshwater larvae. Subsequently, Bradley and Phillips showed that salt was secreted by the posterior rectal segment only (anterior rectum selectively reabsorbs solutes, (Bradley and Phillips, 1977)).

Finally, a recent effort has focused on rectal structure and function in anopheline mosquitoes (Smith et al., 2007; Okech et al., 2008; Smith et al., 2008), including *Anopheles gambiae*, the most important of the world's malaria vectors. Based on the localization patterns of carbonic anhydrase, V-type H^+ ATPase and Na^+/K^+ P-ATPase, Smith and colleagues proposed that anophelines have distinct cell types in the rectum, the dorsal anterior rectum cells (DAR cells) and non-DAR cells, that regulate ions. In response to changing salinity, these cells show significant plasticity in Na^+/K^+ P-ATPase localization.

Because rectal structure in culicines versus anophelines is important to the sections below, we summarize it again here (Fig. 4.13).

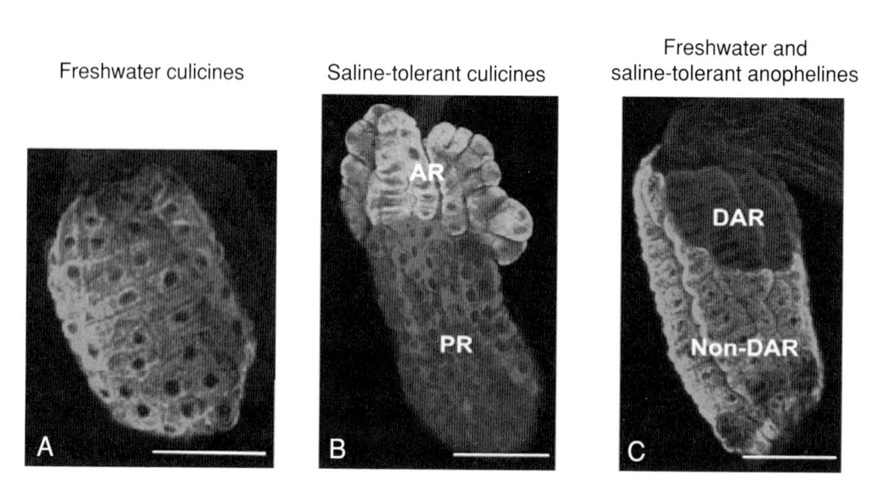

Fig. 4.13 Comparison of the rectal structure of the freshwater culicine *Aedes aegypti* (A), the saline-tolerant culicine *Ochlerotatus taeniorhynchus* (B), and the freshwater anopheline, *Anopheles gambiae* (C) using confocal microscopy of whole mount immunohistochemical preparations. See original paper for color versions. Freshwater culicine larvae have structurally uniform recta, whereas saltwater culcines have recta divided structurally into anterior (AR) and posterior (PR) regions. By constrast, both freshwater and saline-tolerant anopheline larvae have two-part recta, with a patch of cells on the dorsal anterior rectum (DAR) distinct from the rest (non-DAR). Redrawn from Smith et al. (2008), with permission from The Company of Biologists.

Salt-tolerant culicines have evolved segmented recta, with an anterior segment specialized in absorption and a posterior segment specialized for salt secretion; freshwater culicines have an unsegmented rectum that resembles the anterior rectum in the salt-tolerant species. By contrast, all anophelines, regardless of whether or not they are salt-tolerant, have a patch of rectal cells (DAR cells) that is ultrastructurally and biochemically distinct from the rest of the rectal cells (non-DAR cells). Depending on external salinity, the non-DAR cells appear to be able to secrete or to absorb salt.

4.6.3 Short- and Long-term Plasticity in ATPase Function in Response to Environmental Conditions

As discussed in Chapter 2, unlike vertebrate epithelia that power transport primarily, though not exclusively, with Na^+/K^+ ATPases, insects appear primarily to use V-type H^+ ATPases. For the ecological and environmental physiologist, an important question is: how are ion-motive processes modified in response to challenges over different timescales—physiological, developmental, and evolutionary? Compared to the detailed work done on transport mechanisms (Chapter 2), less is known about the question of *response*.

One answer to the question of short-term regulation is that the diuretic and antidiuretic hormones (summarized in Section 4.4) control the ATPases and, more broadly, control the components (resistances, capacitances) of the cellular circuits they energize. This answer is reasonable—but whereas the gross effects of hormones on water movements often are known, the connection between the hormone and the ion-moving machinery often is not. Several exceptions are illuminating. Using *Ae. aegypti*, Beyenbach's group studied the effects of mosquito natriuretic peptide (MNP, a calcitonin-like peptide) and a leucokinin, both of them diuretic peptides, on transepithelial movement of ions and water. MNP stimulated secretion of NaCl and water but not KCl, and did so by increasing the basolateral conductance to Na^+ (return current) (Beyenbach, 2003b). By contrast, leucokinin stimulated secretion of NaCl, KCl, and water, by vastly decreasing the paracellular resistance to movement of Cl^- (Beyenbach, 2003b). Paracellular resistance is controlled by septate junctions between cells, and the work on mosquito tubules provides an example of one of the fastest known instances in any animal in which the conductance of septate junctions changes (Beyenbach, 2003b).

The ion machinery is also plastic over developmental and evolutionary timescales. The developmental timescale is important for mosquitoes because many larvae are subject to changing salinity as they develop—either declining salinity from freshwater flooding (rain) or increasing salinity from evaporation or oceanic intrusion (Clements, 1992). Estuarine mosquitoes are the most likely to experience rapid, profound changes in their ionic environments. The evolutionary timescale is important because mosquitoes have radiated into fresh-, saline- and brackish-water habitats.

A key study illuminating both developmental plasticity and evolutionary change, by Smith et al., used larvae of two culicine and two anopheline species (in both groups there was one freshwater and one saline-tolerant species). Larvae from all four species were reared in either fresh or saline water (25% sea water). Using immunohistochemistry, they mapped rectal locations of carbonic anhydrase (CA), V-type H^+ ATPase, and P-type Na^+/K^+ ATPase. The main result was that larvae of anopheline, but not culicine, species showed major shifts in the localization of Na^+/K^+ ATPase when reared in fresh versus saline water (Smith et al., 2008). In *An. gambiae* and *An. albimanus* (anopheline) reared in freshwater, both species showed similar rectal tissue distributions: in non-DAR cells they had apical V-type H^+ ATPase and basal Na^+/K^+ ATPase. This pattern matched precisely the pattern seen in the unsegmented rectum of *Ae. aegypti* (culicine) reared in freshwater. Similar patterns were observed in larvae (anterior rectum) of the culicine *Ochlerotatus taeniorhynchus*, which is salt tolerant but was reared in freshwater. For capturing dilute ions from freshwater, this arrangement of pumps makes sense. The apical V-type H^+ ATPase hyperpolarizes the membrane, which provides a potential that drives Na^+ into the cell, where it is then pumped from cell to hemolymph by the basal Na^+/K^+ ATPase (Smith et al., 2008).

The picture shifts dramatically when the larvae of these species are reared in saline water (Smith et al., 2008). In the anopheline species, especially the salt-tolerant *An. albimanus*, there was significant upregulation of Na^+/K^+ ATPase in DAR cells and downregulation in non-DAR cells. In non-DAR cells, this shift appears to decouple the functions of Na^+/K^+ ATPase from those of V-type H^+ ATPase so that the cells no longer absorb Na^+ (as they do in freshwater). Instead, the pattern of protein localization comes to resemble the pattern seen in the posterior rectum (the salt-secreting section) in *Oc. taeniorhynchus*. In particular, the apical membranes in both species (non-DAR in *An. albimanus* and posterior rectal segment in *Oc. taeniorhynchus*) contain V-type H^+ ATPase. Smith et al. (2008) propose that the energy is used to *secrete* salts and thereby alleviate the load imposed in saline water. However, because pump activity can be regulated by reversible assembly, it's possible too that non-DAR cells have inactivated pumps when larvae are in saltwater.

The study of Smith et al. (2008) illustrates how developmental plasticity and evolutionary change can be traced to the biochemical level. What remains is to integrate the observed shifts in protein localization back into fuller models of ion and water transport (Harvey, 2009; Beyenbach and Piermarini, 2009; Wieczorek et al., 2009), and to evaluate the mechanisms across broader swaths of insect diversity.

4.7 Why are There No Marine Insects?

Despite their ecological success elsewhere, insects have only a tenuous foothold in the world's oceans. They usually are not major players in terms of biomass or ecological importance; only one book (Cheng 1976, edited volume) has been written on marine insects in the past 40 years. Moreover, even some groups associated with oceans may not be truly marine. For example, in the water-strider genus *Halobates* (Cheng, 1985), 46 species are associated with ocean habitats—but only 5 are truly oceanic (Spence and Andersen, 1994). Moreover, because water striders occur on top of the water rather than in it, it may be more reasonable to consider them terrestrial insects that happen to walk on water.

Why insects should have failed to colonize oceans is unclear, although their failure has generated numerous hypotheses (Buxton, 1926; Usinger, 1957; Van der Hage, 1996; Maddrell, 1998; Vermeij and Dudley, 2000; Ruxton and Humphries, 2008), with a recent evaluation of their merits by Ruxton and Humphries (2008). Below we outline some of the major ideas. Some of them focus on physiology, some on ecology, and some on interactions between the two.

One possibility is that insects are fundamentally unsuited to the rigors of ion and water regulation in seawater. Unlike their close relatives, the crustaceans, which have tissue osmotic pressures similar to those of seawater, insects generally have much lower osmotic pressures and would incur large osmoregulatory costs in seawater (see Maddrell 1998). However, counterexamples abound, including the massive diversification of mosquitoes into non-oceanic but high-salinity water (see previous section). Indeed, like crustaceans, *Culex* larvae avoid costs of osmoregulation by accumulating compatible organic solutes in their blood. Another counterexample is ephydrid flies, which complete much of their lifecycle underwater in hypersaline lakes. Evidently living and reproducing in saline waters is, for some insects, not impossible; they simply haven't been successful globally in terms of biomass and species diversity.

A second set of hypotheses invokes insect respiratory architecture. Insects are characterized by an air-filled tracheal system, which may be fundamentally unsuited to life in water. For example, it may be difficult to transport oxygen rapidly enough from surrounding water into the tracheal system. Or at depth the tracheal system may be prone to flooding. Here too counterarguments are diverse. There are many tracheate insects that live in freshwater. Moreover, some freshwater insects have closed tracheal systems—systems that are air-filled but not linked to the environment via open spiracles—which presumably would resist flooding even at high hydrostatic pressures. These species exchange oxygen across expanded tracheal gills of one kind or another. One interesting way to test the depth-flooding hypothesis would be to examine tracheal structure and function in insect ectoparasites living on diving mammals.

Maddrell (1998) proposed an ecological version of the tracheal hypothesis. He observed that the ecology of aquatic invertebrates is driven by predation pressure, especially from fish, which hunt visually. In response, many invertebrates undergo pronounced diel migrations, avoiding fish by descending into gloomy depths during the day. Maddrell proposed that having an air-filled tracheal system restricts the opportunities for going deep—because high hydrostatic pressure would collapse tracheal tubes and restrict oxygen movement. It's possible to imagine that insects could evolve stiffer tracheal tubes, but perhaps stiff tubes would incur functional costs in shallow depths or in air (e.g. inability to ventilate by tracheal compression). Another possibility is that tracheal systems lead to inevitable buoyancy problems. Because tracheae contain air, insects are naturally less dense than water, making them buoyant. In shallow water, buoyancy may require large energy outputs to overcome; and deep-water insects would not escape buoyancy problems if their tracheal tubes remained air filled.

A separate set of hypotheses focuses on ecological interactions rather than physiological constraints. One hypothesis, proposed by van der Hage (1996), focuses on co-evolution between insects and flowering plants. The logic: the oceans contain few angiosperms because it is difficult to send and receive pollen in seawater; on land, insects are diverse because of co-evolutionary associations with angiosperms; therefore, because angiosperms are rare in the oceans, so are insects. This hypothesis was criticized by Ollerton and McCollin on the grounds that: 1) insects evolved approximately 200 my before angiosperms appeared; 2) one group of angiosperms that has made it into the oceans, the seagrasses, have not been followed by insects; and 3) angiosperms have indeed evolved mechanisms for aquatic pollination (Ollerton and McCollin, 1998).

A broader hypothesis is that no single factor prevents insects from invading the seas. Rather, their long evolutionary history on land means that invading the seas would require coping with an overwhelming number of new physical and biotic problems (Buxton, 1926). A modern twist on this idea was proposed by Vermeij and Dudley (2000), who argued that transitions between marine, freshwater, and terrestrial environments are fundamentally difficult for all taxa, and therefore it is not surprising that it has been for insects too. Their argument builds on Buxton's multivariate idea but adds in a dimension of ecological competition: that the three habitats are so different, and require evolutionary adjustments in so many organismal components, that invaders are usually outcompeted by preexisting taxa in place for longer periods of evolutionary time.

Recent molecular phylogenies of arthropods shed new light on this argument. An older idea, still appearing in textbooks, is that the closest relatives of insects are myriapods. Because myriapods are terrestrial, the evolutionary origin of insects was thought to be fully terrestrial. However, recent papers (Cook et al., 2004; Glenner et al., 2006) have overturned this long-held belief in favor of an alternative: the sister group to hexapods now appears to be the branchiopods (brine shrimp, fairy shrimp, water fleas), or branchiopods and Malacostraca. Branchiopods are common in the world's freshwater systems. This phylogenetic insight suggests that insects (hexapods) are a terrestrial group of an otherwise marine and freshwater clade (see Fig. 1.1). If so, the ecological competition hypothesis for the current lack of marine insects acquires new weight. In particular, the sister clade to insects already dwells in fresh and seawater, and it likely fills many of the niches that would otherwise be available to invading insects.

5
Nutrition, Growth, and Size

5.1 Defining the Problem

Insects are important to virtually all terrestrial ecosystems. They function as major herbivores in natural, urban, and agricultural communities, are critical pollinators for most angiosperms, and play major roles in predation, parasitism, and recycling. In all of these functions, insects eat both plant and non-plant foods and convert them into new insects. This chapter focuses on the general questions of how nutritional factors affect the growth and size of insects. Although top-down effects such as predation are important, bottom-up nutritional effects on organisms are critical and may dominate many systems (Kos et al., 2011). Most of the chapter is about insect herbivores because most studies of insect nutrition have focused on herbivorous species, necessitating only minor attention to predators and parasitoids.

The defining the problem section of this chapter is an exceptionally large part of it. It was written this way as the "problems" associated with nutrition and growth are quite diverse; we cover issues of problems with limiting and excess nutrients and imbalanced diets in this section, as well as covering the problems caused by allelochemicals and some of the solutions offered by microbial symbionts. We then cover behavioral responses to food, post-ingestive responses, acute thermal effects on feeding and growth, and then adaptations to specific diets. We then discuss environmental effects on growth and development, and conclude with issues related to the causes and consequences of body size.

5.1.1 Overview of the Conversion from Food to Tissue and Energy

The ability of insects to convert food efficiently into reproduction is central to their ecological success. Food is complex, potentially comprising water, simple molecules that don't require digestion (e.g. glucose and amino acids or ions like Na^{+} and Cl^{-}), macromolecules utilized by insects (e.g. proteins, lipids and fats), macromolecules that cannot be digested and are excreted (e.g. cellulose), and molecules that are toxic to the insect (e.g. nicotine). Absorbed food can be

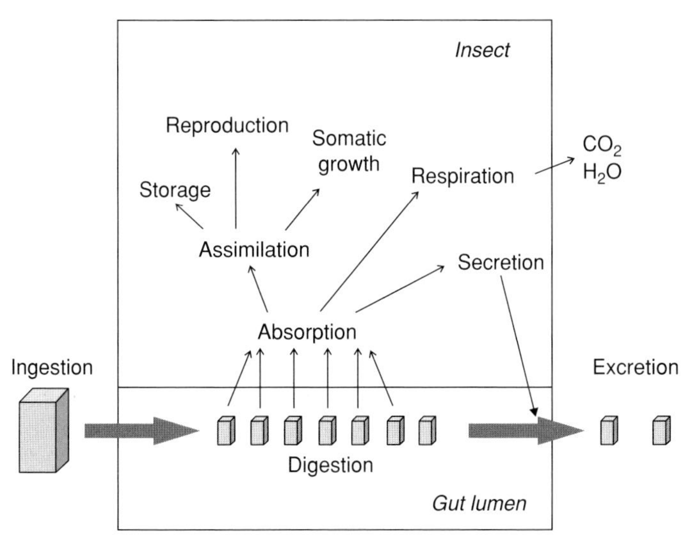

Fig. 5.1 Input/output budgets are key components of analyses of nutritional ecology. Ingested food is digested in the gut into component parts (e.g. proteins into amino acids) which are mostly absorbed, though some ingested material is usually excreted (e.g. undigested cellulose). Components (e.g. amino acids) are absorbed across the gut epithelium and then assimilated into tissues (somatic tissues, gametes, storage tissues, such as lipids), used as energy stores (respiration), or secreted (e.g. saliva). Detailed accounting of inputs, assimilation into tissues, and outputs are critical components of understanding an organism's needs and effects on its community. Such budgets can be calculated for mass, energy, or any individual chemical or element.

metabolized to support ATP production, and can also be assimilated into body tissues, including somatic tissues like muscle and cuticle, reproductive tissues (eggs and sperm), and storage reserves such as glycogen, lipids, or storage proteins (Fig. 5.1). Absorbed material can also be secreted (e.g. saliva, pheromones) or excreted (e.g. nitrogenous wastes). The allocation of assimilated resources can vary developmentally and evolutionarily, as trade-offs exist between investment in reproduction vs. somatic tissues that enable successful foraging, defense, and survival of environmental variation (O'Brien et al., 2008).

As in birds, amphibians, and reptiles, the renal and digestive systems intersect (Chapter 2). The renal system (Malpighian tubules plus hindgut) is responsible for homeostatic regulation of ions and water, and it has a high capacity to remove excess intake of ions or water, or to reduce excretory loss of these to inconsequential levels (Buckner et al., 1990; Buckner and Newman, 1990; Ehresmann et al., 1990; Phillips, 1981). The feces also contain nitrogenous compounds that result from metabolism of absorbed amino acids. If more amino acids are absorbed than required for protein synthesis, then excess amino acids will be deaminated, producing carbohydrates and ammonium. The ammonium can be excreted directly by the hindgut, or converted to complex molecules such as urates or allantoin by the fat body. Urates may be stored in the fat body (Cochran, 1985), or secreted by the Malpighian tubules. Secreted urates are usually converted to uric acid by an acidic hindgut (O'Donnell et al., 1983). Ammonium is secreted by the hindgut and can leave the hindgut and feces as gaseous NH_3 if the material is moist and pH neutral, or as NH_4^+ in more acidic, dry conditions (Harrison and Phillips, 1992; Thomson et al., 1988).

The portion of ingested food that must be used for energy depends on the energy content of the food (joules g^{-1}) and the metabolic rate (joules s^{-1} or watts) of the insect. Metabolic rate depends primarily on the rate of ATP use, as internal stores of ATP are minimal relative to the flux through the ATP cycle (Fig. 5.2). Increased growth or formation of reproductive tissue will require ATP input for formation of synthetic bonds, increasing ATP need and metabolic rate. Increasing skeletal muscular activity, such as walking, running, flying, or chewing, will

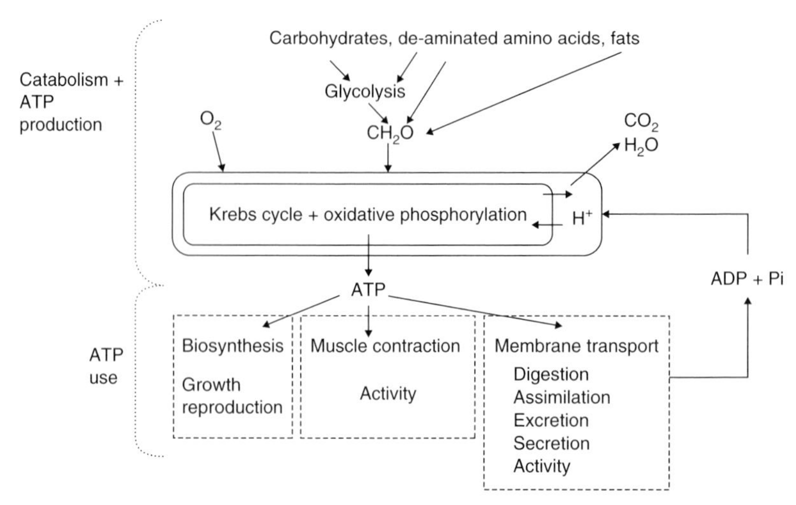

Fig. 5.2 A key concept in organismal physiology is that energy expenditure (ATP breakdown) depends primarily on ATP use rates. ATP utilization by biosynthesis (growth, formation of reproductive tissue), muscle contractions (activity, feeding) or membrane transport, produces ADP and inorganic phosphate, which stimulate mitochondrial fuel uptake and ATP production. Simple sugars, fats, and de-aminated amino acids can be catabolized to acetate, which can be used as fuel by the mitochondria to produce ATP, carbon dioxide, and water. Although there is generally a tight coupling between ATP use and catabolism, H^+ can leak across the inner mitochondrial membrane (futile cycling), which can consume oxygen and use up a significant proportion of catabolic energy without producing ATP. Such processes may help insects rid themselves of excess energy (Zanotto et al., 1997).

require more ATP; for example metabolic rate generally increases linearly with walking speed (Lighton and Feener, 1989), flight can increase metabolic rate 10 to 20 times at a given body temperature (Harrison and Roberts, 2000), and cutting leaves can increase the metabolic rate of leafcutter ants by 30 times (Roces and Lighton, 1995). Increasing active transmembrane processes also will increase metabolic rate. These include the active transport processes that drive absorption in the midgut and excretion in the Malpighian tubules and hindgut, as well as the pumps that create and maintain potentials across plasma and organelle membranes (primarily Na^+/K^+ ATPases, H^+ATPases, and Ca^{++} ATPases).

An interesting question is whether there is a step in the consumption/digestion/absorption/assimilation process that limits the entire process. From a systems perspective, natural selection acts on the integrated output of the digestive/metabolic system, suggesting that evolution should act to ensure matching of capacities of each step of this series, so that no single step will be limiting (Weibel et al., 1998). However, this hypothesis of physiological symmorphosis is not always supported. Grasshoppers and many herbivores surrounded by abundant food are well-known to consume food in bouts interspersed with longer "rest" periods during which digestion and absorption occurs, suggesting that consumption is often not rate-limiting for herbivores (Chambers et al., 1996; Simpson, 1990). Also, the fact that many insects can increase consumption when artificial foods are diluted with cellulose indicates that for these insects, consumption is not limiting. However, this may differ for insects feeding on very hard foods, which can spend very long periods of time chewing (see below). Studies showing that protein digestion is completed in the anterior midgut of *Manduca sexta* suggest that digestion is not limiting, but that absorption or assimilation may be (Woods and Kingsolver, 1999). These findings suggest that it may often be possible to identify rate-limiting steps in the consumption-to-assimilation series for a particular species and environment, speeding identification of key steps in phenotypic or evolutionary adjustments to changing diet.

5.1.2 Theoretical Perspectives on Insect Feeding and Nutrition

Even predaceous or parasitic insects usually have a different body composition from what they eat. However, this is particularly true for herbivores, the most common feeding habit among insects (Table 5.1). Although the ideal food does not necessarily have an identical composition as the consumer (due to certain foods being preferentially metabolized, and variation in digestibility and absorption), many insect foods are dramatically mismatched to nutritional needs. In addition, insects are fairly homeostatic in their body composition, while the composition of plants can vary dramatically with environmental conditions. The mismatch between nutritional needs and food composition drives a suite of behavioral and physiological adaptations that enable insects to survive on

Table 5.1 Average biomass composition of carbon, nitrogen, and phosphorus, and their elemental ratios, for herbivorous insects, terrestrial leaves, and wood (all dried). Herbivorous insects generally have higher nitrogen and phosphorus contents than their diets, and lower and less variable carbon:nitrogen (C:N) and carbon:phosphorus (C:P) ratios, but similar (if more variable) nitrogen:phosphorus (N:P) ratios than leaves (Elser et al., 2000; Mattson Jr., 1980). Wood is extremely low in nitrogen and phosphorus relative to carbohydrate (Haack and Slansky, 1987; Higashi et al., 1992).

	Insects	Leaves	Wood
% carbon	50	46	50
% nitrogen	10	3	0.5
% phosphorus	0.9	0.3	0.03
C:N median (range)	6 (4–14)	32 (7–100)	500
C:P median (range)	73 (40–250)	799 (250–3500)	1700
N:P median (range)	23 (10–44)	27 (5–65)	16

extremely diverse foods. These range from xylem, with extremely low concentrations of nutrients to rotting carrion, and even to plants loaded with enough toxins to make humans hallucinate wildly before dying. Certainly the capacity for insects to grow and reproduce on diverse foods is central to their ecological and evolutionary success.

Energy provided the first major theoretical framework for understanding insect foraging and feeding, and consideration of energetics remains a strong component of understanding the physiological ecology of growth (Karasov and Martinez del Rio, 2007). Foundational studies in ecology focused on energy flow through trophic levels, and some important early studies focused on acquiring energy in an optimal foraging context and extensive mathematical treatments have been developed (Karasov and Martinez del Rio, 2007; Stephens and Krebs, 1986). A major advantage of energy-focused optimal foraging theory is the capacity to use a single currency (energy) to integrate costs and benefits of foraging. Multiple studies, mostly in hymenopterans (bees and ants) have examined whether net rate of gain (rate of energy intake minus rate of energy expenditure, all divided by time spent) or efficiency (energy expenditure/energy intake) best explain insect foraging behaviors, and the answer appears to vary with species and environment (Cartar, 1992; Fewell et al., 1991; Heinrich, 1975; Schmid-Hempel et al., 1985). Behavioral experiments suggest that foraging currency can be influenced by predators and the energy state of a colony (Cartar, 1991a; Cartar, 1991b; Jones, 2010). In the ants *Camponotus rufipes*, *Paraponera clavata*, and *Pogonomyrmex occidentalis*, the energetic value of the resource far exceeds the foraging cost, suggesting that time costs or other nutrients are more likely than energy costs to influence foraging decisions (Fewell et al., 1996; Fewell et al., 1992; Schilman and Roces, 2006). The metabolic theory of ecology is, to a major extent, built on the use of the consideration of energetic turnover to link molecular, organismal, ecological, and evolutionary processes (Brown et al., 2011).

A second theoretical theme of insect nutritional ecology is the idea of limiting nutrients. To grow and function, insects require sufficient amounts of energy, essential amino acids, essential lipids, vitamins, and salts. As an example, growth of grasshoppers eating Romaine lettuce increases when they are supplemented with phenylalanine, suggesting a protein-synthesis-limiting deficiency of this particular amino acid in the diet (Bernays and Woodhead, 1984). Similarly, decreases in vitamin and salt intake below critical limits negatively impacts growth, survival, and fecundity in the insects that have been tested (Dadd, 1970). Thus, one approach to insect nutrition is to attempt to determine which nutrient is limiting to growth and reproduction for a particular organism and diet.

A third major theoretical approach in nutritional ecology focuses on the ratios of nutrients in foods. This approach has been separately formalized as "ecological stoichiometry" by Bob Sterner and Jim Elser (Sterner and Elser, 2002), and "the geometric framework" by David Raubenheimer and Steve Simpson (Raubenheimer et al., 2009; Simpson and Raubenheimer, 1995). These two approaches differ in that ecological stoichiometry traces elements through ecosystems, whereas the geometric framework focuses on behavioral and physiological regulation of nutrients by individuals. Ecologists often measure and discuss elemental concentrations such as N (nitrogen) and C (carbon), often assuming these are reasonable proxies for the nutrients (e.g. proteins and carbohydrates) studied by behaviorists and physiologists. Together, ecological stoichiometry and the geometric framework provide powerful arguments that how organisms respond to the *relative* availability of nutrients is critical to their nutritional ecology. Stoichiometric and geometric frameworks will recur throughout this chapter.

A key finding of the geometric framework studies is that insect feeding systems are adapted to obtain nutrient targets of multiple nutrients, and that consuming excess nutrients can depress fitness (Raubenheimer et al., 2005; Simpson et al., 2004). Thus, most recent studies of insect nutrition focus on responses to multiple nutrients, and consider not just limiting nutrients but also the costs of excess consumption. Even social insects (ant colonies) have been shown to balance consumption of multiple nutrients (Cook and Behmer, 2010; Cook et al., 2010; Dussutour and Simpson, 2009).

The geometric framework is a quantitative approach to evaluate relative intake of nutrients in a multidimensional nutrient space, with as many axes as there are functionally-relevant nutrients (Behmer, 2009). A good starting place to understand geometric frame models is to consider the response of an insect to feeding on a single defined diet (Cheng et al., 2008; Raubenheimer and Simpson, 1993). In these models, foods are represented as rays (often called rails) in a multidimensional nutrient space, because insects will consume nutrients in a defined ratio as they consume more of that food (Figs 5.3 and 5.4). When given a choice in the lab, insects will select a mixture of food that provides an optimal consumed amount and ratio of nutrients, termed the *intake target*, which provides the needed mix of nutrients for that developmental stage (Raubenheimer

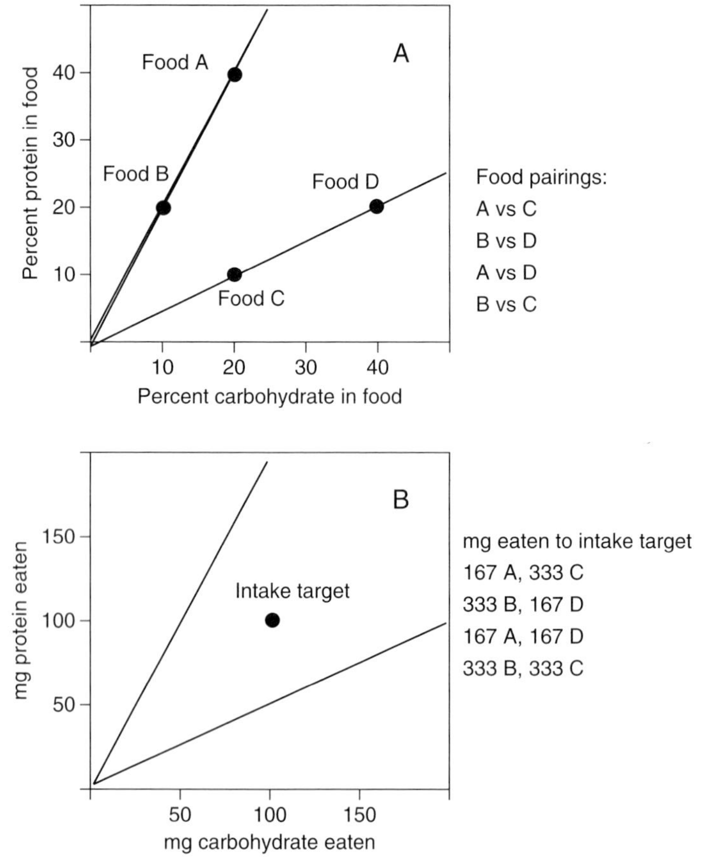

Fig. 5.3 Using the geometric framework to determine the *intake target* ratio for protein and carbohydrate. An organism is offered pairs of diets that differ in the ratio of protein:carbohydrate (Diet A: 20% carbohydrate:40% protein, Diet B: 10% carbohydrate:20% protein, Diet C: 20% carbohydrate:10% protein, Diet D: 40% carbohydrate:20% protein. If an animal eats only one food, it moves "along the rail" consuming protein and carbohydrate according to the ratio in that food (A: upper panel), but mixing allows the animal to attain the intake target (B: lower panel). In this hypothetical case, the organism consumed 100 mg protein and 100 mg carbohydrate regardless of the diet pairings, indicating that the *intake target* is 100 mg for both protein and carbohydrate. From Simpson and Raubenheimer (1995), used with permission from Elsevier.

et al., 2009; Simpson and Raubenheimer, 1995) (Fig. 5.3). There is evidence from field studies too that at least some insects choose among possible foods to approach intake targets.

Although the geometric framework has been focused most frequently on feeding behavior, it can also be used to understand digestive efficiency and assimilation. When feeding on an optimal diet (Fig. 5.4.A), an insect feeds to its intake target for carbohydrate and protein. Of the nutrients ingested, some are used to fuel metabolism; the mix of nutrients assimilated into the tissue of growing animals is termed the *growth target* (Fig. 5.4).

Animals can still achieve their growth targets on imbalanced foods if they are capable of sufficient post-ingestive compensation; for example by utilizing the excess nutrient preferentially for energy generation (Fig. 5.4.B). However, animals that exceed their capacity for such post-ingestive nutrient balancing will not be able to attain the growth targets, and will experience reduced growth (Fig. 5.4.C).

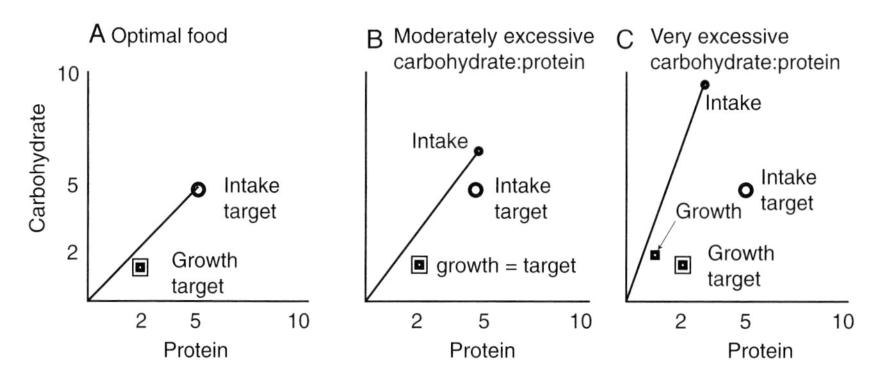

Fig. 5.4 Potential outcomes for animals consuming single foods differing in their carbohydrate:protein ratio during growth. A) An animal consuming an optimal diet can consume that single food to reach its intake target of carbohydrate and protein (in this case, 5 carbohydrate:5 protein). Some of the consumed nutrients are used for energy or secreted, and some are incorporated into the animal tissues. The growth target is the amount of protein and carbohydrate the animal must assimilate to optimize fitness (in this case, 1.5 carbohydrate: 2 protein). B) Consuming a diet with moderately excessive carbohydrate:protein ratio. Animal consumes excess food until it is able to attain sufficient protein; excess ingested carbohydrate can be stored or eliminated via post-ingestive compensation (e.g. carbohydrate excretion or increased metabolic rate), allowing it to achieve the growth target. C) Consuming a diet with an excessive carbohydrate:protein ratio that limits growth. Animal consumes much more food and carbohydrate, but limits to post-ingestive compensatory mechanisms are reached, leading to insufficient protein intake and reduced growth (2.5 carbohydrate: 1 protein). Modified from Raubenheimer, Simpson and Mayntz (2009), with permission from John Wiley and Sons.

5.1.3 Limiting Nutrients and Problems with Excess

Which "functionally relevant" nutrients affect insect performance and fitness? Tests of the geometric framework have primarily focused on the macronutrients protein and carbohydrate; however, in theory, any nutrient required by the insect could be limiting. Nonetheless, certain nutrients are unlikely ever to be limiting for specific insects in particular niches due to their excess concentration in the diet relative to the insect (e.g. water for xylem-feeders, potassium for folivores). Clearly, the relevant nutrients will vary by diet, and also with environmental variation that affects metabolic rate or the need for specific nutrients (e.g. insects preparing for diapause or migration). Predatory insects generally will have fewer chemical mismatches between tissue and diet, suggesting that attaining enough food is the main challenge (however, see data to the contrary below). Herbivorous insects, on the other hand, often have large amount of food available to them, but food that is highly imbalanced relative to their intake targets, necessitating both behavioral and physiological compensation.

The mismatch between consumer and food can also be affected by variation in body composition among species. For example, predators have 15% higher body nitrogen contents than herbivores (Fagan et al., 2002), more derived insect orders (Diptera, Lepidoptera) have 15–25% lower body nitrogen content (Fagan et al., 2002), and larger insects tend to have lower concentrations of phosphorus (Bertram et al., 2008; Woods et al., 2004). There is also considerable intra-specific variation in body stoichiometry (Bertram et al., 2008). In most cases, we do not yet know whether these differences in body composition are reflected in predictable differences in intake targets or limiting nutrients in the field. Guilds of related generalist grasshopper species in a single community consume different ratios of protein:carbohydrate, providing niche separation, but as yet it is not clear whether these different intake targets reflect differences in body composition (Behmer and Joern, 2008).

Another important finding of geometric framework studies is that excess consumption of many types of nutrients can be deleterious (Raubenheimer et al., 2005). We humans are well aware that excess calories can lead to obesity; in insects too, there is increasing evidence that high carbohydrate consumption can lead to obesity. Excess intake of protein and salt is deleterious for insects, and it seems likely that other nutrients can be consumed in excess as well. One important conclusion from a variety of research using geometric approaches is that for many, if not most nutrients, excess consumption has costs.

5.1.3.1 Protein

Protein is the major component of most somatic and reproductive tissues and thus is a critical macronutrient. Insect herbivores generally contain much more nitrogen than do plant leaves (Table 5.1). The protein content of plants varies substantially, with leaves of herbaceous legumes generally having the highest nitrogen, followed by leaves of other forbs, then leaves of deciduous trees and grasses, with evergreens having the least (Schoonhoven et al., 2005). Insects feeding on wood, detritus, or xylem often must cope with extremely low dietary protein content (Table 5.1). The mismatch between dietary and body protein content partially explains why protein limitation is common in herbivorous insects (Mattson Jr., 1980; Scriber, 1984; White, 1993).

For both caterpillars and grasshoppers, if dietary protein levels are sufficiently low, the limits of capacity to consume and process the diet are reached, and body nitrogen, growth, and survival all decrease (Lee et al., 2002; Woods, 1999). For the caterpillar *Pieris rapae* and also several grasshopper species, deficits of protein are worse for growth than are deficits of carbohydrate (Morehouse and Rutowski, 2010; Simpson et al., 2004). Multiple studies have shown that protein levels beyond what are normally available in nature enhance growth, body size, and reproduction of insect herbivores (Huberty and Denno, 2006a; White, 1993). Similarly, the assimilation efficiency of the bug *Leptoterna dolabrata* increases linearly with dietary nitrogen (McNeill and Southwood, 1978). Increases in dietary protein content over the range found in the field increases reproduction in grasshoppers (Joern and Behmer, 1997; Joern and Behmer, 1998). Field fertilization of nitrogen often increases nitrogen levels in plants and insects (Jonas and Joern, 2008), and leads to increases in population density of grasshoppers (Fig. 5.5; (Ritchie, 2000)). However, nitrogen

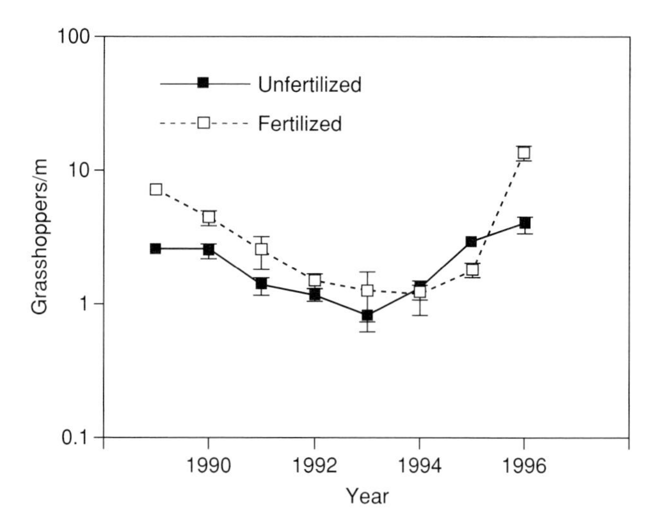

Fig. 5.5 Nitrogen fertilization increased population density of grasshoppers at a grassland site in some but not all years. Nitrogen was added as ammonium nitrate to a Minnesota old-field prairie dominated by native plants. From Ritchie (2000), with permission from Ecological Society of America.

fertilization had no effect on growth of juvenile *Omocestus viridulus* grasshoppers (Berner et al., 2005), and the response to dietary protein varies considerably among grasshopper species in the lab, emphasizing extensive, unexplained variation among species in intake targets (Behmer and Joern, 2008; Joern and Behmer, 1998).

Insects require dietary sources of ten amino acids (methionine, tryptophan, threonine, valine, leucine, isoleucine, lysine, phenylalanine, histidine, and arginine (O'Brien et al., 2005)). The protein deficit situation for insect herbivores can be exacerbated by the fact that plants may have low concentrations of essential amino acids needed by insects. Also, a significant portion of plant nitrogen may be inorganic (ammonium$^+$). Ammonium can be used to produce non-essential amino acids but not essential amino acids (Hirayama et al., 1996). The aromatic amino acids phenylalanine and tryptophan are major components of insect cuticle, and insects seem often to need these amino acids disproportionately (Bernays and Woodhead, 1984; Brodbeck and Strong, 1987). Low dietary protein can impair the immune system (Srygley and Lorch, 2011), perhaps due to deficiencies in immune system proteins. Although low dietary protein can directly limit growth, negative effects on insects can also occur due to correlated effects related to compensatory feeding. Diets with low protein:carbohydrate ratios stimulate compensatory feeding that leads to excess energy intake, which can lead to negative effects on fitness associated with accumulating too much fat (Simpson and Raubenheimer, 2009). Also, compensatory feeding may increase the risk to a herbivore of being detected by a predator or parasitoid (Bernays, 1997).

Shortages of particular essential amino acids may explain why particular diets are poor. Karowe and Martin (Karowe and Martin, 1989) explored the effects of both using the lepidopteran larva *Spodoptera eridania*, fed on artificial diets varying eight-fold in the concentration of casein (a protein), and sometimes substituting zein for casein. When fed diets very low in protein, larvae were extremely small and fewer than half survived; as total dietary protein tripled, body mass increased 40-fold and survival increased to 100%. Body N and protein levels were not completely homeostatic, as larval %N increased by 33% and protein concentration by 80% when dietary protein increased eight times. As dietary protein content increased eight-fold, growth rates only increased about 20%, as increased feeding allowed substantial compensation for low protein levels. Partially substituting zein for casein reduced growth rates by 25%, increased uric acid excretion, and elevated metabolic rates by 25%. Karowe and Martin concluded that growth rates were reduced on the zein-containing diets due to limitations in essential amino acids, and that the costs of converting excess amino acids to uric acid partly accounted for the increase in metabolic rate.

Although more protein usually is beneficial, excess protein can be deleterious for some insects. Scriber's literature survey found 44 studies in which herbivore performance decreased with increasing plant nitrogen (Scriber, 1984). High protein reduced larval survival and production in *Rhytidoponera* ants (Dussutour and

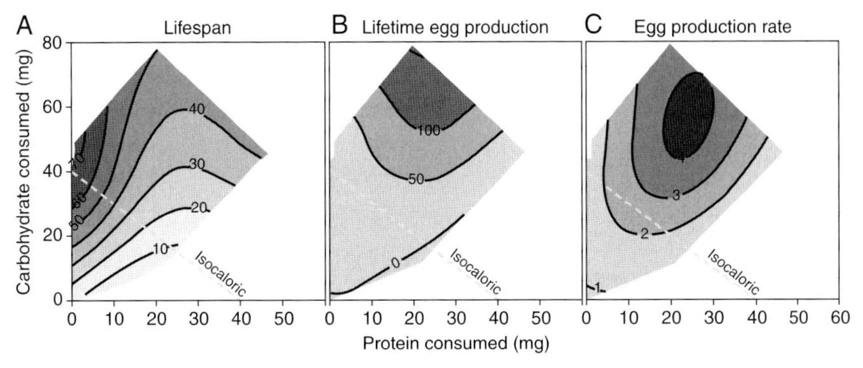

Fig. 5.6 Effects of protein (x axis) and carbohydrate (y axis) intake on survival (left), lifetime egg production (middle), and egg production rate (right) for Queensland fruit flies (*Bactrocera tryoni*) fed single diets varying in carbohydrate:protein ratios. Plotted are fitted surfaces with dark shades indicating higher values. Dashed lines indicate isocaloric intake. High carbohydrate:protein diets maximized lifespan, low carbohydrate:protein diets maximized egg production rate, and intermediate carbohydrate:protein diets maximized lifetime reproduction. Redrawn, with permission, from Fanson et al., 2009.

Simpson, 2009). Higher protein content in the leaves of some plants consumed seems to impede carbohydrate assimilation and growth in the Australian plague locust, *Chortoicetes terminifera* (Clissold et al., 2006). Excess protein intake strongly reduced the lifespan and fecundity of fruit flies and crickets (Fig. 5.6, (Simpson and Raubenheimer, 2009)). High protein:carbohydrate foods reduced lipid stores and increased mortality in fire ants (Cook et al., 2010). Similar results have been found for grasshoppers (Joern and Behmer, 1998). The mechanisms by which high protein affects survival and reproduction are poorly understood but perhaps are related to coping with nitrogenous waste products, increased production of oxygen radicals, or changes in insulin/TOR/AMPK signaling (Simpson and Raubenheimer, 2009). Understanding these mechanisms is a high priority. It is also interesting to consider how such issues might play out in insects that eat very high-protein foods (e.g. blood, meat).

5.1.3.2 Carbohydrate

Carbohydrates are major constituents of leaves and seeds, and often are the major energy source available to herbivores. Many insects that have been tested self-select carbohydrate:protein ratios of near 1:1 (Behmer, 2009), suggesting that carbohydrate is an important nutrient. Grass-feeding grasshoppers and caterpillars tend to have intake targets with even higher carbohydrate:protein ratios than generalist grasshoppers, suggesting evolution of their behavior and physiologies to match the lower protein content of their host plants (Behmer, 2009).

In the lab, insufficient intake of carbohydrate can reduce size, shorten lifespan, and reduce reproduction in caterpillars and grasshoppers (Lee et al., 2002; Simpson et al., 2004) (Fig. 5.6). In the field, relatively poor growth of the Australian plague locust *Chortoicetes terminifera* on the perennial grass *Astrebla lappacea* appears to be due to insufficient carbohydrate absorption and assimilation (Clissold et al., 2006). Insufficient nectar intake and honey storage lead to death during overwintering in temperate honeybee colonies (Winston, 1987). For some predatory insects, diet supplementation with carbohydrates (or lipid) has been shown to increase fitness in the field (Helms and Vinson, 2008; Wilder and Eubanks, 2010). Higher carbohydrate intake increased pheromone production in male cockroaches, leading to increased attractiveness to females and reproductive success (South et al., 2011). However, excess dietary carbohydrate content had negative effects on reproduction in grasshoppers (Joern and Behmer, 1998) and flies (Fig. 5.6). Caterpillars (*Spodoptera littoralis*) over-consume carbohydrate when fed low protein:carbohydrate food. At high levels of carbohydrate over-feeding, animals had reduced survival, suggesting significant costs to the elevated carbohydrate consumption (Raubenheimer et al., 2005; Simpson et al., 2004).

5.1.3.3 Lipid

Insect predators likely obtain considerable calories from ingesting the lipid stores of their prey, and there is evidence that dietary energetic (predominantly lipid) content can be important for enhancing fitness in some predators (Wilder and Eubanks, 2010). Although plant leaves generally do not contain much lipid, insects can manufacture it from excess carbohydrate and protein. Thus, it is likely that plant lipids are not an important macronutrient for most folivorous insects. As noted in Chapter 2, insects have an absolute dietary need for cholesterol, and can only use certain plant sterols (Behmer and Nes, 2003). Male gypsy moth larvae preferentially consume diets high in lipid, possibly in order to build up nutrient stores for adult flight (Stockhoff, 1993). *Manduca sexta* caterpillars also preferred diets high in fat (Thompson and Redak, 2005).

5.1.3.4 Phosphorus

There has been relatively little research on phosphorus limitation in insects, but stoichiometric theory predicts that phosphorus limitation should be common in insects, as it is in aquatic invertebrates (Elser et al., 2000). Some of the available research supports this hypothesis. The tobacco hornworm, *Manduca sexta*, grows and develops faster when consuming leaves of *Datura wrightii* or artificial diets higher in percent phosphorus, with other elements maintained constant (Fig. 5.7 top, (Perkins et al., 2004)). This response occurred over a range of %P relevant in the field; in fact most natural *Datura* leaves had lower phosphorus contents than needed for optimal growth (Fig. 5.7, bottom), suggesting that phosphorus can

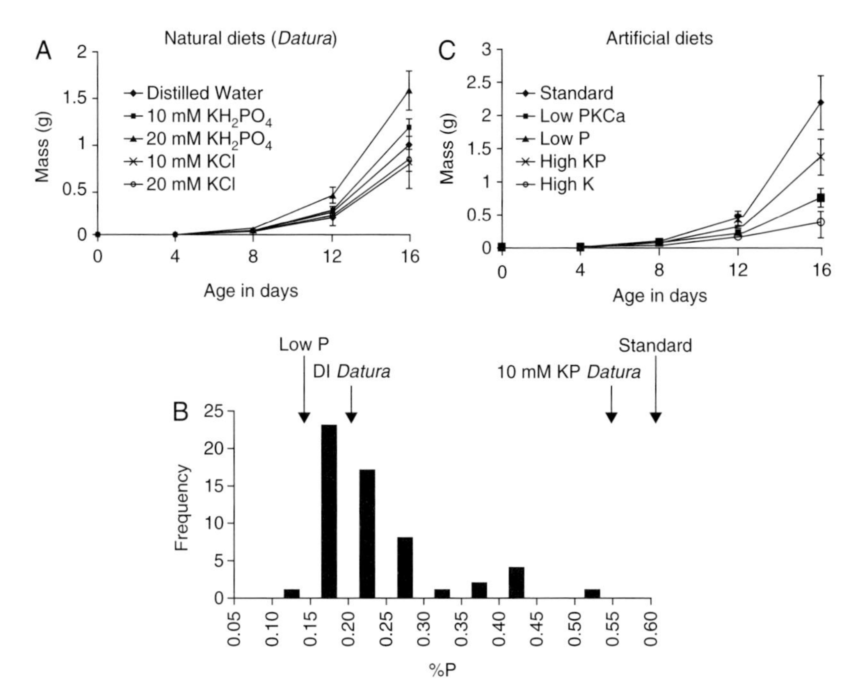

Fig. 5.7 Top left: Growth of *Manduca sexta* caterpillars feeding on *Datura wrightii* leaves whose phosphorus contents were modified by holding (for 2 days) the cut ends of leaf petioles in solutions of different composition. Leaves soaked in salines containing higher phosphate promoted faster growth. Top right: Growth of *Manduca sexta* feeding on artificial diets with varied P content. The diets with the most P (standard and high P) promoted the fastest growth. Bottom: Variation in the percent phosphorus content of *Datura wrightii* leaves in the field in Arizona. Percent P varied widely, and most leaves had lower P contents than required for the most rapid growth. Low P = concentration of P in the low P artificial diets; DI *Datura* = concentration of P in *Datura* leaves soaked in distilled water; 10 mM KP Datura = concentration of P soaked in 10 mM phosphate solution; Standard = concentration of P in the standard artificial *Manduca* diet. From Perkins et al. (2004), with permission from John Wiley and Sons.

limit growth and fitness in the field. Similarly, host trees generally have too little phosphate to support maximal population growth in spruce budworm (Clancy and King, 1993), and mesquite trees in soils with lower phosphate availability had lower %P in their leaves, and supported fewer weevils, which had lower phosphate contents (Schade et al., 2003). Supplementation of normal field diets with phosphorus enhanced fitness of one planthopper, but not a second, more generalist planthopper (Huberty and Denno, 2006a). An elegant example of phosphorus limitation in the field has been documented for noctuid larvae feeding on lupins on Mt. St. Helens (Apple et al., 2009). At higher lupin densities (as occurs later in succession), plant phosphate levels were lower (and C:P higher). Larvae fed on plants with higher phosphate levels or lower carbon levels survived better and grew faster, suggesting that decreased phosphate levels were at least partly responsible for the decreased insect density later in sucession, despite increased availability of food (Apple et al., 2009).

In contrast, higher dietary phosphorus levels reduced growth rates in the grasshopper *Melanoplus bivittatus* (Loaiza et al., 2008), and did not appear to influence choice of food plants by grasshoppers in the field (Loaiza et al., 2010). The diversity of responses to phosphorus again emphasizes the diversity of insect nutritional responses; determining how these relate to variation in body composition, diet, and life history is a significant challenge.

5.1.3.5 Salts

Predatory insects probably have few problems obtaining the correct amounts of Na+, K+, or Cl–, as their diets are well-matched to their body needs. Moreover, insects generally have well-developed mechanisms for excreting excess salts (Phillips, 1981). Because plant leaves are rich in K^+, folivorous insects tend to consume an excess of K^+. However, Na^+ may be in short supply for herbivorous insects. Insects generally maintain body sodium contents that are 10- to 100-fold greater than plants (Kaspari et al., 2008). Ants recruit to NaCl solutions, and this recruitment is greater in herbivorous and omnivorous species than in carnivorous species (Kaspari et al., 2008). In addition, recruitment to NaCl solutions increases with distance from the ocean, perhaps due to the reduced NaCl content of rainfall (Kaspari et al., 2008). Some male moths seek and concentrate Na^+ from puddles, which they pass to females, who allocate this Na^+ to their eggs, strongly suggesting that Na^+ can be a limiting nutrient for the larvae (Fig. 5.8; (Smedley and Eisner, 1996)).

Intake targets have been little investigated for salts, but the data available suggest behavioral regulation of salt intake. Locusts provided a choice of different salt solutions select food mixtures that yield an intake target of 1.8% NaCl. Although one might guess that excess salt intake could be deleterious, there is little evidence for this in grasshoppers (Trumper and Simpson, 1994), perhaps because renal mechanisms for salt excretion are excellent (Phillips et al., 1986).

5.1.3.6 Water

Water is an important limiting nutrient for the growth and survival of some insects. Multiple species of lepidopteran larvae have been shown to grow more rapidly on foliage with higher water content, and manipulative experiments with artificial diets have confirmed that the differential growth on the natural diets was attributable to the specific effect of water availability (Scriber, 1984). Experiments designed to mimic natural variation in plant water content also have shown that caterpillars grow faster on foliage with higher water content (Henricksson et al.,

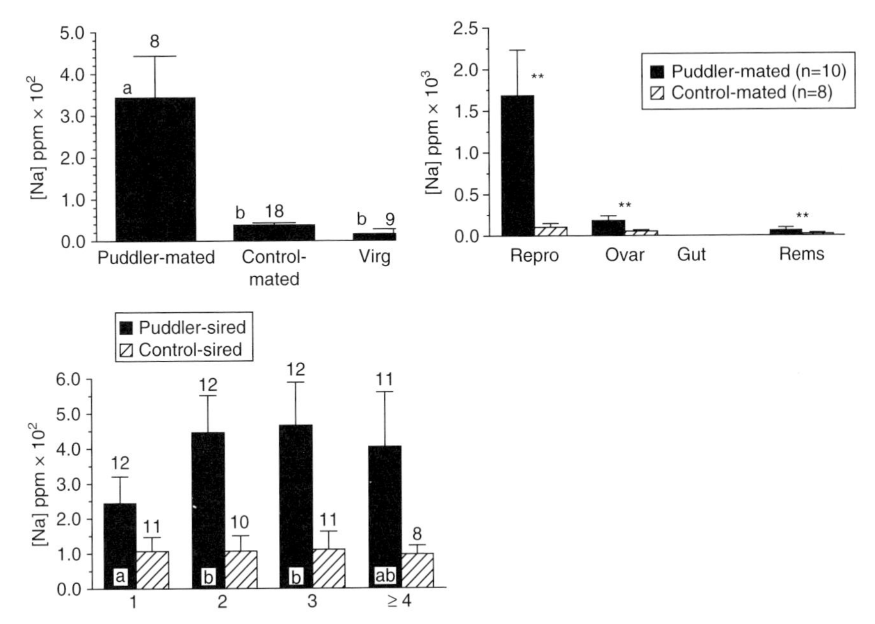

Fig. 5.8 Male moths (*Gluphisia septentrionis*) collect Na^+ from puddles and pass it to females, who allocate the extra Na^+ to their eggs. Top left: Whole body Na^+ content of females mated to puddler males, control males, or no males (virgins). Top right: Na^+ content of different organs of puddler-mated and control-mated females, showing allocation of Na^+ to the reproductive tract. Bottom: Na^+ content of the eggs in puddler-mated vs. control-mated females over four clutches. From Smedley and Eisner (1996), with permission from National Academy of Sciences.

2003; Scriber, 1979). On the other hand, the grasshoppers studied to date can grow more rapidly on water-stressed foliage, perhaps because nutrients are more concentrated (Bernays and Lewis, 1986; Lewis and Bernays, 1985). Grasshoppers prefer dry food if they are well-hydrated but wet food if not, suggesting behavioral regulation toward a target intake of water that aids in maintaining a homeostatic body content of water (Chapman, 1990).

Fluid feeders (nectar, blood, xylem, phloem) face the challenge of extracting nutrients, some in low supply, from a large mass of water. Physiological mechanisms of the control of water transport in response to water loading, and evolved digestive morphologies that allow rapid separation and excretion of water from dissolved nutrients, are described in Chapter 4.

5.1.3.7 Sterols

As noted in Chapter 2, insects and other arthropods lack the capacity to produce sterols such as cholesterol. Thus sterols are potentially limiting, especially since key hormones such as ecdysone are derived from cholesterol. Some insects, such as the leaf-cutting ant *Acromyrmex octospinosus*, obtain sterols from fungi (Maurer et al., 1992). Other insects have fungal symbionts in their guts that produce sterols, including the anobiid beetles *Lasioderma serricorne* and *Stegobium paniceum* (Nasir and Noda, 2003).

5.1.3.8 Micronutrients

In general, insects prioritize intake of macronutrients relative to micronutrients; several studies have demonstrated that insects ignore salts and micronutrients when protein or carbohydrate intakes are limiting (Behmer, 2009). This likely occurs because it is possible for insects to eliminate excess salts and micronutrients using the renal system.

Relatively little is known about which micronutrients insects need and whether their intakes are regulated. In general, the vitamins required by insects seem to be similar to those required by vertebrates (Dadd, 1970). *Drosophila* grow and survive well on folate-free media unless they are given antibiotics, suggesting that they obtain vitamins from endogenous microbes (Blatch et al., 2010). Sometimes micronutrients occur in excess. Host trees generally sustain magnesium levels that reduce population growth rates in spruce budworm (Clancy and King, 1993). Excess iron can damage insects by promoting oxidative damage (Locke and Nichol, 1992; Sohal et al., 1985).

5.1.4 Plant Allelochemicals

Plants often suffer damage from insect herbivores, and so have evolved chemical defenses to dissuade potential consumers. The secondary compounds (as opposed to those involved directly in the primary pathways of metabolism) accumulated by plants that enhance their resistance to herbivory are often termed allelochemicals. These are incredibly diverse, with over 100,000 different compounds discovered so far (Schwab, 2003). Although plant secondary compounds can be difficult to classify, they are often divided into four main types: 1) nitrogen-containing compounds such as alkaloids; 2) terpenoids; 3) phenolic compounds; and 4) acetylenic compounds.

Alkaloids are cyclic nitrogen-containing compounds. Famous examples are nicotine, papaverine, morphine, cocaine, curare, strychnine, quinine, and caffeine. About 20% of angiosperms produce alkaloids, but these are rarely found in gymnosperms (conifers) or cryptogams (ferns) (Schoonhoven et al., 2005). Some plants, particularly grasses, have symbiotic fungi that produce alkaloids that affect insect herbivores or their predators (Faeth and Shochat, 2010). Most alkaloids deter feeding and are toxic to insects at concentrations over 0.1% (Schoonhoven et al., 2005).

Glucosinolates are nitrogen-containing compounds that also contain sulfur, mostly occurring in the Brassicaceae. Glucosinolates such as sinigrin deter feeding and are toxic to many generalist insects, but they also serve as feeding and oviposition attractants to insects that specialize on the species (Nault and Styer, 1972). Another group of nitrogen-containing compounds are the cyanogenics, produced by 11% of all plants (Schoonhoven et al., 2005). These compounds are stored in vacuoles, and produce hydrogen cyanide when the plant is crushed.

Although built mostly from simple isoprene units, terpenoids and steroids are diverse compounds. Major groups include hemiterpenoids, monoterpenoids, sesquiterpenoids, diterpenoids, triterpenoids, tetraterpenoids, and polyterpenoids. Many of these are lipid-soluble and volatile, and isoprene can be emitted from plants at high rates. Terpenoids often impart characteristic flavors or odors to plants. Saponins are a type of terpenoid that can include steroid hormones used by insects such as ecdysone (Schoonhoven et al., 2005).

Phenolics possess an aromatic ring with one or more hydroxyl groups. Phenolics include phenol, hydroxybenzoic acids, coumarins, and flavonoids. Many of these are known to deter feeding (e.g. coumarin) or be toxic (e.g. rotenone). Tannins are common, soluble polyphenolic compounds that bind to proteins, producing insoluble polymers that have less biological activity or cannot be digested as quickly (Schoonhoven et al., 2005).

Plants often store allelochemicals in specific structures that enhance their activity against insect herbivores. Some plants have glandular trichomes that secrete terpene oils; in some cases release is stimulated by touch. Damage to the plant surface releases the volatile terpenes as well as hydrogen cyanide.

Concentrations of allelochemicals vary dramatically among plant parts, and this heterogeneity can influence insect feeding behavior. Young leaves and flowers tend to have higher levels of allelochemicals than older leaves, stems, shoots, or roots; apical and medial parts of leaves tend to have higher levels than basal or edge parts (Schoonhoven et al., 2005). The type of allelochemicals can also vary with location and leaf age. For example, in tropical trees, young leaves tend to have higher levels of alkaloids, phenolics, and terpenes, while more mature leaves have more tannins (Schoonhoven et al., 2005). In leaves of trees, tannins also tend to increase during the season (Forkner et al., 2004). Further compounding this variation, day–night cycles in photosynthesis are associated with major variation in levels of many plant secondary compounds (Schoonhoven et al., 2005).

Environmental effects also create variation in plant secondary compounds. Increasing light availability increases foliar content of condensed tannins in leaves of *Acacia* trees (Mole et al., 1988), whereas low light stimulates cyanogenic compounds in ferns (Cooper-Driver et al., 1977). Such variation can cause insects to feed preferentially on shaded or sunny parts of plants (Schoonhoven et al., 2005). Fertilization with nitrogen reduces the concentration of most carbon-only secondary compounds (Koricheva et al., 1998).

In response to herbivory, plants often ramp up levels of secondary compounds; such changes are termed induced resistance. Induced resistance is widespread in plants, and can produce major effects on insect herbivores (Karban, 2011). Genetic manipulations of tobacco plants have demonstrated the importance of the induced chemicals, and the jasmonate signaling pathways that induce them, in determining herbivore damage (Karban, 2011). Induced chemical defenses are often restricted to regions near damaged areas of plants, which further increases the heterogeneity of plant defenses (Karban, 2011). Many plants use volatile signals to coordinate their induced defenses (Karban et al., 2006), and sometimes this communication can extend to neighboring plants (Schoonhoven et al., 2005). Volatiles can also attract predators and parasitoids of insect herbivores (Turlings et al., 1995).

The effect of allelochemicals on insects depends strongly on food quality, especially on the protein:carbohydrate ratio. Tannic acid can reduce feeding, cause gut lesions, and reduce digestive efficiency in grasshoppers (Bernays, 1981). Negative effects of tannic acid increased with the imbalance of protein:carbohydrate in the diet, and no effects were discernible in diets containing the optimal protein:carbohydrate ratio (Simpson and Raubenheimer, 2001). Deterrence by tannic acid also depended on diet quality, with tannic acid being much more effective at reducing feeding in foods with a low protein:carbohydrate ratio (Behmer et al., 2002).

Plant defensive compounds may be used by insect herbivores to self-medicate. Janzen (1978) suggested that plant secondary metabolites could be used medicinally by herbivores in response to parasites or pathogens, but few experimental data have been presented on this topic for invertebrates. Singer, Mace, and Bernays (Singer et al., 2009) used artificial diets and direct manipulation of parasite load (tachinid flies) to demonstrate that ingestion of pyrrolizidine alkaloids helped woolly bear caterpillars (*Grammia incorrupta*) survive parasites. Pyrrolizidine alkaloids in the diet reduced survival of unparasitized larvae, but improved survival in parasitized larvae. Alkaloids also reduced the number of flies emerging from successfully attacked caterpillars (Fig. 5.9). However, the evidence that caterpillars self-medicate was mixed. In binary food choice tests, surviving caterpillars parasitized with two eggs (but not zero, one, or three eggs) consumed more of the alkaloid-containing food than those that died, and parasitized caterpillars consumed more of cellulose disks treated with alkaloid than did unparasitized caterpillars. However, unparasitized caterpillars consumed the same fraction of diet with the alkaloid as those parasitized with one or three eggs, despite the negative effects of alkaloids on survival. Together, these results suggest that medicinal benefits of alkaloids could select for preference for foods with alkaloids.

In supporting results, Manson et al. examined the effects of alkaloids in nectar on bumblebees (Manson et al., 2010). They found that gelsemine, a nectar

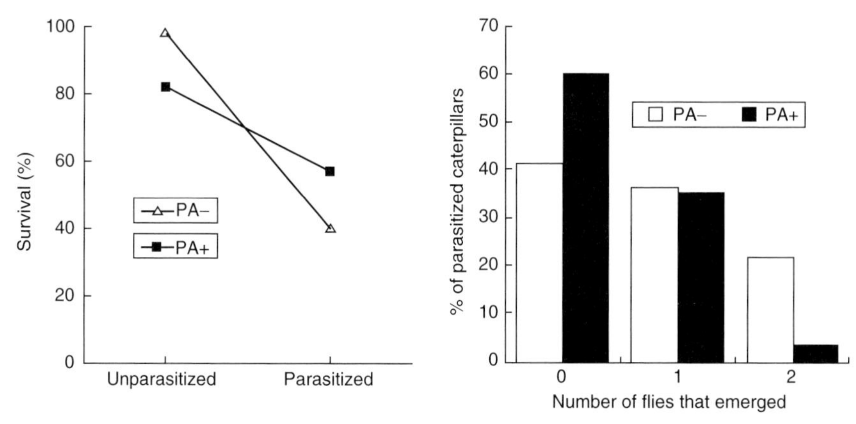

Fig. 5.9 Left: Survival of woolly bear caterpillars (*Grammia incorrupta*) fed pyrrolizidine alkaloids (solid squares) or not, either parasitized by tachinid flies or not. Right: Consumption of alkaloids reduced the number of tachinid flies emerging from the caterpillars. Redrawn with permission from Singer et al. (2009).

alkaloid of the bee-pollinated plant *Gelsemium sempervirens*, reduced pathogen loads in bumble bees infected with the gut protozoan *Crithidia bombi*. The mechanisms of the medicinal benefits of alkaloids, and the nonlinear effects of parasitism on feeding behavior, remain to be elucidated. These exciting results raise the possibility that self-medication in response to parasites or pathogens may be a widespread phenomena in insects.

From the plant perspective, making allelochemicals can carry both costs and benefits. Feeny (1976) suggested that investment in allelochemicals should be greater in "apparent" (large, long-lived, common species) plants than in less apparent (rarer, smaller, or more ephemeral) species. Allocation to chemical defenses appears to be especially high in plants that grow slowly because of limited resources (Fine et al., 2004). Quantitative genetic studies have established that secondary compounds are heritable, that herbivores do indeed exert selection for defense, and that negative genetic correlations often exist between investment in chemical defenses and other plant functions (Mauricio and Rausher, 1997). Another important concept is that plants may adapt to herbivory not only by "resistance" to herbivores (preventing or minimizing attack), but also by "tolerating" of tissue damage, based on an ability to regrow or reproduce (Núñez-Farfán et al., 2007).

5.1.5 The Role of Endo-microbes

Like all animals, insects associate with a diversity of microbes, which can compete for food, attack their tissues, make essential nutrients, and manufacture compounds used in competitive interactions (Dillon and Dillon, 2004). A significant number of microbes are symbionts, which are long-term associations which benefit both the microbes and insects. Symbionts are sources of novel traits, and represent a type of biological alliance (Douglas, 2010). Many of the specialized feeding strategies of insects are possible only due to symbiotic alliances with microbes. This is a large, rapidly expanding area of research. Two recent books and two reviews summarize substantial material on insect nutritional symbioses (Bourtzis and T.A., 2003; Dillon and Dillon, 2004; Douglas, 2009, 2010; Lemos and Terra, 1991). A few of the many interesting and important roles of symbionts in insect nutrition are covered below in the sections related to specific diets. Symbiotic microbes perform many critical functions for insects; the best documented of these are making essential amino acids and vitamins, and digesting cellulose.

Remarkable examples of an obligate insect–microbe symbiosis are mycetocytes (also called bacteriomes), which are found in seven insect orders and several non-insect arthropods (Douglas, 2009). The microorganisms (sometimes multiple species) are intracellular, sometimes living within special structures, and are transmitted from mother to offspring, usually in the egg cytoplasm. Many insects that feed on plant sap (plant-hoppers, aphids, whiteflies, psyllids, scale insects) have bacteriomes, and the mycetocytes appear to provide essential amino acids and vitamins (Douglas, 2010). Mycetocytes have been hypothesized to produce B vitamins for blood-feeders (bedbugs, triatomine bugs, lice) and to be involved in recycling nitrogen from uric acid in termites and in some generalist feeders (cockroaches, (Douglas, 1989, 2007; Mullins and Cochran, 1974, 1975)).The mycetocyte bacteria have reduced genomes and cannot be cultured outside of the insect; conversely, antibiotics that destroy the mycetocyte bacteria usually render the insects sterile (Douglas, 2007).

The best-known mycetocyte symbiosis is between aphids and the bacteria *Buchnera*. The phylogeny of *Buchnera* and its aphid hosts are completely concordant, suggesting a single infection followed by radiation of both aphids and bacteria (Moran, 2007). The obligate symbionts have highly reduced but stable genomes (Douglas, 2010; Moran, 2007). Symbiotic bacteria have lost many metabolic genes, including those for fixing sulfur and for making arginine and tryptophan. The reduced metabolic capacity explains why these bacteria are obligate symbionts, and indicates that they must receive these metabolic products from the host.

Previously it was thought that co-speciation of symbionts and their hosts, and the evolution of reduced genomes in the symbionts, occurred only when the symbionts were intracellular. However, Hosokawa and colleagues (Hosokawa et al., 2006) demonstrated in stinkbugs that mothers pass a gut Proteobacteria to their offspring by depositing a capsule of bacteria with each egg. If the bacteria were removed by antibiotics, the insects were sterile and suffered reduced growth.

These bacteria are closely related to *Buchnera*, the intracellular endosymbionts of aphids. Host and symbiont phylogenies matched perfectly, indicating strict co-speciation (Hosokawa et al., 2006).

Wolbachia are maternally inherited bacteria that commonly spread through host populations by causing cytoplasmic incompatibility (fecundity is strongly reduced when uninfected females mate with infected males). As a consequence of *Wolbachia* infection, infected females are usually less fecund. However, *Wolbachia* in California *D. simulans* evolved over two decades such that infected females are more fecund than uninfected flies, suggesting that *Wolbachia* may now provide some, as yet unknown, nutritional advantage for *D. simulans* (Weeks et al., 2007).

Gut microorganisms change rapidly in response to the insect diet by induction of enzymes and population changes in the microbial community (Dillon and Dillon, 2004). When cockroaches were switched to a low-protein, high-fiber diet, the number of streptococci and lactobacilli inhabiting the foregut and production of lactate and acetate decreased (Kane and Breznak, 1991). A diet rich in cellulose induced increased cellulase activity, methane production, and protozoal populations in the hindgut of the American cockroach, *Periplaneta americana* (Gijzen et al., 1994). When crickets were fed diets with higher protein content, there were changes in the microbial community (detected by community nucleic acids) and a decrease in the production of volatile fatty acids (Santo Domingo et al., 1998), suggesting that microbial release of volatile carbons could be a component of compensation for high carbohydrate diets.

5.2 Behavioral Responses to Food

5.2.1 Finding and Choosing Food

As described above, insect foods are incredibly heterogeneous, and fitness often will depend on an insect being able to find prey, plants, and regions within plants that satisfy their nutritional needs. Flying insects, or insects moving through landscapes (e.g. many insects pupate in the ground and then must locate their food plant as adults), locate their food using visual and olfactory cues. Many insects have excellent vision (caterpillars are one obvious exception), and many have been shown to use shape and color to identify food. For example, adult lepidopterans and aphids have been shown to respond to green colors and/or leaf-like shapes (Schoonhoven et al., 2005). In addition, insects are famously sensitive to olfactory cues, including those of food. Just as we can smell a well-cooked meal, insects respond to the gaseous emissions of their food plants. Plants emit a variety of volatiles (think of cut-grass smells, or the distinctive odor of a pine tree). These volatiles are often similar or identical to chemicals that can serve as feeding deterrents (Schoonhoven et al., 2005). Insects often exhibit upwind flight, tracking plumes of odors to their host plants.

Once on a host plant, insects then choose among leaves and sections within leaves. The mechanisms and specific behaviors vary with the system and the size of the insect but, in general, olfaction and contact chemoreception are believed to be very important. Many insects have chemoreceptors in their mouthparts and on their tarsi that permit them to assess food quality (Chapman, 2003). Special behaviors assist in chemo-analysis of the food quality: many insects make small "test bites" apparently to taste the leaves; grasshoppers vibrate their palps at 10 Hz, increasing surface area of exposure and response of the sensilla (Blaney and Simmonds, 1990). Biting can be either stimulated or inhibited by surface waxes, and many insects can be stimulated to bite by exposure to the waxes of normal host plants (Chapman, 1990).

The ability of insects to attain specific intake targets for multiple nutrients indicates that they can assess many aspects of food quality, including the contents of multiple nutrients. This could occur either by taste, or by post-ingestive mechanisms—whereby changes in body parameters such as protein or lipid levels are sensed and this information is somehow coupled to the neural circuits that control feeding. There is evidence for both types of mechanisms in insects; taste will be covered in this section.

Common hexose and disaccaride sugars stimulate taste receptors in a dose-dependent manner in a variety of insects (Chapman, 1990, 1992; Schoonhoven et al., 2005). These sugars are often at relatively high concentrations (2–10% by dry weight) in green leaves, and even higher in fruits and nectar (Schoonhoven et al., 2005). Although protein content is important for growth, there is little evidence that protein itself stimulates insect taste receptors. Amino acids, sugar alcohols, nucleotides, starches, some phospholipids, and some minerals and vitamins can be stimulating, while salts and a variety of plant secondary compounds can be aversive (Bernays and Simpson, 1982; Chapman, 1990, 2003; Schoonhoven et al., 2005; Simpson and Abisgold, 1985).

For many herbivorous insects, plant secondary compounds are critical in stimulating or deterring feeding (Bernays and Chapman, 2000). A vast literature has demonstrated that plant secondary chemicals can deter insects from feeding (Morgan and Mandava, 1990). These include sinigrin (a glucosinolate), linamarin (a cyanogenic glycoside), chlorogenic acid (a phenolic acid), phlorizin (a flavonoid), strychnine (an alkaloid), caffeine (alkaloid), ajugarin (a diterpenoid), and azadirachtin (a triterpenoid) (Schoonhoven et al., 2005). On the other hand, for many herbivores, plant secondary compounds serve as "token stimuli" for feeding or oviposition (Schoonhoven et al., 2005). Plants often contain both stimulants and repellents of feeding, and the response often depends on the balance of these effects (Bernays and Chapman, 2000; Chapman, 2003).

The ability to respond via taste or olfaction to these compounds is conferred by odor and taste receptors, which in both insects and mammals are seven transmembrane-domain receptors encoded by highly diverse gene families. Much of our knowledge of the genetics and function of these receptors comes from work with *D. melanogaster*, interested readers are urged to consult recent reviews (Hallem et al., 2006; Vosshall and Stocker, 2007).

5.2.1.1 Olfactory Receptors

In many insects, olfactory receptor neurons are found in pairs of olfactory organs, the antennae and the maxillary palps. The surfaces of the olfactory organs are covered with sensory hairs called sensilla, which contain the olfactory receptor neuron dendrites (Fig. 5.10). The structure of the olfactory sensillum usually consists of a cuticular wall containing multiple pores through which odors can enter (Blaney and Simmonds, 1990; Shanbhag et al., 2000). Olfactory sensilla typically contain the dendrites of between one and five olfactory receptor neurons (Hallem et al., 2006). The axons of the olfactory receptor neurons project to functional processing units called glomeruli in the antennal lobes of the brain (Galizia and Rössler, 2010; Hildebrand and Shepherd, 1997). In addition to olfactory receptor neurons, many insect olfactory organs also contain smaller

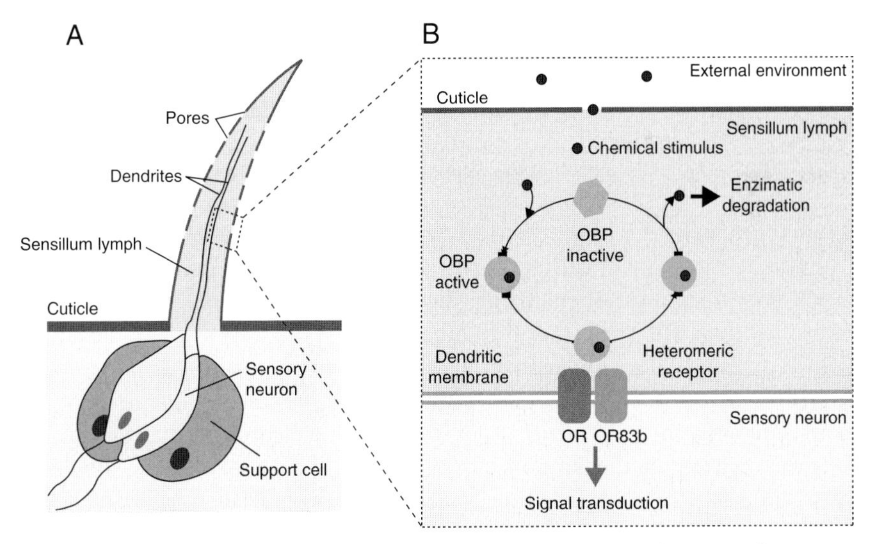

Fig. 5.10 (A) Schematic of an insect olfactory hair. Gustatory sensilla have a similar structure, with only a single pore at the top of the sensory hair. (B) The first molecular steps of the insect chemosensory signal-transduction pathway. Chemicals dissolve in the lymph and then bind to the odorant binding protein (OBP), and then activate the odorant receptor (OR). Enzymes degrade and remove the chemical from the OBP, allowing repeated stimulations. From Sanchez-Gracia et al. (2009), with permission from Nature Publishing Group.

numbers of mechanosensory, thermosensory, hygrosensory, and gustatory neurons (Hildebrand and Shepherd, 1997).

Odor information is coded in the insect brain in a sequence of steps, ranging from the receptor cells, via the neural network in the antennal lobe, to higher order brain centers, especially the mushroom bodies and the lateral horn. Across all of these processing steps, coding logic is combinatorial, in the sense that information is represented as patterns of activity across a population of neurons, rather than in individual neurons. Because different neurons are located in different places, such a coding logic is often termed spatial, and can be visualized with optical imaging techniques such as *in vivo* calcium imaging (Galizia and Szyszka, 2008).

Electrophysiological studies have permitted examination of the responses of single insect olfactory receptor neurons to odors. Such recordings from many insect species including moths, honeybees, mosquitoes, and flies have revealed that different receptor neurons respond to different odors with different response dynamics (de Bruyne et al., 2001; Laurent et al., 2002; Stensmyr et al., 2003; van den Broek and den Otter, 1999). In *Manduca*, one subset of receptor cells exhibited an apparently narrow molecular receptive range, responding strongly to only one or two terpenoid odorants. The second subset was activated exclusively by aromatics, while a third had a broad molecular receptive range and responded strongly to odorants belonging to multiple chemical classes (Shields and Hildebrand, 2000).

In *D. melanogaster*, about 60 Odor receptor (*Or*) genes have been identified that encode 62 Or proteins by alternative splicing (Robertson et al., 2003). The Or proteins are diverse, with many sharing only 20% amino acid similarity. The *D. melanogaster* Or genes do not share sequence similarity with mammalian odor receptor genes. The *Or* genes are widely distributed throughout the genome, and many are found in small clusters of two or three genes (Robertson et al., 2003). Genes within a cluster often share a higher degree of sequence similarity with each other than with the rest of the *Or* genes, suggesting that some of the ancestral *Or* genes may have undergone recent duplication events to give rise to clusters of *Or* genes. Each olfactory receptor neuron expresses only one or a few *Or* genes, primarily in the dendrites. Olfactory receptor neurons expressing similar *Or* genes converge on the same olfactory glomeruli in the antennal lobe (Bhalerao et al., 2003), a similar organization as seen in mammals (Hallem et al., 2006).

In *Drosophila*, it is possible to manipulate the levels of the various Or proteins via gene dilution or overexpression, and these studies have revealed the functional properties of a variety of these *Or* genes, including sensors for cyclohexanone, cyclohexanol, benzaldehyde, and benzyl alcohol—all of which are found in fruits and other natural odor sources (Thorne et al., 2004).

Both olfactory and gustatory organs also express odorant-binding proteins, which are secreted proteins that function in the lymph fluid that fills the sensilla lumen (Galindo and Smith, 2001). Odorant binding proteins may chaperone ligands through the lymph to neuron dendrites where they can be bound by odorant or gustatory receptors. Odorant binding proteins may also concentrate odorant molecules in the sensillum lymph, remove or detoxify odorants, and co-initiate signal transduction with the receptor molecules (Hallem et al., 2006).

5.2.1.2 Taste Receptors

Taste receptors generally come into contact with the stimulus in a solid or liquid form. Typically, contact occurs via a single terminal pore in the receptor (Blaney and Simmonds, 1990; Chapman, 2003). Taste sensilla occur as taste hairs, bristles, or pegs on the palps (Fig. 5.11). Taste hairs are also present on the surface of the tarsi, the wing margins, on the ovipositor of females, and on the internal mouthparts. Larvae have fewer taste sensilla, and, these are located in several organs near the mouthparts, and within the pharynx (Fig. 5.12) (Schoonhoven et al., 2005; Vosshall and Stocker, 2007). Larvae also appear to have more gustatory than odor receptor cells, in contrast to adults (Vosshall and Stocker, 2007). Typically, each taste hair is innervated by unbranched dendrites of multiple chemosensory neurons as well as a single mechanosensory neuron (Falk et al., 1976). Over 15,000 gustatory sensilla occur on adult *Locusta* grasshopper mouthparts (Chapman, 1982).

Gustatory receptor neurons have been analyzed by the tip-recording technique, which involves making contact with the pore at the tip of the sensillum with a solution containing an electrolyte as well as the taste stimulus (Morita and Shirais, 1968). Experiments with various taste stimuli in flies have revealed the presence of four types of neurons—a sugar-sensitive neuron, a water-sensitive neuron, a neuron sensitive to low concentrations of salts, and a neuron sensitive to high concentrations of salts (Dethier, 1976; Shanbhag and Singh, 1992). The low-salt neuron also responds to bitter aversive tastes such as monoterpenes (Ozaki et al., 2003. Salt reception is thought to function via amiloride-sensitive DEG/ENaC sodium channels. Two *D. melanogaster* DEG/ENaC genes, Pickpocket11 and Pickpocket19, are expressed in taste sensilla, and disruption of these genes results in a diminished behavioral response to salt but not to sucrose (Liu, 2003). In locusts, caterpillars and some beetles, sensilla sensitive to amino acids, gamma aminobutyric acid, and sugar alcohols (inositol) have also been identified.

Approximately 68 gustatory receptor genes (*Gr*) have been identified in *Drosophila*. Gr proteins are extraordinarily divergent in sequence, sharing as little as 8% amino acid identity (Clyne et al., 2000; Scott et al., 2001). Phylogenetic analysis suggests that the *Gr* gene family is an ancient family of chemoreceptors from which a branch of *Or* genes subsequently evolved (Scott et al., 2001).

Specific *Gr* genes define the functional properties of taste receptors, but there is no one-to-one matching of genes to neurons as occurs in olfaction. Unlike olfactory

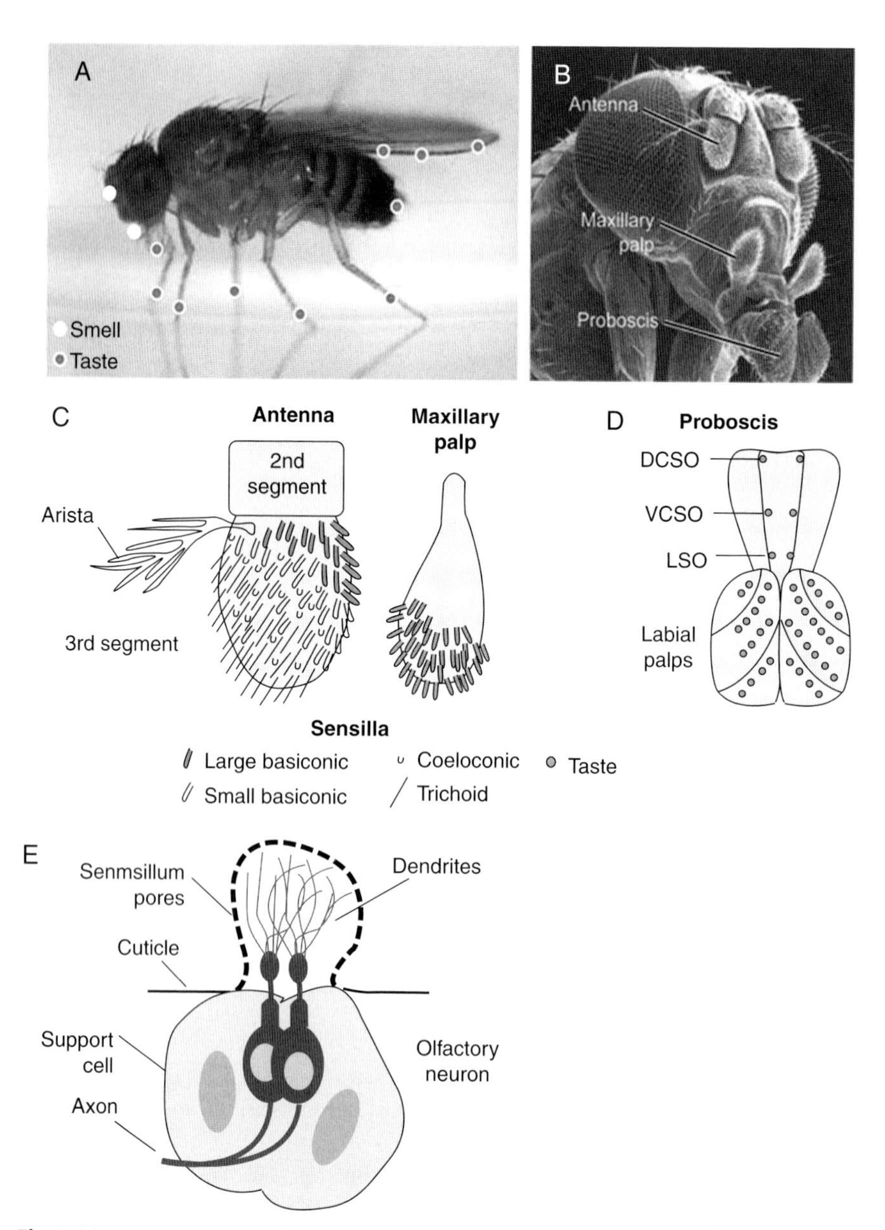

Fig. 5.11 Neuroanatomy of the peripheral chemosensory system of a fly. (*A*) Schematic indicating the position of olfactory (*white circle*) and gustatory (*dark circle*) neurons on the body of the fly. (*B*) Scanning electron micrograph of a fly head, indicating the major chemosensory organs. (*C*) Schematic of the exterior surface of the olfactory organs. (*D*) Schematic of the proboscis. Abbreviations: LSO, labral sense organ; VCSO, ventral cibarial sense organ; DCSO, dorsal cibarial sense organ. (*E*) Schematic of a typical olfactory sensillum, housing two Olfactory Receptor Neurons. From Vosshall and Stocker (2007), with permission from the *Annual Review of Neuroscience*, Volume 30 © 2007 by Annual Reviews www.annualreviews.org.

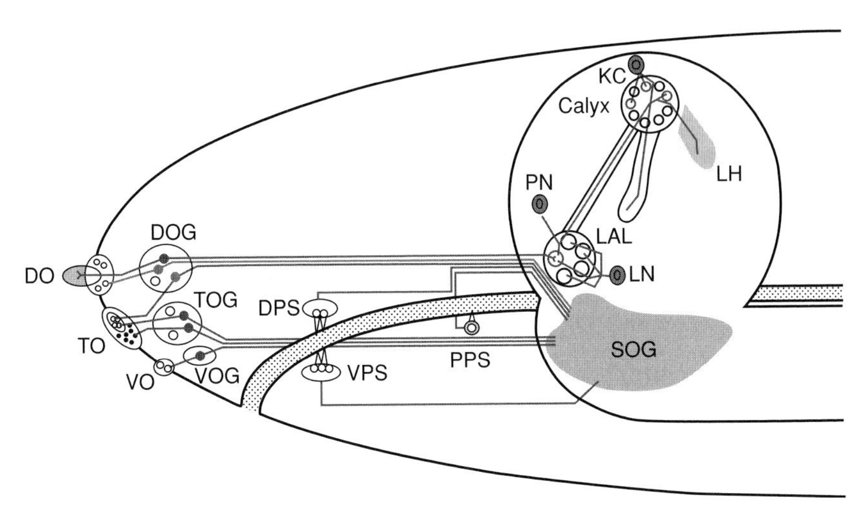

Fig. 5.12 Neuroanatomy of the larval chemosensory system of *Drosophila*. From the three external chemosensory organs, the mixed dorsal organ (DO) comprises the olfactory "dome" (*gray*) and a few putative taste sensilla (*small circles*). The terminal organ (TO), the ventral organ (VO), and the dorsal, ventral, and posterior pharyngeal sense organs (DPS, VPS, PPS) include mainly taste sensilla. The sensory neurons' cell bodies are collected in ganglia below each sense organ (DOG, TOG, VOG). Some neurons innervating the TO are located in the DOG. Olfactory receptor neurons (ORN) project into individual glomeruli of the larval antennal lobe (LAL), which are interconnected by local interneurons (LN). Projection neurons (PNs) link the LAL with two higher olfactory centers, the mushroom body (MB) calyx and the lateral horn (LH). Other abbreviations: KC: Kenyon cell; GRN: afferents. The pharynx is shown stippled. From Vosshall and Stocker (2007), with permission from the *Annual Review of Neuroscience*, Volume 30 © 2007 by Annual Reviews www.annualreviews.org.

receptor neurons, gustatory receptor neurons are more likely to be "generalists" (Blaney and Simmonds, 1990) and to express multiple *Gr* genes (Hallem et al., 2006). In mutant *D. melanogaster* that lack Gr5a, sugar neurons show a severely reduced trehalose response but a normal sucrose response, suggesting that Gr5a mediates the response to trehalose, and a second receptor co-expressed in these neurons mediates the response to sucrose (Dahanukar et al., 2001). Gr5a shows a dose-response to trehalose, but does not respond to other sugars (8). The gene *Gr66a* seems to define a population of bitter-sensitive neurons (Thorne et al., 2004).

Although specific receptors for token stimuli (secondary compounds) have yet to be identified, specific sensilla for this task have been identified in some insects. For example, in the butterfly, *Pieris brassicae*, two cells respond to glucosinolates, the primary stimulus for feeding for this species (Schoonhoven, 2005).

Similarly, specific sensilla that respond to feeding deterrents have been identified, but not yet the individual receptors. Some of these cells are specialized while others respond to a wide range of unrelated deterrent compounds. For example, *Pieris brassicae* and *P. rapae* caterpillars have specialized deterrent cells that respond to very low concentrations of cardenolides (a steroid), and more generalized deterrent cells that respond to higher concentrations of phenolic acids and flavonoids (Van Loon and Schoonhoven, 1999).

How do insects assess concentrations of nutrients? Many taste receptors exhibit concentration-dependent responses, with increased changes in membrane polarization, and modulation of action potential frequency in downstream neurons (Blaney and Simmonds, 1990).

A major region for integrating information about gustation is the subesophageal ganglion (SOG). Gustatory receptor neurons from different peripheral tissues project to different regions of the SOG and tritocerebrum in the brain (Hallem et al., 2006). Different classes of gustatory receptor neurons or gustatory neurons from different regions project to non-overlapping regions of the SOG (Thorne et al., 2004; Wang et al., 2004).

Functional changes in taste sensilla are associated with feeding behavior. At the end of a meal, pores at the tips of gustatory sensilla may close, reducing their sensitivity to stimuli (Bernays et al., 1972). Gustatory sensilla are also less sensitive after meals high in protein, or when levels of amino acids or hemolymph osmolality are elevated (Abisgold and Simpson, 1988; Simpson and Raubenheimer, 1993). Similarly, the sensitivity of sucrose detectors in antennae and tarsi decrease following a nectar meal and then later rise (de Brito Sanchez et al., 2008).

The number of sensilla on the palps and antennae correlate positively with the complexity of the diet on which the insect is reared (Bernays and Chapman, 1998). For example, locusts reared on two complementary foods had 20% more chemosensilla on the maxillary palps than locusts reared on a single optimal diet (Opstad et al., 2004). The locusts reared on two foods also had shorter latencies to feed and were more likely to reject test foods, suggesting that the variation in chemosensilla number is functionally relevant to food choice behavior.

5.2.2 Behavioral Responses of Insects Feeding on Single Foods

When feeding on single-food diets, many insects eat more when dietary protein levels drop, allowing them to obtain sufficient protein for growth (Simpson and Abisgold, 1985; Woods, 1999). This type of compensatory feeding probably is common, allowing insects to cope with variation in plant fiber content, which often increases with plant age. If nutrient dilution becomes too great, the capacity for the insect to consume and process food is exceeded, resulting in a drop in intake of digestible protein and a decrease in growth and survival (Simpson and Abisgold 1985; Woods 1999; Simpson et al., 2004).

5.2.3 Foraging by Social Insects

Social insects are interesting for the study of nutrition and growth as the sterile workers obtain most of the nutrients for the growing larvae. Thus, foraging workers generally select foods that serve the colony's needs rather than their own. There is a vast and fascinating literature on foraging by social insects, and only a few aspects will be covered here. Interested readers are referred to excellent books and reviews on this topic (Beattie, 1985; Holldobler and Wilson, 1990, 2011; Hunt and Nalepa, 1994; von Frisch, 1993).

When given choices among solutions of different amounts of sugars, the ant *Rhytidoponera metallica* forages so that it approaches an intake target for carbohydrate (Dussutour and Simpson, 2008). Free-living ant colonies select among foods differing in protein and carbohydrate content as predicted by a geometric framework. In particular, prior exposure to protein reduces recruitment to protein-rich foods, consistent with nutrient balancing (Christensen et al., 2010). Fire ants prefer relatively balanced carbohydrate:protein foods (Cook et al., 2010). These ants create hoards of collected food, and then extract carbohydrate from these hoards using extracellular digestion (Cook et al., 2010).

Honeybees are known to use multiple cues (visual, olfactory) to find flowers. Honeybees can also learn to associate traits like color, shape, or odor with sugar rewards. When exposed to nectar of different concentrations, there is a threshold level of sugar that elicits proboscis extension. Higher levels of sucrose promote better associative learning of colors (Grüter et al., 2011).

Honeybee foragers tend to specialize on either pollen or nectar, and this preference has a genetic basis (Page and Robinson, 1991). These foraging preferences are associated with differences in sucrose responsitivity, with nectar foragers requiring higher levels of sucrose to stimulate feeding than pollen foragers (Scheiner and Erber, 2009). Temperate-adapted, European honeybees tend to collect more nectar and accumulate honey to survive through the long winter, while tropical-adapted African honey bees tend to collect more pollen, allowing faster population growth (Winston, 1992).

5.3 Post-ingestive Responses of Individual Insects to Food

5.3.1 Responses to Plant Physical Defenses

Leaf toughness has been found to be the best predictor of interspecific variation in rates of feeding (Coley, 1983). Tough plants abrade the mandibles, which can reduce feeding efficiency. Plants can magnify the rate of abrasion by incorporating silica, which makes up to 15% of dry weight of grasses, sedges, and horsetails (Schoonhoven et al., 2005). Increasing the silica content of grasses by silica fertilization caused irreversible wear in the mandibles of the caterpillar *Spodoptera exempta*, reducing the efficiency of assimilation of ingested food and nitrogen extraction from leaves, and reducing growth rates (Massey and Hartley, 2009). As a countermeasure, some insects can harden their mandibular cuticle with extra deposits of manganese and zinc (Schoonhoven et al., 2005). When fed hard grasses, caterpillars (*Pseudaletia unipuncta*) can have head capsules up to twice as large as similarly-sized individuals feeding on soft diets, with commensurate increases in chewing power (Bernays, 1986). The morphology of the mandibles changes developmentally in some caterpillars that feed on woody leaves, changing from toothed mandibles that may aid feeding between leaf veins to incisor-like mandibles that allow vein-cutting (Bernays, 1986).

Clissold et al. recently tested the effects of plant toughness by comparing growth and assimilation of protein and carbohydrate for grasshoppers fed on a turgid Mitchell grass, dried Mitchell grass (higher toughness), and finely chopped Mitchell grass (low toughness). Grasshoppers grew slower, had prolonged development, and reached smaller sizes on the tougher grass (Fig. 5.13). This was primarily due to smaller meals, longer gut passage times, and reduced assimilation of both protein and carbohydrate on the tougher food (Clissold et al., 2009).

Differences in the structure of grasses appear to be critical in determining whether the Australian plague locust *Chortoicetes terminifera* exhibits outbreaks (Clissold et al., 2006). Populations of *C. terminifera* expand after rainfall, as animals consume an annual grass *Dactyloctenium radulans* and a perennial grass *Astrebla lappacea*. The annual, *D. radulans*, is a high quality food, supporting

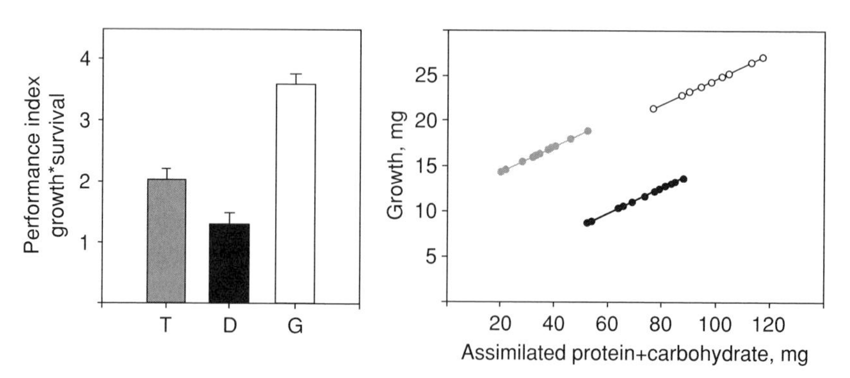

Fig. 5.13 Grasshoppers (*Chortoicetes terminifera* nymphs) show reduced growth and survival on tougher grasses (left), and grow less in relation to the amount of protein plus carbohydrate eaten (right). All animals were fed *ad lib* on Mitchell grass, which was either turgid (T, gray), dried (D, high toughness, black), or ground finely (G, low toughness, open bars/symbols. From Clissold et al. (2009), with permission from Ecological Society of America.

greater consumption and survival, faster growth, and larger body sizes. A key finding was that *C. terminifera* assimilated twice the carbohydrate (per unit carbohydrate consumed) when eating *D. radulans*. Differences in growth and assimilation of *C. terminifera* on the two plants were eliminated if plants were dried and powdered, indicating that the anatomical structure (toughness?) of *A. lappacea* was critical in reducing carbohydrate assimilation and grasshopper growth.

5.3.2 Post-ingestive Compensation for Unbalanced Food

When insects consume diets with unbalanced protein:carbohydrate ratios, they may not be able to attain their intake target by mixing foods. In this situation, insects often over-feed, leading to excess intake of the common macronutrient (carbohydrate or protein). To compensate for eating imbalanced foods, insects may alter strategies of digestion or absorption, store the excess nutrient, funnel different substrates into metabolic networks, and secrete excesses via the renal or exocrine glands (Fig. 5.1).

Generally in mammals, 100% of ingested protein is digested and absorbed, with all regulation occurring post-absorptively. In contrast, insects seem able to incompletely digest and absorb nutrients when dietary contents are high. When the caterpillar *Heliothis virescens* was fed on diets with widely varying protein:carbohydrate ratios, similar growth and incorporation of nitrogen and carbohydrate into tissues occurred despite widely varying protein and carbohydrate ingestion (Telang et al., 2001). At low dietary concentrations of carbohydrate, less than 10% was excreted. However, as carbohydrate intake increased, increasing fractions of it appeared in the feces (Fig. 5.14), suggesting that it wasn't digested or absorbed (or it was absorbed and secreted via the renal system–which seems unlikely). Similar patterns were observed for protein; high dietary protein levels were associated with increased fecal excretion of protein and amino acids (Fig. 5.14). Similarly, *Oncopeltus fasciatus*, which feeds on seeds that contain large amounts of triglycerrides, regularly excretes large oil droplets (Woodring et al., 2007).

Telang and colleagues also provided strong evidence for post-absorptive compensation. As carbohydrate intake increased, an increasing fraction of it appeared in neither the tissues nor the frass (Fig. 5.14), strongly suggesting that caterpillars burned it off with high metabolic rates. As protein intake increased, uric acid excretion increased dramatically, consistent with deamination of absorbed, unassimilated amino acids (Fig. 5.14). Quantitatively, the largest response to increased protein intake (representing up to 50% of the total N budget) was the increased appearance of unidentified nitrogen compounds in the feces (Fig. 5.14). One possibility is that this was mostly ammonium, resulting from deamination of absorbed amino acids and subsequent catabolism of the formed carbohydrates (Telang, et al., 2003), again emphasizing the important role of increased metabolic rate in coping with imbalanced diet.

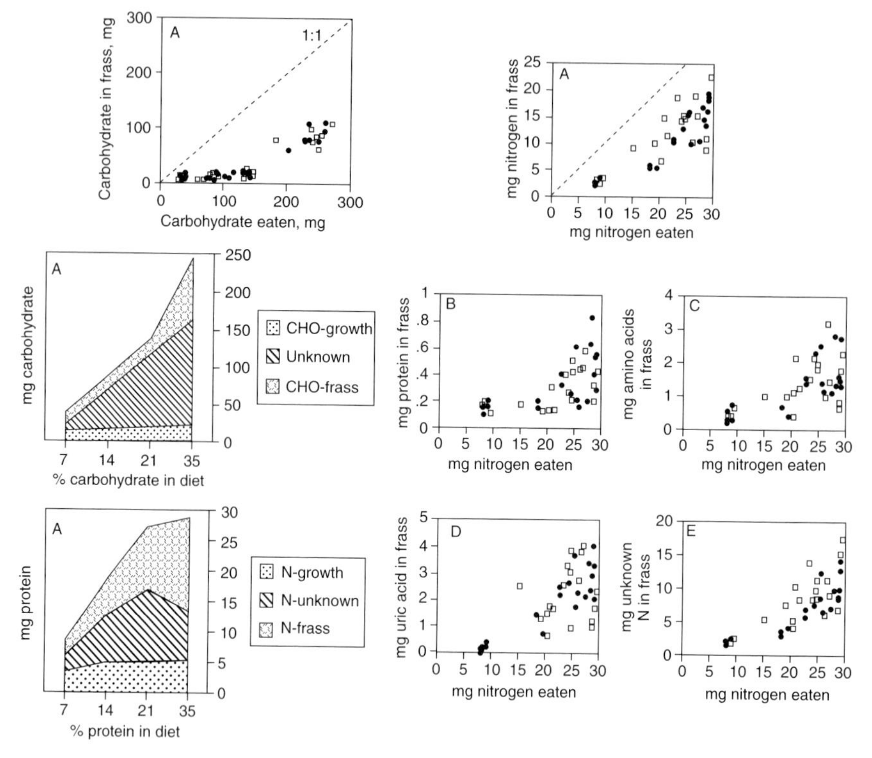

Fig. 5.14 Post-ingestive responses of the caterpillar *Heliothis virescens* (Telang et al., 2003). Top left: On low-carbohydrate diets, virtually all carbohydrate is digested and absorbed, but with excessive consumption, large amounts are excreted (circles: females; squares: males). Lower left: Amount of carbohydrate and protein incorporated into tissues (growth), appearing in the frass, or being unmeasured, plotted vs. the percentage of carbohydrate or protein in the diet. Unknown carbohydrate is likely respired. Unknown nitrogen may be lost as unmeasured ammonia/ammonium in the frass. Right box: As the protein eaten increases, caterpillars excrete an increasing amount of protein, amino acids, uric acid, and other nitrogen. From Telang et al (2003), with permission from University of Chicago Press.

Compensatory post-ingestive responses to imbalanced diets also occur in locusts. In locusts, there is evidence that high caloric intake stimulates metabolic rates, allowing disposal of excess carbohydrate intake (Zanotto et al., 1993). Measurements of gas exchange revealed that locusts fed high-carbohydrate (low protein:carbohydrate ratio) diets had about 25% higher rates of CO_2 emission than animals fed balanced diets (Zanotto et al., 1997). These differences could be driven in part by changes in respiratory quotient (not measured), as it seems possible that locusts shift from catabolizing fat and amino acids toward catabolizing carbohydrate when eating diets with excess carbohydrate. Metabolic rates definitely increased on the high carbohydrate diets, but the mechanisms responsible remain unclear (Zanotto et al., 1997). Possibilities include increased activity, increased costs of absorption/assimilation, and various forms of futile cycles (e.g. increased mitochondrial proton leak or increased substrate cycles such as inter-conversions between glucose and trehalose).

Nutrient storage may also be a common response to excess intake. *Heliothis virescens* caterpillars accumulated higher levels of storage proteins as they consumed more protein, potentially benefitting adult reproduction (Telang et al., 2002). As predicted by the geometric framework, compensatory feeding of insects on diets with low dietary protein caused storage of excess lipid (Fig. 5.15). Thus, differences in the protein:carbohydrate content of diets lead to changes in body composition, with an inverse relationship between body protein and lipid stores (Karowe and Martin, 1989).

The energetic demands on males and females often differ, with females usually having higher total nutrient requirements, especially protein, due to the demands of oogenesis. Not surprisingly, juvenile females often show strategies to accumulate more nutrients, including extra juvenile molts, higher feeding rates, and longer development times (Slansky and Scriber, 1985). Testing within the geometric framework has demonstrated that female *Heliothis virescens* caterpillars had higher protein intake targets than males; they ate more than males, but only if the diet was high in protein. The sexes also differed in post-ingestive responses to diet, with females utilizing protein more efficiently than males (Telang et al., 2003).

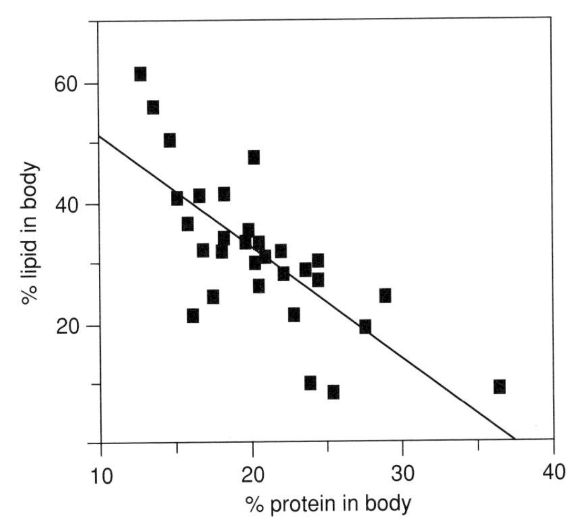

Fig. 5.15 Protein and lipid content of the body varied inversely in caterpillars of *Spodoptera eridania* fed diets varying in protein content. Caterpillars consuming diets lower in protein were fatter. From Karowe and Martin (1989), with permission from Elsevier.

5.3.3 Post-ingestive Responses to Plant Allelochemicals

As described above, many plants, particularly trees and forbs, defend themselves with a variety of secondary allelochemicals that can deter feeding, interfere with digestion or absorption, or poison tissues directly. How do insects survive and grow in the face of this chemical onslaught?

Plants defend themselves with both standing and induced defenses. Wounding of plant tissue by insect feeding induces secretion of plant hormones such as jasmonic acid (JA), which have a variety of defensive effects, including the production of defensive chemicals. Some of these allelochemicals can have dramatic effects; for example, the L-quisqualic acid in zonal geraniums causes paralysis in beetles for several hours by blocking the action of the excitatory neurotransmitter, L-glutamate (Ranger et al., 2011).

One approach for herbivores is to interfere with the induction of plant defensive compounds. Saliva from caterpillars such as *Helocoverpa zea* contains glucose oxidase (Musser et al., 2005; Musser et al., 2002). This saliva reduces nicotine levels in tobacco leaves, and improves the growth of caterpillars feeding on those leaves (Fig. 5.16). Caterpillars that are experimentally blocked from secreting glucose oxidase have lower survival and growth (Musser et al. 2005), and experimental application of glucose oxidase to these leaves reduces leaf nicotine levels, suggesting this enzyme is interfering with allelochemicals induction. However, the exact mechanisms by which glucose oxidase inhibits nicotine production in the plant remain to be determined.

Another approach to overcoming plant defenses is to interfere with their mode of action in the gut. A lovely example of such a defensive strategy is revealed by Konno et al.'s studies (e.g. Konno et al., 2010) of the relationships between privet leaves (*Ligustrum obtusifolium* (Oleaceae)) and specialist caterpillars that feed on them. Privet leaves contain up to 3% oleuropein, a phenolic secoiridoid. When leaves are damaged by herbivory, β-glucosidase and polyphenol oxidase from the plant tissues activate oleuropein into a strong protein denaturant (structurally related to glutaraldehyde) that can covalently bond to proteins, removing their lysine. Ten of eleven privet specialist caterpillars tested maintained high concentrations of glycine (up to 160 mmol l^{-1}), GABA (up to 60 mmol l^{-1}), or β-alanine in the gut lumen, and the concentrations of these amino acids in the gut are significantly higher than in other caterpillars that do not specialize on privet (Konno et al., 2010). An injection experiment with ^{15}N-labeled glycine showed that glycine is actively transported from hemolymph to the midgut lumen against the concentration gradient in one caterpillar species (Konno et al., 2001). Konno et al. have shown *in vitro* that any amino acid, in sufficient quantities, can protect lysine from oleuropein. Further, feeding glycine to non-specialist caterpillars prevents the decrease in lysine concentrations and allows normal growth (Konno et al., 2009). Together, these results suggest that the elevated

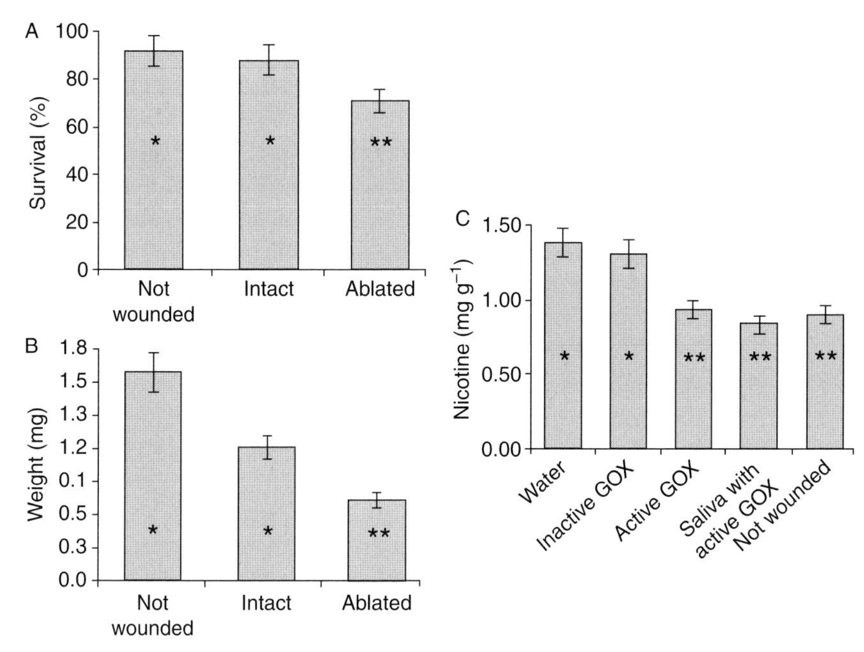

Fig. 5.16 *Helicoverpa zea* caterpillars secrete saliva via spinnerets. (A) Proportion of *Helicoverpa zea* caterpillars surviving after being reared on leaves that were either: "not wounded"—never previously exposed to caterpillars, previously fed on by caterpillars with "intact" spinnerets, or previously fed on by caterpillars with surgically ablated spinnerets. The reduced survival on plants previously fed on by caterpillars with ablated spinnerets suggests that the plants respond to feeding by inducing a defense mechanism, and that caterpillar saliva is able to reduce that defensive induction. (B) Weights of surviving caterpillars fed on leaves of the three categories. (C) Nicotine levels in wounded leaves: exposed to water, inactivated glucose oxidase enzyme (GOX), active GOX, or salivary-gland extracts containing active GOX. Nicotine levels in an intact, non-wounded leaf are shown on the right. From Musser et al. (2002), with permission from Nature Publishing Group.

levels of glycine and GABA in insect digestive juices and oleuropein in privet leaves represent the outcome of a co-evolutionary arms race (Konno et al., 2010) (Fig. 5.17).

Another defensive approach is to avoid absorbing the allelochemicals. Some protection of this kind is provided by the peritrophic membrane (Barbehenn, 2001). As described in Chapter 2, the peritrophic membrane is a sheath of chitin microfibrils, proteins, and glycoproteins that forms a layer between the ingested food and the epithelium of the alimentary canal. The peritrophic membrane is able to retain considerable amounts of tannins and allelochemicals within the gut lumen. Although individual tannin molecules are usually small enough to pass through the pores of the peritrophic membrane; tannin forms complexes that

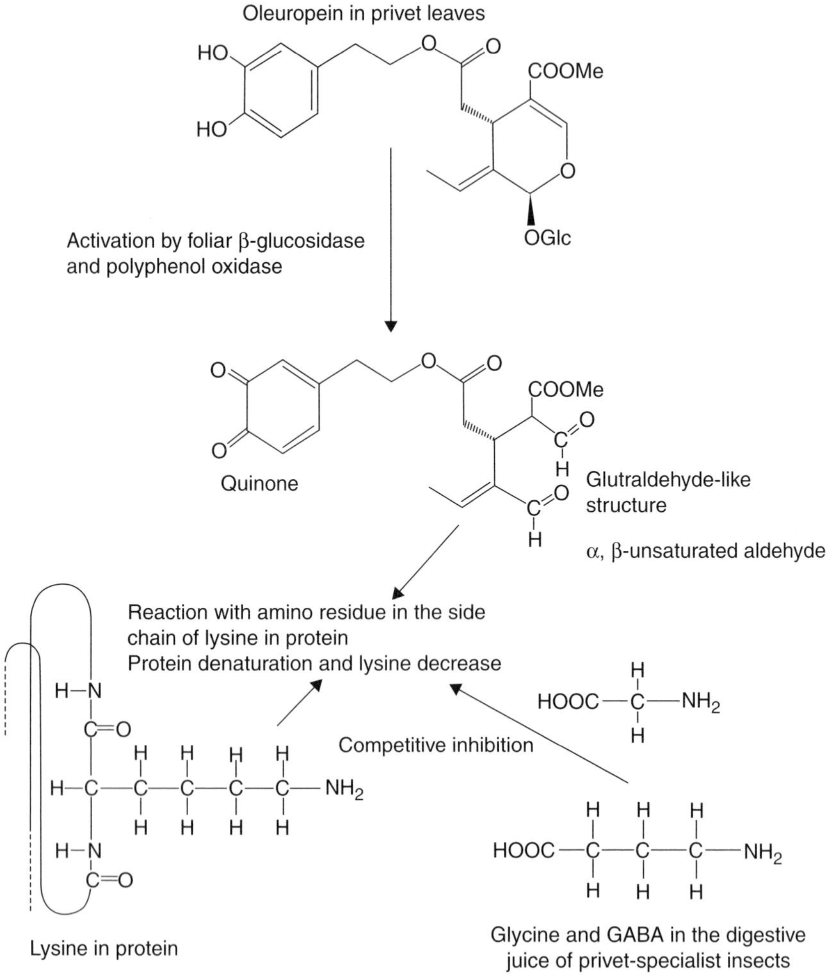

Oleuropein in privet leaves

Activation by foliar β-glucosidase and polyphenol oxidase

Quinone

Glutraldehyde-like structure

α, β-unsaturated aldehyde

Reaction with amino residue in the side chain of lysine in protein

Protein denaturation and lysine decrease

Competitive inhibition

Lysine in protein

Glycine and GABA in the digestive juice of privet-specialist insects

Fig. 5.17 Proposed model for antinutritive effects of oleuropein, a glycoside produced by privets which, when activated, can bind to lysine in the insect gut, reducing the availability of this essential amino acid. Privet-specialist caterpillars (multiple species) can secrete large amounts of glycine and GABA into their midgut, which competitively interfere with the lysine-binding reaction, allowing the insects to achieve higher uptake of lysine. From Konno et al (2010), with permission from Springer Science + Business Media.

often are large enough to be retained (Barbehenn, 2001). Such complexes (which may include tannins, proteins, lipids, and polyvalent metals) may prevent a variety of allelochemicals from passing through the peritrophic membrane (Barbehenn, 2001). Any hydrophilic molecules for which there are no protein-based uptake mechanisms (e.g. glycosides) are likely to be excreted without absorption into the body. The peritrophic membrane can also adsorb allelochemicals (such as tannins) by non-covalent binding, which reduces movement of the toxin from the gut to the hemolymph (Bernays, 1981). Allelochemicals that do not cross the peritrophic membrane are excreted rapidly due to the rapid gut throughput of most insect herbivores.

Absorbed allelochemicals can be metabolized to non-toxic forms and then secreted in some insects. For example, nicotine, often in solanaceous plants, is an easily absorbed toxin that binds to acetylcholine receptors. Some species such as the caterpillar *Manduca sexta* and the grasshopper *Melanoplus differentialis* metabolize more than 90% of ingested nicotine to the non-toxic cotinine, and then secrete these metabolites via the Malphighian tubules in less than 2 hours. In contrast, nicotine loaded into the house fly remains in the fly, unmetabolized, for many hours (Self et al., 1964).

A major set of players in detoxification are the mixed function oxidases, also called cytochrome P450 monooxygenases (P-450s). These are heme-thiolate proteins which can use NADPH to oxidize a wide variety of natural and synthetic hydrophobic toxins. The genes encoding these enzymes constitute one of the largest gene superfamilies known, and some insect genomes can contain over 100 P-450 genes (Feyereisen, 2005). P-450s are known to be induced by, and metabolize, a wide variety of plant secondary compounds (Berenbaum, 2002). Oxidized allelochemicals are then generally excreted by the Malphigian tubules.

5.3.4 Sensory Systems and Neuroendocrine Regulation of Feeding and Digestion

The nervous system regulates feeding in response to external factors such as stimulants and deterrents in food, and to internal factors such as the degree of fullness and hemolymph nutrient levels. Stimulants and deterrents in food, and the optical, olfactory, and gustatory receptors that sense these, have been discussed above. Internally, gut fullness is sensed by stretch receptors that provide input to the central nervous system, inhibiting feeding (Bernays and Chapman, 1973). In addition, elevated levels of hemolymph amino acids and osmolality inhibit feeding, at least partially by inhibiting the sensitivity of gustatory sensilla (Abisgold and Simpson, 1987, 1988).

Feeding, digestion, and absorption are regulated by a variety of nutritional and neuroendocrine controls. Positive and negative inputs from food to the central nervous system are modified by feedbacks from peripheral systems such as gut stretch receptors and blood composition (Simpson and Bernays, 1983). A recent review by Audsley and Weaver (2009) provides extensive detail on the current state of knowledge of neuroendocrine regulation of feeding and digestion, only some of which will be covered here. In general, it is thought that exposure to phagostimulants, or consumption of food, causes cocktails of neurohormones to be released directly into the gut as neuromodulators, or into the hemolymph, where they act as hormones.

The major neuroendocrine structures involved in the control of feeding and digestion are the stomatogastric nervous system (SNS) and the suboesophageal ganglion (SOG). The SNS consists of a number of interconnected ganglia lying directly above the esophagus and below the brain. The most cranial ganglion of the SNS is the frontal ganglion, which lies directly on the dorsal surface of the esophagus, innervating the buccal region. The next caudal ganglion of the SNS is the hypocerebral ganglion, which is linked to the corpora cardiaca and the proventricular ganglion. Caudally, the SNS bifurcates into an enteric plexus that envelopes the proventriculus (Audsley and Weaver, 2009). The SOG is the first ganglion of the ventral nerve cord, and is the only ventral nerve ganglion in the head. There is evidence from a variety of insects that the SNS and SOG can both stimulate and inhibit gut function and feeding, primarily via secretion of different neurohormones.

Surgical studies support a role for the SNS in stimulating foregut activity. Although peristalsis of the insect gut is myogenic, the SNS stimulates peristaltic waves, particularly in the foregut (Schoofs and Spieß, 2007). Removal of the frontal ganglion, or severing the recurrent nerve, results in accumulation of food in the foregut, slower food ingestion, and slower growth (Audsley and Weaver, 2009).

Many RFamide peptides associated with the SNS have been shown to regulate feeding and the myoactivity of the visceral gut muscles. These peptides, defined by their carboxy-terminal arginine (R) and amidated phenylalanine (F) residues (hence RFamide), have been identified in the nervous systems of animals within all major phyla, and are often associated with the regulation of feeding and digestion (Bechtold and Luckman, 2007). RFamide peptides and RFamides have been identified in a variety of insects, and have been shown to inhibit gut peristalsis and feeding (Audsley and Weaver, 2009). RFamides occur in nerves from the SNS to the gut.

In vertebrates, the RFamide peptides cholecystokinin (CCK), neuropeptide Y, galanin, and bombesin are known to be involved in the control of food intake. Insect sulfakinins, peptides which display substantial sequence similarities with the vertebrate gastrin/CCK peptide family, can be secreted by the SNS and by the central nervous system (Audsley and Weaver, 2009). Sulfakinin injections inhibit food consumption in the locust, *Schistocerca gregaria*. No effect on the tip resistance of the gustatory sensilla was found after injection with sulfakinin, suggesting that sulfakinins do not reduce food intake by decreasing the sensitivity of taste receptors (Wei et al., 2000). Sulfakinins have now been shown to inhibit food consumption and stimulate secretory activity by the digestive tract in a variety of insect species, suggesting that these are general "satiety hormones" in insects (Audsley and Weaver, 2009). Interestingly, sulfakinins appear to suppress carbohydrate feeding but not protein feeding in blow flies, supporting the idea that the two macronutrients are regulated by independent mechanisms (Downer et al., 2007).

Other hormones associated with the SNS, and believed to regulate feeding and digestion, are the tachykinins, proctolin, allatoregulatory peptides, and myoinhibitory peptides. Tachykinin-related peptides can be secreted by the SNS, the central nervous system, and by the gut itself (Audsley and Weaver, 2009). Tachykinins appear to be released after a meal, and they stimulate contractions of both the hindgut and foregut in various insects (Nassel et al., 1998). Proctolin is released from the SNS and the central nervous system, and it stimulates gut peristalsis in some but not all insects (Audsley and Weaver, 2009). Allatostatins and allatotropins (well-known for their ability to regulate secretion of juvenile hormone from the corpora allata) are released by multiple regions of the SNS and from endocrine cells of the midgut in some insects (Audsley et al., 2005). At least in *Manduca sexta*, allatostatins tend to inhibit gut peristalsis while allatotropins stimulate it (Matthews et al., 2007). Myoinhibitory peptides are secreted by the SNS and central nervous system, and have been shown to inhibit contractions of the fore- and hindgut, as well as feeding, in several insect species (Aguilar et al., 2006).

The SOG also secretes RFamides and particular FXPLRamide peptides which have homologies with neuromedin U-8, which regulates feeding and body weight in rodents (Audsley and Weaver, 2009). In *D. melanogaster*, the *hugin* gene encodes several neuropeptides, including hugin γ (Meng et al., 2002). The *hugin*-expressing neurons are localized exclusively to the SOG and modulate feeding behavior in response to nutrient signals. Over-expression of *hugin* results in suppression of growth and feeding, while blocking synaptic activity of *hugin* neurons results in an increase in feeding, suggesting *hugin* neurons generally inhibit feeding (Melcher and Pankratz, 2005). If fly larvae are reared on low-protein diets or starved (which should increase hunger), expression of *hugin* is downregulated (Melcher and Pankratz, 2005). Axons from *hugin*-expressing cells arborize with the axon terminals of gustatory sensory neurons within the SOG, suggesting that these cells may mediate sensillar sensitivity (Bader et al., 2007).

Eating often means consuming excess water and salts. Thus it is not surprising that hormones that activate the insect renal system also are activated by feeding. Two important diuretic hormones in *Manduca sexta*, both homologous to mammalian corticotropin releasing factor (CRF), activate secretion of the Malpighian tubules (Coast, 2007). These peptide diuretic hormones are generally released from the brain or ventral nerve cord. Injection of diuretic hormones or consumption of these hormones on food disks suppress feeding in *Manduca*, though the mechanisms are not clear (Audsley and Weaver, 2009). Another CRF-related diuretic hormone, Lomci diuretic hormone, appears to be important in inducing satiety (reducing meal size and increasing time between meals) by decreasing the sensitivity of the gustatory sensilla (Goldsworthy et al., 2003).

Serotonin (5-hydroxytryptamine) is released after feeding, and it plays a role both in stimulating diuresis and renal function (Coast, 2007) and in a variety of feeding-related events in insects (Orchard, 2006). Serotonin is generally released from the suboesophageal ganglion or ventral nerve cord, and can be released locally or into the hemolymph as a circulating hormone. At least in *Rhodnius prolixus*, serotonin also induced salivary secretions, fore- and hindgut peristalsis, fluid transport into the crop, plasticization of the abdominal cuticle (to allow the stretching that accompanies a meal), and stimulation of heart beat (Orchard, 2006). Injected serotonin inhibited feeding on either protein or carbohydrate in the flesh fly (Haselton et al., 2009). Serotonin appears to be involved in regulation of the intake target in the cockroach *Rhyparobia madera* (Cohen, 2001). This roach self-selected an intake target of 25% protein:75% carbohydrate. Injection of serotonin-related drugs inhibited carbohydrate feeding, while injections of serotonin antagonists promoted carbohydrate over-feeding (Cohen, 2001).

Octopamine is often considered the fight or flight hormone of insects and thus might be expected to suppress feeding. However, octopamine has been shown to increase the sensitivity of gustatory receptors to stimuli in several insects, increasing feeding rate (Farooqi, 2007). At present it is not clear whether this modulation is distinct from the effects that octopamine has on activity and metabolism, or whether the effects are related, that is because higher activity and metabolic rates require higher energetic intake.

5.4 Thermal Effects on the Rate and Efficiency of Feeding and Growth

The generally positive effects of higher body temperatures on feeding, digestion, and growth over moderate temperature ranges have been described in Chapter 3. As described there, many insects use behavior to achieve relatively high, stable body temperatures, thus promoting increased rates of feeding and food processing (Harrison and Fewell, 1995). How does body temperature affect the efficiency of growth? Like many other grasshoppers, *Locusta migratoria* generally selects a high body temperature in a gradient (38 °C). As had been shown previously, between 26 and 38 °C, higher temperatures strongly increased intake, development, and growth rates (Miller et al., 2009; Fig. 5.18). However, the efficiency of conversion of digested food to tissue was highest at the intermediate temperature of 32 °C, suggesting that these grasshoppers sacrificed efficiency to maximize growth rate (Miller et al., 2009). The mechanism responsible for this increased efficiency at 32 °C relative to lower or higher temperatures remains unknown, and does not seem to be related to simple exponential effects of temperature on metabolic rate (Miller et al., 2009).

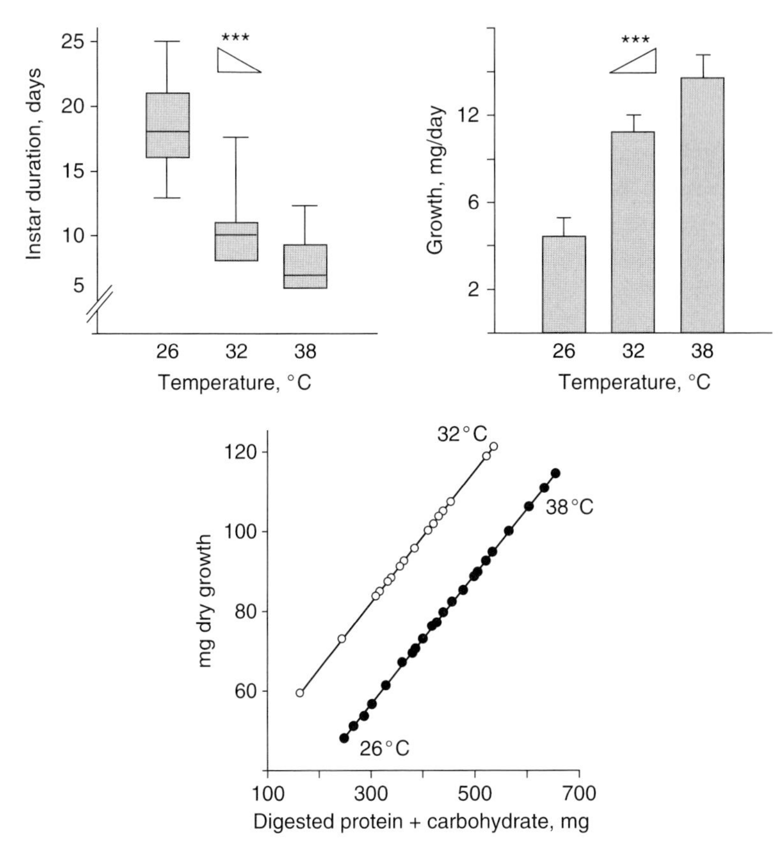

Fig. 5.18 Higher temperatures speed growth (top right) and reduce development time (top left) in grasshoppers (*Locusta migratoria*) but reduce the efficiency of conversion of digested food to tissue (bottom). From Miller et al. (2009), with permission from The Royal Society.

How does temperature interact with the macronutrient content of food? Lee and Roh explored this question by raising caterpillars of the moth *Spodoptera exigua* Hübner (Lepidoptera: Noctuidae) at a range of temperatures and protein:carbohydrate ratios (Lee and Roh, 2010). Feeding and growth were best at an optimal, intermediate ratio, and growth rates were higher at higher temperatures (Fig 5.19). Extreme diets (high or low protein:carbohydrate levels) constrained thermal increases in growth and feeding. However, survival strongly decreased at higher temperatures, with high prepupal mortality observed especially when animals were consuming foods with extreme protein:carbohydrate ratios (Fig. 5.19). Thus the optimal temperature for feeding and growth was higher than the optimal temperature for survival and lipid storage. The Miller

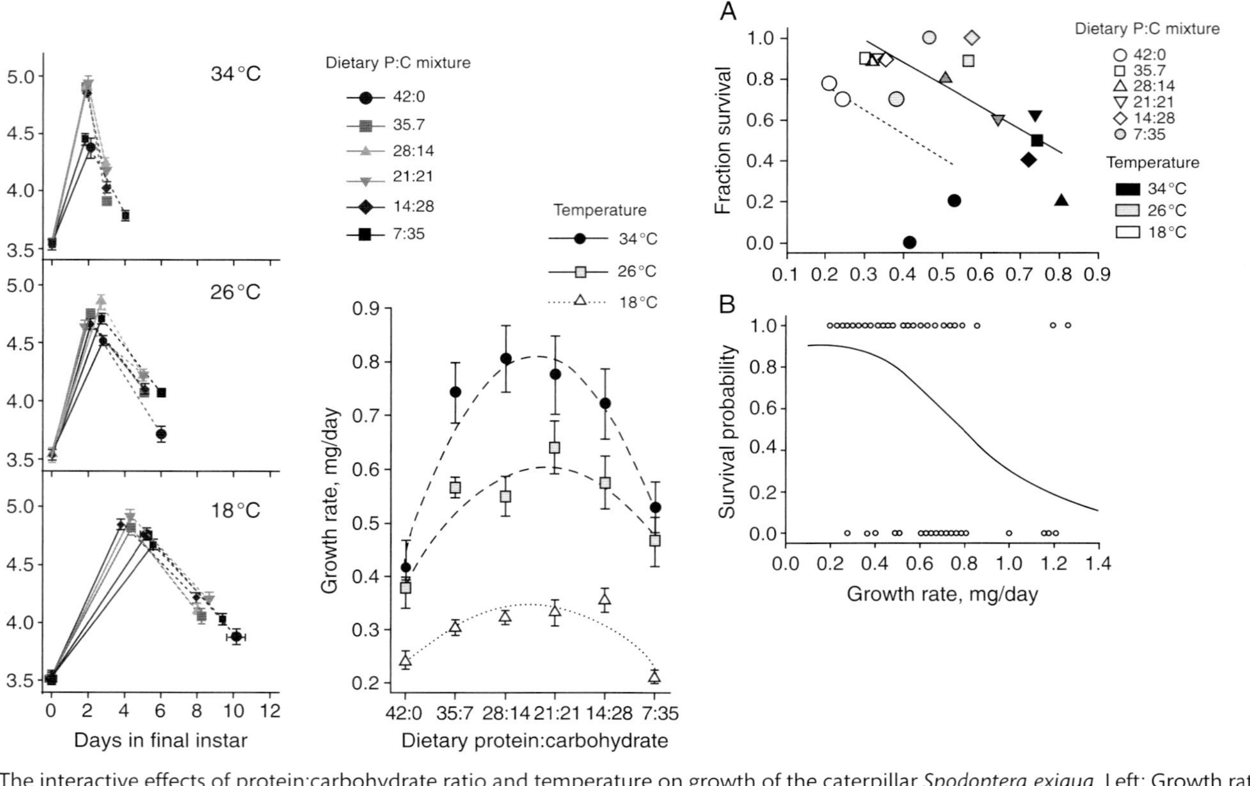

Fig. 5.19 The interactive effects of protein:carbohydrate ratio and temperature on growth of the caterpillar *Spodoptera exigua*. Left: Growth rate vs. protein:carbohydrate ratio at three temperatures. Note that at extreme protein:carbohydrate ratios that the temperature effect on growth is muted. Right: Higher growth rates and temperatures were associated with lower survival rates. From Lee and Roh (2010), with permission from John Wiley and Sons.

et al. (2009) and Lee and Poh (2010) studies together support the hypothesis that fitness may sometimes be maximized at lower temperatures than those that optimize rates of growth and feeding.

5.5 Adaptations of Feeding and Digestive Systems in Populations and Species

Insects ingest food in various ways depending on their ecological niche and morphology. The type of food (liquids, leaves, grasses, carrion, detritus) correlates strongly with phylogeny at the level of Family, suggesting that much of the major diversification in feeding strategies occurred early in evolution. On the other hand, there is increasing evidence that populations of insects can evolve quantitative shifts in diet utilization in response to changes in diet.

A key factor underlying evolution of diet may be evolution of the sensory system. In *Drosophila*, quantitative genetic studies of chemosensory responses to benzaldehyde (a volatile often emitted from rotting fruit) suggest that substantial variation exists in natural populations, especially among larvae (de Brito Sanchez et al., 2008). Both within and across species, key chemosensory genes appear to exhibit "birth and death evolution" (Sanchez-Gracia et al., 2009). *Or* and *OBP* genes that are physically close to each other on the chromosome are generally related, suggesting birth by tandem gene duplication, followed by independent evolution of these gene families. Loss (death) of functional genes also occurs due to mutations such as deletions.

5.5.1 Within-population Evolution of Feeding and Digestion

Laboratory selection experiments indicate that insects can evolve behaviors and physiologies specialized for particular dietary ratios of protein:carbohydrate. Warbrick-Smith et al. studied the performance of *Plutella xylostella* caterpillars that had been reared for over 350 generations on a single artificial diet that differed strongly in protein:carbohydrate ratio from their natural diet (Warbrick-Smith et al., 2009). When given a choice, the larvae self-selected protein: carbohydrate ratios similar to the rearing diet (and different from the natural diet), and deviations in the protein, carbohydrate, or water content away from the levels in the rearing diet strongly reduced growth and survival.

Interestingly, *Plutella* moths kept for multiple generations on high carbohydrate foods evolved mechanisms to dispose of carbohydrate, as they had less fat at any given level of carbohydrate intake than did moths reared on high protein:carbohydrate food. Although the mechanisms are unknown, moths likely evolved either higher metabolic rates or lower absorption of dietary carbohydrate.

Field studies of different populations also suggest population-level evolution of feeding and digestive systems. For example, using the grasshopper *Melanoplus sanguinipies*, Fielding and Defoliart (2008) compared second-generation lab populations derived from wild populations in either Alaska or Idaho. Those from Alaska had shorter instars and lower lipid stores than did those from Idaho when reared on identical diets (Fielding and Defoliart, 2008).

Most phytophagous insects feed on only one or a few hosts. Host races are important examples of within-species variation in insect nutritional ecology, and are often thought to provide a first step toward sympatric speciation, though this remains controversial (Drès and Mallet, 2002). Host-races show genetic differentiation associated with different plant hosts; this is often associated with the fact that mating occurs on the host plant (Drès and Mallet, 2002). This is a large area of research which will only briefly be covered here with a few examples.

Oshima (Ohshima, 2008) studied the leaf-mining moth, *Acrocercops transecta*, which has races associated with *Juglans ailanthifolia* and *Lyonia ovalifolia*. Females usually oviposit on their natal host plant. Transplant experiments demonstrated that *Juglans*-associated individuals failed to survive on *Lyonia*, but *Lyonia*-associated moths could grow on *Juglans*. The populations on *Juglans* and *Lyonia* were clearly separated by a phylogeny based on mitochondrial DNA, which indicated that the *Lyonia* populations evolved from *Juglans*-related populations. The two host races mated and produced eggs that hatched successfully. However, hybrids could not grow on *Lyonia*. Together these data suggest that evolution of a race that can survive on *Lyonia* required adaptations in morphology or physiology.

Among insects, the apple fly, *Rhagoletis pomonella*, is one of the best-studied examples of host races. The apple fly shifted within the past 200 years from native hawthorn trees to domesticated apple trees in North America, and population variation among races has been extensively studied (Bush, 1994; Jiggins and Bridle, 2004). Host choice and host-specific mating are critical aspects of differentiation of these biotypes. Forbes and Feder recently showed that apple and hawthorn flies have a strong sensory bias for their host-plants, preferentially responding to volatile blends from their own plants, and strongly preferring host-plant associated visual cues (Forbes and Feder, 2006). Such data demonstrate the critical role of changes in sensory processing that must underlie host race formation.

The tephritid fly *Tephritis conura* feeds on at least nine different thistle (*Cirsium*) species, and there is evidence that populations on these different host species are genetically distinct (Diegisser et al., 2007). Different races have different relative ovipositor lengths, and these differences are maintained even when the races are reared on identical hybrid species, indicating a genetic basis for the variation (Diegisser et al., 2007). Phylogenetic data based on mitochondrial DNA indicates that *C. heterophyllum* is the ancestral race, with populations moving to a variety of thistle species including *C. oleraceum*. Larvae transplanted from

C. heterophyllum to *C. oleraceum* grew poorly and had low survival compared to larvae associated with *C. oleraceum*, providing evidence for morphological or physiological adaptations to the new host (Diegisser et al., 2008).

5.5.2 Specialists and Generalists

Many insects exhibit fairly extreme dietary specialization, choosing only one or a few species of plants within a genus. Conversely, other insects are extremely polyphagous, consuming hundreds of plants species (Schoonhoven et al., 2005). Sometimes a species is polyphagous, but individuals are relatively specialized. The degree of specialization varies widely among taxonomic groups (Fig. 5.20), for reasons that are unclear. Even insects that are polyphagous generally consume a limited number of plants, and exhibit distinct preferences.

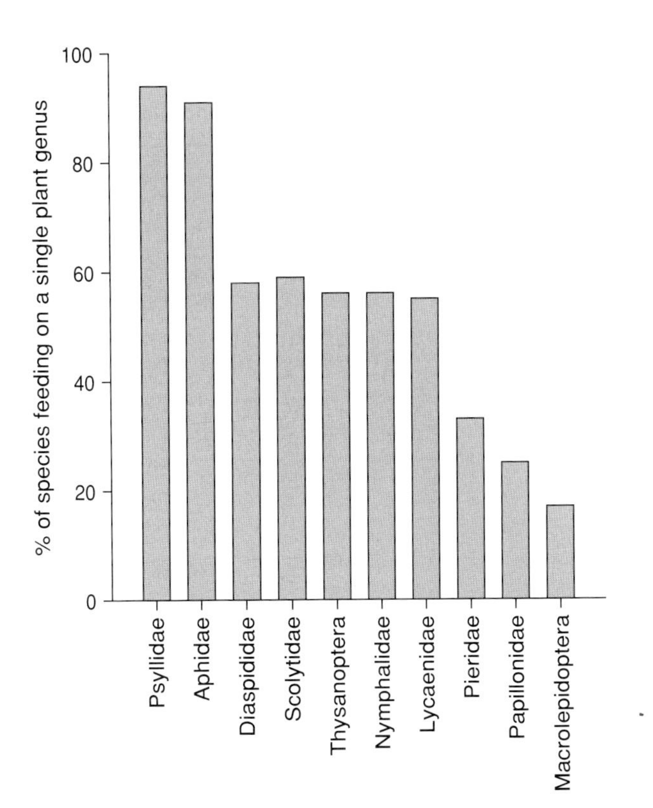

Fig. 5.20 Variation in degree of specialization among some herbivorous insect families. Psyllidae, Aphidae, and Diaspididae are Hemiptera (bugs), Scolytidae is a family of Coleoptera (beetles), and Nymphalidae, Lycaenidae, Pieridae, and Papillonidae are families of Lepidoptera (butterflies). Drawn from Table 2.3 of Schoonhoven et al. (2005).

Closely-related insects are more likely to feed on the same or closely related plants, a pattern observed for chrysomelid beetles (Kopf et al., 1998), butterflies (Janz et al., 2001), leaf-mining moths (Lopez-Vaamonde et al., 2003), and walking sticks (Crespi and Sandoval, 2000). Phylogenetic conservation of host choice is also observed in pollen-foraging bees (Sipes and Tepedino, 2005). The most likely explanation for this phylogenetic conservation of host choice is similarity in morphology and chemistry of closely-related plants. Despite the tendency for phylogenetic conservation of host choice, it does not appear that speciation is generally associated with host switches (Sipes and Tepedino, 2005).

Polyphagous grasshopper species appear to have more gustatory sensilla than specialist feeders (Chapman, 1982). Oligophagous grass-feeding grasshoppers, such as locusts, are strongly deterred from eating artificial diets containing significant levels of plant secondary compounds, while polyphagous species such as *Schistocerca gregaria* are not, and may even be stimulated to feed (Blaney and Simmonds, 1990; Chapman, 1990).

Host specialization is likely to require, and perhaps is caused by, changes in the chemosensory system of the insect. Specialization might indicate that insects acquired a new or greatly increased attraction to particular chemicals or lost the repulsion to the new host-plant's odor and taste. Most *Drosophila*, including the widely-studied *D. melanogaster* and *D. simulans*, are relative generalists, feeding on a wide variety of decaying fruit and vegetation. However, a few species are specialists. One is *Drosophila sechellia*, a narrow ecological specialist that feeds on the fruit of *Morinda citrifolia*, and is attracted to volatiles of that fruit. The odorant receptors of *D. sechellia* are particularly sensitive to hexanoate esters, which are commonly emitted by *M. citrifolia*, and the glomeruli for processing the hexanoate signals are enlarged, demonstrating that evolution has acted on multiple aspects of the olfactory system (Dekker et al., 2006). Six odorant receptor and thirteen gustatory receptor genes have been lost in *M. sechellia*, suggesting that specialization is associated with a narrowing of chemosensory capacity (McBride, 2007; Nozawa and Nei, 2007). There also appears to be more purifying selection across all odorant and odorant-binding proteins of generalist compared to specialist *Drosophila* species, suggesting that the function of many of these is more critical to fitness in generalist species (Sanchez-Gracia et al., 2009). Kropp et al. found that the expression of a number of odorant receptor and odorant binding proteins was strongly upregulated in the antennae of *D. sechellia* when compared to *D. melanogaster* or *D. simulans*, suggesting a possible mechanism for the gain in hexanoate sensitivity (Kopp et al., 2008).

5.5.3 Adaptations to Herbivory

Plant species are highly variable in the stoichiometry of their tissues (Table 5.1). C:N content of leaves varies from 5–100, C:P content from 220–3400, and N:P content from 2–60 (Elser et al., 2000). Plants also vary tremendously in shape, form, toughness, and allelochemical content. What adaptations have occurred to allow different insect species to feed and grow successfully on different types of plants?

5.5.3.1 Co-evolution

Terrestrial biodiversity is dominated by plants and the herbivores that consume them. Nearly half of extant insects feed on living plants, with at least 400,000 species of phytophagous insects feeding on 300,000 species of plants (Schoonhoven et al., 2005). Macroevolutionary hypotheses have suggested that much of this diversity is driven by reciprocal evolution of plants and insect herbivores. A classic article by Ehrlich and Raven (Erlich and Raven, 1964) suggested that in response to herbivory a plant species may evolve a novel, highly effective chemical defense that enables escape from most or all of its associated herbivores, enabling the plant lineage to radiate into diverse species (hence, related plants tend to share similar chemistry). After some time, one or more insect species colonize and adapt to this plant clade. These insects then undergo adaptive radiation, as new species arise and adapt to different, but related, plants. Erlich and Raven proposed that such a pattern could explain why related insects tend to use related plant hosts.

Evidence from molecular cladograms suggest that shifts to new plant clades is a major force associated with increases in insect diversity, emphasizing the key role of plant–insect interactions in driving the tremendous diversity of insects (Winkler et al., 2009). Mitter et al. showed that herbivorous clades are significantly more species-rich than non-herbivorous sister clades, also supporting an important role for plant–insect interactions in driving speciation (Mitter et al., 1988). Farrell used phylogenetic data to examine the evolution of beetles (Farrell, 1998). He showed that primitive, Jurassic fossil species represent basal lineages that are still associated with cycads and conifers. Repeated origins of angiosperm-feeding lineages were associated with enhanced rates of diversification, supporting the hypothesis that the incredible species diversity of beetles is explained by their diverse and specialized feeding strategies (Farrell, 1998).

5.5.3.2 Evolution of Insect Feeding Structures

Insect mouthparts are dramatic examples of the power of evolution to transform morphology. The type of food ingested is strongly correlated with the structure of the feeding apparatus. For example, fluid feeders like aphids, mosquitoes, and butterflies have hollow-tube mouthparts, with cibarial pumps for generating the negative pressures necessary to drink.

The physical toughness of leaves provides a major deterrent to herbivores. Plants can evolve tougher leaves by increasing the content of sclerophyllous tissues including lignified veins or by adding silica. Palm fronds are particularly tough, as are the leaves of woody plants and the blades of C4 grasses, although new leaves are often softer than mature ones (Bernays et al., 1991). Gramnivorous grasshoppers have relatively large but short, mandibles with chiseled ends and a molar region; these have evolved multiple independent times from more generalized species (Bernays et al., 1991). Caterpillars that feed on the leaves of woody plants or grasses have smooth "scissor-like" mandibles, whereas caterpillars feeding on softer dicot leaves tend to have overlapping cusps on the inside edge of the mandibles (Bernays et al., 1991).

5.5.3.3 Holding on to the Leaf

Many plants defend themselves by physical means, including trichomes or hairs. To overcome trichomes, many insects have specialized tarsal claws for gripping on to these leaves, and such claws appear to be critical adaptations in many cases of host specialization (reviewed by (Bernays et al., 1991)). For example, dock beetles (*Gastrophysa viridula*) use specialized setae to adhere to a small smooth area of their host leaf (Bullock and Federle, 2011).

5.5.3.4 Adaptations for Overcoming Plant Chemical Defenses

A central tenet of Ehrlich and Raven's theory is that evolution of plant chemical defenses is followed closely by biochemical adaptation in insect herbivores, and that newly evolved detoxification mechanisms result in adaptive radiation of herbivore lineages. This tenet has recently been explored by Wheat et al. (2007), focusing on family Pieridae (white and sulfur butterflies). Pieridae use three major host plant groups: the Fabales (including legumes Legumes), the Brassicales (glucosinolate-containing plants such as cabbages), and Santalales (mistletoes); phylogenetic analysis suggests that feeding on Fabales is ancestral (Wheat et al., 2007). Feeding on Brassicales presented a radical new chemical challenge known as the glucosinolate-myrosinase system. When larvae damage leaf tissues, the formerly compartmentalized myrosinase enzyme comes into contact with non-toxic glucosinolates, which are hydrolozyed into toxic products such as isothiocyanates (the so-called "mustard oil bomb"). There are two independent lepidopteran mechanisms for detoxifying the glucosinolate-myrosinase system. The first is glucosinolate sulfatase, which desulfates glucosinolates, producing metabolites that no longer act as substrates for myrosinases (Ratzka et al., 2002). The second mechanism is production of nitrile-specifier proteins by the larval midgut, which promotes the formation of nontoxic, excretable nitrile breakdown products (Wittstock et al., 2004). Only pierids that feed on glucosinolate-containing plants express nitrile-specifying proteins (Wheat et al., 2007). Using a variety of molecular clock approaches, Wheat et al. showed that nitrile-specifying proteins evolved shortly after the evolution of the Brassicales, and that this was followed by an adaptive radiation of these glucosinolate-feeding Pierinae.

Perhaps the most detailed example of co-evolution of plant defenses and insect herbivores involves swallowtail butterflies, P-450s, and the furanocoumarins in Apiaceae (Berenbaum, 1983; Berenbaum, 2002). Furanocoumarins are toxic to many organisms because, in the presence of ultraviolet light, they covalently bind to thymine, cross-linking DNA and inactivating proteins (Berenbaum, 2002). Hundreds of different furanocoumarins exist (Berenbaum, 2002). Furanocoumarin synthesis in plants can be induced by insect herbivory via jasmonic acid signaling cascades (Zangerl and Berenbaum, 1998). Only a few insects feed on furanocoumarin-containing plants, mostly in a few families of Lepidoptera (Berenbaum, 2002). These include over 75% of swallowtail butterfly species in the genus *Papilio*; and they can rapidly detoxify these toxins via P450-mediated systems (Berenbaum, 2002). P-450s that catabolize furanocoumarins are found in related generalist species. However specialists such as the parsnip webworm, *Depressaria pastinacella*, exhibit P-450s highly specialized for the particular furanocoumarins found in their host (Mao et al., 2006). Phylogenetic and comparative analyses suggest that the P-450s have evolved from a low-activity form that catabolizes the wide variety of substrates found in generalists to a high-activity form specialized to catabolize furanocoumarins in specialists (Li et al., 2003; Wen et al., 2006).

A completely different strategy for coping with plant allelochemicals is illustrated by leaf-cutting ants (Holldobler and Wilson, 2010). Leaf-cutting ants depend on microbial symbionts in their nests to overcome toxins in the plant leaves they collect. The microbial partners are fungi of the genus *Leucoagaricus*. Higher attines, such as *Atta* and *Acromyrmex* species, harvest leaf fragments which they chew and deposit on the fungus in their nests. This physical disruption promotes access by the fungal hyphae, which subsist on the plant material, detoxifying plant secondary compounds. The fungal mycelium is the only food consumed by ant larvae and also contributes to the diet of adults.

5.5.4 Wood and Detritus Feeders

5.5.4.1 Wood Feeders

Although wood is a super-abundant food, it is also a challenging food for insects. It is extremely low in nutrients such as nitrogen and phosphorus, and most of the carbon is locked up in cellulose, which is indigestible by most insects (Haack and Slansky, 1987; Higashi et al., 1992).

The first problem faced by wood-feeders is the large quantity of cellulose in their diet. Wood-feeders cope with cellulose in a variety of ways. Some feed indirectly on wood by eating the fungi that digest the cellulose. Some feed only on the sugars and proteins in the wood. And some make their own cellulases capable of digesting cellulose. Cellulose-digestion was previously thought to require microbial symbionts, but clear genetic and biochemical evidence has revealed that some animals including insects possess endogenous cellulases. In insects (termites and cockroaches), cellulase genes have been identified with classic eukaryotic traits (upstream TATA boxes, polyadenylation and cleavage signals in appropriate sites, eukaryotic *cis* regulatory sites (Watanabe and Tokuda, 2001)). This finding explains the observation that antibiotic treatments are unable to block cellulose digestion in many termites and cockroaches (Watanabe and Tokuda, 2001). Since animal cellulase has been found in crayfish as well insects, but is lacking in the nematode *C. elegans*, an ancient horizontal transfer of this gene from prokaryotes is hypothesized, followed by inheritance from stem arthropods (Watanabe and Tokuda, 2001).

All termites yet tested possess intrinsic cellulases, as do Thysanura and some beetles and cockroaches (Douglas, 2009). Some cockroaches and a few termites possess protists in their guts that appear to assist in cellulose digestion (Douglas, 2009). Some insects, including some higher termites and leaf-cutting ants, cultivate cellulose-digesting microbes in their nest (Martin, 1977).

A second challenge posed by feeding on wood is to obtain sufficient protein, specifically the required essential amino acids. Some cerambycid beetles have symbiotic yeast-like endosymbionts located in special structures of the midgut that provide essential amino acids and vitamins, and detoxify plant allelochemicals (Walczynska, 2009). One area of controversy is whether wood-feeding insects use nitrogen-fixing microbes to overcome the low protein content of wood. Nitrogen-fixing bacteria have been demonstrated to occur in some termite species, as well in one wood-feeding beetle (and the medfly) (Nardi et al., 2002). However, the nutritional significance of such nitrogen fixation is uncertain, as no physiological experiments have yet demonstrated that such fixed nitrogen is incorporated into insect tissues.

Detritus feeders are also widely-considered to be nutrient-limited (Moe et al., 2005). Insect detritivores likely selectively forage on nutrient-rich subcomponents of the detritus (e.g. bacteria). Nonetheless, obtaining sufficient protein and essential amino acids may be challenging. Collembolans feeding on low-nitrogen fungus strongly increase their body carbon:nitrogen ratios (body nitrogen content decreasing to a very low 7%), probably due primarily to accumulation of lipid as they cope with the excess carbohydrate they must consume to gain sufficient protein (Larsen et al., 2011). Larsen et al. have also provided stable isotope evidence suggesting that these soil collembolans may possess gut bacteria that contribute essential amino acids when these insects feed on detritus with very low nitrogen content. Microbial symbionts are likely of substantial importance in all detritivores. As an example, hindgut bacteria have been shown to significantly increase the capacity of crickets to digest soluble plant storage carbohydrates (Kaufman and Klug, 1991).

5.5.5 Fluid Feeders

5.5.5.1 Phloem

Phloem sap is also a challenging diet that is utilized by insects in the order Hemiptera, such as aphids, as well as some beetles (Gündüz and Douglas, 2009). Phloem poses four major challenges for the feeding insect: 1) high water content; 2) high sugar concentration; 3) deficits of essential amino acids; and 4) deficits of sterols (Behmer and Douglas, 2011). The first two can severely stress the osmoregulatory physiology of sap feeders.

All sap-feeding insects consume excessive water. Morphological adaptations that permit rapid extraction and excretion of the water load are described in Chapter 4. As described in more detail there, osmoregulation in aphids also requires a functional aquaporin for water transport across the filter chamber; inhibition of aquaporin expression by RNAi causes osmotic disruptions (Shakesby et al., 2009). How does this work? Shakesby et al. demonstrate that the aquaporins are localized in the stomach and distal intestine of the aphids, and propose that water flows from the intestine to the stomach through these aquaporins, helping to dilute the high osmotic pressure of phloem.

The osmotic pressure of phloem sap is often over four times that of insect hemolymph, reaching 1.5 molar, primarily due to high concentrations of sucrose. Such high osmotic pressures in the gut lumen should extract water from the hemolymph, causing sap feeders to shrivel as they feed. Aphids use sucrase and transglucosidases in their gut to link the sucrose molecules together into oligosaccharides (Price et al., 2007). When these enzymes are inhibited, the osmotic pressure in the hemolymph rises, and aphids die within hours (Karley et al., 2005).

The deficiencies in essential amino acids in phloem are solved by microbial symbionts, which appear to be universal in phloem-feeders (Macdonald et al., 2011). Perhaps the best-studied system for understanding the nutritional interactions between insect host and symbionts is that of the pea aphid, *Acyrthosiphon pisum*. Many essential amino acids in phloem are insufficient to meet the aphid's needs, so they must derive supplementary amino acids from their symbiotic bacteria, *Buchnera aphidicola* (Gündüz and Douglas, 2009). The pea aphid lacks other gut microbes, simplifying analysis of these interactions (Oliver 2010). Feeding experiments with artificial diets demonstrate that production of essential amino acids by the symbionts is sufficient to meet all the amino acid deficiencies in the phloem of the broad bean (Gündüz and Douglas, 2009).

The reduced *Buchnera* genome contains most of the genes necessary for making essential amino acids, and the aphid has the complementary genes to complete amino acid metabolism (Moran, 2007). For example, the *Buchnera* in the pea aphid has most of the genes for the biosynthesis of amino acids but lacks the genes for steps in the synthesis of phenylalanine, the branched chain amino acids, and methionine. Remarkably, the pea aphid has genes for enzymes mediating these reactions, but no other reactions in these biosynthetic pathways, indicating that the capacity to synthesize these five essential amino acids can be shared between the host and bacterium (Macdonald et al., 2011). Moreover, the pea aphid has lost some of the genes for purine recycling, but these are possessed by *Buchnera*, suggesting that the symbionts perform aspects of purine metabolism for the aphids (and vice versa), providing yet another facet of this obligate alliance (Ramsey et al., 2010). The interconnectedness of aphid and *Buchnera* metabolism may directly benefit the aphid as growth of the bacteria may be limited by supply of substrates from the aphid (Macdonald et al., 2011). Ensuring that microbial symbionts do not overgrow may be a key aspect of successful use of such symbionts by insects (Chandler et al., 2008).

Aphid clones differ in their host-plant species and varieties, and also in their need for particular essential amino acids. MacDonald and colleagues tested whether variation in the need for particular essential amino acids could be explained by genetic variation in the *Buchnera* symbionts. This was not the case; there was relatively little genetic variation among symbiont lineages in the genes for making essential amino acids (Macdonald et al., 2011). Metabolic modeling suggests that, in principle, the different aphid strains could regulate differential production of essential amino acids by supplying substrates at different rates (Macdonald et al., 2011).

Another role for *Buchnera* in aphids appears to be production of the vitamin riboflavin. Aphids do not require riboflavin in the diet to grow unless they are treated with antibiotics to remove *Buchnera* (Douglas, 2009).

Other phloem feeders in the Auchenorrhyncha utilize a different bacterial symbiont, *Sulcia muelleri*. The *Sulcia* phylogeny closely matches that of the hosts, consistent with maternal transfer (Fig. 5.21, (Moran, 2007)). Phylogenetic analyses are consistent with the hypothesis that this symbiont evolved from a gut-feeding Bacteroidetes when the Auchenorrhyncha evolved in the Permian (Moran, 2007).

5.5.5.2 Xylem

Xylem is an even more abundant and challenging liquid food than phloem, as it is even lower in vitamins and macronutrients. Xylem-feeding insects such as sharpshooters, plant-hoppers, and spittlebugs have evolved a number of important adaptations to be able to grow and feed on this challenging diet.

A key aspect of such adaptation may relate to incorporation of new symbionts that bring new metabolic capabilities. Sharpshooters (a type of leaf-hopper, Hemiptera: Auchenorrhyncha) are xylem-feeders that evolved from phloem-feeding ancestors (Cicadomorpha). Cicadomorpha possess *Sulcia muelleri*. Sharpshooters have an additional symbiont, *Baumannia cicadellinicola*, which occurs in different cells within the bacteriome. Phylogenetic analyses suggest that *B. cicadellinicola* was acquired in the Eocene, consistent with the first appearance

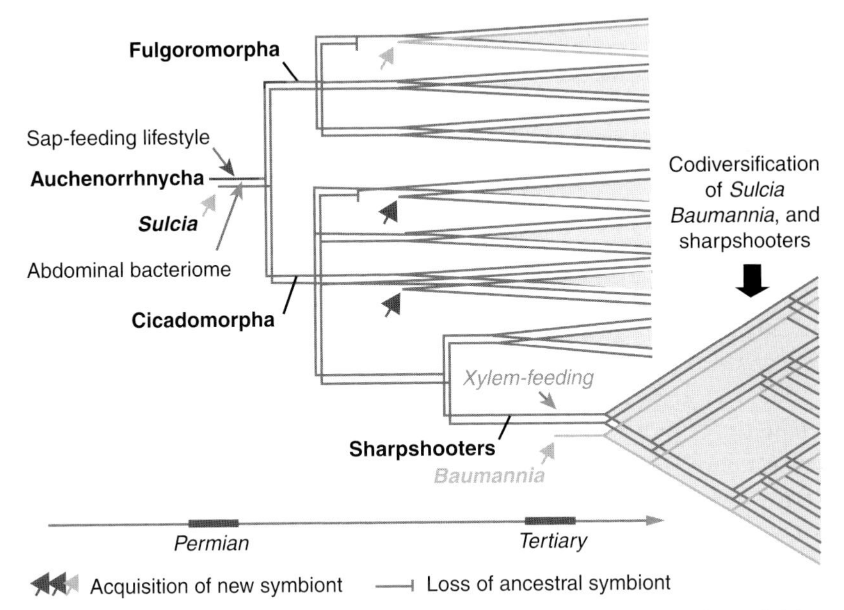

Fig. 5.21 Ancestors of the Auchenorrhyncha (Hemiptera) acquired the symbiont *Sulcia* coincident with acquisition of the sap-feeding lifestyle; thereafter, evolutionary diversification in *Sulcia* and the host is closely matched. (Moran, 2007). Sharpshooters acquired *Baumannia* symbionts co-incident with the evolution of xylem-feeding in the Tertiary, and evolutionary diversification in *Baumannia*, *Sulcia*, and the hosts are thereafter tightly linked in this clade. From Moran (2007), with permission from National Academy of Sciences.

of fossils of this type of insect (Moran, 2007). After acquisition of the *B. cicadellinicola*, the sharpshooters and bacterial phylogenies are completely congruent, suggesting that acquisition of this second symbiont was critical to evolution of xylem-feeding in this guild (Fig. 5.21). Analysis of the *B. cicadellinicola* genome shows that the symbiont has multiple metabolic pathways not present in *S. muelleri*, including the synthetic pathways for several vitamins (Moran, 2007).

5.5.5.3 Blood Feeders

Feeding on blood of vertebrates (haematophagy) has evolved in more than 14,000 insect species, appearing independently in several insect orders (Lehane, 2005; Ribeiro, 1995). The ancestors of haematophagous Cyclorrhapha were likely coprophagous, with licking or sucking mouthparts. For example, the tsetse fly, Glossina (Diptera: Glossinadae) feeds by ingesting blood from wounds, later cutting the skin surface, and finally penetrating the dermal layer with a modified piercing proboscis (Balashov, 2006). Kissing bugs (Hemiptera, Triatominae) derived from plantsap-sucking forms via a group that later developed predatory habits (Schofield, 2000). Anoplura (lice) probably developed from already parasitic insects which fed on the feathers, hair, or skin of their vertebrate hosts (Lehane, 2005).

Feeding on the blood of vertebrates is a dangerous endeavor, and selection has favored the evolution of morphologies, behaviors, and salivary chemicals that reduce feeding time by the insect and detection by the host. Most hematophagous insects studied to date produce at least one anticoagulant, one anti-platelet, and one vasodilatory protein in their saliva (Ribeiro and Francischetti, 2003). Most of the anticoagulants target one or more of the serine proteases in the clotting cascade. Of these, the most commonly isolated are anti-thrombins (Ribeiro and Francischetti, 2003). RNAi knock-down of the genes for these proteins has confirmed their anti-thrombin effect, and their efficacy at increasing ingestion rates.

The salivary glands of triatomid bugs (hemipterans which vector Chagas disease) contain multiple proteins called nitrophorins that can constitute up to 50% of the salivary fluid. These are heme-containing proteins that bind and deliver nitric oxide (NO) to the host, increasing local vasodilation (Champagne et al., 1995). The release of NO from the nitrophorins is triggered by the pH change that occurs when the saliva is injected into the host (Champagne et al., 1995). Some nitrophorins also bind histamine, preventing swelling and pain, and others are anticoagulants. Reducing nitrophorin concentrations by RNAi extends feeding times and reduces ingestion rates (Araujo et al., 2009).

Insects that are blood-feeders throughout their lifecycles, including the tsetse flies *Glossina*, the anopluran sucking lice, and the cimicids (bed bugs) have mycetocyte symbioses. It has often been suggested that these symbionts provide B vitamins (Akman et al., 2002), and there is genomic evidence for this (Douglas, 2009). However, as yet there is as yet no direct experimental support for the hypothesis (Douglas, 2009).

Blood-feeding insects must cope with digestion of large amounts of heme, a powerful pro-oxidant and cytotoxin. Paiva-Silva et al. have shown that in *Rhodnius prolixis*, the blood-feeding carrier of Chagas disease, that heme degredation occurs by a previously unknown pathway, in which the heme molecule is first modified by addition of cysteinylglycine residues before cleavage of the porphyrin ring (Paiva-Silva et al., 2006). Heme is degraded by heme oxidase to biliverdin, iron, and carbon monoxide, with the green biliverdin being an important camouflage molecule in plant-feeders, and likely being an important antioxidant in all insects (Paiva-Silva et al., 2006).

5.5.5.4 Nectar and Fruit Feeders

Nectar is a high-water content, sugar-rich food, usually containing relatively low amounts of amino acids and vitamins (Baker and Baker, 1983). Nectar sugars are mostly converted by insects to the disaccharide trehalase; this compound (and fat body glycogen) can serve as a significant carbohydrate source for activity (Schilman and Roces, 2008). Most nectar feeders must also consume other foods, such as pollen, dung, mud, or carrion to obtain protein, phosphate, and vitamins (Beck et al., 1999; O'Brien et al., 2003). However, some insects have preference for nectars that contain higher levels of amino acids, and this preference can be increased by growing larval plants in elevated CO_2, and decreased by fertilizing larval plants with nitrogen, suggesting that larval protein intake affects adult nectar preference (Mevi-Schütz et al., 2003). Some insects, including many Lepidoptera, forage primarily for nectar as adults, and so must utilize protein consumed by the larvae to produce eggs (O'Brien et al., 2005). Proteins consumed by the larvae are stored as essential amino acid-rich storage proteins that are then utilized in adult reproduction (Pan and Telfer, 1996, 2001; Wheeler et al., 2000). In contrast, the carbohydrate used in reproduction often comes primarily from the nectar procured by the adults (O'Brien et al., 2002).

Like nectar, fruits tend to also be very high in sugar content and low in protein. Up to 25% of the nitrogen in fruit can be non-proteinaceous, making it of no value to animals, and the proteins in fruits often are relatively deficient in some essential amino acids required by insects (Fischer et al., 2004). Many Lepidopterans forage on fruits as adults and leaves as larvae. Fisher et al. have shown that fruit consumption is essential to oviposition in the tropical frugivorous butterfly, *Bicyclus anynana*, and that carbohydrates from the fruit are the major carbon source for the eggs; however protein is mostly obtained from the larval diet (Fischer et al., 2004).

The nutritional quality of fruit may be improved by rotting, as microbes concentrate the protein in the fruit by respiring carbohydrate. Rotting fruit is a major source of adult food for many adult Lepidoptera, and Diptera such as *Drosophila*. Rotting produces anerobic endproducts such as ethanol and acetic acid, which can be used as a volatile cue to locate the rotting fruit (Becher et al., 2010). In addition, *Drosophila* can metabolize ethanol and use it as a metabolic fuel. Low concentrations of ethanol enhance longevity reproduction, while high concentrations inhibit reproduction and reduce survival in *Drosophila* (Etges and Klassen, 1989; Parson, 1989). Some flies that feed on rotting fruit (or dung) appear to have gut pH and enzymes specialized for bacterial digestion, including acid-active lysozyme and cathepepsin (Lemos and Terra, 1991).

5.5.6 Pollen Feeding

Pollen is a major source of protein for many insects, especially bees. The diet breadth of pollen feeders is quite variable, with some species being extreme generalists, and others specializing on only a single species (Sipes and Tepedino, 2005).

The powdery nature of pollen makes it impossible for basic insect mandibles to handle, so pollen feeders exhibit a diversity of morphological specializations. In honeybees, the pollen grains are first captured on the body hairs as the bee brushes by the anthers. Then, during hovering flight, the bee captures the pollen grains with special combs on the pro- and mesothoracic legs, mixes the pollen with saliva, compresses it into a pellet with the hind legs, and then transfers the pollen pellet onto the corbicula, a ring of hairs on the hind tibia. The bee rubs its hind legs together to transfer further pollen onto the pellet.

Pollen odors appear to be important in attracting pollinators. Pollen odors are more pronounced in insect-pollinated flowers (Dobson and Bergstrom, 2000). Pollen volatiles include repellents such as methyl alcohols and ketones, reflecting the likely need for plants to minimize pollen removal by non-pollinating insects and damage from pathogens while promoting dispersal and pollination (Dobson and Bergstrom, 2000). Pollen odors mostly derive from the pollenkitt, the oily covering of pollen that usually accounts for about 5% of the pollen by weight (Dobson and Bergstrom, 2000).

Pollen is highly nutritious, containing minerals, proteins, lipids, carbohydrates, and fiber. However, the nutritional value of pollen can be quite varied, with protein content ranging from 2.5 to 61% (Roulston and Cane, 2000). Pollen is challenging to digest. The outer wall of pollen (exine) is particularly hard and unreactive. Complete dissolution of the exine requires heating at 97°C in monoethanolamine for 3 hours (Stanley and Linskens, 1974). Enzymatic degradation of the exine is known to occur only in certain Collembola that secrete an enzyme called exinase (Scott and Stojanovich, 1963). Some insects (flower thrips) can pierce the grain with their mandibles and suck out the contents (Kirk, 1987). Regurgitated nectar can be used to cause pollen grains to swell and leach their contents, a method that has been observed in butterflies and nectar flies (Gilbert, 1972; Johnson and Nicolson, 2001). Flower-feeding scarab beetles are able to digest the contents of the pollen grains, excreting the empty shells, suggesting that their digestive enzymes can penetrate into the pollen grain (Johnson and Nicolson, 2001).

5.5.7 Mutualisms in insect–plant interactions

Although this chapter and book focuses primarily on the physiology and behavior of insects, such traits evolved in the context of their ecological communities, with associated competition, predation, parasitism, and commensal and mutualistic interactions. Particularly relevant to this chapter are the many mutualistic interactions that have been documented for insects and their food plants. Mutualisms are defined as reciprocally positive interactions between species, with benefits usually quantified in terms of fitness or population dynamics.

A variety of insect–plant mutualisms occur, but most fit into the categories of: 1) pollination; 2) protection of plants from herbivores and enemies; and 3) seed dispersal. Nutritional symbioses with microbes are also often considered to be mutualisms; however, this topic is covered elsewhere within this chapter. Mutualisms have been covered in a variety of excellent books and review articles (Beattie, 1985; Bronstein, 2009; Bronstein and Holland, 2008; Stadler and Dixon, 2008). Often both species involved derive both costs and benefits from the interaction. For example, *Manduca sexta* moths are the primary pollinators of jimsonweed (*Datura wrightii*) in the American southwest, clearly providing a benefit to this plant (Bronstein et al., 2009). During pollination, the moth consumes large quantities of nectar that support the energy cost of flight. However, female moths also lay eggs primarily on *Datura*, and the plants can be completely denuded by foraging caterpillars, which clearly is a major cost to the plant. From the perspective of the adult insect, there are costs associated with foraging on intermittently available flowers, which the moth minimizes by also foraging on other plants (Bronstein et al., 2009). For the caterpillar, it seems likely that there are costs associated with feeding on a plant which is heavily defended by alkaloids, although the vast majority of these are excreted and even injection of the alkaloids into the blood does not have detectable effects (J. Hildebrand, personal communication).

Pollination by insects is believed to have evolved even before the origin of angiosperms, but to have exploded in diversity and complexity during the Cretaceous and Tertiary (Grimaldi and Engel, 2005; Labandeira, 1997, 1998). Recent studies suggest that assemblages of plants and pollinators are rather generalized; most plants are pollinated by multiple species, and most pollinators visit multiple plants (Waser and Ollerton, 2006). Insect preference for certain plant characteristics, such as flower size, can drive evolutionary change in those traits (Galen, 1996). Less is known about how insect behavior, morphology, and physiology is adapted for their mutualistic interactions. However, as one example, the long tongues of some butterflies, moths, and bees are clearly adaptive for extracting nectar from flowers with long corollas (Nilsson, 1988).

Most pollination is passive and generalized, with insects picking up pollen on their bodies while foraging for nectar or pollen, with that pollen subsequently being brushed onto plant ovules. However, some insects exhibit behavioral and morphological traits that actively enhance pollination. Each of these cases involves a highly-evolved mutualism in which both pollination and seed predation play roles.

Fig–fig wasp interactions are particularly striking as they involve some wasp species that are active pollinators and some that are not. Fig wasps pollinate figs, and lay eggs on the ovules. The larvae subsequently eat the seeds. Actively pollinating species have hairy pockets (corbicula) on their front coxae that transfer pollen to thoracic pockets. When arriving at a new flower, they actively transfer pollen to the ovules with their forelegs (Cook et al., 2004). Active pollination benefits the plants, and fig species with actively pollinating wasp species have lower anther:ovule ratios (Cook et al., 2004). The costs and benefits of active pollination to the wasp are less clear, but evolution of this behavior may be driven by benefits to the larvae of successful fertilization of the seeds. The behavior and morphology of active pollination has evolved multiple times within a single genus of fig wasps, and in all cases the anther:ovule ratio of the host fig is matched to the pollinator, suggesting that the evolutionary forces exerted by such mutualisms are strong (Cook et al., 2004).

A striking evolutionary novelty observed in yucca moths that facilitates their mutualistic interaction with the yucca plant are the tentacular mouthparts that can move yucca pollen with great precision (Pellmyr and Krenn, 2002). Tentacles appear to develop from the galea that form the proboscis in most moths. The tentacles are controlled by a hydraulic extension mechanism similar to that used for control of the proboscis in most moths, suggesting that these structures could evolve with relatively few evolutionary steps (Pellmyr and Krenn, 2002).

5.6 Regulation of Growth and Development

5.6.1 Embryogenesis

Study of insect embryonic development is an extremely active area of research, especially using molecular approaches in *Drosophila*. Many specific aspects of pattern formation, the role of morphogens in creating tissue variation, and specific transcriptional pathways are now known (Jaeger, 2009). In addition, the ability to visualize gene expression inside *Drosophila* eggs provides an exquisite tool for examining the effects of environmental stresses on development (Vlisidou, 2009). For example, fly embryos have been used to understand molecular targets of methylmercury pollution on neural system development (Rand et al., 2009) and hypoxia effects on tracheal development (Mortimer and Moberg 2009). Developmental biologists have created models that help explain canalization in the face of environmental variation that have the potential to be extremely useful for understanding the effect of environmental variation on major developmental processes (Manu et al., 2009).

5.6.2 Hormonal and Genetic Control of Juvenile Growth

An insect's body size depends on growth rate and development time in the juvenile stage. Genes affect growth rate by influencing capacities to consume, process, and assimilate food, and the hormonally-regulated transitions to a sexually mature state that determine development time. These intrinsic factors are modulated by environmental factors such as food availability, food quality, and temperature.

A fundamental problem is to understand the mechanisms that cause an animal to grow to a certain size. In recent years, a variety of research has provided new insights into this process in insects. The hormonal mechanisms responsible for controlling the transition from feeding to molting are now being elucidated.

Recent studies have shown that cellular growth in *D. melanogaster* is regulated by a large number of signaling pathways including Myc/Max, Cyclin/Cdk and Ef2, Wnt, P38 MAPK, FRAP/TOR, and JAK/STAT (Bjorklund et al., 2006). The TOR/AKt/FOXO pathway has been identified as a key signaling pathway that links availability of amino acids to protein synthesis, cellular proliferation, and body size (reviewed by Mirth and Riddiford, 2007). Organ and body size can also be intrinsically regulated during development by the Hippo kinase cascade, which when activated leads to the inactivation of the transcriptional co-activator Yorkie, suppressing cellular proliferation and increasing apoptosis, thus reducing organ and body size (reviewed by Dong et al., 2007; Pan, 2007). Organ and body size are also regulated by temporal changes in Dpp signaling, which alter the ratio of cellular proliferation to apoptosis (Wartlick et al., 2007; Schwank and Basler, 2010). Additional factors, such as mechanical forces on organs, may also be important in determining organ and body size (Aegertner-Wilmsen et al., 2007). Most of these growth-signaling pathways have been shown to also function in mammals, and thus are likely to occur broadly in animals.

Insects grow to a certain point, after which they molt to the next instar or adult stage; thus controlling the transition from the growth to the reproductive adult stage is critical in determining size. For holometabolous insects, the first step in initiating the transition toward metamorphosis occurs when the juvenile attains a critical weight, defined experimentally as the size at which starvation no longer delays the timing of the molt (Nijhout et al., 2006). After a juvenile passes its critical weight, levels of juvenile hormone (JH) begin to decline due to decreased secretion of hormone and increased hemolymph levels of juvenile hormone esterase (Nijhout et al., 2006). The fall in JH triggers secretion of ecdysone from the prothoracic gland, which causes the transition from feeding to wandering stage and then ecdysis. Experimental elevation of JH extends the duration of the terminal juvenile instar and tends to increase body size, whereas blocking the release of JH shortens development time and reduces size (Nijhout and Williams, 1974). An additional factor in triggering molting may be growth of the prothoracic gland in response to insulin-like peptide signaling, leading to increased ecdysone production by the prothoracic gland (Colombani et al., 2005b; King-Jones and Thummel, 2005; Mirth et al., 2007). Ecdysone appears to inhibit growth and to be a key factor in the transition from the growing larval stage to the subsequent stages (Colombani et al., 2005b; King-Jones and Thummel, 2005; Fig. 5.22).

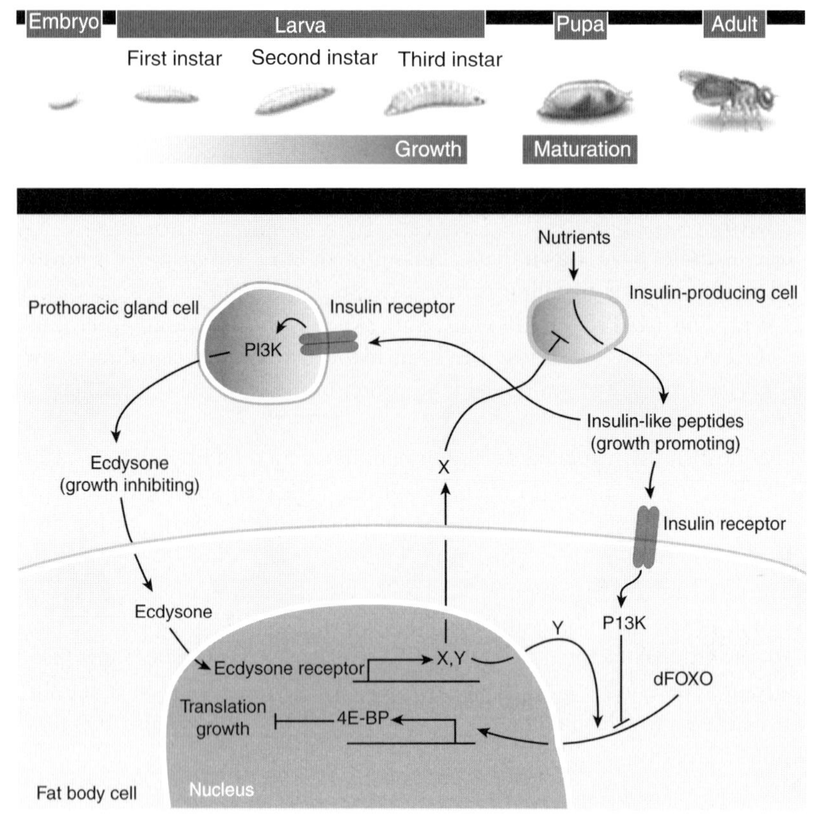

Fig. 5.22 (Top) The four major stages of the *Drosophila* lifecycle: embryo, larva, pupa, and adult. Growth occurs during larval stages in response to insulin signaling and basal levels of the steroid hormone ecdysone. This is followed by sexual maturation during metamorphosis. (Bottom) The prothoracic gland releases ecdysone that activates the ecdysone receptor in fat body cells, producing an unknown factor X. This factor may suppress growth by inhibiting the release of insulin-like peptides from insulin-producing cells in the brain. Insulin-like peptides activate the insulin receptor and PI3K signaling pathway that blocks nuclear translocation of the transcription factor dFOXO. The ecdysone receptor may also induce expression of a factor Y that directs nuclear translocation of dFOXO, activating genes that inhibit growth, including that which encodes the 4E-BP protein translation inhibitor. From King-Jones and Thummel (2005), with permission from AAAS.

Insulin-like peptides (ILP, seven currently identified in *Drosophila* (Brogiolo et al., 2001)) are important circulating regulators that link growth and development rate to nutritional status (Wu and Brown, 2006). Although there is evidence for ILP in a diverse variety of insects, most of the functional characterization of ILP signaling has been done with *Drosophila* (Wu and Brown, 2006).

Some ILPs are produced, stored, and released into the hemolymph from neurosecretory cells or glands in the brain, while others are produced as local growth factors (Wu and Brown, 2006). ILP are transported in the hemolymph by binding proteins that reduce their availability to receptors (Arquier et al., 2008; Honegger et al., 2008).

At target tissues, ILP bind to receptor tyrosine kinases (and perhaps other receptors), activating a variety of intracellular pathways that stimulate cellular growth and development (Fig. 5.23). Binding of ILP to the receptor activates a catalytic subunit that phosphorylates insulin receptor substrate proteins (*chico* in *Drosophila*), which in turn activates PI3K (phosphatidylinositol-3-kinase),

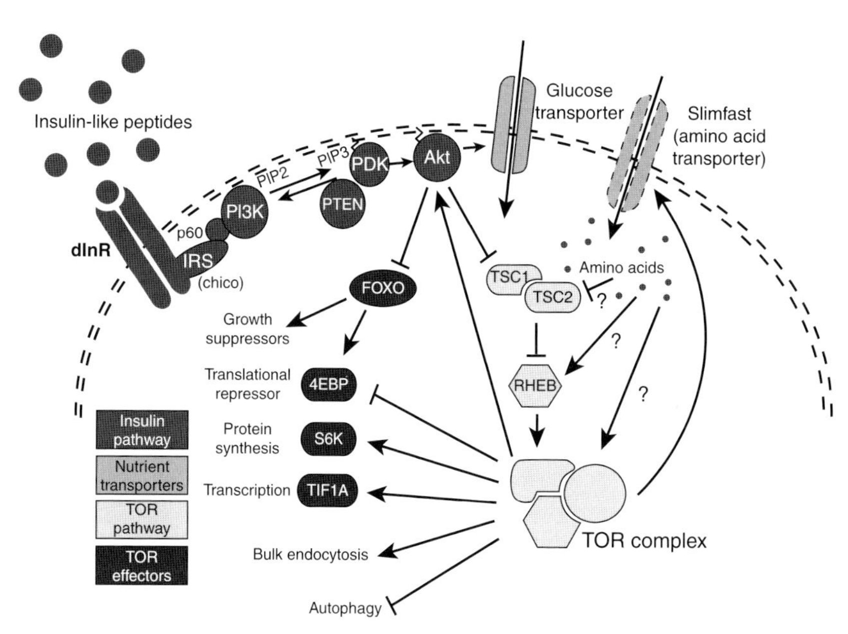

Fig. 5.23 Signaling pathways involved in nutrient-dependent growth of *Drosophila* (Mirth et al., 2007). Insulin-like peptides (ILP) bind to the drosophila insulin receptor (dInR), which in turn activates a signaling cascade via PI3K and Akt that enhances cellular growth by inhibiting repressors of growth and translation (FOXO), and activating nutrient uptake and protein synthesis. Higher levels of nutrients activate growth via RHEB and TOR. Abbreviations: IRS: insulin-receptor substrate (encoded by *chico*); PI3K: phosphatidylinoside 3-kinase; PIP2: phosphatidylinositide 4,5 bisphosphate; PIP3: phosphatidylinositide 3,4,5 triphosphate; PTEN (reverses insulin signaling by catalyzing PIP3 to PIP2); PDK: phosphoinositol-dependent kinase; Akt (protein kinase B); FOXO: forkhead box; TSC1/TSC2: tuberous sclerosis dimmer; RHEB: Ras Homolog Enhanced in Brain; TOR: Target of Rapamycin; 4EBP: 4E Binding Protein; S6K: S6 kinase; TIF1A: Transcriptional Intermediary Factor 1A. From Mirth et al. (2007), with permission from John Wiley and Sons.

leading to activation PDK1 (phosphatidylinositol dependent kinase1) and AKT/protein kinase B (a serine/threonine kinase).

Activation of ILP/PI3K/AKT signaling increases growth and size by at least three mechanisms (Fig. 5.23). AKT improves the cell's energy status by causing insertion of glucose transporters into the membrane and increasing glucose uptake. Activated AKT phosphorylates and thereby inactivates FOXO (forkhead box containing protein), a transcription factor which reduces cell proliferation and number (Junger et al., 2003). Activated AKT also leads to increased signaling in the target of rapamycin (TOR) pathway. Activated TOR increases cell growth (size) by at least three mechanisms, by activating protein synthesis initiation via phosphorylation of elongation factors, by activating translation elongation (mediated by phosphorylation of the ribosomal protein S6 kinase, (Wang and Proud, 2006), and by promoting amino acid uptake to the cell (Hennig et al., 2006).

Other hormones have been shown to interact with ILP and reduce ILP stimulation of growth. Increased ILP signaling increases juvenile hormone secretion via insulin receptors in the corpora allata, and perhaps via effects on secretion of neuropeptides that regulate the juvenile hormone synthesis (Tu et al., 2005). Juvenile hormone suppresses growth of the primordial imaginal discs during larval starvation, thus antagonizing ILP (Truman et al., 2006). Growth is also inhibited by ecdysone, which antagonizes ILP signaling at the level of FOXO and translation elongation factors (Colombani et al., 2005a, 2005b; Mirth et al., 2007; Mirth and Riddiford, 2007). Another mechanism by which ecdysone reduces growth is inhibition of feeding (Liu et al., 2010).

Recent evidence suggests that oxygen plays an important role in developmental control of insect size (Harrison and Haddad, 2011). Key early findings were that hypoxia reduced the body size of mealworms and increased tracheal diameters, while hyperoxia extended instar durations (Loudon, 1988, 1989; Greenberg and Ar, 1996). More recently it has been shown that the body size of *D. melanogaster* is linearly reduced by development in more severe hypoxia (Peck and Maddrell, 2005), but unaffected by hyperoxia (Klok et al., 2009). Moderate hypoxia (10 kPa) tends to reduce growth rate and extend development time (Frazier et al., 2001; Klok et al., 2009), likely due to a reduction in feeding rate (Frazier, 2007). Rearing for multiple generations in progressively lower oxygen levels leads to an even greater developmental suppression of size by hypoxia (Zhou et al., 2007, 2008). Moderate hypoxia's suppressive effect on body size is likely mediated by at least two mechanisms, as hypoxia suppresses growth rate, especially during the final juvenile instar, but also affects cell and body size during the early to mid pupal phase (Heinrich et al., 2011). Hypoxia has been shown to reduce TOR signaling, providing a plausible signaling mechanism that lowers growth rate (Brugarolas et al., 2004; Dekanty et al., 2005). Moderate hypoxia also reduces feeding rates in *D. melanogaster*, which clearly can reduce growth

(Frazier, 2007). The tendency for higher oxygen levels to extend development time (Greenberg and Ar, 1996; Frazier et al., 2001; Klok et al., 2009; Heinrich et al., 2011) suggests that oxygen level might be involved in the sensing of critical weight. As insects grow, spiracles and major tracheae are likely fixed in size, while increases in body mass lead to increased demand for oxygen, and a strong decrease in the safety margin for oxygen delivery (Greenlee et al., 2004, 2005). Possibly, decreasing internal oxygen levels are a component of the trigger for critical weight and the ecdysone cascade that triggers molting.

5.6.3 Hormonal and Genetic Control of Molting and Metamorphosis

Molting is under complex neuroendocrine control (Figs 5.24 and 5.25). The process begins when an insect passes its critical weight, apparently sensed by growth of the prothoracic gland to a threshold size (Mirth et al., 2007). Variation in

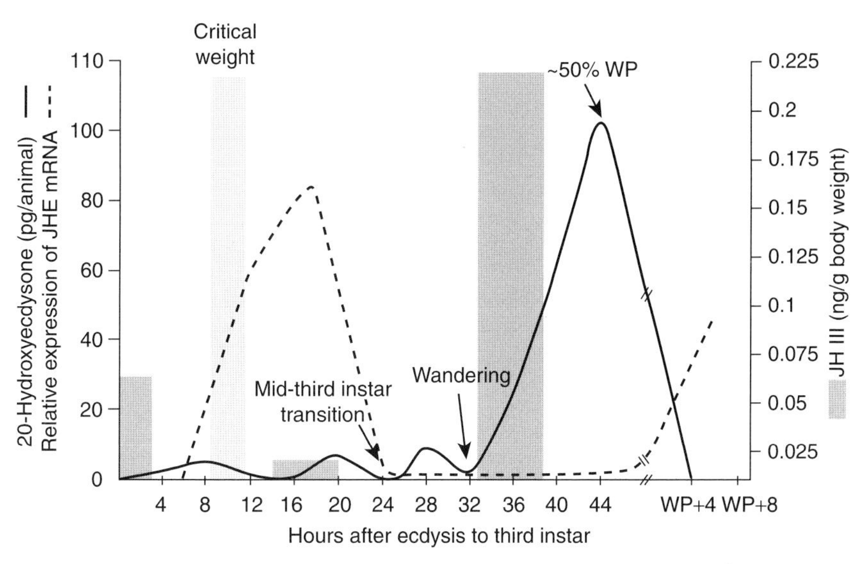

Fig. 5.24 Hormonal control of molting and ecdysis in third instar *Drosophila* (Mirth et al., 2007). Gray bars: relative levels of juvenile hormone (JH); light bar: time period when critical weight obtained; solid line: 20-hydroxyecdysone levels, dashed line: relative expression of juvenile hormone esterase. A hypothesized mechanism for how larvae sense the critical weight: after the critical weight is reached, JH levels fall due to decreased synthesis and a rise in JH esterase. Decline in JH below a threshold causes secretion of prothoracotropic hormone (PTTH) which stimulates the prothoracic gland to secrete ecdysone that triggers the transition from feeding to wandering. From Mirth et al. (2007), with permission from John Wiley and Sons.

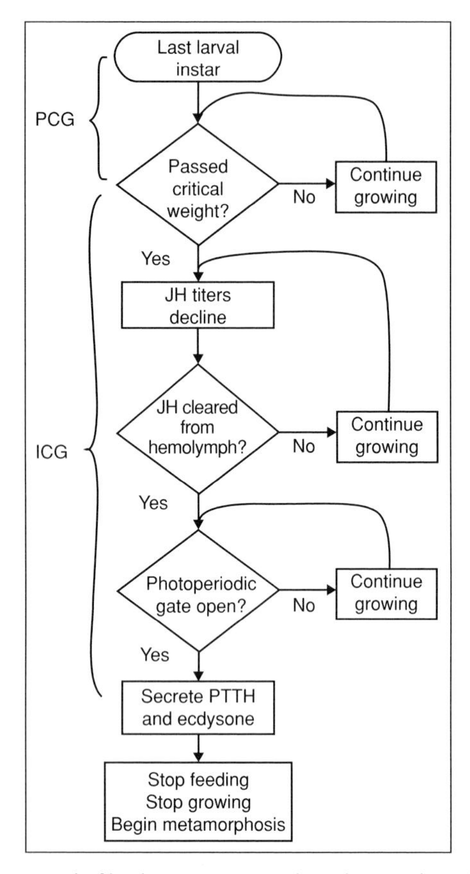

Fig. 5.25 Model for control of body size in insects, based on studies of *Manduca sexta* (Nijhout, 2006). About half of body mass is gained pre-critical weight (PSG), after which juvenile hormone (JH) levels fall until below a threshold and the correct time of day (photoperiodic gate) is reached. The fall in JH triggers the hormonal events that cause wandering behavior and ecdysis. About half of the last instar's mass gain occurs during the interval to cessation of growth (ICG).

critical weights is strongly correlated with variation in maximum larval and adult size across populations of *Manduca sexta* (Fig. 5.26). Growth of the prothoracic gland is stimulated by ILP signaling, and ILP-stimulated growth of the prothoracic gland to a threshold level of ecdysone production is hypothesized to provide a mechanism for sensing of critical weight by the brain (Mirth et al., 2007).

Molting itself is triggered by major peaks of ecdysteroid secretion, stimulated by secretion of prothoracotropic hormone (PTTH) from brain neurosecretory cells.

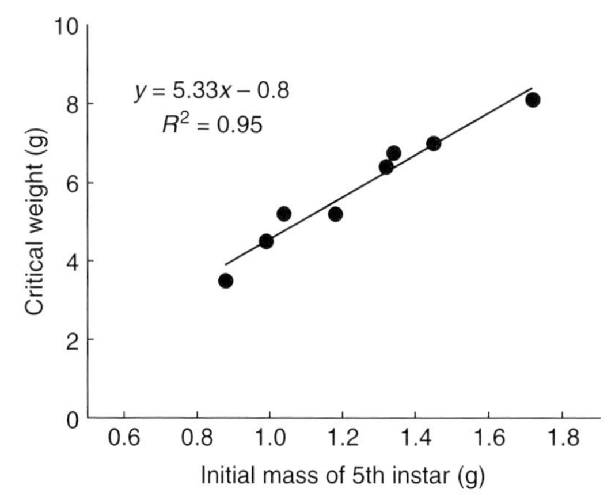

Fig. 5.26 Variation in body size among different genetic strains of *Manduca* are positively correlated with differences in their critical weights (Nijhout, 2006).

Ecdysteroids are converted to 20-hydroxyecdysone (20E) by a variety of tissues. Together, elevations in ecdysone and 20E levels, via the ecdysone receptor (EcR) initiate and coordinate a cascade of transcription factors that trigger the processes of molting; first apolysis (separation of the epidermis from the cuticle), secretion of molting fluid that digests the old cuticle, and development of the new cuticle. Ecdysis is the process of the insect splitting and emerging from the old cuticle, and is also under complex hormonal control (Zitnan et al., 2007). Elevation in ecdysteroids levels trigger the production of both ecdysis triggering hormone (ETH) in Inka cells adjacent to tracheae and the receptors to ETH in the central nervous system. A fall in ecdysone levels is necessary to allow ETH secretion from the Inka cells, which occurs in response to the brain neuropeptides corazonin and eclosion hormone (Zitnan et al., 2007). ETH then triggers the complex pattern of motor behaviors of ecdysis.

Larval–larval molts are accompanied by elevated levels of juvenile hormone, which likely act via receptor proteins termed *ultraspiracle* and *methoprene-tolerant* (*Met*, Riddiford, 2008). In holometabolous insects, elevated juvenile hormone prevents the transcription of *broad*, a transcription factor that causes formation of pupal cuticle and metamorphic differentiation of imaginal discs.

The metamorphic molt occurs when levels of juvenile hormone are low. Decline in juvenile hormone levels below a threshold allows the secretion of PTTH at the next photoperiodic gate (Mirth and Riddiford, 2007). PTTH induces the ecdysone secretion that causes cessation of feeding and larval wandering, and a second, later peak of PTTH and ecdysone secretion trigger pupariation. Multiple other brain neuropeptides in addition to PTTH are also involved in the control of ecdysone secretion, at least in the silkworm, *Bombyx mori* (Yamanaka et al., 2006).

5.6.4 Developmental Effects of Temperature on Size

Insects, like 80% of ectotherms studied, tend to be smaller when reared at higher body temperatures, the so-called "temperature size rule" that "hotter is smaller" (Atkinson, 1994; Kingsolver and Huey, 2008; Angilletta, 2009). This pattern of phenotypic plasticity is reversed in a few orders of insects, including orthopterans (crickets and grasshoppers; Atkinson, 1994). These patterns only hold within moderate temperature ranges; at extreme, stressful temperatures, body size decreases (Karan et al., 1998). Limited data suggest that the "hotter is smaller" rule may also apply to eggs (Honek, 1993).

Body size can change by several distinct mechanisms: by changes in the size of cells, the number of cells, or the amount of interstitial space and connective tissue. In a variety of organisms, cell sizes decrease when they are reared at warmer temperatures (Arendt, 2007; van Voorhies, 1996). In all 12 studies of the thermal effects on size in insects, changing cell size was a major contributor to changing body size (Arendt, 2007). The data for insects are difficult to generalize since so far only two studies have been done with species other than *D. melanogaster* (Arendt, 2007). However the observation that crickets have larger cells when reared at colder temperatures, despite being larger at higher temperatures, supports the link between temperature and cell size for insects (Mousseau, 2000).

Higher temperatures tend to increase growth rates and development rates; the smaller size of insects reared at warmer temperatures may reflect the fact that development time is reduced more than growth rate is increased (van der Have and de Jong, 1996). However, the mechanisms responsible for greater effects of temperature on differentiation than growth are unclear. Interestingly, although larval *Manduca sexta* attain a smaller maximal size when reared at warmer temperatures, the critical weight did not vary with temperature (Davidowitz et al., 2003). The smaller body size at higher temperatures was associated with a shorter duration of the period from the critical weight to the initiation of wandering behavior, when larvae cease feeding (Davidowitz et al., 2004; Davidowitz and Nijhout, 2004). During this period, juvenile hormone levels are declining to a level sufficient to permit initiation of the hormonal cascade associated with molting. Thus, at least in *Manduca*, a plausible mechanism for the reduction in development time at higher temperatures is more rapid progression of neuroendocrine events such as the clearing of juvenile hormone from the hemolymph (Davidowitz et al., 2004; Davidowitz and Nijhout, 2004).

Whether temperature-related changes in body size are beneficial (adaptive) or are a passive consequence of temperature remains controversial (Deere and Chown, 2006; Stillwell, 2010; Wilson and Franklin, 2002; Woods and Harrison, 2002).

5.6.5 Evolutionary Effects of Temperature and Latitude on Size

To test for evolutionary effects of latitude on body size, populations of insects from different locations are reared in the lab under common-garden conditions, ideally for several generations to eliminate the possibility of parental effects. When this is done, some insect species collected from lower latitudes are smaller (consistent with Bergmann's rule) but others show the opposite trend (reverse Bergmann's), or no pattern at all (Blanckenhorn and Demont, 2004; Hawkins, 1995).

There is strong evidence that low temperature drives the evolution of larger *Drosophila* at higher latitudes. First, this pattern is observed across multiple continents (Van't Land et al., 1999), and the pattern evolves quickly during invasions (Gilchrist et al., 2001; Huey et al., 2000). Second, *Drosophila* evolves smaller body sizes when reared at warmer temperatures in the laboratory (Partridge et al., 1994). Most convincingly, studies artificially selecting for large body size demonstrate that a larger body size is favored when selection occurs at lower temperatures (McCabe and Partridge, 1997; Reeve et al., 2000).

Seasonality also has strong effects on latitudinal patterns in size. Small, multivoltine species such as *Drosophila melanogaster* are more likely to show patterns consistent with Bergmann's rule (Blanckenhorn and Demont, 2004). Larger species with longer development times are more likely to exhibit a reverse Bergmann's trend, suggesting that animals with long development times are likely to experience selection associated with the end of the growing season to mature more rapidly at a smaller size (Blanckenhorn and Demont, 2004; Chown and Gaston, 1999). Significant work in other invertebrates suggests that organisms with higher growth (including those selected for high growth rate by short growing seasons) require higher P contents due to their need for higher levels of ribosomal RNA to support higher rates of protein synthesis (Elser et al., 1996).

Other factors, such as starvation and desiccation resistance, may also contribute to the evolution of body size across latitudes (Stillwell, 2010). Dry conditions or long dearth periods may select for larger body size, overwhelming effects of season length or temperature for some species (Stillwell, 2010).

5.6.6 Bigger is Better

In the lab, there is generally a strong, often exponential, relationship between body size and fecundity (Roff, 2002). A substantial number of studies have now considered the relationship between body size and fitness in the field by examining the evidence for directional selection on body size (recently reviewed by Kingsolver and Huey, 2008; Kingsolver and Pfennig, 2004). Kingsolver and colleagues have examined linear directional selection, measured as the variation in the trait (in units of standard deviations) relative to variation in relative fitness, an approach that allows comparisons across animals of different sizes and systems. Their meta-analyses found strong positive relationships between body size and measures of fitness (fecundity, survival, mating success). Thus, the data available suggest that there should, in many systems, be selection for insects to achieve larger body sizes.

Given that "bigger is better," why are so many insects small? Blanckenhorn (2000) pointed out that a variety of countervailing selection pressures may lead to smaller body size, including decreased survival associated with longer exposure to dangers such as predation or parasitism during the growth phase of the lifecycle, reduced maintenance and repair in fast-growing organisms, reduced agility, or increased detectability. However, there is little support in the literature for such stabilizing selection (Kingsolver and Huey, 2008). One possibility is that ecological conditions limit the ability of larger animals to achieve their potential fecundity. Gottard et al. (2007) recently demonstrated this in the butterfly *Pararge aegeria*. In the lab, when thermal conditions are suitable throughout the day, oviposition rates increased strongly with body size (Fig. 5.27). However, if the time when temperature was suitable for flight and oviposition was restricted to 2 hours per day, oviposition rates were lower and there was little effect of body size on fecundity (Fig. 5.27). Over the natural temperature range, no oviposition occurred below 15°C, and oviposition rates reach an asymptote above 25°C (Fig. 5.27). This thermal pattern for oviposition was compared to natural conditions for *P. aegeria*. Modeling suggested that under natural conditions, there would be little fecundity benefit for larger butterflies in the field. Such limitations could occur for other reasons (e.g. predator- or parasite-free time), and could provide an important mechanism for stabilizing selection on body size.

5.6.7 Nutritional Regulation of Growth and Development

Insects given low-quality food, or small quantities of higher-quality food, grow more slowly, and generally achieve smaller adult sizes, as described above. What mechanisms are responsible for these effects? Obviously one possible explanation is simple mass-action effects; shortages of building blocks could slow anabolic pathways. Thus low sugar intake could lead to low hemolymph trehalose levels, leading to low intracellular glucose levels, causing slowing of glycolysis due to substrate shortages. While such mechanisms are plausible, there is little direct evidence for this occurring. By contrast, we have substantial evidence for other signaling mechanisms that link nutrition to growth.

In response to nutritional variation, the rate of growth in insects is regulated by the insulin/insulin growth factor signaling pathway and the target of rapamycin (TOR) pathway (Bohni et al., 1999; Oldham and Hafen, 2003). One current model proposes that the fat body senses nutritional status (specifically, levels of amino acids). Signals from the fat body cause variation in how much insulin is

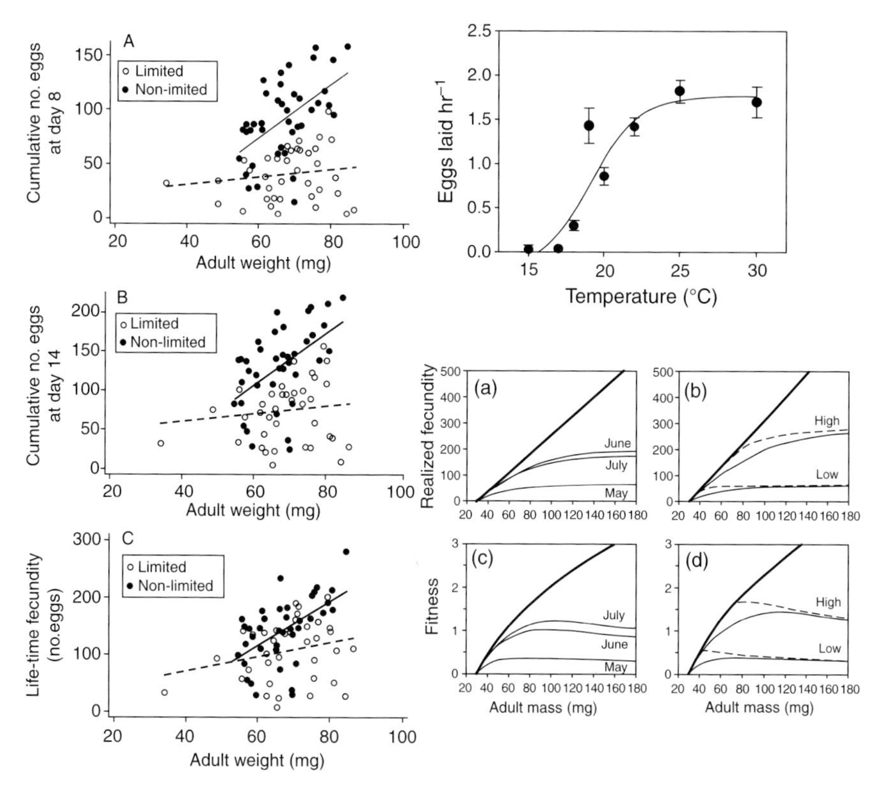

Fig. 5.27 Left Cumulative number of eggs laid (top, 8 days of age, middle, 14 days of age, bottom, lifetime) vs. female mass for the butterfly, *Pararge aegeria*. Filled symbols (solid line) show results when the females can lay eggs for eight hours/day; open symbols (dotted line) show the results with two hours per day oviposition time (limited treatment). Upper Right: oviposition rate as a function of temperature. Lower right: Predicted size-fecundity (*a*, *b*) and size-fitness (*c*, *d*) relationships for *Pararge aegeria*. In each case, bold lines indicate predictions of lifetime fecundity assuming no time limitation during oviposition, whereas thin lines indicate predictions assuming time constraints for various thermal conditions. *a* and *c* show mean effects of natural variability in temperature and sunshine hours for the months May to July; *b* and *d* show mean effects of constant (*dashed lines*) versus natural variability (*solid lines*) in mean daily temperatures. For the field estimates, female body temperature was 4.25°C above ambient during all periods of sunshine. From Gotthard et al. (2007), with permission from the American Society of Naturalists.

release from the brain, leading to altered growth. Geminard et al. studied this recently in *Drosophila* (Géminard et al., 2009). They found that low-protein diets or starvation caused insulin to accumulate in brain neurosecretory cells (detected with antibodies), and to decline in the hemolymph, suggesting reduced secretion. Refeeding with a high-protein diet caused insulin to disappear from the brain, suggesting secretion. Single amino acid diets indicated that branched-chain amino acids (leucine or isoleucine) were responsible for stimulating insulin secretion.

How is starvation sensed? Starvation leads to inactivation of the insulin/TOR pathway in the fat body (Géminard et al., 2009). Evidence suggests that this leads to a change in hormone secretion by the fat body, leading to reduced insulin secretion by the brain and general reduction in growth (Géminard et al., 2009). Thus, the availability of nutrients is sensed by the fat body, which controls insulin release from the brain via a secreted humoral factor.

5.7 Body Size

Insect body sizes range from egg-parasitizing wasps only 140 μm long to 0.55 m walking sticks and scarab beetles that exceed 50 g. Within species there can also be considerable (orders of magnitude) variation in body size, both during ontogeny and within populations. Body size affects virtually every aspect of insect physiology, ecology, and evolution including metabolic rates (Bartholomew and Casey, 1978; Chown et al., 2007), water loss rates (LeLagadec et al., 1998), heat balance biology (Bartholomew and Casey, 1977; Stone, 1993), range size (Gaston and Blackburn, 1996), fecundity (Honek, 1993), and flight behavior (Lehmann, 2002), and thus is one of the most important parameters affecting the environmental physiology of insects. Many of these patterns in life history can be related to size related variation in metabolic rates, otherwise known as the metabolic theory of ecology (Brown et al., 2011). This broad-scale approach linking metabolism to a wide variety of processes has been extremely influential as scientists have worked to connect phenomena occurring across different scales of biological organization.

5.7.1 Body Size and Metabolic Rate

Like other animals, insects have inactive metabolic rates that *scale*, such that larger insects have lower mass-specific metabolic rates, at least for broad interspecific comparisons. On a log-log plot of metabolic rate versus mass, the slope is approximately 0.75, after phylogenetic correction (Chown et al., 2007). However, the results are more varied for intraspecific comparisons, particularly during ontogeny. Honeybee larvae increase in mass by more than 400-fold in only four days, and their metabolic rates scale with mass 0.9 (Petz et al., 2004). Larvae of the tobacco hornworm (*Manduca sexta*) span four orders of magnitude in body mass and exhibit CO_2 emission rate scaling with a mass exponent of 0.98 across the entire larval stage (Greenlee and Harrison, 2005). However, individual instars show different patterns of metabolic rate scaling; as larvae grow within an instar, the mass-specific CO_2 emission decreases with age or size among early instars, but it increases with size in the final larval instar (Greenlee and Harrison 2005). Similarly in grasshoppers, the pattern of CO_2 emission rate scaling varies across instars with exponents ranging between 0.45 and 0.91, while across its entire development, metabolic rate scales with the exponent 0.73 (Greenlee and Harrison, 2004a; Greenlee and Harrison, 2004b).

Interestingly, social insect colonies also appear to exhibit metabolic rates that scale approximately with mass 0.75. Honeybee clusters maintain a relatively constant core temperature when air temperature falls; mass-specific metabolic rates and mass-specific heat loss from the cluster falls in larger clusters due to a reduced surface area-to-volume ratio (Southwick et al., 1990). In the polymorphic ant *Pheidole dentata*, lower mass-specific colony metabolic rates arise from larger colonies having a greater fraction of larger "major" workers (Shik, 2010). In a harvester ant species with monomorphic workers, *Pogonomyrmex californicus*, lower mass-specific metabolic rates of larger colonies may be due to larger colonies having a lower fraction of active workers (Waters et al., 2010). Similar hypometric scaling patterns with disparate mechanisms suggest common underlying ecological or evolutionary forces that can be addressed by varied mechanisms in different species. Social insect colonies may be particularly useful for investigating mechanisms responsible for metabolic scaling due to the ease of manipulating and measuring specific components of a superorganism.

5.7.2 Ecological and Evolutionary Factors Affecting Body Sizes

Many ultimate factors affect the evolution of animal size (Allen et al., 2006). In a variety of insects, reproductive success, including fecundity in females and competitive ability of males, increases with individual size, due to positive correlations between adult size and egg size and number, foraging ability, fighting ability, and capacities to survive resource shortages (Brown and Mauer, 1986). Nonetheless, most species are small, as smaller animals tend to be more viable and require less space and time and fewer nutrients (Blanckenhorn, 2000). Ecological factors such as community interactions (competition and predator–prey) and the match between the size of animals and their food also affect distributions of body size (Allen et al., 2006). Taxon-specific physiological or biomechanical constraints can influence body size distributions; certainly, insects are usually smaller than vertebrates, a pattern that remains to be definitively explained.

5.7.3 Constraints on Maximum Size

Terrestrial arthropods in general, and insects in particular, are usually smaller than vertebrates. It is not unreasonable to say that insects dominate terrestrial niches for animals smaller than 1 g, while vertebrates dominate niches for animals larger than 1 g. Why is this?

At least four hypotheses have been put forward. The first is that exoskeletons are incompatible with large body size. This argument is based on the engineering principle that, to resist buckling, a cylinder must have wall thickness that increases faster than cylinder diameter, assuming constant material properties. In support, all terrestrial arthropods are relatively small (spiders and insects have similar maximal sizes) and marine crustaceans which need not support their weight reach much larger sizes than terrestrial ones. Although this would appear to provide a fundamental limit on insect size, little data has yet been published to support this compelling hypothesis, which is often stated as fact (e.g. Price, 1997). It is true that the thickness of the exoskeleton of the grasshopper leg increases with age/size (Hartung et al., 2004; Katz and Gosline, 1992, 1994). However, the only broad interspecific comparison of investment in the exoskeleton across insect body size found that exoskeleton mass scaled isometrically with body mass, which is unexpected if larger insects must have thicker exoskeletons to survive breakage (Lease and Wolf, 2010). As pointed out by Price (1997), it is possible that exoskeletal problems primarily arise during molting, when the soft cuticle could potentially turn into a "great soft pancake."

A second hypothesis is that small size is required by insects having tracheal respiratory systems. Unlike vertebrates and most other invertebrates, which transport oxygen to their tissues via a circulatory system, the insect tracheal system delivers oxygen via blind-ended tracheoles through which diffusion is likely an important mechanism of gas transport. The time for molecular diffusion increases exponentially with distance, and this distance may increase in larger insects. As described in Chapter 6, virtually all large insects use convection within large tracheae to enhance oxygen transport, so the simplistic vision of insects having completely diffusion-based gas exchange should be exterminated. Nonetheless, diffusion is ultimately responsible for gas exchange in the blind-ended tracheoles, and all insects examined so far (up to 20-g beetles) can recover from exposure to anoxia despite that exposure imposing a paralyzed state that should inhibit musculo-skeletal induced convection.

The oxygen-limitation hypothesis for the small size of insects has been invigorated by the realization that the gigantic insects of the late Paleozoic occurred in hyperoxic conditions (Berner, 2006; Dudley, 1998; Graham et al., 1995; Ward, 2006). Some insects of the Carboniferous were up to 10-fold larger than those in similar groups alive today (Grimaldi and Engel, 2005; Shear and Kukalova-Peck, 1990), and certainly this correlation between historical hyperoxia and insect gigantism supports the notion that insect size is limited by oxygen delivery system. However, the paucity of the insect fossil record and the complex interactions between atmospheric oxygen level, organisms, and their communities make it impossible to definitively accept or reject this hypothesis.

Nonetheless, a variety of recent empirical findings support oxygen-limitation of insect size, including: 1) most insects develop smaller body sizes in hypoxia, and some develop and evolve larger sizes in hyperoxia; 2) insects develop and evolve lower proportional investment in the tracheal system when living in higher atmospheric PO_2, suggesting that there are significant costs associated with tracheal system structure and function; and 3) larger insects invest more of their body in the tracheal system, potentially leading to greater effects of aPO_2 on larger insects (Harrison and Haddad, 2011; Harrison et al., 2010). In addition, *D. melanogaster* evolves larger body sizes when reared in hyperoxia (Klok et al., 2009), and progressive rearing in higher oxygen levels for many generations increases the body size of flies (Zhao et al., 2010). Together, these studies provide a wealth of plausible mechanisms by which tracheal oxygen delivery may be centrally involved in setting the relatively small size of insects and for hyperoxia-enabled Paleozoic gigantism.

Instead of maximal insect size being limited by some biophysical mechanism, it is possible that this pattern results from adaptive responses to ecological interactions. Evolution of the Paleozoic giants has been hypothesized to have occurred as a response to stable, optimal environmental conditions (Briggs, 1985), and the availability of coal-swamp forests as a new habitat with little competition or predation (Briggs et al., 1991). Such improved environmental conditions could directly increase insect abundances or shift body size distributions. The large size of some of the Paleozoic insects could have been the result of size-selection by predators, or the need for greater force generation in ground litter filled with tree branches (Shear and Kukalova-Peck, 1990). Temperature could be involved, as lower temperatures tend to cause insects to be larger (Kingsolver and Huey, 2008), and global temperatures were low during the late Carboniferous when some insects were gigantic (Royer et al., 2004). Possibly the Paleozoic giant insects were displaced by the evolution of larger, more successful flying vertebrates.

5.7.4 Trade-offs in Investments Among Body Parts

One of the most interesting aspects of allometry, now a focal area in developmental and evolutionary biology, is the relative investment in body proportions. One of the best-studied examples of such trade-offs are the morphological trade-offs observed between developmentally plastic morphotypes utilized for dispersal vs. sedentary morphotypes specialized for feeding and reproduction. Reduced investment in flight apparatus (wings and flight muscle) can allow increased investment in other structures and systems that enhance feeding, growth, and reproduction. For example, short-winged forms of planthoppers have greater investment in feeding musculature than long-winged forms, both within and between species (Huberty and Denno, 2006b). An elegant series of papers have demonstrated the many biochemical and morphological trade-offs associated with winged vs. wingless morphs in crickets (Zera, 2006; Zera and Zhao, 2003a, 2003b, 2006a, 2006b).

Allometric development of dung beetle horns is tied tightly to mating strategy (Rowland and Emlen, 2009). The largest males have extremely large and well-developed horns that allow them to defend tunnels containing hornless females. Smaller males have reduced horns and resort to sneaking strategies to obtain access to females, such as slipping past guarding males or digging side tunnels. Horn size depends on both genetic and nutritional factors, with variation among species. In some species, males exceeding a size threshold develop horns, in others, there is a positive relationship between horn and body size in all males, but above a threshold body size, horns are larger, and in many species both mechanisms exist and there are functionally three male horn phenotypes (Fig. 5.28). At least in some of these species, the smallest males may mimic females (Rowland and Emlen, 2009).

The developmental mechanisms that result in altered development of body parts are beginning to be clarified. In beetles, differential expression of

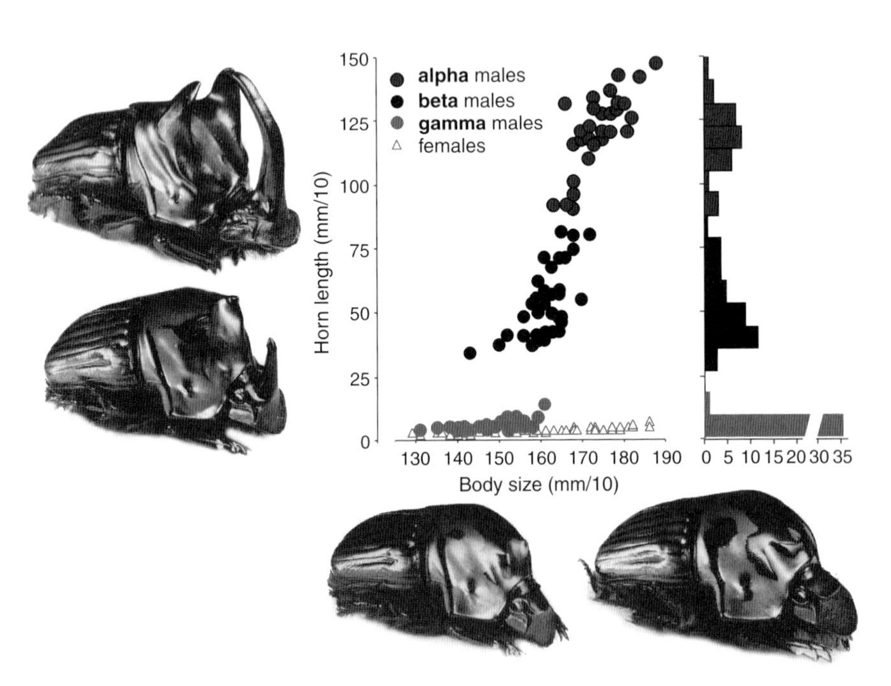

Fig. 5.28 Male trimorphism in scarab beetles. Alpha and beta males both produce horns, but the relative size of these differs. Gamma males and females lack horns. From Rowland and Emlen (2009), with permission from AAAS.

limb-patterning genes and insulin receptors appears to modulate differential growth of horns in beetles (Emlen et al., 2006). Moszek and Rose (2009) recently used RNA interference to provide evidence that the limb-patterning genes *Distal-less* and *homothorax* regulate differential horn expression in beetles of different sexes, sizes, and species. Thus we are beginning to move toward understanding both the proximate and ultimate mechanisms for understanding both shape and size.

6 Oxygen

6.1 Defining the Problem

Oxygen's dual nature was unintentionally captured by Peter Mere Latham's old quote: "Poisons and medicine are oftentimes the same substance given with different intents." Like all aerobic organisms, insects must have oxygen to run their metabolisms. But too much oxygen can be poisonous. The central respiratory problem for insects is to regulate oxygen delivery to tissues under variable internal and external conditions. This chapter focuses on the physiological mechanisms that insects use to do so—and the molecular, ecological, and evolutionary consequences of oxygen deprivation and surplus. As in previous chapters, we focus on a few selected topics that highlight recent progress, at the expense of encyclopedic coverage.

The organizing idea of this chapter is that the instantaneous oxygen level, in any tissue, represents a quasi-equilibrium between oxygen consumption by the tissue and delivery by respiratory structures; the problem for an insect is to maintain adequate flux of oxygen from the environment to its tissues when both tissue demand and environmental supply vary. Problematic tissue oxygenation occurs whenever there is gross mismatch between supply and demand. Insufficient tissue oxygenation may occur when an insect transitions from rest to vigorous activity, especially flight (increased demand), or when insects encounter environmental hypoxia (decreased supply). Because oxygen supports energy and material movements through an insect—which are used for fitness-related processes like growth, reproduction, and activity—the link between oxygen physiology and insect ecology likely is profound, although at present under-explored. The first part of the chapter will focus on problems of tissue *hypoxia*. But an insect at rest can also experience the opposite problem—tracheal delivery exceeding tissue demand, leading to tissue *hyperoxia*. The resulting problem, production of reactive oxygen species, can wreak havoc on the basic molecules of life.

6.1.1 Principles of Gas Exchange

Gas exchange can become unbalanced under any of three broad conditions: from rapidly changing oxygen demand by tissues, from changes in environmental oxygen, or from changes in the resistance to gas flux by the respiratory system. Understanding the causes and consequences of these conditions depends on understanding mechanisms of transport. Unfortunately, the relative importance of mechanisms, even at a fundamental level—that is diffusion versus convection—is unknown for most insects. However, this area has seen vigorous debate and substantial progress in the past few years. We start with the basic equations describing the diffusion and convection of both oxygen and carbon dioxide.

6.1.1.1 Diffusion

Diffusive transport is the net movement of O_2 or CO_2 molecules from areas of high concentration (or partial pressure) to areas of low concentration, driven by random thermal motion. Because individual molecules go only on random walks, bulk transport is the statistical outcome of many such random walks. The diffusion coefficient, $D(m^2s^{-1})$, is a measure of how rapidly diffusing gases move. Denny (1993) demonstrates how to derive values of D from underlying molecular kinetics. D depends on multiple factors, among them temperature (faster diffusion at higher temperatures), molecular weight of the diffusing molecule (slow diffusion of larger molecules), and identities and concentrations of other molecules present. D also depends on whether the gas is diffusing in liquid or gas; in liquids, gases bump into neighbors much more often than they do in gases, and therefore diffusion is many-fold slower in liquids than in air. For example, the diffusion coefficient of O_2 is approximately 10,000 times slower in water than in air (Denny, 1993). When more than two gases are present, the situation becomes complicated and may give counterintuitive results (for example, see Brès and Hatzfeld, 1977).

These observations explain why insects can exchange gases, at least in part, by diffusion—because transport occurs in air-filled tracheae (rather than blood in vertebrates or the related crustaceans). The complementary observation is that insects are small, meaning that gases don't have far to go. Small size is disproportionately important because diffusive transport and distance are not related isometrically. To wit, the average distance an individual molecule travels (x_{rms}, the root mean square distance traveled by a molecule) increases with the square root of time (t):

$$x_{rms} = \sqrt{2Dt}. \quad (1)$$

In tissues, which are essentially water, a common diffusion distance for oxygen is 10 μm, which requires ~0.01 s; for 1 mm, which is 100 times as far, it takes 100 s, which is 10,000 times as long; for 1 m, it takes 10^8 s (almost 3 years) (Schmidt-Nielsen, 1997). Similar scaling, though with much faster transport times, applies to oxygen diffusing in air-filled tracheae. Clearly, although diffusion is speedy on microscopic scales, it rapidly becomes inadequate for supplying oxygen over organism-sized distances.

At steady-state, diffusive flux is given by Fick's first law:

$$J_i = -D_i \frac{\partial C_i}{\partial x}, \tag{2}$$

where J_i is the flux density of gas of species i, the number of molecules crossing an arbitrary point in the x direction per area per unit time, C_i is gas concentration gradient, and x is distance. The negative sign indicates that the gas (oxygen) moves from high concentrations to low. Equation 2 is one-dimensional, but two- or three-dimensional forms can be used (Rashevsky, 1960).

As stated, Equation 2 is a partial differential equation that is quite general; simpler versions can be derived for special cases. The usual way forward is to assume fixed concentrations at the boundaries of some region of interest. In the steady state, this assumption gives a linear concentration gradient—that is $\partial C_i/\partial x$ is rewritten as $\Delta C_i/\Delta x$, or even more explicitly as $(C_a - C_b)/x$, where C_a and C_b are gas concentrations at either end of some structure of length x. In addition, physiologists often do not describe *flux density*, J_i, but rather total flux (molecules per time), J_i, which can be obtained by multiplying both sides of Equation 2 by the area, A, through which the flux occurs. Together these modifications give the more familiar version:

$$J_i = -AD_i \frac{\Delta C_i}{x}. \tag{3}$$

Yet another rearrangement may be desirable in some situations, especially when one has little idea of the values of A, D, or x: all of the parameters on the right side except ΔC can be collapsed into a single term called diffusive condwuctance, G, so that:

$$J_i = G_i \Delta C_i. \tag{4}$$

Conductance can be thought of as "how easily oxygen moves down a given concentration gradient," and is the inverse of resistance. This equation is also a molecular analogue of Ohm's law, $I = V/R$, with the mapping $J \rightarrow I$, $G \rightarrow 1/R$, and $\Delta C \rightarrow V$.

An important assumption in Equation 4 is that diffusion along the entire path of x occurs in the same fluid phase, either air or water. But such an assumption may not capture situations of biological interest. For example, one may be interested in fluxes from air-filled tracheae to mitochondrial surfaces inside cells.

Here, using gas concentration, C_i, is inappropriate and partial pressure, P_i, should be used instead, which leads to the concepts of pressure and solubility.

The total pressure (P) of a mixed gas is the sum of the partial pressures of its component gases (P_i), a relationship known as Dalton's Law. Thus, for n different component gases:

$$P = P_1 + P_2 + \ldots + P_n \,. \tag{5}$$

The utility of biological thinking about partial pressures becomes apparent when oxygen moves between air and water. Imagine a water bottle half full with water and the remaining head space with ambient air (21% oxygen, which at sea level is about 21 kPa). If the bottle is shaken vigorously, to ensure close contact between air and water, oxygen in the air and oxygen in the water come into equilibrium. Afterward, on average, no net oxygen moves in either direction. And yet, the *concentration* of oxygen (mol m^{-3}) is much higher (about 30 times) in the air space than in the water. In fact, what is in equilibrium between the head space and the liquid is oxygen's *partial pressure*. Partial pressure acts like chemical activity (or fugacity): no net movement of molecules occurs when oxygen's tendency to escape is the same from air into water as it is from water into air.

To convert between partial pressure and concentration, Henry's Law must be invoked: at a constant temperature, the equilibrium concentration of gas dissolved in a liquid is proportional to the partial pressure of gas above it, or:

$$P_i = k_i C_i \,, \tag{6}$$

where k is the Henry's Law constant. For oxygen in water at 25°C, $k = 7.7 \times 10^{4} 1$ kPa mol^{-1}. In respiratory physiology, Equation 6 usually is rewritten in terms of a solubility coefficient (α), that is, in terms of concentration of gas in liquid obtained per unit partial pressure of the gas above the liquid: $C_i = \alpha_i P_i$. For oxygen in water at 25°C, $\alpha = 1.3 \times 10^{-5}$ mol L^{-1} kPa^{-1}. Because CO_2 is more polar than O_2, the solubility of CO_2 in water (and biological tissues) is higher than that of O_2; α for CO_2 at 25°C is 3.4×10^{-4} mol L^{-1} kPa^{-1}.

A further complication is that physically dissolved O_2 and CO_2 may react with water or bind to molecules in solution; such products diffuse independently with their own diffusion constants and concentration gradients. In particular, the total content of oxygen in a fluid will be the sum of physically dissolved and chemically bound oxygen (and there has been much recent interest in insect hemoglobins (Burmester and Hankeln, 2007)). The total content of CO_2 also includes other chemical forms, as CO_2 reacts with water to form carbonic acid, and then bicarbonate and carbonate, depending on the pH:

$$CO_2 + H_2O \leftrightarrow H_2CO_3 \leftrightarrow H^+ + HCO_3^- \leftrightarrow 2H^+ + CO_3^{2-} \,. \tag{7}$$

Carbonic acid is present only fleetingly; it rapidly falls apart into bicarbonate and a proton. Also, in most insect tissues, carbonate occurs at very low concentrations. Therefore, when CO_2 dissolves in aqueous solutions, the primary products are bicarbonate and protons. The latter depresses the pH of body fluids. Insects can, in some cases, actively drive the chemical equilibrium to the left or right by adding or subtracting protons from body compartments. The total CO_2 content of a solution is the sum of physically dissolved CO_2 and all chemical products (e.g. bicarbonate, carbonate, and CO_2 bound to proteins).

The capacitance coefficient, β, is a measure of the increase in total content of gas in the fluid as the partial pressure increases (analogous to Henry's law, Equation 6), including all physically transformed (e.g. bicarbonate) and chemically bound (e.g. oxyhemoglobin) forms. $\beta_i = \Delta C_i / \Delta P_i$, where ΔC_i and ΔP_i are changes in concentration and partial pressure, respectively (Dejours, 1975). For insect fluids, the flux equation above (Equation 3) can be rewritten in terms of partial pressures as:

$$J = -AD_i\beta_i \frac{\Delta P_i}{x}. \tag{8}$$

When using Equation 8 across air–liquid boundaries, which occur between tracheal air and tracheolar water, allowance must be made for the different values of D, β, and x in air versus water. This can be done by modeling flux as separate steps, in series, and sharing a common partial pressure of gas at their interface. For a steady-state, pure-diffusion system, these fluxes in series must be equal:

$$J = -AD_{air}\beta_{air}\frac{(P_a - P_b)}{x_{air}} = -AD_{water}\beta_{water}\frac{(P_b - P_c)}{x_{water}}. \tag{9}$$

For oxygen, for example, P_a could be the Po_2 at the spiracles, P_b the Po_2 at the air–water interface in the tracheoles, and P_c the Po_2 at the mitochondria.

Finally, for a steady-state system, the diffusive flux of oxygen, J, must match the metabolic consumption of oxygen, $\dot{V}O_2$, so that:

$$\dot{V}O_2 = -AD_{air}\beta_{air}\frac{(P_a - P_b)}{x_{air}} = -AD_{water}\beta_{water}\frac{(P_b - P_c)}{x_{water}}. \tag{10}$$

Because insects are small, a steady-state assumption will often be reasonable—that is, changes in tissue metabolic rates will show up quickly as changes in gas emission rates and ΔPo_2(see Förster and Hetz, 2010). However, when the conductance of the system changes (e.g. if all spiracles open or close), there is always a period of non-steady-state equilibration. Also, during rapid metabolic changes, gas exchange will not be steady state, although usually the timescales of change in metabolism will be much longer than the equilibration time, so that insects will usually be in quasi-equilibrium. Equilibration time will be extended when internal gas pools are large relative to the net flux (e.g. large air sacs, significant buffering of CO_2 by the tissues).

6.1.1.2 Convection

Krogh (Krogh, 1920a) did a set of analyses and experiments showing that small insects could, in principle, meet oxygen demands entirely by diffusion. This early finding tended later to be over-interpreted to mean that insects *in general* transport gases by diffusion—even though a number of early workers, including Krogh himself (Krogh, 1920a, 1920b), clearly observed tracheal compression and ventilation (Baudelot, 1864; Macloskie, 1884; Demoll, 1927; Gunn, 1933). Subsequently, however, the literature has come to contain spectacular examples of convective air flow in insects, with a variety of physiological mechanisms employed to drive and control flow (summarized below). These data show that many insects, of virtually all sizes, exhibit some convection in addition to diffusion in their respiratory systems. And, counterintuitively, convection may be an important respiratory mechanism that helps small insects to avoid excessive water loss (Kestler, 1985). Active respiratory convection in insects in the gas phase of the tracheal system occurs most often due to tidal compression and expansion of air sacs or tracheae (Weis-Fogh, 1967; Westneat et al., 2003). Most commonly (see Section 6.2.1), these result from contraction of intersegmental muscles that draw cuticular plates together, compressing body volume and increasing pressure.

Convection is the bulk flow of a fluid (air or liquid, liters min^{-1}), driven by a pressure gradient. Gas delivery by convection is the product of fluid flow and the total gas content of the fluid. For insects that are transporting O_2 primarily due to convection through the spiracles (e.g. hypoxia-exposed grasshoppers: (Greenlee and Harrison, 1998)), oxygen consumption $(\dot{V}O_2)$ can be related to breathing frequency (f), tidal volume (V_T), and difference in oxygen fraction between inspired and expired air $(F_{I_{oxygen}} - F_{E_{oxygen}})$:

$$f V_T (F_{I_{oxygen}} - F_{E_{oxygen}}) = \dot{V}O_2 \,. \tag{11}$$

This equation can be framed in terms of inspired and expired partial pressures by taking into account the relationship between partial pressure and concentration of gases in air ($\beta = 1/RT$ for ideal gases in air):

$$f V_T \beta (Po_{2air} - Po_{2\exp}) = \dot{V}O_2 \tag{12}$$

This equation may be applied to insects that breathe tidally—that is, inspire some volume of air through the spiracles and then expire a similar volume back out. In some species, however, there is coordination of spiracular opening and pumping such that air moves unidirectionally through parts of the tracheal system, especially large longitudinal tracheae (Weis-Fogh, 1967). In these cases, Equation 11 may also be applied, with inspired and expired air occurring via different spiracles.

Mechanistically, total air flow through insect tracheae also depends on several more basic properties of the air stream, including whether air flow in the tracheae is laminar or turbulent. Turbulence depends on a fundamental property of flowing fluids, the Reynolds number (Re):

$$\mathrm{Re} = \frac{U \rho L}{\mu} \qquad (13)$$

where U is fluid velocity, ρ is fluid density, L is characteristic length (i.e. diameter of the tracheal tube), and μ is dynamic viscosity. Re is dimensionless and expresses the relative magnitudes of inertial versus viscous forces on fluid flow. High values indicate that objects in a flowing fluid (or moving through a fluid) will themselves continue to move even when propulsion stops and, likewise, that fluid packets diverted by objects behave relatively independently from other packets of fluid (i.e. flow is turbulent). Conversely, low values indicate that viscous damping dominates inertial components.

As Equation 13 indicates, larger objects (in this discussion, larger tube diameters) and faster flows give higher Re. For flow in smooth tubes, a rule of thumb is that Re = 2000–3000 is a transition point between turbulent and laminar flow; Re values greater than this range are likely to result in turbulent flow (Vogel, 1994). The Reynolds number of tracheal air matters because the volumetric flow rate per unit pressure differential depends strongly on whether flows are laminar or turbulent (Vogel, 1994). Miller's study of giant African beetles provides an informative example (Miller, 1966). His species had very large longitudinal tracheae that originated at an anterior spiracle, traveled along the metathorax supplying secondary tracheae going to flight muscles, and exited through a posterior spiracle. In the cerambycid, *Petrognatha gigas*, average tracheal diameter was 2.4 mm (and was ~ 20 mm long), and air flowed at ~ 30 cm s^{-1}. Together with values for air density and viscosity (Vogel, 1994), these values give an Re of ~ 50 (i.e. Re << 2000), indicating that flows in *Petrognatha* are laminar. This result strongly suggests that convective flows in *all* insect tracheae are laminar, as the large tracheal diameters and air velocities in *Petrognatha* represent an extreme upper end of size and speed; most other insects have smaller tracheae and slower flows, both giving lower Re.

When fluid flow is laminar, relationships among pipe (i.e. tracheal) morphology, pressure differentials, and volumetric flow rates are described (in steady state) by the Hagen-Poiseuille equation:

$$V = \frac{\pi \Delta \mathrm{P} r^4}{8 l \mu}, \qquad (14)$$

where V is volumetric rate of air flow, ΔP is pressure difference between the beginning and end of the tube, r is tube radius, l is tube length, and μ is dynamic viscosity. The dependence of flow on the 4th power of radius is one of the more remarkable aspects of this equation (and contrasts with diffusion's dependence on the 2nd power of radius). It means that the volumetric flow depends strongly on small changes in r—for example for a given ΔP, halving the tracheal radius results in 1/16 the flow.

Recent video images of insect tracheal systems *in vivo*, taken with x-ray synchrotron imaging, provide a complication: they indicate that many parts of the tracheal system compress simultaneously (Westneat et al., 2003; Socha et al., 2007). These images suggest that rather than thinking of insect tracheae as rigid tubes through which air is driven by muscular pumps, that insect tracheae often can be analogs of peristaltic pumps, collapsing in response to local elevations in hemolymph pressure, and opening due to elastic properties of the taenidia, which are arranged like supporting hoops around the tracheal tubes. With such a system, volumetric flow will be a simple multiple of the volume of the tubes compressed and the frequency of compressions. However, Equation 14 may still be important for insects if the collapsing tracheae (or air sacs) drive air through rigid tracheae, and Equation 14 may apply to spiracles, which likely have quite rigid walls.

Gas transport via convection in the hemolymph has traditionally been considered unimportant in insects due to the lack of O_2-binding pigments; however, recent findings require a reassessment. For cells that lack a supply of tracheae, such as hemocytes, O_2 transport through hemolymph must be important, and hemocyte "docking" with tracheal lungs has been reported (Locke, 1998). It is now clear that many insects have respiratory pigments in the hemolymph (Burmester and Hankeln, 2007), and so significant convective delivery of oxygen via the hemolymph may well occur, though this has never been quantified. Convective transport of CO_2 through the hemolymph may be relatively more important than O_2, as the hemolymph contains relatively high levels of CO_2, and the higher solubility of CO_2 than O_2 in tissues yields a proportionally greater conductance across the walls of the major tracheae (Kestler, 1985; Schmitz and Perry, 1999). Moreover, if carbonic anyhdrase is present in the walls of the major tracheae, this could be a major route of CO_2 elimination (Slama, 1994). As yet there have been no quantitative studies of O_2 or CO_2 transport via the blood in any insect; the difficulty of quantifying convective flow in open circulatory systems, and the distributed nature of the gas transport system, will make this challenging.

6.1.1.3 Mixed Diffusion and Convection

For most insects, gas exchange likely occurs by a mixture of diffusion and convection. Such a mixture can take two forms. First, in any specific domain of the tracheal system, both convection and diffusion may contribute to the movement of oxygen and carbon dioxide. Mathematical descriptions of such mixed diffusion–convective systems remain challenging. Second, transport in different domains may be dominated by different mechanisms, so that the correct overall model invokes multiple mechanisms in series. For example, oxygen transport across spiracles might occur by convection, followed by mixed diffusive and convective transport through air to the tracheoles, and then pure diffusive transport through air and then liquid as oxygen moves from tracheoles to mitochondria.

Regardless of how gases are transported, the two equations—for diffusion (Equation 10) and convection (Equation 12)—provide key conceptual frameworks for the rest of the chapter: they provide quantitative descriptions of how various parameters *must* be matched to one another. For example, if the metabolic rate goes up and the spiracles and tracheae do not change in any way, either convection or ΔP must increase. Conversely, for ΔP to remain constant when metabolic rate rises, convection or spiracular opening must increase in direct proportion to metabolic rate. The following sections can each be considered an analysis of physiological problems related to the parameters in the two equations. Specifically, Section 6.1.2 analyses potential problems associated with high $\dot{V}O_2$, Section 6.1.3 potential problems of ΔP stemming from low ambient O_2 levels, and Sections 6.1.4 and 6.1.5 consider issues related to low A/x. Problems of solubility also will appear during discussion of tracheolar fluid levels.

6.1.2 Inadequate Delivery During High Oxygen Demand

Insects—at least some species—are famous for displaying large metabolic scopes, defined as the ratio of maximum sustained aerobic metabolism to basal or resting metabolism. Metabolic scope is important in two ways. From a *power perspective*, metabolic scope represents power available to an insect for fitness-enhancing activities, including foraging, migrating, territorial establishment, mate finding or guarding, and predator escape (Rogowitz and Chappell, 2000). Conversely, scope is a means of conservation during periods of quiescence—what we might call an *efficiency perspective*. In other words, insects with large metabolic scopes conserve energy and water better when inactive, which may aid during periods of deprivation.

Typical vertebrate metabolic scopes are 5–15. By contrast, those of endothermic flying insects can be much higher (Harrison and Roberts, 2000). For example, Bartholomew and Casey (1978) showed that 1-g sphingid moths had metabolic scopes of around 170 and 1-g saturniid moths around 125 (Fig. 6.1). For large flying insects, a major contributor to scope is flight-driven differences in body temperature. Although most moths, wasps, and bees are ectothermic at rest (i.e. cool), many are endothermic and warm during flight, maintaining thoracic temperatures of 35–42°C (see Chapter 3). Small flying insects, like *Drosophila*,

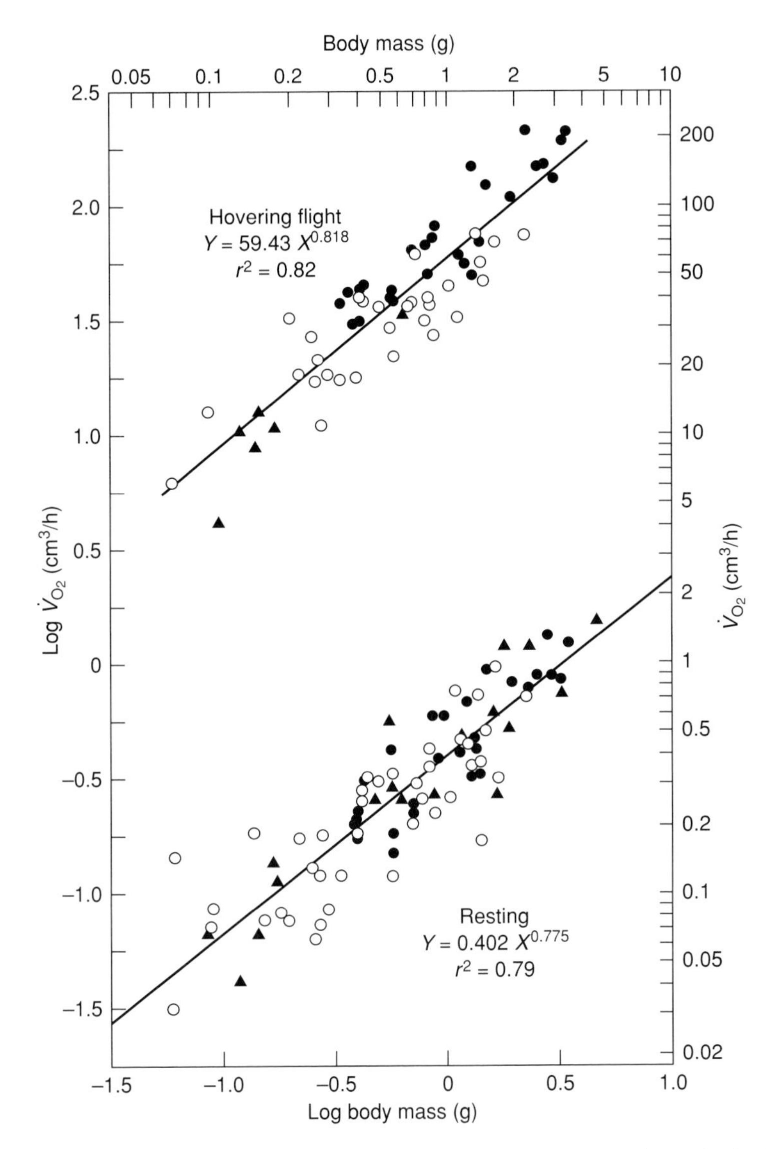

Fig. 6.1 Oxygen consumption, at rest and during hovering flight, for moths in the families Sphingidae (closed circles) and Saturniidae (open circles). Moths in other families are indicated by closed triangles. Redrawn from Bartholomew and Casey (1978), with permission from The Company of Biologists.

which generally warm up little during flight, have aerobic scopes similar to those of vertebrates (Lehmann et al., 2000).

Large increases in oxygen consumption and carbon dioxide production during activity challenge the respiratory system. From Equation 10, it is clear that a ten-fold increase in $\dot{V}O_2$ entails either a 10-fold increase in ΔP, G, ventilation, oxygen extraction, or some combination that provides the 10-fold increase in flux. Another respiratory challenge is the increased demand for gas exchange associated with growth. During their lifetimes, individual insects may grow by spectacular factorial amounts, and larger, older insects in general will have higher metabolic rates than their earlier, smaller incarnations.

Few studies have examined whether increases in $\dot{V}O_2$ are accompanied by increases in ΔP (larger drops between ambient and tissue Po_2); so far, studies with flying insects suggest the answer is no, but for jumping grasshoppers it is yes. Komai (Komai, 1998, Komai, 2001) showed that in a tethered hawkmoth, *Agrius convolvuli*, there is only a minor drop in muscle Po_2 during flight. In bumblebees, flight muscle Po_2 was lower and varied with abdominal pumping at rest, but also did not decrease significantly during flight (Komai, 2001). Measurements of tissue oxygen during light emission by elaterid larvae also suggest that ΔP is relatively constant during increases in metabolism (Timmins et al., 1999). By contrast, measurements of tracheal gases in hemolymph suggest that O_2 falls dramatically and CO_2 rises in the legs of jumping grasshoppers (Krolikowski and Harrison, 1996).

One approach to determining the match between tracheal capacity and tissue demand involves manipulating external oxygen supply and measuring the responses of metabolism or performance. The conclusion of such studies is that resting adult insects are relatively unaffected by short-term hypoxia but that other life stages may be more susceptible (Harrison et al., 2006). Moreover, *flight* performance often falls in even moderate hypoxia. The critical partial pressure of oxygen for metabolic rates depends strongly on activity in the locust, *Schistocerca americana* (Fig. 6.2). Resting locusts had low metabolic rates, which were further depressed by ambient Po_2 only when it dipped below 1 kPa O_2, whereas flying locusts exhibited a critical Po_2 between 10 and 21 kPa O_2 (Rascón and Harrison, 2005). Similar experiments have been done with dragonflies (*Erythemis simplicicollis*, (Harrison and Lighton, 1998)) and honeybees (*Apis mellifera*, (Joos et al., 1997)), and their critical oxygen levels were 30 and 15 kPa O_2—that is *E. simplicicollis* appeared to be oxygen-limited even in normoxia.

How well does tracheal capacity match increases in metabolic rate that occur with growth? Greenlee and Harrison (2004a) showed that critical Po_2 for metabolic rate declined substantially in older instars of locusts (*S. americana*), indicating that (at least at the beginning of the instar) tracheal capacity more than matches increases in oxygen needs. This pattern reflects both proportionally greater convection, which depends on proportionally greater content of tracheae and air

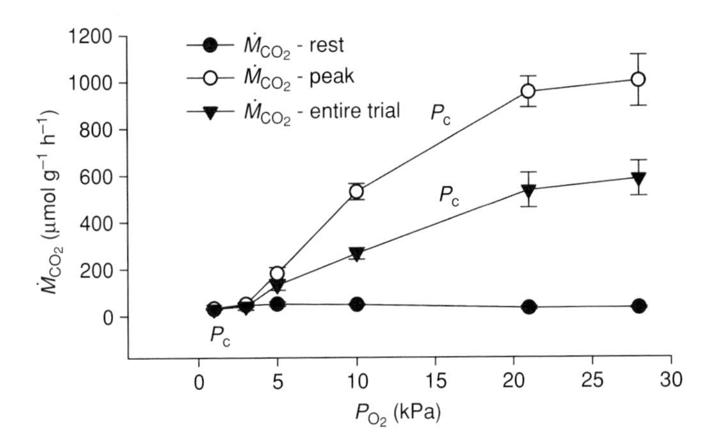

Fig. 6.2 Metabolic sensitivity of grasshoppers (*Schistocerca americana*) to ambient oxygen level (Rascón and Harrison, 2005). Sensitivity depended strongly on grasshopper activity–those at rest were not nearly as sensitive to oxygen availability as those engaged in vigorous activity. P_c denotes critical partial pressure of oxygen. Redrawn from Rascón and Harrison (2005) with permission from Elsevier.

sacs, and higher breathing frequencies in older instars and adults (Greenlee et al., 2009). The critical Po_2 for metabolic rate remained constant across instars in the caterpillar *Manduca sexta*, suggesting that tracheal capacity closely matches tissue needs during growth, at least at the beginning of the instar (Greenlee and Harrison, 2005).

In contrast to cross-instar studies, studies of critical Po_2 *within* instars suggest that tracheal capacity does not keep up with increasing need for oxygen. In both grasshoppers and caterpillars, critical Po_2 rose strongly for animals tested near the end of the instar (Greenlee and Harrison, 2004b, 2005). Within instars, body mass can more than double, whereas major branches of the tracheal system remain unchanged, their size being modified only during molts. Thus, higher body mass may place higher demands on a static delivery system, or may even reduce O_2 delivery capacity by compressing tracheae and air sacs. These findings suggest that one impetus for molting is to rematch oxygen supply to demand.

6.1.3 Inadequate Oxygen Delivery Stemming from Environmental Hypoxia

Although many of the most obvious insects around us—houseflies, dragonflies, bees, butterflies—live in oxygen-rich environments (normoxic = 21 kPa O_2 at sea level), less obvious species, or hidden stages within a species, inhabit environments with intermittent or chronic hypoxia (defined as $Po_2 < 21$ kPa).

These environments include hypoxic water, soils and litter, flooded soils, fermenting fruits, and high altitudes (Hoback and Stanley, 2001; Schmitz and Harrison, 2004). Reduction in external oxygen reduces the gradient driving oxygen into tissues, requiring an increase in ventilation or diffusive conductance ($AD\beta/x$) if $\dot{V}O_2$ is to remain constant (Equation 10). Species that live in hypoxic environments such as flooded soils exhibit a strong capacity for anaerobic metabolism and the ability to strongly reduce metabolic rate and behavior (Hoback and Stanley, 2001).

6.1.4 Inadequate Oxygen Delivery Through High-Resistance Structures

Limits to oxygen delivery can also arise from high-resistance (= low conductance) oxygen delivery systems. Here we consider structures whose resistance is naturally high and subsequently the exogenous factors that can alter resistance.

In all insect *eggs*, oxygen flux occurs across multiple layers of the eggshell, which bear little resemblance to a tracheal system (Fig. 6.3) (Hinton, 1981).

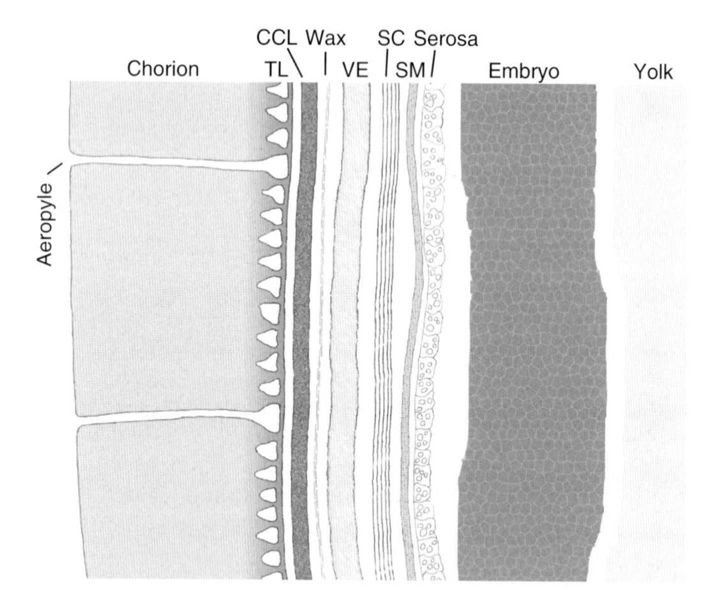

Fig. 6.3 Eggshell structure of a sphingid moth, *Manduca sexta*. The outer proteinaceous chorion is traversed by many air-filled tubes (aeropyles), which exchange gases between the atmosphere and the air-filled trabecular layer (TL). Below the trabecular layer is a crystalline chorionic layer (CCL), a wax layer, and the vitelline envelope (VE). Further inside is an extraembryonic layer of cells, the serosa, which secretes the serosal cuticle (SC) during the first half of development. The embryo is adjacent to the serosa and draws nutrients and energy from the yolk. Redrawn from Woods (2010), with permission from Elsevier.

Oxygen diffuses down tubular aeropyles into a film of air in the trabecular layer; from there it diffuses through a crystalline layer of protein and an associated wax layer, then through the vitelline envelope and the serosal layers before reaching the embryo. The two functional differences between eggshells and tracheal systems are 1) that eggshell transport occurs entirely by diffusion and 2) that diffusion occurs across multiple liquid or semi-solid spaces rather than in air, as occurs through most of the tracheal system.

Eggshells constitute high-resistance structures in the sphingid moth *Manduca sexta*. Both increasing age and higher temperatures caused the Po_2 in central regions of *Manduca* eggs to decrease (Woods and Hill, 2004). The oldest, warmest eggs had broad areas that were nearly anoxic. In all treatments, the steepest drop in Po_2 was across the eggshell, demonstrating that it is the high-resistance component of the delivery pathway. Temperature and ambient Po_2 also interacted to affect metabolic rates. At low temperatures (22, 27°C), egg metabolic rate was not strongly affected by ambient Po_2, whereas at high temperatures (32, 37°C) egg metabolic rate decreased linearly with declining Po_2. Moreover, at high temperatures, hyperoxia stimulated egg metabolic rates, implying that eggs may have been oxygen limited *even in normoxia*. Most of the total resistance arose from either (or both) of two layers, the crystalline chorionic layer or the wax layer, which lies just under the crystalline layer (Woods et al., 2005a).

We know little about the consequences for eggs of reaching or exceeding limits to oxygen delivery across the eggshell. Outstanding questions include: What other evolutionary pressures bear on insect eggshells? Are eggshells highly impermeable to oxygen as a means of conserving water? How does eggshell structure and function change during adaptation to novel egg environments? These questions are ripe for study, especially given our growing understanding of eggshell formation and microstructure (Trougakos and Margaritis, 2002), comparative morphology (Hinton, 1981), and physiology (Woods et al., 2009).

6.1.5 Inadequate Oxygen Delivery from Disturbance to, or Pathologies of, the Tracheal System

In tracheal systems, two classes of disturbance can decrease tracheal conductance: physical changes in the tracheal environment and tracheal infections. The latter is especially interesting, as most biologists working on insect respiration study healthy, vigorous individuals. We know little about the natural prevalence of respiratory pathologies in nature nor about the degree to which they impair respiratory function.

One recently discovered problem, which leads to low conductance, is ice formation in tracheoles, the distal-most tubes of the tracheal system. Tracheoles normally contain liquid in their tips. In the cold, however, this liquid can freeze, causing two problems: oxygen moves very slowly through ice, and tracheal ventilation depends on movement, which is impossible if muscles are partially frozen. How do frozen insects withstand hypoxia? Morin et al. explored this problem using *Eurosta solidaginis*, the goldenrod gall fly (Morin et al., 2005). In summer, larvae form galls in which they feed and grow. Last-instar larvae then spend the winter in their galls, which exposes them to very low temperatures. Much of the water, including tracheal water, in a mid-winter larva may be frozen. Morin et al. (2005) focused on *Eurosta*'s hypoxia-inducible factor 1α (HIF-1α), a molecule known from studies of mammals and a few invertebrates to trigger and coordinate gene expression in response to hypoxia (see Section 6.3 for molecular details). Levels of *hif-1*α transcripts were stimulated by both experimental anoxia and by cold. Morin et al. (2005) suggested that chill-induced expression *anticipates* freeze-induced hypoxia. The more general implication is that insects can evolve sophisticated mechanisms for dealing with predictable disturbance to tracheal conductance.

One might also expect that freezing of other body fluids could compress or collapse tracheal tubes, even if the tracheolar water remained liquid. However, using x-ray synchrotron imaging, Sinclair et al. (2009) examined the details of freezing of larvae of two species of Drosophilidae (*D. melanogaster* and *Chymomyza amoena*). Although freezing of different body compartments was readily visible and caused some displacement of tracheal tubes, there was no evidence that tracheae were collapsed by the expansion of external ice (Sinclair et al., 2009).

The second class of tracheal disturbance is respiratory infection, which can include bacteria, fungi, mites, and other parasites. Little is known about such infections, with the exception of those in honeybees (*Apis mellifera*). What is known about tracheal blockage in bees—a research focus because of their economic importance (Delfinado-Baker, 1988; Finley et al., 1996; Fore, 1996)—suggests that respiratory occlusion in insects generally could be an important selective force on tracheal function.

Honeybees are susceptible to infection with the tracheal mite, *Acarapis woodi*, almost throughout their range. Individual bees are normally infected within the first few days of adult life, and the mites feed on hemolymph via tracheal walls and reproduce in major thoracic tracheae (Fig. 6.4) (Delfinado-Baker, 1988). Harrison et al. (2001) studied effects of tracheal mites (*A. woodi*) on safety margins for oxygen delivery in flying honeybees. They initially measured flight metabolic rates of bees in different levels of atmospheric oxygen (5, 10, 21, and 40 kPa oxygen). In normoxia (21 kPa O_2) and hyperoxia (40 kPa O_2), infected bees had flight metabolic rates that were as high as those without mites, but at lower O_2 levels (10 and 5 kPa) increasing total mite load was correlated with lower flight metabolic rate. These results suggest that bees with tracheal mites had smaller safety margins for O_2 delivery during flight. Harrison et al. (2001) suggested that functional effects of reduced safety margins may appear during very high metabolic loads—for example when bees are flying in cold conditions.

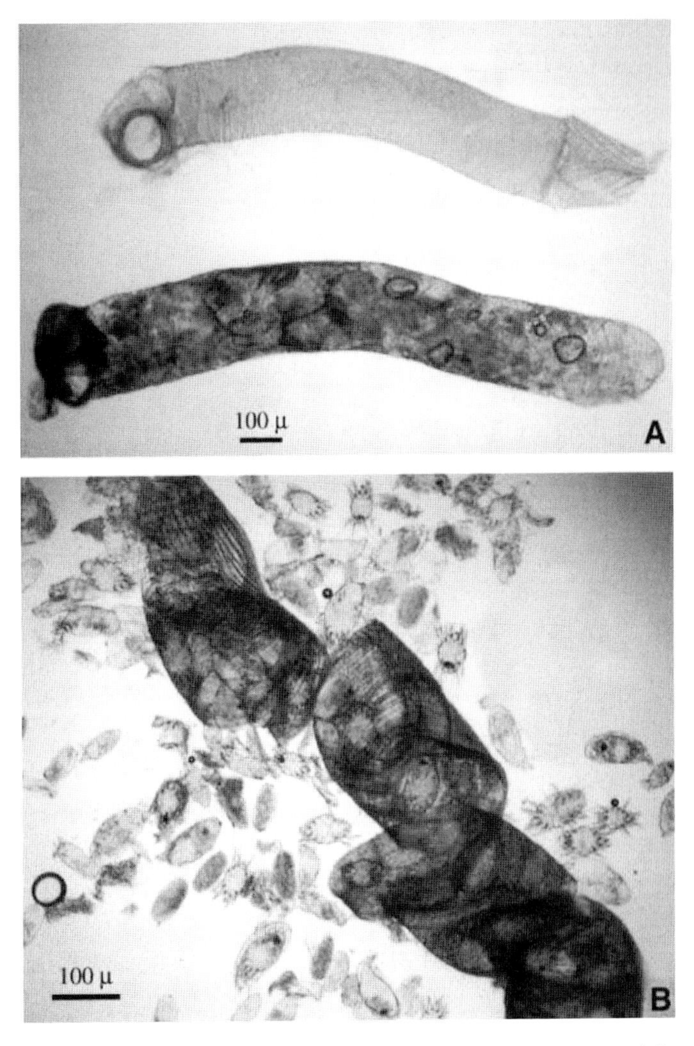

Fig. 6.4 Prothoracic tracheal trunks of honeybees, *Apis mellifera*, uninfected (A) and infected (B) with mites (*Acarapis woodi*). The infected trunk is ripped open to show the mites. Reprinted from Harrison et al. (2001) with permission from The Company of Biologists.

A follow up experiment tested this idea. The authors examined mite loads of bees returning from mid-winter flights versus those grounded on the snow. Grounded bees had, on average, twice as many tracheal mites. The prevalence and effects of tracheal mites in other flying insects are poorly known, although initial reports suggest that prevalence can sometimes be high and effects dramatic

(Husband and Shina, 1970; Otterstatter and Whidden, 2004; Korner and Schmid-Hempel, 2005).

Another kind of parasitic infection, by trypanosomes, reduces oxygen delivery by altering tracheal morphology. The bloodsucking bugs *Rhodnius prolixus* and *Triatoma infestans* (Reduviidae) host several trypanosomes. Eichler and Schaub (1998) experimentally infected *R. prolixus* and *T. infestans* with the trypanosome *Blastocrithidia triatomae*, and measured tracheal supply to different internal organs. In *T. infestans*, but not *R. prolixus*, infection substantially reduced the density of tracheoles going to the rectum, small intestine, and Malpighian tubules. This effect may be explained by vitamin B deficiencies induced by the trypanosome, based on other studies involving symbionts. Specifically, bugs reared without symbionts also had reduced tracheolar densities, but this effect could be rescued by supplementing the diet with vitamin B. Bugs with reduced tracheolar densities also showed other physiological pathologies, including reduced excretion of fluid after blood meals. These effects were hypothesized to stem from inadequate oxygen delivery, but direct tests of this idea have not been done.

Other kinds of tracheal infections—such as microsporidian fungi (Sokolova and Lange, 2002) or baculoviruses (Engelhard et al., 1994)—may also affect oxygen delivery capacity, but no work has been done on them. We suggest that tracheal pathologies deserve renewed attention, as they may be widespread in nature and could have far-ranging consequences for insect performance in ecologically relevant contexts.

6.1.6 Ecological Consequences of Tracheal Safety Margins

Despite sparse data on the oxygen sensitivity of metabolism and performance, we suggest that it is time to look beyond physiological experiments per se to evaluating ecological and evolutionary consequences of approaching limits to oxygen delivery. For example, one could manipulate Po_2 of air available to interacting organisms: males fighting over females (Rogowitz and Chappell, 2000), dragonflies escaping from predators (Harrison and Lighton, 1998), honeybees collecting nectar and pollen from flowers (Joos et al., 1997). Manipulating Po_2 in arenas large enough to allow realistic interactions would be difficult but not insurmountable. As is well known, researchers have managed to manipulate CO_2 levels in entire forests (Strain, 1991). Alternatively, one could manipulate an insect's tracheal system (e.g. by spiracular occlusion; (Heymann and Lehmann, 2006)), release it back into its natural environment, and measure performance under demanding metabolic conditions. Finally, one could use chemical or molecular techniques, artificial selection experiments, or changes in oxygen rearing environment, to manipulate tracheal size and morphology (at the risk of manipulating other, unknown variables). To our knowledge, such experiments have never been done on free-flying insects in nature.

6.2 Responses to Hypoxia

The full suite of insect responses to inadequate O_2 delivery is dazzling, spanning timescales from milliseconds to generations, and organizational levels from molecular to organismal and even supra-organismal (see Chapter 1). Figure 6.5 summarizes currently known, or hypothesized, structures and molecules involved in Po_2 homeostasis. We use the figure to organize the following discussion.

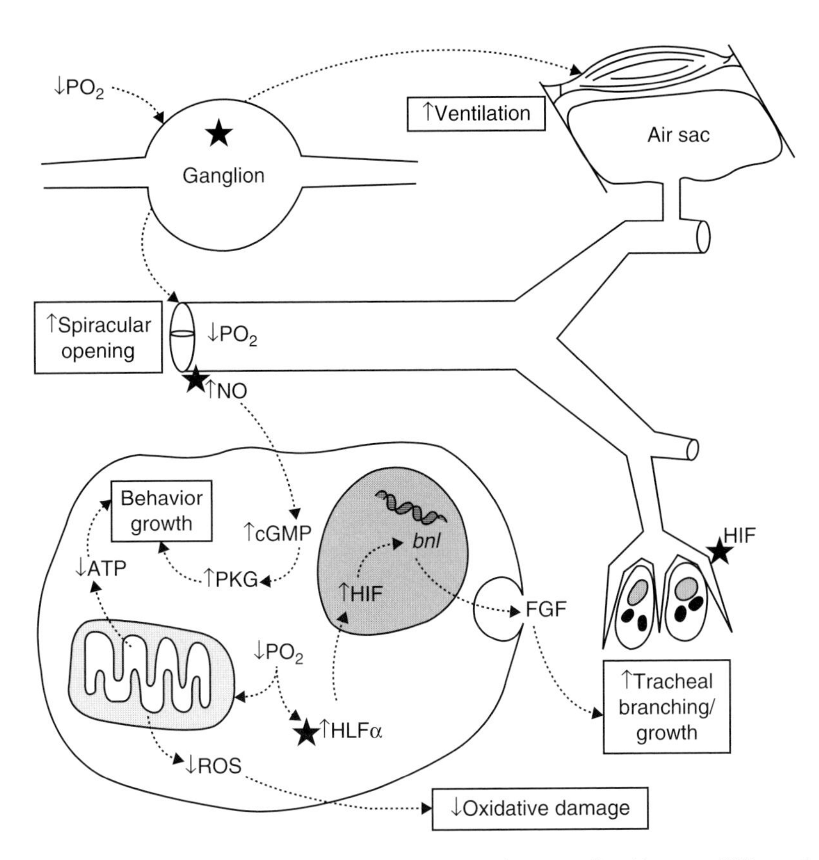

Fig. 6.5 Schematic of the insect respiratory system and a generalized insect cell illustrating pathways of oxygen homeostasis. CO_2- and O_2-sensitive ganglia in the central nervous system control both spiracular valves and muscles that drive convective air flow. Putative oxygen sensors are indicated by stars. Longer-term tissue hypoxia activates secretion of nitric oxide (NO) from some cells and expression of hypoxia inducible factor (HIF) in others. These factors lead to downstream effects such as secretion of a fibroblast growth factor (FGF) homolog, which stimulates tracheal growth and branching. Changes in internal Po_2 can have other direct effects on long-term physiological function by altering production of mitochondrial ATP or reactive oxygen species (ROS). Reprinted from Harrison et al. (2006), with permission from Elsevier.

6.2.1 Short-term Responses

6.2.1.1 Movement

Many insects encounter intermittent or chronic environmental hypoxia (Hoback and Stanley, 2001). It stands to reason that such insects would have evolved behavioral means of positioning themselves in oxygen levels to their liking. However, although literature in this area suggests that insects do respond behaviorally, the field is poorly developed, especially using comparative approaches. Recent advances in understanding mechanisms of hypoxia sensing (see Section 6.3) may soon change this.

The best example is by Wingrove and O'Farrell (1999), who studied responses of *Drosophila* to hypoxia. Larvae of *D. melanogaster* forage on foods rich in bacteria and yeast. Respiration by microbes and fungi must occasionally draw down oxygen to levels that are uncomfortably low for larvae. Wingrove and O'Farrell (1999) exposed third-instar larvae feeding on yeast paste to 1% oxygen. Within seconds of exposure, larvae stopped feeding and exited the yeast, and this behavior seems to be at least partially mediated by nitric oxide signaling (Wingrove and O'Farrell, 1999). Morton et al. (2008) have uncovered further genetic and neural mechanisms underlying the rapid larval response (Morton et al., 2008). Outstanding questions include how O_2 gradients shape feeding patterns of *Drosophila* in nature, whether common mechanisms underlie hypoxia responses across insects, and whether behavioral responses to hypoxia are important in aquatic insects (Apodaca and Chapman, 2004).

6.2.1.2 Tracheoles and Spiracles

Barriers to oxygen flux occur at both ends of the tracheal system: at the spiracular openings to the ambient atmosphere and in the tracheolar tips, which contain liquid. Clearly, oxygen moves less rapidly through occluded than open spiracles, and through liquid than air. Thus, two ways to increase flux, in response to higher tissue demand or lower ambient oxygen, are spiracular opening and fluid removal (Wigglesworth, 1983). Both mechanisms appear to be widely used by insects.

Early work on tracheal respiration focused on spiracles as oxygen gateways. Wigglesworth (1935) subjected fleas (*Xenopsylla cheopis*) to atmospheres containing modified amounts of oxygen, finding that spiracles were open more during hypoxia and less during hyperoxia. This basic finding, that spiracles respond to ambient oxygen levels, is now well established (Burkett and Schneiderman, 1974b), and it is undoubtedly important for insects living in hypoxic environments, and whenever oxygen demand varies.

A good recent example is from flying *Drosophila* (Lehmann, 2001). In flies, force production, CO_2 emission (a measure of metabolic rate), and H_2O emission (a measure of relative spiracular opening) were highly correlated (Fig. 6.6). As flies worked harder, they expended more power, which gave higher instantaneous

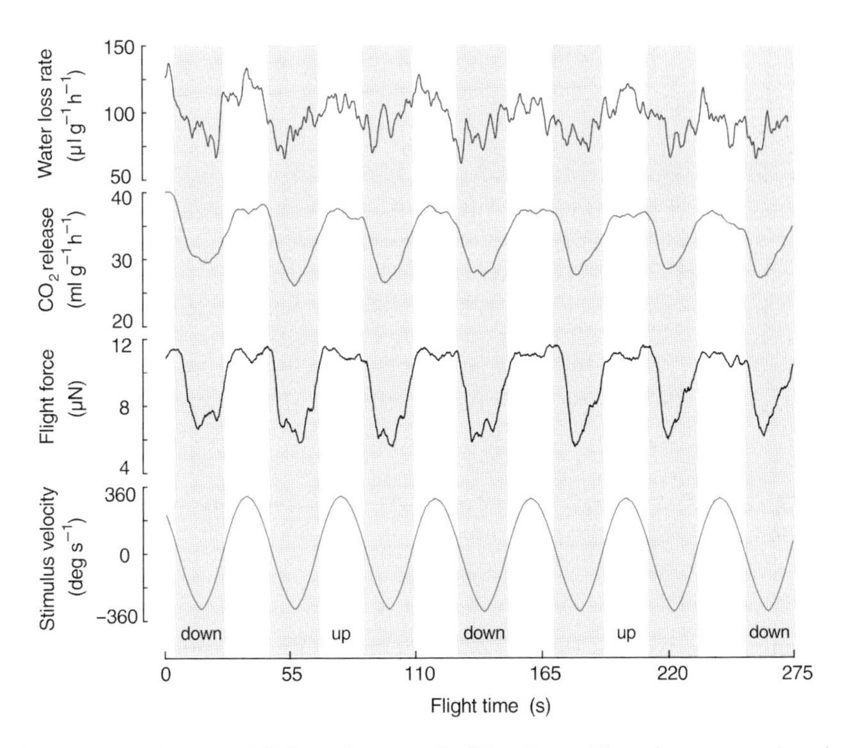

Fig. 6.6 Gas exchange and flight performance by flying *Drosophila melanogaster* tethered in a virtual-reality arena. "Stimulus velocity" refers to the velocity of an image projected in the arena. Flies attempted to stabilize the image in their visual fields and, in doing so, produced upward and downward flight forces. Carbon dioxide and water emission, measured concurrently, showed that gas exchange was highly correlated with flight forces produced. Reprinted from Lehmann (2001), with permission from AAAS.

rates of CO_2 emission. Water loss rates increased in phase with CO_2 emissions, which occurred because flies opened spiracles wider to obtain O_2. Lehmann concluded that during flight spiracular opening is matched to metabolic demand, and that this matching lowers desiccation risk during flight in dry conditions.

Variation in tracheolar fluid levels is also an important mechanism for modifying tracheal conductance (Wigglesworth, 1930, 1931, 1935). In initial observations on larval *Aedes argentus* (now *Ae. aegypti*), Wigglesworth (1930) showed that resting larvae had tracheoles filled with liquid. When larvae were asphyxiated (by blocking the respiratory siphon), however, liquid rapidly disappeared and air entered finer branches. The effect could be reversed by unblocking the siphon. In 1935, Wigglesworth performed a definitive set of experiments with the flea, *Xenopsylla cheopis*, showing that exposure to progressive levels of hypoxia (21, 10, 5, 0.8% O_2) resulted in tracheoles filled with air for proportionately greater

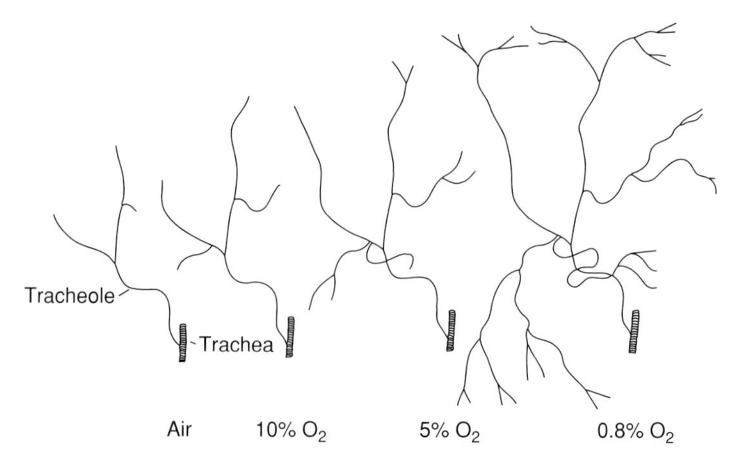

Fig. 6.7 Extent of gas filling of tracheoles of a flea (*Xenopsylla cheopis*) exposed to different ambient levels of oxygen. From Wigglesworth (1935), with permission from The Royal Society.

fractions of their volume (Fig. 6.7). Substitution of gas for liquid in the tracheolar path would support much higher O_2 fluxes to tissues, because of much higher O_2 diffusion and capacitance coefficients in gas; however, the quantitative effect of tracheal fluid levels on insect tracheal conductance has never been measured.

Wigglesworth (1930) also suggested a mechanism controlling tracheolar fluid levels: an equilibrium between capillary pressure (drawing fluid into and along tracheoles) and osmotic pressure of tracheolar cells (higher tissue osmotic pressures tending to draw fluid back into surrounding cells). In such a balance, increases in tissue osmotic pressure during activity—for example from incomplete oxidation of metabolites—would reduce tracheolar fluid levels. He supported this contention experimentally by showing that injection of hypertonic solutions into living larvae rapidly increased extent of tracheolar air. However, direct measurements of hemolymph osmolytes during hypoxia have not been made, and the actual mechanism by which insects control tracheolar fluid levels remains unclear.

One of the most extensive analyses of the functional importance of tracheolar fluid levels is by Timmins et al., who examined mechanisms controlling bioluminescent flashing in fireflies (*Photinus* sp.) (Timmins et al., 2001). It has been known for almost a hundred years that exposing fireflies to hyperoxia can induce light emission. This observation suggests that fireflies control patterns of flashing by controlling oxygen access to photocytes. Timmins et al. used a set of gas manipulations, coupled with direct electrical stimulation of the lantern organ, to calculate distances over which oxygen must diffuse in both gaseous and aqueous phases. Their observations, and others from the literature, led the authors to suggest a multipart mechanism for control of flashing: 1) O_2 access to photocytes is modulated by tracheolar fluid levels (for an alternative, see (Aprille et al., 2004)); 2) changes in tracheolar fluid level occur after nervous stimulation of tracheal end cells; and 3) nerve pulses stimulate ion movement into adjacent cells, which rapidly draw fluid osmotically out of the tracheoles. This last part builds on Wigglesworth's (1935) proposed osmotic mechanism. In fireflies, the osmotic hypothesis gained direct experimental support from Maloeuf (1938), who showed that constant glowing could be induced by injecting hypertonic solutions into the abdomen.

6.2.1.3 Ventilation

From studies on diverse species, it is now clear that air often *flows* through tracheal systems. Air flow is driven by pressure gradients between different tracheal domains or between tracheae and external air. The gradients themselves can arise by either passive or active processes (though the energy for both is traceable back to the insect's metabolism). Passive mechanisms include Bernoulli ventilation, in which pressure gradients between anterior and posterior spiracles, arising from different flow speeds past them, drive flow longitudinally through the tracheal system. Miller (1966), for example, demonstrated passive unidirectional air flow through giant cerambycid beetles. In wind speeds comparable to what a *P. gigas* would generate in flight, air entered the second spiracle, flowed through the main thoracic trunks, and exited via the third spiracle (Miller, 1966). A second passive process is suction ventilation, in which air flows into the system when spiracles open after a period of closure. During closure, tracheal pressure declines as oxygen is consumed by metabolism, and the resulting carbon dioxide does not entirely take oxygen's place, as it dissolves disproportionately into hemolymph (Hetz and Bradley, 2005).

Active processes move air by compressing tracheae or tracheal air sacs. Compression can occur by either of two mechanisms: directly by muscles or indirectly by high hemolymph pressure, itself caused by muscular contraction. The most familiar of the indirect mechanisms is abdominal pumping, which is obvious in bees freshly returned from foraging, and in grasshoppers. Indirect compressions may also be coupled to the circulatory system (Wasserthal, 1996; see Chapter 2). We consider these two mechanisms below and also briefly discuss 1) an unusual mechanism of ventilation by flying *Drosophila* and 2) new insight into tracheal ventilation stemming from recent advances in x-ray synchrontron videography.

6.2.1.4 Autoventilation Driven by Flight Muscles

Oxygen consumption rises dramatically during flight, and one way for the respiratory system to exchange gases is to harness a portion of the flight motor's energy for pumping. The functional importance of in-flight thoracic pumping was first recognized by Weis-Fogh and Miller (Weis-Fogh, 1964a, 1964b, 1967; Miller, 1966). In detailed experiments on *Schistocerca gregaria*, Weis-Fogh observed that notal plates of the thorax deformed substantially, in phase with wing movements, causing changes in thoracic volume. He proposed that thoracic pumping directly deformed the primary thoracic tracheae and air sacs, driving air through them.

Wasserthal (2001) worked out the details of autoventilation in flying *Manduca sexta*, a large (1.5–4-g) hawkmoth, which is of interest given the large aerobic scopes of hawkmoths (Bartholomew and Casey, 1978); see Fig. 6.1. Using sensitive transducers, he measured pressures at both the anterior (first) thoracic spiracle and the mesoscutellar air sac (which connects directly to the posterior, or second, thoracic spiracle), during tethered flight. Pressures in both spaces varied cyclically with wingbeat and were likely coupled to spiracular movements to generate unidirectional airflow.

6.2.1.5 Body Compression by Intersegmental Muscles

Abdominal pumping may be coordinated with patterns of spiracular opening in a way that produces unidirectional air flow (Bailey, 1954; Duncan and Byrne, 2002). The taxonomic prevalence of such unidirectional flow is unknown, although it has now been demonstrated in Lepidoptera (Wasserthal, 2001), Hymenoptera (Bailey, 1954), Coleoptera (Miller, 1966), and Orthoptera (Weis-Fogh, 1967). Additional work has explored unidirectional airflow in flightless dung beetles (Duncan and Byrne, 2002, 2005).

Abdominal pumping rates respond to insect oxygen status (Miller, 1960; Arieli and Lehrer, 1988; Gulinson and Harrison, 1996; Miller, 1960; Arieli and Lehrer, 1988; Gulinson and Harrison, 1996), with tracheal Po_2 playing a central role. Gulinson and Harrison (1996) perfused tracheae of the desert locust, *Schistocerca americana*, with different levels of oxygen (and carbon dioxide), causing rapid changes in tracheal gas levels. The locusts showed correspondingly rapid changes in rates of abdominal pumping, increasing it during hypoxic perfusion and decreasing it during hyperoxic perfusion. These results suggested that locusts regulate oxygen levels in the tracheal system at ~18.8 kPa O_2 (Weis-Fogh; 1967; Gulinson and Harrison, 1996). Furthermore, many insects show elevated rates of abdominal pumping during flight (Bailey, 1954; Miller, 1966).

In addition to these examples, there is also evidence for low amplitude pressure fluctuations, probably driven by intersegmental muscle contractions, that move air in many insect larvae and pupae (Slama, 1988; Slama, 1999; Tartes et al., 2002). In these cases, although the insect appears motionless to the human eye, micro-compressions of the body drive convective flow through spiracles. These micro-compressions may also be related to recent observations of tracheal compression using x-ray videography (see below).

6.2.1.6 Convection Driven by Proboscis Pumping in *Drosophila*

Lehmann and Heymann demonstrated that flying *Drosophila* show large, cyclic releases of CO_2 (at about 0.37 Hz), even when generating constant mechanical power (Lehmann and Heymann, 2005). Using video analysis, they showed a strong correlation between *proboscis extension* and CO_2 release (Fig. 6.8), leading them to suggest that proboscis extension promotes oxygen supply to the head. The visual apparatus—retina and optic lobes—accounts for more than 20% of the resting metabolic rate. Oxygen to support this metabolism, however, comes from the thorax. The authors suggest, "[T]he hypothetical benefit of the proboscis-induced ventilation for breathing might be to circumvent this bottleneck for

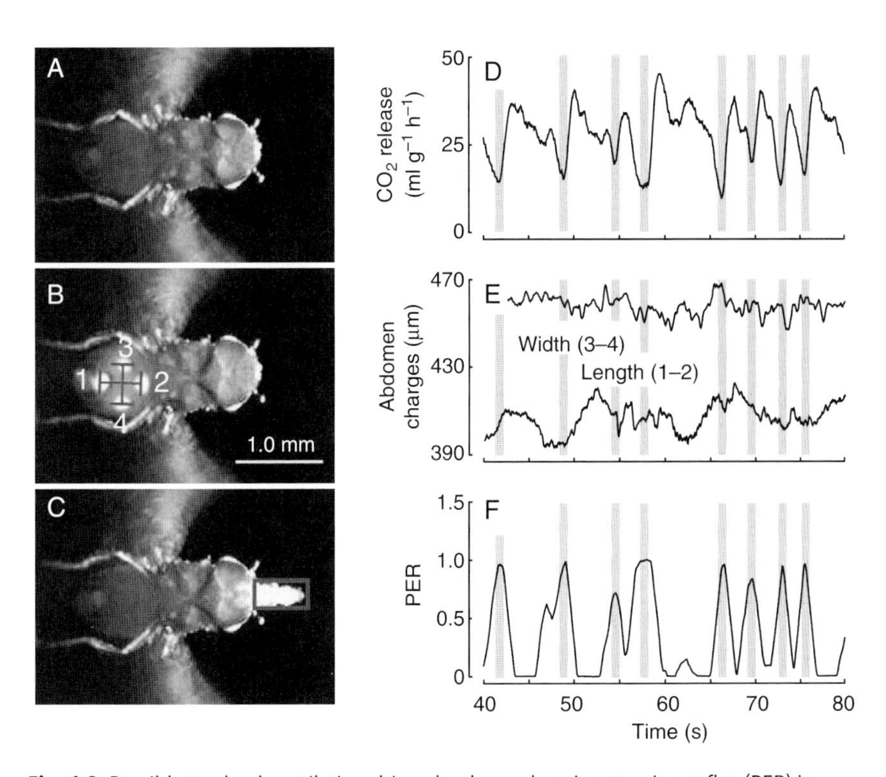

Fig. 6.8 Possible tracheal ventilation driven by the proboscis extension reflex (PER) in flying *Drosophila melanogaster* (Lipp et al., 2005). (A) View of fly from below. (B) Tracking of fluorescent spots on the abdomen allows calculation of abdominal deformation, which may drive airflow. (C) Proboscis extension reflex (PER) during flight, an alternative mechanism that may drive airflow. (D) CO_2 release by a flying fly corresponds much better with degree of PER (F) than of abdominal deformation (E). From Lehmann and Heymann (2005), reprinted with permission from The Company of Biologists.

diffusive respiration, in order to ensure evacuation of CO_2 from—and the supply of oxygen to—the fly's brain." It will be fascinating to learn whether this behavior can be adjusted to modify oxygen delivery in different oxygen environments, or under different demands, and whether it occurs in other insects.

6.2.1.7 Insights from X-ray Videography

In 2003, Westneat et al. used synchrotron x-rays to observe tracheal systems inside living beetles (*Platymus decentis*), carpenter ants (*Camponotus pennsylvanicus*), and house crickets (*Achaeta domesticus*) (Westneat et al., 2003). It was clear from videos that tracheae were compressed cyclically, with cycle duration on the order of 1 s. They proposed that compression was driven by contracting jaw or limb muscles and that re-expansion depended on recoil by taenidial rings in the tracheal wall. Compressions likely move air into different parts of the tracheal system, or to and from the insects via their spiracles. Another possibility is that they raise the total pressure (and with it, the partial pressure of O_2) in the distally trapped air, which would increase the driving gradient for O_2 into tissues.

The x-ray method has been increasingly refined (Socha et al., 2007; Westneat et al., 2008), and several questions raised by Westneat et al. (2003) are now being answered. One was whether the observed compressions in fact exchange gases between insect and environment. Using the beetle *Pterostichus stygicus*, Socha et al. approached this question with a combination of x-ray videography and flow-through respirometry, showing that cyclic pulses of CO_2 emission were associated with cyclic tracheal compression (Socha et al., 2008). A second question was whether the rate of tracheal compression was sensitive to an insect's oxygen status. Using the grasshopper *Schistocerca americana*, Greenlee et al. (2009) showed with the x-ray method that insects in hypoxia (~ 5 kPa O_2) compressed tracheal air sacs at significantly higher rates than did insects in normoxia.

Collectively, these results suggest that tracheal compressions are taxonomically widespread and can be important for ventilation even when gross body movements are not seen. A key outstanding question is *how* insects effect compression. In x-ray videos, some tracheae, or sections of tracheae, may compress while others do not, perhaps due to local mechanisms (muscles) or differences in tracheal compressibility (Socha et al., 2008).

6.2.2 Mid- to Long-term Responses: Developmental Plasticity in Respiratory Systems

An insect considers its tracheae to be part of its exoskeleton—tracheae are shed along with the exoskeleton at each molt. Molts thus represent opportunities to expand and upgrade. The obvious reason for enlarging is to rematch oxygen supply to growing demand from tissues (Greenberg and Ar, 1996). A series of papers dating from the 1980s suggests that the cue driving both enlargement of tracheae and tracheolar proliferation is tissue Po_2.

6.2.2.1 Patterns of Tracheal Plasticity

Loudon measured tracheal size and morphology in larval *Tenebrio molitor* (mealworms) reared in different levels of oxygen. The number and position of major tracheal branches were unaffected by oxygen treatment, but their cross-sectional areas underwent marked enlargement in hypoxia—cross-sectional area was 40% higher (compared to normoxic larvae) in larvae exposed to 15 kPa O_2 and 120% higher in larvae exposed to 11 kPa O_2. She concluded that tracheal hypertrophy was an adaptive mechanism for maintaining adequate diffusive fluxes of O_2 to tissues in the face of reduced driving gradients (Loudon, 1989). Similar but less pronounced changes in tracheal diameter were found in *D. melanogaster* (Henry and Harrison, 2004).

Level of atmospheric oxygen also affects the number and morphology of tracheoles (Locke, 1958; Wigglesworth, 1954; Centanin et al., 2010). Jarecki et al. exposed larvae of *D. melanogaster* to 20 hours of 5, 21, or 60% O_2 in N_2, and then examined their tracheoles (Jarecki et al., 1999). Larvae subjected to hypoxia (5% O_2) had 68% more tracheoles compared to siblings reared in normoxia, and the tracheoles were longer and more tortuous (Fig. 6.9). Tracheole proliferation began within one hour of hypoxic exposure. Conversely, larvae in hyperoxia had fewer, straighter tracheoles.

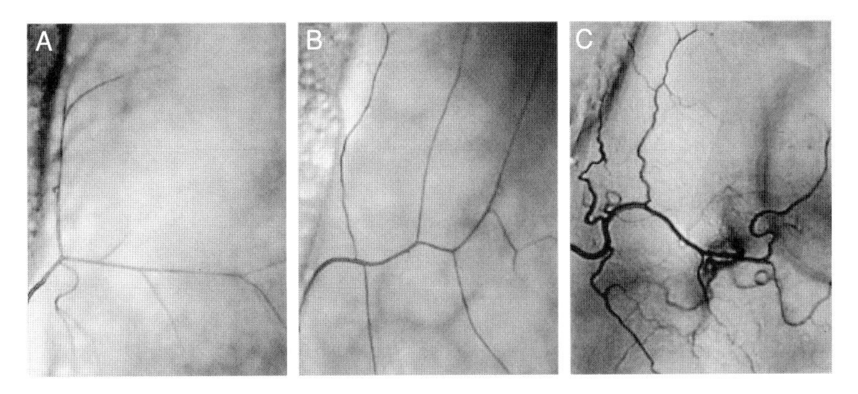

Fig. 6.9 Effect of ambient oxygen on tracheation in third-instar larvae of *Drosophila melanogaster* (Jarecki et al., 1999). Larvae were reared in room air and then transferred to 60% (A), 21% (B), or 5% O_2 (C) for 40 hr. Lower oxygen levels gave more tracheal branches and increased tortuosity. From Jarecki et al. (1999), with permission from Elsevier.

6.2.2.2 Cellular and Molecular Mechanisms of Tracheal Plasticity

Drosophila melanogaster has become the model system for understanding tracheal morphogenesis (Ghabrial et al., 2003; Kerman et al., 2006; Affolter and Caussinus, 2008). The *Drosophila* tracheal system develops in four steps: formation of epithelial sacs in each body segment of the early embryo, followed by sequential sprouting of primary, secondary, and terminal branches.

The cellular details of how this occurs are increasingly well known (Affolter and Caussinus, 2008). Starting about 5 hours after egg laying, the epithelial sacs form from invaginated clusters of ectodermal cells. These tracheal precursors undergo a final cell division during invagination to generate around 80 cells in each sac. Remarkably, all subsequent events forming the larval tracheal system occur by cell migration and deformation—no additional cell division occurs (Samakovlis et al., 1996). At 7 hours, primary branches begin budding, in fixed positions, from each sac. Cells in each bud organize themselves into a tube, each with 4–20 cells. Later, secondary branches, each formed by a single cell, arise from the primary tubes. These single cells form tubes as they extend. Over the subsequent few days, these secondary branch cells give rise to terminal branches (tracheoles), all of which form as cytoplasmic extensions (Ghabrial et al., 2003).

The genetics underlying each step are also known. We do not provide exhaustive coverage here, as several reviews are available (Ghabrial et al., 2003). However, control of the three branching steps (primary, secondary, and terminal) involves the same genes, and interactions among these genes determine how oxygen alters tube morphology, at least in the terminal tracheolar step.

Two primary genes involved in tracheal branching are *branchless*, which encodes a homolog of mammalian fibroblast growth factors (FGF), and *breathless*, which encodes a homolog of mammalian FGF receptors (FGFR) (Fig. 6.10). Before onset of primary branching, tracheal cells express Breathless FGFR. Concurrently, small clusters of cells around each tracheal sac secrete Branchless FGF. Tracheal cells then migrate toward the FGF signal. *Branchless* appears to turn off when the migrating tracheal cells reach the FGF source and, in some cases, is expressed by other groups of cells farther away. The upstream events controlling initial *branchless* and *breathless* expression are partially known (Ghabrial et al., 2003).

Like primary branching, secondary branching is also controlled by *branchless* and *breathless* (Fig. 6.10), but correct patterning requires a second molecular mechanism. Primary cells that arrive closest to the original Branchless FGF signaling centers are induced to express a set of genes that control secondary sprouting. The second part of the mechanism involves *sprouty*, which encodes a protein that antagonizes FGF (Hacohen et al., 1998). *Sprouty* expression is induced as part of the secondary branching gene set, but it acts to stop neighboring cells

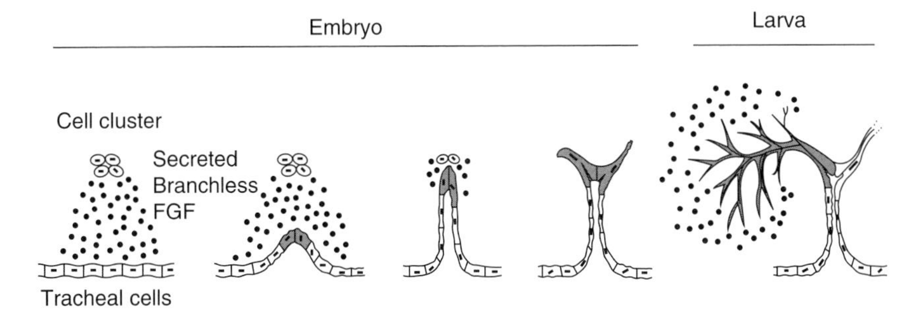

Fig. 6.10 Branchless FGF pathway controls each branching step during tracheal morphogensis (Ghabrial et al., 2003). Clusters of cells secrete Branchless FGF, with secretion enhanced in hypoxia. The secreted growth factor acts as a chemoattractant that guides outgrowth of primary branches. During larval life, branchless turns back on again, but in a completely different pattern, to control outgrowth of terminal branches. The gene is expressed yet again during pupal life where it controls budding of adult air sacs (not shown). Modified from Ghabrial et al. (2003) with permission from *Annual Review of Cell and Developmental Biology*, Volume 19 © 2003 by Annual Reviews www.annualreviews.org.

from inducing the secondary branch genes. Thus, although many primary cells potentially receive FGF signals, only a few of them go on to form secondary branches. *sprouty* mutants respond vigorously to FGF, forming multiple ectopic secondary and terminal branches (Hacohen et al., 1998).

Terminal branching too is controlled by *branchless* and *breathless*, under yet another form of control (Jarecki et al., 1999). In this final stage, low oxygen stimulates expression of *branchless*, and cells outside the tracheal system secrete Branchless FGF. In turn, Branchless FGF acts as a terminal branch inducer (by binding to breathless FGFR in secondary branches) and a chemoattractant that guides extending terminal branches. This control mechanism—induction and attraction of tracheoles by hypoxic tissues—likely plays a central role in distributing oxygen throughout the body to locations of high demand. In summary, the tracheal branching program uses a core FGF pathway expressed repeatedly at different stages and under different kinds of control.

A key outstanding question is how cells sense hypoxia and transduce that information into tracheal plasticity. Several lines of evidence point toward the hypoxia-inducible factor (HIF) system (Jarecki et al., 1999; Ghabrial et al., 2003; Gorr et al., 2006) (see Section 6.3.1), which is strongly conserved and widely used. Centanin et al. showed in *Drosophila* that in mild hypoxia, the HIF-1α homolog Sima accumulates in both tracheal and non-tracheal cells (Centanin et al., 2008). In tracheal cells it induces *breathless*, whereas in non-tracheal cells (the targets for oxygen delivery) it contributes to induction of *branchless*.

The tracheal cells are sensitized to respond to branchless ligands emanating from nearby cells. Centanin et al. suggest that this complementary induction of *breathless* and *branchless* is a core process that distributes oxygen, via terminal cell branching and recruitment, to tissues where it is needed most.

Although a genetic model for hypoxia responses is coming into focus for terminal branching, the details of hypoxia-driven plasticity at higher levels (primary and secondary branching) are still obscure. Nonetheless, there are known developmental effects of oxygen on primary and secondary tracheal diameters in both *Drosophila* (Henry and Harrison, 2004) and *Tenebrio* (Loudon, 1989). Perhaps oxygen acts via other sets of recently discovered genes that control primary and secondary tracheal tube morphology (Llimargas and Lawrence, 2001; Tonning et al., 2005; Luschnig et al., 2006). For example, over- or under-expression of some *DWnt* genes in *Drosophila* changes dorsal trunk patterning and size (Llimargas and Lawrence, 2001). Mutants of *krotzkopf verkehrt*, a chitin synthase gene, and of *serpentine* and *vermiform*, show altered tracheal morphology, including length, diameter, and tortuosity (Tonning et al., 2005; Luschnig et al., 2006).

The O_2 sensitivities of these genes are poorly understood. Arquier et al. suggested that morphogenesis of primary and secondary tracheae is controlled by the HIF-1 signaling cascade, perhaps via these genes (Arquier et al., 2006). More recently, Mortimer and Moberg confirmed this suggestion, showing that primary and secondary branching in embryos of *Drosophila* is affected by hypoxia in a way that depends on Sima, the HIF-1α homolog (Mortimer and Moberg, 2009). The effects of hypoxia depended on timing. Exposure to hypoxia early in embryogenesis caused severe disruption of tracheal morphology. Exposure to milder hypoxia later in development caused increased growth of primary tracheal branches and increased secondary branching, both of which require *sima*. These results suggest that soon we will have a more complete picture of how oxygen levels lead to remodeling at all levels of tracheal morphogenesis, at least in *Drosophila*. Understanding these genetic mechanisms across a broader swath of the insects remains a distant goal.

6.2.2.3 Plasticity in Ventilation

Other more behavioral forms of plasticity are also possible. Harrison et al. measured effects of developmental hypoxia on ventilation rates of adult grasshoppers (*Schistocerca americana*). Individuals were reared, from the first instar, in 5, 10, 21, of 40 kPa oxygen. In adulthood, all grasshoppers showed statistically indistinguishable abdominal pumping rates at their rearing Po_2. However, when tested at 21 kPa oxygen (normoxia), those reared in hypoxia (5 or 10 kPa O_2) had much lower rates of ventilation than those reared in normoxia. Harrison et al. (2006) suggested that developmental hypoxia led either to sensory adaptation of Po_2 sensors that drive ventilation or to developmental increases in number or size of tracheae and tracheoles (Harrison et al., 2006).

6.2.2.4 Plasticity in Eggshell Permeability

All insects go through one life history stage—the egg stage—in which there is no tracheal system, at least early in development. Rather, fluxes of respiratory gases and water are mediated by the eggshell (see Fig. 6.3). Do eggshells respond to either internal or environmental hypoxia?

Zrubek and Woods examined this problem using eggs of the hawkmoth *Manduca sexta*. Batches of eggs were exposed to 15, 21, or 35% O_2 for 1 day, during which time rates of CO_2 and water emission were monitored. Compared to normoxia (21%), hyperoxia (35%) had no detectable effects. However, eggs in hypoxia (15%) had significantly lower metabolic rates (lower CO_2 emission) and higher water loss rates than did normoxic eggs. The authors interpreted these data to mean that insect eggs actively participate in balancing O_2 and water fluxes by modifying eggshell conductance—that to obtain sufficient oxygen, eggs increase eggshell conductance, which exacts the price of higher rates of water loss (Zrubek and Woods, 2006). Although the eggshell response is not nearly as rapid as the tracheal responses described above, it appears to serve the same purpose of matching O_2 supply to demand. A second developmental process in the egg stage may promote oxygen supply to tissues of pharate larvae: about three-quarters of the way to hatching, the embryonic tracheal system undergoes rapid air filling by a process of cavitation and bubble expansion (Woods et al., 2009). Sudden air filling of the tracheal system probably switches it to a functional state capable of rapid oxygen transport.

6.2.3 Trade-offs in the Design of Insect Tracheal Systems

The multiple responses to hypoxia described above suggest that oxygen flux can be limiting even under moderate tissue demand (e.g. during larval development) or moderate environmental hypoxia. Why not allocate enough to the tracheal system that tissues always have the oxygen they need?

The first and most general answer stems from the idea of symmorphosis, developed by Weibel and colleagues (Weibel et al., 1991, 1998). They proposed that structural design of physiological systems is matched to functional demand. In the context of tracheal systems, the symmorphosis hypothesis suggests that tracheal systems should be designed to provide maximal oxygen flux from environment to mitochondria at rates similar to those demanded from, for example, mitochondria operating under maximal rates of ATP production. This argument implies that tracheae are costly, either in terms of materials or perhaps space within the insect. The fact that *D. melanogaster* evolve smaller tracheae when reared for multiple generations in hyperoxia (Henry and Harrison, 2004) supports the hypothesis that tracheae are sufficiently costly that fitness is enhanced by reducing tracheal investment.

Larger insects have lower mass-specific metabolic and gas exchange rates (Chown et al., 2007). Thus we would predict that the tracheal system should scale hypometrically, as observed for mammalian capillaries and fish gills, or perhaps isometrically as observed for mammalian lungs (Weibel et al., 1998). However, the three studies that have investigated the scaling relationship of the tracheal system to date suggest that tracheal investment is hypermetric, with greater proportional investment in larger insects. During ontogeny of *S. americana*, tracheal investment increases in the leg muscle (Hartung et al., 2004) and at the whole-body level, with tracheal volumes and ventilation scaling approximately with mass 1.3 (Greenlee et al., 2009; Lease et al., 2006). Similarly, across four tenebrionid beetle species, tracheal volumes scale with mass 1.29 (Kaiser et al., 2007). Such a trend appears to be general for insects: tiny stick insects have tracheal volumes of around 2% (Schmitz and Perry, 1999), while giant scarabaeid beetles have tremendous air sacs (Miller, 1966). Theoretical calculations suggest that the observed hypermetry is consistent with a need to overcome reduced rates of diffusive gas exchange in longer, blind-ended tracheoles (Harrison et al., 2009).

Increasing animal volume per unit mass, via increases in tracheal volume, could affect many aspects of performance, such as increasing drag, lever arms for locomotion, or niche space required. Based on the scaling of the tracheal system in grasshoppers, a 1-kg grasshopper would have a volume of 3.6 l, greatly increasing the mass-specific need for nutrient investment in the exoskeleton and tracheal system and likely the susceptibility to breakage (Greenlee et al., 2009). Tracheal hypermetry could also directly limit maximal insect size by filling all available space within key regions of the body that cannot be expanded for biomechanical reasons. In interspecific comparisons of beetles, the most dramatic example of hypermetry occurred at the connection between legs and body (Kaiser et al., 2007).

Another factor constraining tracheal design is water loss; obtaining oxygen requires that insects present exchange surfaces to the environment that necessarily lose water. Higher rates of O_2 uptake, or increases in tracheal conductance, may exact costs in terms of increased water loss (Woods and Smith, 2010). Honeybees flying in moderate hypoxia maintain metabolic rates but increase evaporative water loss, likely due to increased ventilation (Joos et al., 1997). Fruit flies vary water loss rates with CO_2 emission rates during flight (Fig. 6.6) in a manner that suggests that spiracle opening is closely matched to oxygen needs—opening them only when the flight motor demanded additional O_2 and simultaneously paying water costs for obtaining that O_2 (Lehmann, 2001). Similar oxygen–water trade-offs appear to constrain the developmental physiology of insect eggs (Zrubek and Woods, 2006).

Finally, too much oxygen is dangerous and the need to maintain tissues at relatively low Po_2 levels may constrain the design and function of the tracheal system (Hetz and Bradley, 2005). This may partially explain why insects close their spiracles at low metabolic rates (Hetz and Bradley, 2005), and why insects evolve smaller tracheae at higher Po_2 (Henry and Harrison, 2004).

6.3 Sensing and Responding to Hypoxia

Almost all responses discussed in this chapter involve sensing and responding to O_2. How do insects do it? Is the same sensing mechanism used over different timescales? Rapid progress on these questions has been made in the past few years. It now appears that insects use two main molecular systems, one based on hypoxia-inducible factors (HIFs) and one on atypical guanylyl cyclases, to sense O_2 and respond over long and short timescales, respectively.

6.3.1 Hypoxia-inducible Factors (HIFs)

The main molecular mechanism mediating hypoxia-driven changes in gene expression is a family of highly-conserved transcription factors—the hypoxia-inducible factors (HIFs)—that occur throughout examined animals, including mammals, teleosts, nematodes, and multiple arthropods (Gorr et al., 2006). HIFs occur as heterodimers that bind to hypoxia-response elements (HREs) within regulated genes. The global effects of HIF expression are only now becoming clear. In humans, for example, Manalo et al. estimated that 2–5% of *all genes* are regulated by HIF (Manalo et al., 2005). In mammals, the two main functions affected by HIF-regulated genes are 1) increasing tissue oxygen during modest hypoxia; and 2) increasing anaerobic carbohydrate metabolism in a cell-autonomous way during severe hypoxia (Ebbesen et al., 2004; Gorr et al., 2006).

HIFs do not interact with O_2 directly; rather their levels and actions are themselves regulated by O_2 at many levels, which fine tune responses to particular tissues and cells (Acker and Acker, 2004; Kaelin and Ratcliffe, 2008). Each HIF is a heterodimer composed of HIF-1α and HIF-1β subunits. HIF-1β is stable whereas HIF-1α is unstable, such that low levels of O_2 promote HIF-1α persistence while high levels promote its degradation. If HIF-1α persists, it binds to HIF-1β and the complex moves into the nucleus, where it acts as a transcription factor activating downstream genes involved in the hypoxia response.

The main (but not only) mechanism by which O_2 regulates HIF-1 activity is O_2-dependent hydroxylation of proline or asparagine residues in HIF-1α—that is, high O_2 allows hydroxylation to occur (demonstrated in *Drosophila* by Arquier et al. 2006; Kaelin and Ratcliffe, 2008). Hydroxylated subunits then interact with another protein, the von Hippel Lindau (VHL) tumor suppressor protein, which instructs the E3-ubiquitin ligase enzyme complex to tag the molecule for proteasomal degradation. Hydroxylation can also alter interactions between heterodimer subunits, reducing their transcriptional activity. In mammals, several other O_2-sensing mechanisms have been proposed, including heme-containing molecules and reactive oxygen species (Acker and Acker, 2004).

Work on *Drosophila* has shown how this signaling cascade may be focused on individual tissues, particularly the tracheal system. Arquier et al. (2006) constructed a fusion protein consisting of the O_2-dependent degradation domain (ODD) of HIF-1α and the green fluorescent protein (GFP), which allows *in situ* visualization of ODD expression. They found that hypoxia (5% O_2) led to marked increases in ODD expression in tracheal-associated cells but no differences in ectodermal cells—the tracheal system but not the ectoderm was sensitive to the hypoxia. Arquier et al. proposed that the differences stemmed from expression patterns of VHL (the von Hippel Landau protein), which is expressed in tracheae but not ectoderm making the former tissue hypoxia sensitive (Arquier et al., 2006). These results demonstrate that hypoxic sensitivity can be highly tissue specific. See Section 6.2.2 for details of physiological and developmental responses of tracheae to HIF expression.

6.3.2 Atypical Guanylyl Cyclases

HIF-driven changes in protein expression involve, at a minimum, the following steps: migration of HIF-1 heterodimer to the nucleus, binding to DNA, recruitment of transcription machinery, transcription, translation, protein folding and modification, and export and localization of protein product. This sequence is slow: it cannot cause physiological change in less than about half an hour. But insects respond in as little as a few seconds to changes in O_2 levels (Wingrove and O'Farrell, 1999), indicating the existence of some other, more rapid, mechanism.

David Morton and colleagues may have discovered it (Vermehren et al., 2006). The intracellular messenger, cGMP, is synthesized by guanylyl cyclases, of which there are two classes. The first, the typical soluble guanylyl cyclases, are heme-containing heterodimeric enzymes which are strongly activated by nitric oxide (NO) but not O_2. The second class, atypical guanylyl cyclases (in *Drosophila* the subunits *Gyc-88E*, *Gyc-89Da*, and *Gyc-89Db*), are potently activated by hypoxia (i.e. *lack* of O_2) but are virtually insensitive to NO.

In *Drosophila*, evidence now indicates that atypical guanyl cyclases (aGC) function as molecular O_2 sensors. Morton (2004b) showed that COS-7 cells co-transfected with *Gyc-88E* and *Cyc-89Da* or *Gyc-89Db* had low levels of cGMP in normoxia but increasingly higher levels of cGMP in increasingly severe anoxia (Fig. 6.11). The response was fairly linear, as would be expected of an O_2 sensor. Moreover, Morton observed significant accumulation of cGMP within 1 minute of anoxia in cells containing the *Gyc-88E/89Db* combination (Morton, 2004a, 2004b). Though they did not show it, the authors hypothesized that,

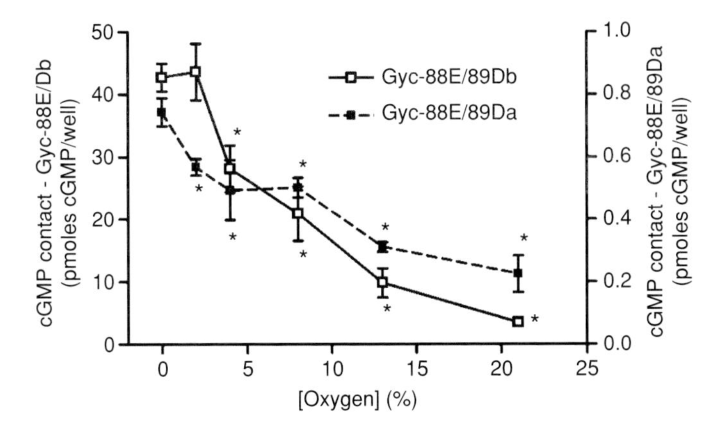

Fig. 6.11 Graded response of atypical guanylyl cyclases to ambient oxygen levels (Morton, 2004a). COS-7 cells were transfected with the subunits shown, exposed for 1 hour to the different levels of oxygen shown, and then assayed for cGMP content. The linear response to oxygen is a key characteristic expected for molecules acting as oxygen sensors. From Morton (2004) with permission from the American Society for Biochemistry and Molecular Biology.

in normoxia, O_2 binds directly to the aGC heme group; in hypoxia, O_2 dissociates from the heme group, activating the enzyme. Support for this idea has been put forward by Huang and Haddad (2007). Morton et al. identified two populations of neurons that express *Gyc-89Da* or *Gyc-89Db*. Using tetanus toxin, they were able to block synaptic transmission of just these neurons. Blocking the *Gyc-89Da* neurons prevented adult eclosion, and blocking the *Gyc-89Db* neurons blocked initiation of the first larval ecdysis (Morton et al., 2008). Altogether, how common is this mechanism across the insects? The answer is unclear, but many sequences for aGCs are now available from *Anopheles*, *Bombyx*, *Manduca*, *Apis*, *Tribolium*, and the nematode *Caenorhabditis elegans* (Morton, 2004b).

An important step in establishing physiological function *in vivo* is to localize expression patterns, as Langlais et al. have done for *Drosophila*. They found Gyc-88E and Gyc-89Db in central and peripheral neurons of both embryos and newly hatched first-instar larvae. Peripheral expression stained most intensely in neurons associated with external sensilla—which are a primary conduit for sensing ambient hygro and chemical environments (Langlais et al., 2004). In one preparation, they observed guanylyl cylcase extending in a single dendrite to the very tip of the sensillum. Furthermore, in abdominal segments A1 and A2, *Gyc-88E* and *Gyc-89Db* were expressed in several neurons innervating the tracheal system. Together, these spatial patterns support the idea that O_2-sensing aGCs mediate rapid nervous responses to changing internal and external O_2 status.

6.3.3 Two Rapid-Response Systems with Unknown Oxygen Sensors

Ultimately, we would like to be able to trace variation in hypoxia responses to a molecular level and, conversely, to predict how molecular variation is manifest in whole organisms, including non-model insects. That day is coming; here we review two rapid-responses systems in insects for which molecular explanations are needed.

The first is oxygen's effects on ventilation (see Section 6.2.1), which can be dramatic. Early work by Miller (1960a) on locusts (*Schistocerca gregaria*) indicated the influence of CO_2 and O_2 on the central nervous system (CNS). By micro-perfusing different parts of the locust and its nervous system with altered gas compositions, Miller showed that head and thoracic ganglia contain CO_2 receptors that alter pumping frequency. Oxygen affected CNS rhythmic activity but less dramatically than did CO_2. Subsequent work defined a central pattern generator (CPG) operating in the locust metathoracic ganglion

Burrows (1996), and more recent work has refined the role of the metathoracic ganglion in gas detection (Bustami and Hustert, 2000; Bustami et al., 2002). Using isolated nerve cords from the lubber grasshopper (*Taeniopoda eques*) and the American desert locust (*Schistocerca americana*), Bustami and colleagues showed that ventilatory burst patterns could persist for several hours and that pulse rate depended on both CO_2 and O_2 partial pressures in the head space above the preparation. Based on similar experiments, gas detectors in the CNS have also been inferred for *Periplaneta americana* (Woodman et al., 2008). What remains is to identify the sensors themselves and their location in the CNS—which likely will involve the atypical guanyl cyclases discussed above.

The second system for which molecular sensors have not been identified is spiracle opening and closing. In general, spiracles open for longer or more often when intra-tracheal CO_2 is high or O_2 low (Harrison et al., 2006). A key question is whether this control is exerted via central or peripheral parts of the nervous system. Certainly there is strong evidence for centralized responses to oxygen (Van der Klook, 1963; Burkett and Schneiderman, 1974a). Several tantalizing clues also suggest at least some local control near spiracles. Langlais et al. (2004), for example, showed in *Drosophila* that some peripheral nerves innervating tracheae expressed high levels of *Gyc-88E* and *Gyc-89Db*, and Wingrove and O'Farrell (1999) demonstrated high levels of nitric oxide synthase in spiracular glands associated with posterior spiracles in *Drosophila*. However, full elaboration of a model, from chemosensation to spiracular control, still remains. Förster and

Hetz propose a new model that connects internal levels of CO_2 and O_2 to spiracle opening and closing (Förster and Hetz, 2010).

6.4 Living at High Altitudes

Living at altitude presents a suite of physiological and ecological challenges, of which low oxygen is only one (Hodkinson, 2005). Others that are O_2-relevant include lower air density and lower temperature (Dillon et al., 2006). Some changes are partially offsetting. For example, in hypo-dense air, lower Po_2 may be partially offset by higher O_2 diffusion coefficients. And other changes are synergistic. For example, flying in hypodense air at altitude requires greater metabolic output, which in turn requires extracting oxygen at greater rates from lower ambient Po_2. Thus, studying effects of declining O_2 availability per se is complicated by systematic changes in other factors. Such factors also include biological effects of changing host-plant, disease, and parasite communities (Hodkinson, 2005). In this section, we describe some of the gas exchange challenges faced by flying insects at altitude and relate those challenges to altitudinal patterns of insect physiology and ecology. Additional work has been done on aquatic insects in high altitude streams (Rostgaard and Jacobsen, 2005; Jacobsen and Brodersen, 2008), which face the twin oxygen problems of low ambient Po_2 and potentially slow oxygen transport through water (see Section 6.5).

6.4.1 Flight in Thin Air

Flight at low air density requires greater energy expenditure (Dudley, 2002). How do insects generate extra force? How large is the metabolic cost? Are there strict altitudinal limits beyond which insects cannot fly, and do these stem from O_2 limitation? Studying insect flight mechanics and physiology across altitudinal gradients is non-trivial. Consequently, much of what we know comes from laboratory studies in which air density, Po_2, and sometimes both, are manipulated experimentally (Dudley and Chai, 1996).

Dudley (1995) studied effects of variable air density on three species of orchid bees (Euglossini) from lowland, tropical Panama. He manipulated air density by flying bees in either air or heliox (21% O_2 in He), which has a density around 66% lower than air. Bees in heliox exhibited dramatic increases in lift production and power output stemming from increases in wing stroke amplitude (from 105 to 140°) and not in wingbeat frequency. Large increases in power output suggested that bees had much higher metabolic rates in heliox, although Dudley did not directly measure gas exchange or heat output.

Using carpenter bees (*Xylocopa varipuncta*), Roberts et al. (2004) varied air density using different ratios of He and N_2 as the inert gas (all treatments had

21% O_2, although changes in air density probably changed the diffusion coefficient of oxygen) and measured a suite of flight kinematics and metabolic rate (as CO_2 emission). In hypodense air, carpenter bees, unlike orchid bees, increased both stroke amplitude and wingbeat frequency, although the frequency effect was small. The metabolic rate of hovering bees also increased by more than 50% in hypodense air. Perhaps the most ecologically relevant finding stems from the large variation in the body size of bees in the study. Body mass varied more than three-fold, from 0.4 to 1.2 g. The proportion of mass devoted to thorax (flight motor) versus abdomen also varied, such that small bees allocated a much larger percentage of total mass (~50%) to thoracic muscle mass than did large bees (~25%). This differential allocation affected the *reserve capacity* (or safety margin) of flight performance traits (Fig. 6.12): small bees could respond to hypodense air by increasing most measured traits, including stroke amplitude, wingbeat frequency, and mass-specific power output. Large bees, by contrast, had little or no reserve capacity: they appeared to be performing at near-maximal levels even in normodense air.

Using wild-caught *Drosophila melanogaster*, Dillon and Frazier (2006) showed that, although flies are capable of flight over a wide range of barometric pressures, low pressure and temperature reduce motivation to fly. Flies subjected to different combinations of temperature and pressure were physically capable of flight in

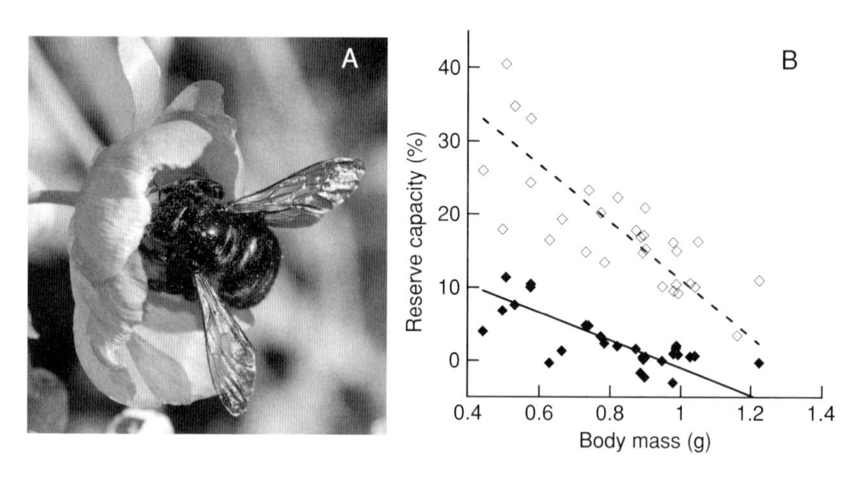

Fig. 6.12 (A) Foraging female of *Xylocopa varipuncta* (photo Kathy Keatley Garvey). (B) Allometry of reserve capacities for hovering carpenter bees (*Xylocopa varipuncta*)–wingbeat frequency (filled symbols) and stroke amplitude (open symbols). Data were obtained by flying bees in normo- and hypo-dense air, obtained by substituting He for N_2. The pattern indicates that larger bees have smaller safety margins for flight performance; the largest bees appear to have non-existent safety margins even in normo-dense air. From Roberts et al. (2004), with permission from The Company of Biologists.

a wide range of conditions; flies failed only at the lowest air pressures (33% of sea level, equivalent to the air pressure at the top of Mount Everest!) and then only at lower temperatures. However, motivation to fly changed dramatically with both temperature and air pressure, and the effects were interactive. In particular, at lower pressures and temperatures, flies were much more likely to fly only when coerced. These findings likely are relevant to *Drosophila* species in the wild, which occur at up to 3000 m in the Sierra Nevada of California (Dobzhansky, 1948) and 5000 m in the Himalayas (Khare et al., 2002).

6.4.2 Multiple Effects of Air Density on Oxygen Supply

The studies above indicate that flight in lower air densities requires more aerodynamic power and higher metabolic rates. In nature, the decrease in air density at higher altitudes is accompanied by decreasing partial pressure of oxygen (Po_2) , which may limit metabolic rates and further restrict flight capacity (note Po_2 in air = 0.21 * total barometric pressure, *P*). However, lower total atmospheric pressure results in higher O_2 diffusion coefficients in air. The diffusion coefficient of a gas can be described by the following equation (Denny, 1993):

$$D = k_1 T^{k_2} \frac{P_0}{P}, \qquad (15)$$

where T is absolute temperature, P is pressure, and P_0 is one atmosphere. For oxygen, $k_1 = 1.13 \times 10^{-9}$ m^2 s^{-1} and $k_2 = 1.724$ (Marrero and Mason, 1972). Therefore, a 50% drop in barometric pressure leads to a doubling of the diffusion coefficient, D. Recall that according to Equation 9, diffusive flux is directly proportional to D and the partial pressure gradient. Thus, if gas exchange through tracheae occurs completely by diffusion through air, increasing altitude should have no effect on oxygen delivery, as increases in molecular diffusion rates will exactly balance the drop in the driving gradient.

However, the effect of altitude on D within *air-filled* tracheoles may differ in very small tracheoles, whose diameters are small enough to affect the molecular process of diffusion (Pickard, 1974). The diffusion coefficient of O_2 in a tube depends both on rates of collision with other gas molecules and with the tube walls. In large-diameter tubes, oxygen's diffusion coefficient depends on pressure because most collisions are with other gas molecules. In small-diameter tubes, however, O_2 movement is increasingly dominated by collisions with tube walls, not other gas molecules, a process that is known as Knudsen diffusion (Bird et al., 2002). The transition point between Fickian and Knudsen diffusion occurs when the tube diameter is similar to the mean free path of O_2 (the average distance traveled between collisions) which is about 0.08 μm at room temperature in air at sea level and higher in low-pressure, high-altitude air. The smallest tracheoles have diameters of 0.20–0.25 μm, suggesting that in these tracheoles, *D* may not increase at higher altitudes.

The effect of altitude on oxygen delivery also depends on two other factors: 1) the relative contributions of diffusive versus convective transport; and 2) the relative resistances of air- versus liquid-phase gas transport. First, the greater the role that convection plays, the less the self-cancelling effects of barometric pressure matter, because a 50% decrease in barometric pressure (and therefore Po_2) causes a 50% decrease in the molar amount of oxygen transported by a packet of moving air (see Section 6.1). Second, the self-cancelling effects of barometric pressure on O_2 transport apply only to diffusion in *air phase*. But the distal steps in oxygen transport—through tracheolar walls, cell membranes, and cytoplasm to the mitochondria—all occur in liquid, and diffusion coefficients of oxygen in liquid are essentially unaffected by barometric pressure; therefore, flux will be decreased by a lower Po_2 gradient. Clearly, it is difficult to quantitatively predict the effect of altitude on the oxygen delivery capacity of insect tracheae, and empirical studies are needed.

Altitude probably does have strong effects on gas exchange by insect eggs. Throughout most of embryonic development, the main resistance to O_2 flux occurs in the insect eggshell and liquid (yolk) around the embryo. Although the chorion and trabecular layers of the eggshell are usually air-filled, Woods et al. (2005b) showed that in eggs of the hawkmoth *Manduca sexta* the chorion provided negligible resistance to O_2 movement. Instead, resistance was localized to solid or semi-solid layers between the chorion and embryo. Consequently, low Po_2 at altitude should be a real and pressing problem for insect eggs; we predict that, among insect life stages, eggs will be most sensitive to high altitude hypoxia. Again, this idea has never been tested.

6.4.3 Oxygen–Water Trade-Offs at Altitude

As hikers know, high-altitude air feels dry, and one must drink fluid at above normal rates to stay hydrated. Dehydration occurs because high-altitude air contains less water vapor, even though its *relative humidity* may still be high. Vapor density is a measure of the concentration of water per unit volume of air (g m^{-3}) whereas relative humidity is a fractional measure of how much water the air contains compared to how much it could contain if vapor-saturated. Because cold air contains much less water vapor at saturation, cold, high-altitude air contains little water even at near-saturation humidities.

The driving force for insect water loss is the vapor density gradient between body tissues and environment. One might expect that low ambient vapor density would magnify the gradient, always leading to greater water loss at altitude. However, again, there is an interesting and ecologically-relevant subtlety stemming from the temperature-dependent vapor capacity of air. Insects with low body temperature—those that are small, resting, or cryptic (i.e. most high-altitude insects most of the time)—will have very low *internal* vapor densities because they are cold. In these cases, the driving gradient for water loss may also be low, leading to low rates of water loss. By contrast, large flying insects that maintain high thoracic temperatures in flight—such as bees and hawkmoths, which can maintain 35°C in the thorax—may produce very high *internal* vapor densities, as hot air flowing through the thorax rapidly becomes vapor saturated. Thus, hot-bodied insects flying at altitude will produce *enormous* water vapor gradients between themselves and the environment. These water loss rates will be further increased by high-altitude hypoxia as insects likely increase tracheal conductance (e.g. ventilation) to obtain sufficient oxygen (Joos et al., 1997). A possible mitigating factor is metabolic water production, which can compensate for a significant fraction of total water lost during flight (Lehmann, 2001). Again, there is little empirical data on altitude effects on water loss in insects to evaluate these predictions.

6.4.4 Morphological Characteristics of High-altitude Insects

As a group, high-altitude insects show more brachyptery (short wings) and aptery (absence of wings) (Hodkinson, 2005). In some species, brachyptery appears as a polymorphism, with higher frequencies at altitude (Hodkinson, 2005). Alternatively, related species may show fully-winged forms at low altitude and, in congeners, brachypterous forms at altitude. Ecological conditions leading to the evolution of brachyptery are disputed (Zera and Harshman, 2001). Constraints on O_2 delivery may play a role, but no empirical data are available to separate this hypothesis from other potential reasons for not flying (high rates of water loss, cold temperatures, wind, etc.).

The second altitude-associated morphological pattern, seen in high altitude species and populations that retain wings, is an increase in wing area and a decrease in wing loading (the ratio of body mass to wing area). Insects with lower wing loading reduce the power (and oxygen consumption) necessary for flight (Dudley, 2002; Dillon and Frazier, 2006). Stalker and Carson (1948) showed that *Drosophila robusta* had higher wing area at the high end (1400 m) of an altitudinal transect in Tennessee; thorax lengths did not change across the transect, suggesting that neither did body mass. Hepburn et al. (1998) showed that high-altitude (2300 m) honeybee colonies had workers with approximately 10% lower wing loading, resulting primarily from increases in wing area. Dillon and Frazier (2006) measured thorax and wing widths on museum specimens of the bumblebee *Bombus festivus*, which are found naturally across a broad altitudinal gradient (400–5200 m). Workers, but not queens and males, had lower wing loading—wing length declined with altitude, but body mass (as estimated from thoracic measurements) declined even more.

Is increasing wing area (and decreasing wing loading) due to developmental plasticity or genes, and is it driven by oxygen, temperature, air density, or some other biotic variable? No studies have yet reared insects at low air pressures and measured wing areas or loading, but developmental plasticity of wing areas in response to temperature and oxygen has been demonstrated for fruit flies. Rearing at cold temperatures increases wing area and improves flight performance at cold air temperatures, suggesting that temperature rather than oxygen is the primary variable affecting altitude-related changes in wing area (Frazier et al., 2008). Genetically-based differences in wing area have been demonstrated across altitudes among populations of *D. buzzatii* (Norry et al., 2001).

6.5 Living in Water

> "For a member of a terrestrial group entering freshwater, respiration probably poses a greater problem than does any other vital function."
>
> G. Evelyn Hutchinson (1981)

Unlike the sparse literature on high-altitude insects, the literature on aquatic insects is immense, reflecting their importance as food sources for commercial and sport fishes, as nutrient recyclers and agents of trophic transfer, and as bioindicators of water quality. Also, they are more readily accessible to biologists, as there are numerous freshwater habitats throughout the world. Not surprisingly, much is known about their O_2 biology, both structures and physiologies (see review by Resh et al. (2008)) and the effects of O_2 on distribution and abundance. Below we highlight only a few selected and (mostly) recent studies of aquatic respiratory physiology.

Living in water fundamentally reconfigures the problems of gas exchange. The two most difficult problems are low O_2 content of water and vastly smaller O_2 diffusion coefficients—2.1×10^{-9} m^2 s^{-1} in water versus 2.0×10^{-5} m^2 s^{-1} in air, at 20°C (Denny, 1993). Most aquatic adaptations therefore involve behaviors, morphologies, and physiologies aimed at extracting scarce, recalcitrant oxygen from large volumes of water, or maintaining direct connections with the atmosphere. One problem that plagues terrestrial insects—loss of respiratory water—disappears for aquatic insects. This release allows the evolution of large surface areas for exchange across high-conductance structures.

In this section, we take a broad view of the gas exchange mechanisms that support aquatic insects. In particular, we consider "aquatic" to mean *any* liquid habitat—from commonly considered habitats like streams and ponds to less traditional habitats like rotting fruit (*Drosophila* larvae) and caterpillar blood (parasitoids). This definition obscures finer distinctions made by other authors but emphasizes evolutionary diversity of functional solutions to the common problem of gas exchange in liquids.

6.5.1 Open Versus Closed Tracheal Systems

Tracheal systems of most insects are *open*, with spiracles connected directly to the atmosphere (or any other suitable source of O_2). The first four sections below describe various elaborations of open systems used by aquatic insects. By contrast, other aquatic insects have *closed* tracheal systems, which lack functional spiracles. Gas exchange in these groups occurs across high-conductance, high-surface-area structures, either the whole body cuticle or projections from the tracheal system called lamellae or gills. Closed tracheal systems are described in a subsequent section.

6.5.2 Strategy 1: Stay Directly Connected to the Atmosphere

Many aquatic insects, though submerged, maintain direct connections between their tracheal systems and the atmosphere using tubes (snorkels or respiratory siphons). A modest but familiar example is the *Drosophila* larva, which feeds head-down in various forms of muck. Throughout larval development, most O_2 is supplied only by the two posterior-most spiracles (Manning and Krasnow, 1993). Mosquito larvae use a similar arrangement in water, though usually with the addition of an elongated tube (siphon). Many larvae divide their time between underwater activities (swimming, feeding) and surface activities (resting, suspension feeding), during which they hang from the surface tension with valves in the siphon open to the atmosphere to allow gas exchange. More dramatic examples of snorkels (Resh et al., 2008) include long respiratory tubes emerging from the posterior tracheal systems of rat-tailed maggots (Diptera: Syrphidae; e.g. *Eristalis tenax*) (Fig. 6.13A) and water scorpions (Hemiptera: Nepidae; e.g. *Ranatra*) (Fig. 6.13B). Eggs in liquid or semi-liquid media (e.g. *Drosophila*) often bear respiratory horns, which may also facilitate gas exchange using plastrons (Fig. 6.13C, D; see below) (Hinton, 1981). Eggs of some parasitoids have respiratory horns that emerge from their hosts (Hinton, 1981).

The consequences of respiratory tubes, for both physiology and ecology, are poorly known but almost certainly important. In air, most terrestrial insects obtain O_2 from multiple sets of spiracles arranged along the thorax and abdomen. In this arrangement, the problem is mostly one of radial O_2 transport into the body, which, because insects are elongate, minimizes transport distances. By contrast, respiration through one or a few tubes restructures the geometry so that longitudinal transport is more important. Such an arrangement may require mechanisms for convective air movement, although these are completely unstudied. In addition, one-point O_2 entry may require a rearrangement of the relative proportions of the tracheal system. For example, the main longitudinal tracheae in *Drosophila* larvae are almost three times wider at the posterior end (near the functional spiracles) than at the anterior end (Henry and Harrison, 2004).

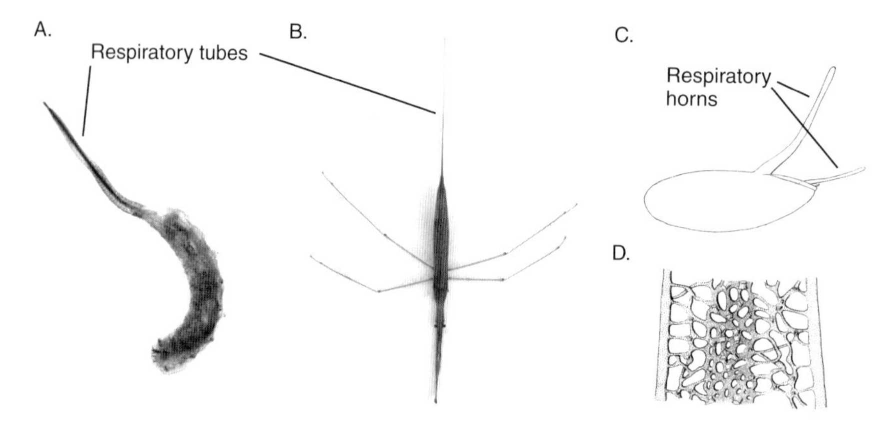

Fig. 6.13 Respiratory tubes of (A) rat-tailed maggots (*Eristalis tenax*) and (B) water scorpions (*Ranatra linearis*). (C) Respiratory horns of *Drosophila gibberosa* (from Hinton 1960) show projecting from egg. (D) Optical section of the posterior respiratory horn of *D. gibberosa*, about 100 μm from its base (from Hinton 1960). The mesh is hydrophobic and filled with air (a plastron). (Photo credits: (A) Brian Jones; (B) Piet Spaans; (C) and (D) from Hinton (1960) used with permission from The Royal Society.)

Moreover, in hypoxic atmospheres, the diameter at the posterior end showed plastic increases in size whereas the diameter of the anterior end did not (Henry and Harrison, 2004).

6.5.3 Strategy 2: Take Air Along for the Ride

Many insects also maintain open, air-filled tracheal systems, *without* direct connection to the atmosphere. They do so with gas gills, which are gas volumes bounded by a liquid–gas interface serving as an exchange organ. Gas gills come in two basic types: compressible and incompressible (the latter is also known as plastron respiration). Although superficially similar, these two types have different consequences for gas supply.

Compressible gills consist of air bubbles or films trapped against the insect's body and connected with its tracheal system. Many diving beetles and bugs use trapped air to support respiration (Resh et al., 2008). Trapped air bubbles can act as O_2 reserves, but a more important function is to increase effective surface area for gas exchange with surrounding water. The externally-trapped air multiplies the functional surface area of the combined spiracles to which it is connected. Thus, during its lifetime, a bubble usually extracts much more O_2 from the water than it originally held when first formed. A drawback of compressible gills is that they have a limited useful lifetime, because N_2 diffusion into surrounding water gradually causes the bubble to collapse (Rahn and Paganelli, 1968). Thus, insects using this mechanism must return periodically to the surface to refresh the bubble, which can expose them to predators (Hutchinson, 1981).

Tsubaki et al. (2006) describe the reproductive consequences of compressible gas gills used by the damselfly *Calopteryx cornelia* (Odonata: Calopterigidae). Females oviposit underwater in forest streams, normally staying submerged for 20 to 120 minutes. Tsubaki et al. (2006) showed that the number of eggs laid was linearly related to time spent underwater. While underwater, females derive O_2 from a film of air covering the body and wings, presumably trapped by the hydrophobic external surface. The authors measured time to asphyxiation under different conditions, including hypoxic versus normoxic water, stirred versus unstirred water, and immobilized versus freely moving females. In addition, in some treatments they removed the air film by coating wings with Vaseline. The largest negative effects on survival time were obtained with coated wings in unstirred water (survival time < 25 min) in hypoxic water (50% of air saturation). The largest positive effect was in the semi-submerged treatment, in which wings could renew trapped air. Tsubaki et al. (2006) suggest that oviposition time in high-oxygen streams is not limited by respiratory physiology, although this may not be true in poorly oxygenated water or higher temperatures.

The second type of gas gill—the incompressible gill or plastron—differs from the compressible type in that it can persist indefinitely. Its extended life arises from direct mechanical support of the air film by hydrofuge hairs or other hydrophobic structures (Thorpe and Crisp, 1947a, 1947b). The surface area of the supported film provides a large surface for obtaining O_2 from the surrounding water. O_2 enters the insect tracheal system via open spiracles that are in contact with the plastron film. Many larval and adult insects use plastrons (Resh et al., 2008), and they are especially common in eggs of aquatic species and of terrestrial species subjected to periodic flooding (Hinton, 1981). Insects using plastrons may no longer have to return to the surface, and whole stages or even lifecycles can take place under water.

The ecological consequences of using gas gills are largely unknown. A few authors have suggested broader hypotheses linking respiratory mechanisms to ecology. In particular, Hutchinson (1981) noticed that within the Corixidae (Hemiptera) the larger Corixinae occur primarily in cool, temperate waters, whereas smaller Micronectinae are primarily tropical. Both species use compressible gills. Hutchinson suggested that warmth reduces compressible gill effectiveness in two ways: by accelerating rate of collapse and by raising metabolic rates. Thus, warm water may force species using compressible gills to be small to avoid demanding rates of O_2 delivery in excess of what the gill can deliver. Rigorous functional and comparative tests of this hypothesis have not been done.

6.5.4 Strategy 3: Mine Photosynthetic Oxygen from Plants

Another alternative is to gather O_2 produced by photosynthesis. Two general schemes can be envisioned. The first is for insects to position themselves in water enriched with O_2 from nearby plants. For example, the pond-dwelling damselfly *Enallagma cyathigerum* (Odonata: Coenagrionidae) usually oviposits near green stems of water plants, regions that have high O_2 levels (Miller, 1994).

The second scheme is to tap internal O_2 stores directly from plants. Many aquatic plants maintain buoyancy by storing O_2–for plants a waste product—in vacuoles, and they also have aerenchyma, a spongy tissue with intercellular air spaces through which O_2 can move. For aquatic plants, O_2 movement in aerenchyma supports cellular respiration in roots, which often occupy anoxic sediments. Some aquatic insects have evolved means of piercing plant tissues to tap this O_2 source directly, including mosquito larvae (Diptera: Culicidae) of *Mansonia*, *Coquilletidia*, and *Taeniorhynchus* (Resh et al., 2008).

One of the most economically important root piercers is the rice water weevil, *Lissorhaptrus oryzophilus* (Coleoptera: Curculionidae). It is the most destructive early-season rice pest in the USA (Zhang et al., 2006) and has recently been introduced into rice-producing areas of Asia (Chen et al., 2005; Saito et al., 2005). Larvae have specially modified spiracles for piercing rice roots and reaching aerenchymal O_2 (Fig. 6.14A). At the end of the last instar, larvae find a root, make a hole in it, and form the pupa next to the hole (Fig. 6.14B), with the cocoon chamber connected directly to root aerenchyma (Zhang et al., 2006).

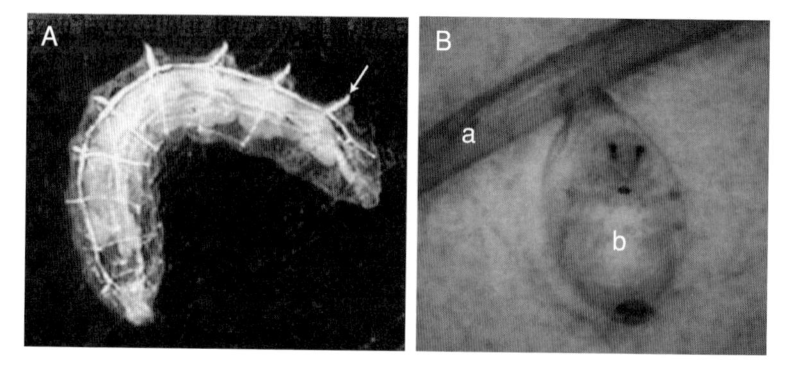

Fig. 6.14 Root tapping by the rice water weevil *Lissorhoptrus oryzophilus* (Zhang et al., 2006). (A) Larvae have specially modified spiracles (arrow) for piercing rice roots and reaching aerenchymal oxygen. (B) At the end of the last instar, larvae find a root (a), make a hole in it, and form the pupa (b) next to the hole. Redrawn from Zhang et al. (2006) with permission from the Kansas Entomological Society.

The potential consequences of depending on photosynthetic O_2 are diverse. First, aerenchyma may be hyperoxic, especially in or near plant leaves (Sand-Jensen et al., 2005), leading to hyperoxic stress. Second, root-piercing insects experience diurnally fluctuating O_2 supply, with potentially very low Po_2 at night. For example, root Po_2 in the halophytic stem succulent *Halosarcia pergranulata* had daytime values in the range of 5 kPa O_2 whereas dark values were less than 0.5 kPa O_2 (Pedersen et al., 2006). Thus, root-piercers seem likely to require euryoxic physiologies able to cope with intermittent anoxia. Third, patterns of insect activity and growth seem likely to be directly linked to seasonally active periods of photosynthesis (Houlihan, 1969, 1970).

6.5.5 Closed Tracheal Systems Exchange Gases Across Large Surface Areas and High-conductance Barriers

Some aquatic insects have dispensed entirely with functional spiracles, having so-called closed tracheal systems. In these species, oxygen must cross the cuticle before entering the air-filled tracheal system. Not surprisingly, the location at which oxygen crosses the cuticle, and the cuticle's characteristics, bear strongly on oxygen flux. Living in water relieves the dangers of desiccation, which probably facilitates the evolution of this strategy in the first place. In general, a distinction is made between species that obtain oxygen across the body wall (cutaneous respiration) and those that have evolved elaborations of the cuticle (tracheal gills) that give higher surface areas for gas exchange. The two modes of transport, however, are not mutually exclusive—even species with tracheal gills may obtain, under some circumstances, the majority of their oxygen across the body surface (Resh et al., 2008). Indeed, some authors have rejected the idea that tracheal gills are important for oxygen uptake at all and are instead used for other functions, such as ionic regulation (Mill, 1974). For example, Morgan and O'Neil (1931) experimentally removed gills from larval *Macronema* (Trichoptera) and found no effect on oxygen consumption. However, experiments on gill-less specimens in stressful conditions (higher temperature, lower oxygen) have clearly shown the importance of tracheal gills to respiration (Resh et al., 2008).

6.5.6 Parasites Inside Hosts

Insects living in other animals (parasites) face similar problems to those living underwater: O_2 diffusion through fluid is slow, and host body fluids may be hypoxic. Moreover, for insects living in other *insects* (e.g. parasitoids) the problem is worse, because O_2 in the host is delivered primarily via the tracheal systems, with the main body fluid, hemolymph, carrying little or none. For some parasites, the solution is analogous to the O_2-delivery mechanism discovered by root-piercers. They tap directly into the host tracheal system.

Ichiki and Shima (2003) describe such a tapping mechanism used by the highly polyphagous parasite, *Compsilura concinnata* (Diptera: Tachinidae), which parasitizes primarily caterpillars (Lepidoptera). Female *C. concinnata* inject fully developed first-instars, still in the egg, into the host (ovovivipary), which in Ichiki and Shima's experiments were silkworms, *Bombyx mori*. The tachinid larva hatches and migrates to the cavity between peritrophic membrane and midgut epithelium of the host, where it uses anal hooks to attach itself to a host tracheole. Apparently the hooks pierce the tracheole and establish a direct connection between the host tracheole and the parasite's main longitudinal tracheae. Later instars pull additional tracheae and tracheoles through the host midgut epithelium and attach these also via anal hooks.

Unlike *C. concinnata*, most other tachinid parasites obtain O_2 via a respiratory funnel (Clausen, 1940; Salt, 1968), a breathing tube derived from host defensive cells that the parasite appropriates for itself (Stireman et al., 2006). Respiratory funnels do, however, allow tachinid larvae to maintain direct access to the atmosphere through their posterior spiracles. Stireman et al. (2006) argue: "The ability to capitalize on the immune response by forming respiratory funnels may allow tachinids flexibility to ecologically 'explore' new hosts more easily, resulting in dynamic evolution and diversification of host associations." The hypothesis is supported by the observation that species remaining free in the hemolymph without direct connections to outside air have, in general, smaller host ranges (Stireman et al., 2006).

7
Techniques and Applications

There is no substitute for visiting, or at least talking with, an expert on the various methods in insect environmental physiology. However, even getting started requires finding relevant papers. Here we provide a compendium of selected methods in the field, with a few key references that cover the method, usually specifically in insects. In the remainder of this chapter, we cover a few specific methods in insect environmental physiology that are sometimes difficult to find in the methods sections of scientific papers.

7.1 Insect Molecular Biology

Over 40 insect species now have genomic data included in NCBI's BLAST search, and more than 25 have searchable protein databases. The seven-volume *Comprehensive Molecular Insect Science* (edited by K. Iatro, S.S. Gill, and L.I. Gilbert) provides a broad, deep overview of the field, and a recent text entitled *Insect Molecular Biology and Biochemistry* (edited by L.I. Gilbert) provides a concise, up-to-date summary of these topics.

Drosophila researchers have an especially wide set of genetic tools available to them, including the Flybase website, which covers the genetics and molecular biology of *Drosophila melanogaster*. In addition to genomic maps and a variety of query tools (such as being able to search for genes associated with particular structures or physiological functions), there are specific lists of vectors and transposons, as well as easy ways to find people conducting *Drosophila* research on particular topics.

There is a growing number of *Drosophila* stock centers (Bloomington, Vienna, University of California, San Diego, Kyoto) that maintain useful lines of flies. It is possible to order flies with deficiencies or excesses of particular genes. Gal4/UAS drivers permit controlled expression of particular genes in specific tissues or developmental stages, and these can be coupled with RNAi stock that are now available for silencing many *Drosophila* genes. Mosaic animals can be created in which specific cells lack or over-express particular genes using FRT or FLP insertions. Recent excellent texts reviewing *Drosophila* molecular biology include

Drosophila Protocols (by M. Ashburner and R.S. Hawley, 2008) and *Drosophila: Methods and Protocols* (by C. Dahmann, 2010).

One of the key aspects of identifying new genes involved in specific processes has been development of high throughput screens that enable testing of large numbers of flies for defects. This so-called forward genetics approach utilizes mutagenesis (e.g. irradiation) or P-element strains with deletions of particular genes to identify genes that affect particular functions. Such approaches have yielded many new insights. Examples of such high-throughput approaches relevant to environmental physiology include vertical baffled tubes that allow "fraction collecting" of flies that drop out at different upper or lower temperatures (Huey et al., 1992) and automated flight trackers (Marden et al., 1997).

7.2 Working with Insects–Behavior and Physiology

7.2.1 Gluing Things to Insects

Attaching devices to insects or, conversely, attaching insects to devices (e.g. tethers) is often problematic, because insects are small, are often covered in scales or hairs, and have hydrophobic cuticles. Moreover, many are good at escaping even quite devious constraints, either by brute force (they can be strong for their body size) or by slipping away.

There are myriad ways of restraining insects and attaching things to them. For any particular problem, we encourage the reader to think non-linearly and to try as many approaches as possible. Nevertheless, there are set of approaches that have been used repeatedly by different workers. Many commercial adhesives are available, although few of them are suitable, especially those that have toxic components.

7.2.1.1 Waxes

Beeswax is often a successful adhesive for insects. To deliver small amounts of melted beeswax to a specific location, wrap nichrome wire around a metal dissecting pin. Control the heat of the nichrome wire with a variable AC current. We often find it useful to have the current switchable by a foot pedal, allowing both hands to be free for handling the insect. Another method is to wrap a fine wire around the end of a soldering iron, which is then heated. One can modify the wire temperature by altering the length of the wire that projects away from the soldering tip. With either wire-heating method, with the wire at the right temperature, it's possible to melt small quantities of wax from a stick of beeswax. With the wire held vertically, the wax flows down the wire and forms a drop at the tip, which can be applied quickly to the insect or device without causing significant thermal damage.

7.2.2.2 Glues and Epoxies

Glues come in many forms. However, most of them are unsuitable for insects because they contain organic solvents. Superglue rarely works, and often spreads a toxic shell over the insect. In general, organic solvents provide fast drying times but are too toxic during initial exposure.

The more suitable glues are those that harden by some other method than evaporation of organic solvents. One slow-drying, non-toxic glue is Elmer's mucilage glue (water based), which we have used to attach insect eggs to leaf surfaces (Potter et al., 2010). It damaged neither the insect nor the plant, unlike other solvent-based glues. Other glues activated by UV irradiation may be suitable, as are dental cements (Bullock et al., 2008). In addition, epoxies are potentially suitable, as they generally involve chemical reactions between two components that are mixed together, which doesn't necessarily involve organic solvents. Heymann and Lehmann (Heymann and Lehmann, 2006) describe both UV-activated glue and an epoxy that work on *Drosophila* with no apparent toxicity. Another approach is to use heated glue (glue guns); these provide strong connections and adhere well to most cuticles. To deliver smaller amounts of heated glue in a controlled fashion, use the nichrome-wire or soldering tip approach described above for beeswax.

7.2.2 Marking Insects

For many behavioral experiments, it is very useful to be able to mark individuals. We have found Testor's™ enamel paints to be non-toxic, and to stick well to insect cuticles. Many colors are available, so quite a few insects can be individually coded. If the insect is large enough, the numbered tags and glue provided by Opalith-Zeichenplattchen (Germany) or The Bee Works (Canada) are excellent, as is the glue that is provided.

Fluorescent powders can be dusted on insects and are particularly useful for tracking small cryptic insects in the field. Some studies have reported effects of such powders on survival or behavior, so this should be tested before use (Warner and Bierzychudek, 2009).

In studies of Lepidoptera, it is common to mark wings using a Sharpie. Often one just needs to write an identifying code on the wings. Although wings can sustain lots of wear over a lifetime, which may obliterate the number, often it persists readably for a few days or more. Some studies have used Sharpies for more extensive modification of wing phenotypes. For example, Kingsolver (Kingsolver, 1996) used black Sharpies to manipulate the wing phenotypes of western white butterflies (*Pontia occidentalis*), experimentally mimicking a natural seasonal polyphenism, and measured selection on the manipulated butterflies in the field.

7.2.3 Rearing Insects

There are multiple excellent books on rearing insects and making suitable artificial diets. The techniques are quite variable, depending on species. A large fraction of the environmental physiologists working on insects use one of the major model species, including the fruit fly, *Drosophila melanogaster*, the tobacco hornworm, *Manduca sexta*, the honeybee, *Apis mellifera*, the locusts *Schistocera* or *Locusta*, the mealworms *Tenebrio* or flour beetle *Tribolium*, the mosquito *Anopheles*, or the kissing bug *Rhodnius prolixus*. For such species, and a few others (such as the butterflies reared in zoos), there are excellent resources available on the web to help with rearing. However, the vast majority of insect species have never been reared in the lab, partly due to their specialized dietary requirements. In these cases, start by thinking about the insect's diet and lifecycle. Many zoos are now rearing insects and can be a good source of information on rearing.

One aspect of insect rearing that is often not considered is prevention of human allergies, which can be serious. Many insects create extensive small particles that can serve as antigens. Good ventilation systems and use of masks and gloves are highly recommended for any decently-sized insect colony.

7.2.4 Permits for Insect Collection, Transport, and Rearing

In some parts of the world (e.g. the northern countries of Europe), the long winters ensure that insects transported in from the south cannot establish themselves and become pests. However, in most countries, including the USA, prevention of invasive pest insects is an important goal of agricultural programs, and federal permits will be needed to transport or rear insects. In the USA, a permit from the United states Department of Agriculture (USDA) is required to move most insects across state lines, and for rearing of many insects, especially those related to species that have historically caused significant agricultural damage. A separate permit is required for each species, and scientists should check with their governmental agricultural and plant protection agencies before importing or beginning to rear insects.

7.2.5 Virtual Environments and Computer-assisted Data Collection

New advances in computers and digital screens in particular have permitted the construction of virtual environments for insects that can be used to explore many aspects of their sensory and behavioral systems. For example, Lehmann (Lehmann, 2001) put tethered *Drosophila* in a virtual-reality flight arena, in which he could modify the forces they produce during flight by altering the angular velocity of a set of projected black stripes. The use of virtual reality flight arenas is becoming more common, and they have been used to study flight control and odometry in grasshoppers, blow flies, *Drosophila*, and honeybees.

It's also possible to use image processing software to keep track of individual insects and to automatically estimate parameters. Such an approach can allow large-scale collection of data with little effort (that is, *after* the device is working). For example, see Tammero and Dickinson's (2001) work on flying *Drosophila*. A camera-based method for tracking multiple animals simultaneously has recently been proposed by Straw et al. (Straw et al., 2011).

7.2.6 Telemetry and Remote Monitoring of Insects

There has been rapid miniaturization of telemeters. Naef-Daenzer et al. (2005) describe 0.2-g telemeters, which may be appropriate for insects, and they discuss problems of attaching them to animals. Wang et al. (2008) attached 0.23-g telemeters to the hawkmoth *Agrius convovuli* (mass 0.94 g), and tracked them in laboratory arenas; the effective radius of the device was 3 m. Hedin and Ranius (2002) studied dispersal of the beetle *Osmoderma eremite* with around 0.5-g transmitters, and obtained signals from up to 330 m away. Sword et al. (2005) used 0.45-g telemeters to track individual Mormon crickets (*Anabrus simplex*) in migratory bands. Wikelski et al. (2007) attached 0.3-g telemeters to dragonflies (*Anax junius*) and followed their migration for up to 12 days. More recently, Wikelski et al. (2010) mapped foraging ranges of large, neotropical orchid bees (*Exaerete frontalis*), with readings obtained from 100–300 m away. With extensive surveying, they were able to locate bees up to 5 km away from their core ranges.

In the longer term, it would be enormously productive to track insects on global scales by remote sensing. Wikelski et al. (2007) have proposed an initiative to develop a small-animal, satellite-based tracking system.

For insect telemetry, a key problem is that the radio-emitting devices are quite heavy. A large proportion of the mass comes from the battery, and one mass-reducing strategy is to dispense with the battery altogether. A promising approach is passive radio-frequency identification (RFID). RFID tags are powered by incoming radio waves from *interrogaters* and they put out unique identifying codes, also in the radio spectrum. These devices have been developed with an eye toward tracking commercial goods and allowing rapid payments. They have recently been adapted for use in tracking insects. For example, Moreau et al. (2011) used passive RFID tags to track individual ants over long periods of time. Other recent papers using RFID techniques are (Ohashi et al., 2010; Stelzer et al., 2010; Sumner et al., 2007; Vinatier et al., 2010).

7.2.7 Acoustic and Vibrational Communication

Insects produce a tremendous diversity of sounds for an equivalently large diversity of purposes. Many sounds, especially those that humans can hear, are involved in sexual communication. Those audible to us, however, are just the tip of a sonic iceberg. A few examples: background noise can mask the courtship songs of male *Drosophila montana* (Samarra et al., 2009); caterpillars that are social parasites on ant colonies make sounds that mimic those made by their hosts (Barbero et al., 2009a, 2009b); and the Asian corn borer *Ostrinia furnacalis* rubs specialized wing scales together to make extremely low-intensity songs for sexual communication (Nakano et al., 2008). There is also much known now about the mechanical and neurological mechanisms used by insects to detect sounds. For example, female *Drosophila* listen to male courtship songs using specialized "antennal ears" (Riabinina et al., 2011). Experimentally playing sounds to insects, and recording faint sounds, requires relatively high-quality speakers, microphones, and amplifiers (Riabinina et al., 2011). Bertram and Johnson (1998) describe a technique for simultaneously monitoring up to 256 individuals indefinitely using relatively inexpensive electronics.

The examples above involve mechanical disturbance of air (i.e. sound waves). However, it has become increasingly clear that most insects use another kind of mechanical communication (Cocroft and Rodrídguez, 2005): vibrations of their substrates. Cocroft and Rodrídguez (2005) estimate that "[of] the insect families in which some or all species communicate using mechanical channels, 80% use vibrational signals alone or in combination with other mechanical signals, and 74% use vibrational signals alone." Because plant-associated insects are so diverse, much of insect vibrational communication occurs through plant parts. The mechanical differences between host plants, or between parts within a plant, can affect how vibrations (signals) propagate, and therefore can constrain communication or lead to evolution of signaling systems. Cocroft and Rodrídguez (2005) summarize and discuss four main methods for recording insect vibrations: laser vibrometry, accelerometers, phonograph cartridges, and guitar or bass pickups.

Insect communication (and more broadly, animal communication) has attracted substantial theoretical work (Romer et al., 2010). Gillooly and Ophir (2010) develop a model that relates acoustic communication to body size and temperature, and they apply it to diverse taxa including insects. For good book-length reviews of insect communication, see (Greenfield, 2002) and (Drosopoulos and Claridge, 2005).

7.3 Metabolism, Gas Exchange, Internal Gases, Air Flow

7.3.1 Metabolic Rate/Respirometry

Metabolic rate (watts) is a measure of the net summation of all energy-transforming processes in the body. Some physiological processes such as anabolism consume energy, but, in animals, the energy to generate these decreases in entropy comes from catabolism of fuels, with a portion of the energy captured in the form of ATP before use (Fig. 5.2). The most direct way to measure metabolic rate is to measure heat production using a calorimeter. Calorimeters used for chemistry can be very useful for measurement of metabolic heat production of small, diapausing insects, and are critical for assessing anaerobic metabolism (Downes et al., 2003; Harak et al., 1999; Neven and Hansen, 2010). These techniques are very sensitive to evaporation, and so generally require maintenance of near 100% humidity.

Most commonly, metabolic rate is estimated by respirometry, the measurement of the rate at which moles or volumes of oxygen are consumed or carbon dioxide emitted. This indirect approach is generally suitable for insects which are usually aerobic (the metabolic rate of many insects goes to near-zero when oxygen is removed) (Downes et al., 2003; Kolsch et al., 2002). Catabolism of fuels is accomplished by oxidizing hydrocarbons, and the rate of oxygen consumption and carbon dioxide production measured, with metabolic rate calculated from tables in the literature. These techniques sound simple, but it is relatively easy to make simple errors that yield data that may appear reasonable but can be spectacularly wrong. A comprehensive, up-to-date reference on techniques is provided by Lighton (2008), and Sable Systems, Inc. provides highly recommended classes for scientists wishing to apply this technique for the first time.

In general, there are two main classes of techniques. The first is closed-system respirometry, in which the insect is held in a closed volume for some time. Its metabolism alters the composition of the air in that volume, depleting the oxygen and increasing the carbon dioxide. If the carbon dioxide is absorbed (for example, by the NaOH in soda lime), then oxygen consumption rate can be calculated from the decreasing pressure, the system volume and time of closure (assuming temperature and humidity in the chamber is constant). More commonly now, the chamber is flushed into gas analyzers, for measurement of carbon dioxide and oxygen levels. Over the years, many insect physiologists have utilized Warburg apparatus adapted from chemists to conduct closed system respirometry, as these have many glass chambers designed to be sealed, a water bath for regulating temperature, and various systems for measuring chamber pressure and volume. Those with gas analyzers tend to prefer syringes for closed system respirometry chambers, as these are cheap, available, and it is easy to inject a portion of the chamber volume into an air current for analysis. Closed-system respirometry gives an integrated measure of metabolism over relatively long periods of time, averaging whatever variation is occurring due to behavioral variation. If the metabolic rates of the insects are too low to be measured by flow-through respirometry (perhaps due to small cold bugs being of interest), closed system respirometry can still be used to make accurate measurements if the closure time is extended (and leaks are avoided).

The second approach is open-system, or flow-through, respirometry, in which a gas stream of known composition is directed past an insect and through gas analyzers. From the change in composition of the stream and its flow rate, one can calculate rates of gas exchange by the insect. Because the gases flow past the insect, one can measure relatively rapid changes in metabolic rate, especially if the insect chamber is small and the flow rates of gas are high. Key technical aspects of flow-through respirometry include: 1) knowledge of the flush time of the chamber (calculated from flow rate and chamber volume); 2) accurate measurement of air flow; 3) prevention of contamination of the air stream by outside air; and 4) well-calibrated analyzers. Positive-pressure respirometry systems (air flow through the chamber generated by an upstream tank or pump) are generally recommended as small leaks will result in the loss of some of the flushing air (not a problem if the air flow is measured just upstream of the chamber), rather than negative-pressure systems with flow through the chamber generated by a downstream pump, which will pull atmospheric air in through small leaks.

A common question is whether to assess metabolic rate by measuring oxygen depletion or carbon dioxide enrichment (Schmidt-Nielsen, 1990). Both gases have advantages and disadvantages. In general, instantaneous oxygen consumption is more closely allied with instantaneous aerobic metabolic rate than is carbon dioxide emission, for two reasons. First, compared to oxygen, carbon dioxide is far more soluble and reactive in insect tissues, which can lead to the build-up of pools of CO_2 and bicarbonate that don't appear immediately in the gases that one might analyze. Second, anaerobic metabolism often produces acids that drive bicarbonate from the tissues, causing increased carbon dioxide emission rates from the animal. While not many insects have been shown to exhibit anaerobic metabolism, we do know that this occurs in jumping grasshoppers, some caterpillar intersegmental muscles, during drowning, and during hypoxic exposure in larval *Drosophila* (Gäde, 1985; Hoback and Stanley, 2001; Kölsch, 2001).

The complications associated with carbon dioxide suggest that, whenever possible, one should measure oxygen consumption. Unfortunately, the theoretical advantages of measuring oxygen consumption rates are offset by great technical advantages of measuring carbon dioxide emission rates when the investigator is concerned with small insects. Compared to oxygen analyzers, modern CO_2 analyzers are at least an order of magnitude more sensitive and generally more robust to laboratory and field environments. For example, although the performance characteristics of essentially all electronics are sensitive to temperature, our own experience is that oxygen analyzers are far more sensitive to temperature than are CO_2 analyzers. Temperature sensitivity magnifies the problems of baseline drift. Except for relatively large, warm and/or active insects, the low-sensitivity, high noise, and drift of current oxygen analyzers makes it impossible to accurately measure oxygen consumption with high temporal resolution using flow-through respirometry. In contrast, except for extremely small insects, it is possible to assess changes in carbon dioxide emission associated with behavioral variation for individuals. Thus, the majority of studies of insect respirometry focus on carbon dioxide emission.

Many authors measure carbon dioxide emission and report this as metabolic rate. However, this is incorrect, because the relationship between carbon dioxide emission rate and aerobic metabolic rate depends strongly on the fuel being oxidized. If it is carbohydrate, the exchange is 1:1, or one molecule of CO_2 produced per molecule of O_2 consumed. However, if the fuel is protein or lipid, the respiratory exchange ratio can drop to as low as 0.7. This ratio, called the respiratory exchange ratio or the respiratory quotient, must be known to accurately calculate metabolic rates from gas exchange. Thus, conversion of carbon dioxide emission rates to metabolic rates requires measurement of the respiratory exchange ratio. Given the technical difficulties with measuring oxygen consumption by flow-through respirometry with most insects, this often requires a separate experiment using closed system respirometry.

Analyses of metabolic rate patterns in physiology and ecology rely on standardized conditions for measurement. In the field of mammalian biology, basal metabolic rate is relatively well-defined as the metabolic rate of resting, non-digesting (post-absorptive) animals within their thermoneutral zone (Hulbert and Else, 2004). Insect physiology lacks such a well-accepted set of criteria for standardizing metabolic rate measurements. First, it can be challenging to get many insects (e.g. ants) to be still in a respirometer! Metabolic rates increase linearly with speeds up to 10 times that of resting (Full, 1997), thus locomotion can produce considerable variation. The best solution for this problem is to be sure to assess motion using direct observation or movement sensors to separate periods of motion from those of inactivity (Vogt and Appel, 1999). A variety of studies have used respiratory patterns (exhibition of discontinuous gas exchange) as an index that insects are in a "resting" state (Davis et al., 2000; Klok and Chown, 2005; Lachenicht et al., 2010). Lower metabolic rates do increase the likelihood of discontinuous gas exchange (Contreras and Bradley, 2009), but some insects can exchange gases discontinuously when active and others never exhibit discontinuous gas exchange. It is also challenging to achieve a well-defined post-absorptive state in insects, as is commonly done for vertebrates. Insects do exhibit a strong increase in gas exchange rates with feeding, so this can introduce considerable variation (Bradley et al., 2003; Gouveia et al., 2000). Some insects tolerate starvation very well, while in other species (e.g. honeybees), high metabolic rates lead to rapid depletion of nutrient stores and death after only a few hours of starvation at 20°C. Investigators should assess the relationship between metabolic rate and time after feeding for each species if the goal is to achieve a stable, post-absorptive state for comparative purposes.

7.3.2 Doubly Labeled Water

This technique for measuring metabolic rate uses water for which both the hydrogens and oxygens are replaced with heavy isotopes. These isotopes can be radioactive or stable; more recent studies have used largely stable isotopes. A known amount of doubly labeled water is injected into the animal and, after period in which the tracer equilibrates into blood spaces, a blood sample is taken to determine initial concentrations. The animal is released and then, days to weeks later, is recaptured for taking a second sample. The technique hinges on the differential routes, and therefore rates, of washout of the heavy isotopes. Because carbonic anhydrase exchanges oxygen between pools of water and bicarbonate, oxygen in body water is in isotopic equilibrium with the body stores of carbon dioxide (bicarbonate). Consequently, some of the heavy oxygen is breathed out as carbon dioxide, with the rest disappearing via water that leaves the body. Heavy hydrogen, by contrast, leaves the body only via lost water. Therefore, the difference in washout rates between oxygen and hydrogen isotopes allows one to estimate the rate at which carbon dioxide is exhaled, which can be converted to a metabolic rate if the respiratory exchange ratio is known. The pro and cons of using doubly labeled water are reviewed by (Butler, 2004), and a full treatment is given by (Speakman, 1997).

Doubly labeled water has been an important technique for measuring field metabolic rates of terrestrial vertebrates (Nagy et al., 1999). However, its application to insects has been relatively rare (Buscarlet et al., 1978; Cooper, 1983; King and Hadley, 1979; Voigt et al., 2005; Wolf et al., 1996, 1999). The technique certainly can work. Wolf et al. (1996) used doubly labeled water to estimate metabolic rates of tethered flying bumblebees, and did extensive cross-validation using respirometry. The values provided by the two techniques differed by less than 4%. Doubly labeled water was then used in free-foraging bumblebees in a later study (Wolf et al., 1999). This is a promising technique for application to questions of field metabolic rates of insects.

7.3.3 Methods for Measuring Internal Gases

A variety of techniques is available for measuring concentrations of oxygen or carbon dioxide in different body compartments. They differ by all the usual axes: ease of use, expense, and spatial and temporal scales of resolution.

7.3.3.1 Electrodes

Techniques for making and using both oxygen and carbon dioxide electrodes are now well developed (Fatt, 1982), and they can be purchased from a variety of suppliers or made in-house. Oxygen electrodes put out tiny currents proportional to local oxygen levels, which require a high-quality amplifier (a picoammeter) to detect. Oxygen electrodes can be made with very small tips (down to a few microns), which consume little of the nearby oxygen, thus allowing one to measure oxygen levels with high spatial resolution. The downside is that such electrodes are quite fragile and require special handling. Miniaturized Clark-type oxygen electrodes have been developed by Hetz and colleagues for measuring oxygen levels directly inside the tracheal systems of lepidopteran pupae (Hetz and Bradley, 2005; Hetz et al., 1994). Fiber optic oxygen electrode systems are also commercially available. In these systems, a pulsed light at an excitation frequency is sent through a fiber optic cable. Fluorescence at the tip responds to local P_{O2}. These systems are sturdier than glass micro-electrodes, but to date are not smaller than 100 microns in diameter, and so are only suitable for very large insects.

7.3.3.2 Electron Paramagnetic Resonance (EPR)

This technique exploits the detectability and oxygen sensitivity of electron spin resonance in biologically acceptable compounds. Recent work in this area has used crystals of lithium phthalocyanine, which are reasonable stable in biological contexts (James et al., 1997; Liu et al., 1993). Timmins et al. injected crystals of lithium phthalocyanine into the prothorax of elaterid beetle larvae (*Pyrearinus termitilluminans*) and used EPR to measure oxygen levels in ambient hypoxia, normoxia, and hyperoxia, and during periods of time when larvae were bioluminescing (Timmins et al., 2000). Kirkton used the technique to measure oxygen levels in the femoral hemolymph of grasshoppers, both at rest and after jumping (Kirkton, 2007).

7.3.4 Air Flow and Ventilation

Quantitative measurement of convection within insects (air or blood flow) is challenging and has only rarely been accomplished. Weis-Fogh developed a system for measuring air flow from spiracles by tracking the movement of a water bead within a glass tube glued to the spiracle (Weis-Fogh, 1967). This is a challenging technique due to the difficulties in gluing to spiracles and the great sensitivity of such systems to temperature and pressure. Sláma has developed elegant systems for measuring bidirectional flow of air through spiracles of pupae by gluing tubes containing two thermocouples (Slama, 1988, 1988). Tracheal air flow has also been estimated from the volume changes of the abdomen during grasshopper pumping using three-dimensional measurement of tracheal dimensions with solar cells (Greenlee and Harrison, 1998). Ventilation has also been estimated qualitatively from pressure changes measured at the spiracle (Wasserthal, 2001). X-ray synchrotron imaging permits direct measurement of volume changes in internal tracheae, though it can be challenging to use these data to assess whole-animal ventilation (Socha et al., 2008, 2010).

7.4 Water, Ions, Transport

7.4.1 Hemolymph Collection

Collection of hemolymph is most commonly done by puncturing the intersegmental membranes and drawing fluid up with a microcapillary or micro-syringe. However, the incredible diversity of insects means that there are many ways to suck an insect's blood. Hemolymph can be collecting by cutting one of the fleshy abdominal prolegs of caterpillars like *Manduca sexta*, or by cutting the proboscis of pupal *Manduca*. Hemolymph will rapidly melanize (turn black) when exposed to oxygen, due to the effect of phenoloxidase with consequent effects on protein content and buffering capacity. Melanization can be inhibited by keeping the hemolymph in an oxygen-free environment, or with phenylthiourea (Fisher and Brady, 1983). Hemolymph clots quickly, making it difficult to cannulate the hemolymph (place a plastic tube into the blood). Cooling of the insect, or addition of chelating agents are often useful to reduce clotting. Clotting mechanisms are of intrinsic interest in understanding the ability of insects to cope with damage or infection, and are beginning to be studied in earnest (Bidla et al., 2005; Haine et al., 2007; Scherfer et al., 2004; Theopold et al., 2004). Although it is challenging to obtain samples of hemolymph that have not been exposed to air, it has been done (Hadley and Draper, 1969), and other studies have shown that the lack of carbonic anhydrase in insect hemolymph means that if hemolymph is sampled within 30 seconds of collection and clotting is prevented, that blood pH and bicarbonate levels appear unaffected (Harrison, 1988a).

7.4.2 Salines

Many studies use relatively simple (e.g. NaCl) salines for insect studies, but most studies that have examined this carefully have found that tissue function is improved by using salines that more closely match actual *in vivo* chemistry (Hanrahan et al., 1984). Insect hemolymph generally has higher organic acid content than vertebrate salines, and is particularly rich in organic acids, sugars (trehalase), and amino acids (Florkin and Jeuniax, 1974; Hanrahan et al., 1984).

7.4.3 Water Content, Cuticle Content

The water content of insects is usually calculated by weighing insects, and then drying them to a constant weight. The drying should be done at a temperature low enough to reduce the likelihood of loss of volatile fats (generally below 50 °C). The cuticular (non-protein) content of insect tissue can be measured by weighing the insect or segment and then soaking the material for several days in a high-concentration (e.g. 2 molar) solution of sodium or potassium hydroxide, followed by rinsing, blotting, and a second weighing. The hydroxide solution dissolves everything but the chitin in the cuticle.

7.4.4 Hemolymph Volume

Determining hemolymph volume is best done by injecting a tracer of known volume and concentration, allowing time for it to equilibrate throughout the hemolymph (generally on the order of minutes to a few hours, although this assumption should be checked), then assaying the tracer concentration in the hemolymph. The hemolymph volume is then calculated from the magnitude of tracer dilution between the injected solution and the assayed hemolymph. The tracer requirements are stringent: the tracer should not be metabolized, broken down, or transported (it should stay in the hemolymph). One good candidate is radiolabeled inulin. For discussion and technique, see (Buckner and Caldwell, 1980; Harrison, 1988b; Levenbrook, 1958; Wharton et al., 1965).

7.4.5 Hemolymph Osmolarity

The key determinant of a suitable method is the volume of hemolymph available. For large insects, containing a milliliter or more of hemolymph, essentially all of the techniques are useable. Easy ones include vapor pressure osmometry, which take advantage of the fact that solutes depress the vapor pressure above solutions; numerous companies (e.g. Wescor) make devices that measure this vapor pressure and convert it to an estimate of solution osmolarity.

Working with small insects, containing small volumes of hemolymph, may require specialized techniques. For samples with sample volumes of 1 μl or less, estimates of osmolarity can be obtained by measuring the sample's freezing point depression. Commercial devices, known as nanoliter osmometers (e.g. Otago Osmometers), provide electronics for controlling this process carefully. The sample is floated in oil, rapidly frozen using the Peltier effect, and then slowly warmed until the ice crystals just disappear. The hardware is arranged so that one can observe the droplet and crystals continuously via a stereomicroscope. The temperature increment below zero that the crystals disappear at is the freezing point depression, which is related in known ways to solution osmolarity (Ge and Wang, 2009).

7.4.6 Regulation of Humidity

There are two major methods for obtaining known relative humidities. The first is to use solutions of water saturated with salts (Winston and Bates, 1960). Depending on the salt used, it is possible to generate almost the entire range of relative humidities in the boundary layer above a solution. For these humidities to be useful, the solution has to be held in a closed container, possibly with slight agitation or stirring to avoid formation of stratified layers of non-saturated water at the liquid surface. Such an approach is an excellent method for holding insects for long periods at fixed humidities.

A more dynamic approach is to bubble streams of air through water of controlled, or at least known, temperature. If the bubbles are small (i.e. from an airstone) and stay in contact with the water more than fleetingly, the exiting air will contain water vapor at the saturation vapor pressure for that temperature. If one has a temperature controlled water bath, it is possible to rapidly generate a wide range of relative humidities this way (RH = saturation vapor pressure at water temperature/saturation vapor pressure at temperature around the insect; these are available from online calculators or from graphs of saturation vapor pressures available in references such as Campbell and Norman, 1997) by simply immersing a bubbling flask into the bath. Ideally, one would use a thermometer of some kind (thermocouples are convenient) to monitor water temperature in the flask; that water may be distinctly colder than the water bath, as the stream of air cools it by evaporation. This approach has been commercialized by several companies into a highly flexible and accurate device called a dewpoint generator (e.g. Li-Cor LI-610).

7.4.7 Epithelial Transport

A fundamental tenet of animal physiology is that good compartments make good organs. Not surprisingly, insects regulate traffic to and from compartments carefully, by modifying the permeability of epithelia and by regulating the density and activity of specific transporters. Several key techniques for examining epithelial transport, described below, have been enormously productive over the past 80 years.

7.4.7.1 Ramsey Preparation

In insects, urine is formed by the Malpighian tubules, which secrete it into the junction between the mid- and hindgut. To study the composition of urine, one needs a way to sample this liquid. Ramsay invented the key technique, now known as "the Ramsay prep." He dissected insects to remove segments of Malpighian tubule and introduced them into a droplet of saline covered with paraffin oil. The cut end of the Malpighian tubule is then carefully pulled out into adjacent oil. The secreted urine then collects as a separate droplet, which can be easily sampled and analyzed. To ensure adequate oxygenation of tubules, Ramsay also pressed a small bubble of air against the saline. Tubules could be kept alive by this technique sometimes for more than 24 h (Ramsay, 1955a, 1955b).

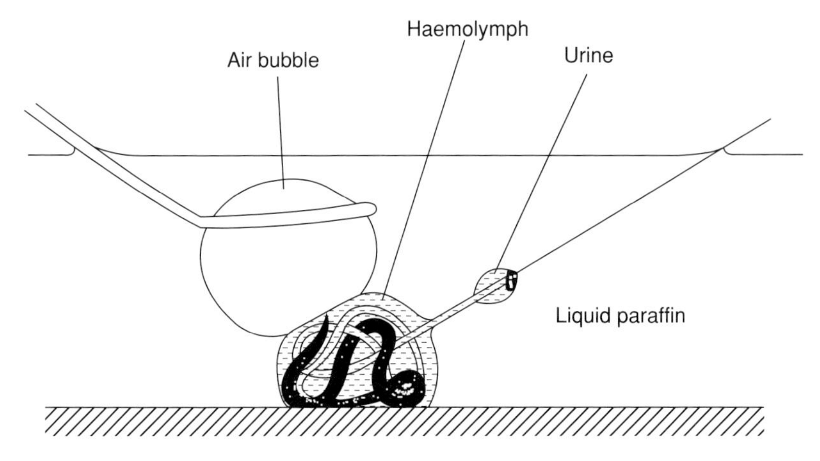

Fig. 7.1 Schematic of a Ramsay preparation. A coiled piece of Malpighian tubule is held in a droplet of saline and the cut end is pulled out into the surrounding oil, so that the secreted urine accumulates as a separate droplet. To ensure adequate oxygenation, a bubble of air can be held against the saline. From Ramsay (1954), with permission from The Company of Biologists.

7.4.7.2 Everted Rectal Sacs

The hindgut (including the rectum) is a functional complement to the Malpighian tubules: the secreted urine and digested gut contents pass over the rectal epithelium before they are excreted, and the epithelium absorbs ions and water that the insects want to keep. Understanding the absorptive properties of the rectum, and the mechanisms underlying it, has been a focal area in insect physiology. Recta, however, are not as convenient to work with as Malpighian tubules. To facilitate sampling of transported substances, Goh and Phillips invented the everted rectal sac technique and applied it to locusts (Goh and Phillips, 1978). The insect is pinned and opened, and a small tube of polyethylene is pushed through the anus and rectum and tied gently to the tips of the rectal pads. After cutting away more material, the tube is pulled back out, thereby everting the rectal sac (lumen side out; hemolymph side in). The everted sac can then be tied shut and placed in experimental salines. Fluid accumulating in the sac can later be sampled and assayed, or water transport estimated by increases in mass of the sac.

7.4.7.3 Flat-sheet Ussing Preparations

In principle, one could measure transport across the midgut also using everted sacs or sleeves. Indeed, this technique (Karasov and Diamond, 1983) has been widely used to measure nutrient transport into or across sections of vertebrate guts (however, there subsequently have been serious concerns about tissue damage arising from this technique; see (Starck et al., 2000) for a discussion). Indeed, one of us (Woods) attempted during his dissertation to use the everted sleeve preparation to measure amino acid transport across *Manduca* midgut, and never got reliable results.

Instead, measurements of nutrient transport across midgut epithelia have tended to use Ussing preparations (Hanrahan et al., 1984). Woods ended up saving his dissertation by collaborating with Mary Chamberlin to use Ussing chambers to measure proline transport across *Manduca* midgut (Woods and Chamberlin, 1999).

7.4.8 Water Loss

Water loss at the whole-body level can be measured using flow-through or closed system respirometry using various commercial water sensors (Lighton, 2008). Measurement of water loss rates requires special care in the choice of materials for tubing and chambers, as many plastics reversibly bind water (stainless steel, glass, and Bevline are good). Water loss has often been measured gravimetrically; this is a reasonable approach if humidities are low and the duration of the measurement is short enough that significant depletion of stored fuels is minimal. Often, investigators have wished to separate the various routes of water loss (excretory, respiratory, cuticular). Excretory water loss usually occurs episodically in insects and so is often relatively obvious if a continuous recording of water loss is available. Cuticular water loss can be distinguished from respiratory water loss in insects that exhibit discontinuous gas exchange with a clear closed phase (Williams et al., 2010). When this does not occur, sometimes insects can be induced to close their spiracles with exposure to hyperoxia, allowing assessment of cuticular water loss rates (Lighton et al., 2004). Another elegant approach is to utilize intra-individual variation in water loss and carbon dioxide emission rates. If these are linearly related, the intercept (at zero carbon dioxide emission) represents cuticular water loss rate (Gibbs and Johnson, 2004). Finally, ventilated capsules glued onto insects have been used to directly assess water loss rates across sections of cuticle (Machin et al., 1992).

7.5 Temperature

7.5.1 Thermocouples and Thermometers

Contact thermocouple thermometry has been used for decades to measure the temperature of individual insects or physical models of insects. Long term monitoring of an insect's temperature is possible with an implanted type-T (copper-constantan) wire thermocouple. For example, small thermocouples implanted within or in contact with extremely small insects, such as *Drosophila* larvae, can measure the heat of freezing, the slight release of heat due to ice crystal formation when an insect freezes (Sinclair et al., 2009). Implanted thermocouples are most appropriate for physiological preparations, but are of limited use in behavioral and ecological studies because the insect's movement is restricted. If the insects under study are large and abundant, then this problem can be overcome by a "grab-and-stab" approach, in which individual insects are captured, quickly restrained on an insulating surface (such as in an insect net or on closed-cell foam), and a narrow-gauge needle thermocouple is rapidly inserted into it. For most insects this causes fatal injury, but on certain larger ones the needle thermocouple can be slipped underneath the cuticle and the resultant injury to the cuticle will self-seal or can be sealed with wax. We have found the best thermocouple for this technique (when working with honeybees and larger insects) to be the MT-29/1 needle thermocouple probe from Physitemp Instruments, Inc. (USA). Stone and Willmer discuss the assumptions and methods of the grab-and-stab technique (Stone and Willmer, 1989). Newly-developed miniature wireless thermocouples and non-contact infrared thermometers with focal areas of just a few millimeters will permit remote temperature monitoring of large insects. Both of these technologies hold great promise for applications in insect physiological ecology.

7.5.2 Operative Temperatures and Heat Budgets

Because insects interact in complex ways with their natural thermal environments, physical models such as operative temperature models are often useful for assessing the mechanisms of heat balance. Operative temperature is the temperature of the animal in the absence of metabolic heat production or evaporative cooling, and thus reflects radiative and conductive/convective heat exchange (Bakken et al., 1985; Dzialowski, 2005). Operative temperature is measured with physical models (most often dead, dried insects), and these can be highly accurate when calibrated (Dzialowski, 2005). Operative temperature models placed in various locations and orientations are generally better indices of insect thermal environments than are shaded air temperatures. Studies using operative temperature models have demonstrated the thermal role of wings in butterfly basking (Kingsolver and Moffat, 1982) and of body orientation in grasshoppers (Chappell and Whitman, 1990).

7.5.3 Infrared Thermography

In infrared thermography, one uses an infrared camera (still or video) to assess surface temperatures of insects or their environments. This technique is becoming widely used because the price of cameras is coming down and their resolution and accuracy is going up. For a good recent example of examining scarab beetles interacting with thermogenic plants, see (Seymour et al., 2009). The technique depends on Wien's displacement law, which states that the distribution of infrared wavelengths emitted from a black-body surface (a perfect absorber and emitter) has a characteristic shape, with wavelengths displaced by an amount that depends on the surface temperature.

The technique is appealing because one can rapidly gather large amounts of data at fine spatial scales. Nevertheless, there are several caveats to using the technique. The first is that it provides information just on surface temperature, not core body temperature, and these may differ. For example, many large flying insects sustain thoracic temperatures in excess of 40°C, but also are insulated with furry scales, such that their surface temperatures are less than 30°C.

Second, the technique contains an important assumption about surface emissivity. A perfect black body has an emissivity of 1 and a reflectance of 0 (all incoming radiation in absorbed and none reflected; in this case, the emitted distribution of IR is predicted by Wien's displacement law). However, most surfaces are not perfect black bodies, meaning they have of emissivity less than 1 and reflectance of less than 0. The farther the emissivity is from 1, the more error is potentially introduced, for two reasons: the camera assumes some particular emissivity (which on many cameras is a settable parameter), and the higher the reflectance the greater the proportion of IR coming from the insect is contaminated by IR bouncing off it from the environment. You should therefore establish the emissivity of your insect surfaces before proceeding. This can be done in the lab by comparing the camera-interpreted surface temperature of the insect (preferably dead and dried, to avoid problems with evaporative cooling) with the camera-interpreted temperature of a known black body, both held in strictly isothermal conditions, such as in an insulated box. A good, cheap "perfect" black body is black electrical tape.

Reflected environmental IR is more likely to be problematic when measuring surface temperatures in the field during the day time, due to intense IR from the sun and sky. The best method to minimize these errors is to take IR images directly after shading the insect from direct sun. If the IR camera has video capability, take a short video and use the first fully-shaded frame for estimating surface temperature; of course, if the object is shaded too long, it will cool significantly. Plant biologists have done a thorough job of grappling with issue in the context of leaf temperatures (Leigh et al., 2006).

7.6 Cardiac Physiology

Although insect hemolymph generally doesn't distribute oxygen in the insects that lack hemocyanin, it does move nutrients, hormones, and heat. Flows through the hemocoel are driven by a dorsal heart and can be supplemented with accessory pulsatile organs. Even though the hemocoel is "open"—meaning hemolymph flows through relatively large spaces rather than through arteries and veins as it does in vertebrates—insects are increasingly understood to have sophisticated mechanisms for controlling patterns of blood flow.

Studying insect heart function and hemolymph flow is complicated by the small size of insects. In general, there are several classes of approaches. These involve at their most simple just arranging appropriate lighting (e.g. strong light directed through the body) so that the heartbeat is clearly visible. This approach does not allow one to trace blood flow, but often it's possible to discern heart rate and whether the heart is beating with retrograde or anterograde pulsations. "Watching" has been taken to very refined level, using optoelectronic techniques to record heartbeat patterns automatically, by Sláma and colleagues (Slama and Farkas, 2005) and by Wasserthal (Wasserthal, 2007). To assess patterns of blood flow, researchers often inject some visible tracer and follow its subsequent flow path. Older studies used insects with semi-transparent cuticle and injected India ink into the hemocoel. More recently, Glenn et al. watched blood flow in mosquitoes by injecting a red fluorescent microsphere (2 μm diameter), and Lee and Socha visualized blood flow in grasshoppers using x-ray synchrotron radiation, with density contract provided by injecting stabilized microbubbles of air (Glenn et al., 2010; Lee and Socha, 2009).

A second approach is to trace blood by applying small amounts of heat and then following where it goes (Wasserthal, 1980). This technique resembles the heat-marking technique used by tree physiologists to measure the speed of xylem and phloem flow in tree trunks. The two technical problems are to apply heat rapidly and locally and to measure very small pulses heat passing by other locations of the body. Wasserthal (1999) used brief pulses of 30-W laser light to heat locally (+1.5°C) and sensitive thermistors in contact with the cuticle to measure subsequent movement of the heat bolus (Wasserthal, 1999). This technique was also used successfully by Sláma and colleagues (Slama, 2010). A similar approach was used by Heinrich in the 1970s to infer patterns of blood flow in adult sphinx moths and bumblebees, except that the heat was provided by the thoracic flight motor (Heinrich, 1971, 1973). During shivering warm-up and flight, these insects produce enormous amounts of heat, and they employ sophisticated patterns of blood flow to control where that thoracic heat goes. Heinrich implanted sensitive thermocouples at various locations in the thorax and abdomen and was able, from pulses of heat associated with each heartbeat, to deduce patterns of flow.

Insect heart function can also be assessed electrically by measuring the impedance of current flow between paired electrodes placed on either side of the heart. These data have been used to measure heart rate, propagation velocity, and direction of flow (Smits et al., 2000).

7.7. Issues with Using the Comparative Method in Insects

Although the diversity of insects provides vast fodder for the comparative method, the diversity also poses challenges for statistical analyses of comparative data. Molecular trees with branch lengths are available only for a small number of insect clades. Because the molecular data is scattered, it is often necessary to create phylogenies using supertrees (assemblies from smaller phylogenies). Outside of these groups, investigators are forced to rely on tree topology alone. Fortunately, most tools for phylogenetic analysis can operate without branch lengths, though with reduced power.

Because of the great age of insect orders and many families, a great deal of the variation in insect physiology occurs at the order and family level. Thus for many interesting biological questions, phylogeny and behavior/physiology are confounded. For example, grass- vs. forb-feeding is highly correlated with subfamily in grasshoppers (Caelifera). Thus tests of what anatomical or physiological factors correlate with grass- vs. forb-feeding must use multiple subfamilies.

8 Conclusions and Future Directions

Ecological and environmental physiology of insects has an extraordinarily bright and interesting future. As a discipline, physiology is undergoing a renaissance, because biologists are increasingly realizing that physiology is a key interface between phenotype, genetics and molecular biology, and the evolution of life. However, the lines delineating physiological problems of the past will increasingly blur, with the result that physiology, molecular biology, and genetics will become increasingly indistinguishable, and that physiological systems and processes will increasingly be folded into interesting ecological and evolutionary contexts.

The list below outlines a series of big problems and questions in insect biology that we think deserve attention. Prognosticating carries risks, and we hope that readers will derive inspiration, not irritation, from these thoughts.

8.1 Insects and Anthropogenic Perturbation of the Environment

Climate change, including alterations in mean and variation in temperature, rainfall, solar insolation, and a wide variety of pollutants, is undoubtedly going to cause major changes in the environments of insects. Although the magnitude of environmental changes in the past, over the several hundred million year history of the insects, has been enormous, the rate at which changes are happening now is unprecedented. Many habitats are rapidly being lost due to increasing human populations and development. New crops, including genetically-altered species, also create new environments for insects. Major questions remain concerning how these anthropogenic perturbations of the environment will affect insects.

Will higher temperatures increase the growth, abundance, and impacts of insect pests?

To what extent will shifts in insect behavior and distribution ameliorate the effects of environmental change?

To what extent will physiological plasticity and evolution ameliorate the effects of anthropogenic change?

Will invasive species exploit the rapid changes? Will they be more likely to eliminate or replace native species?

How do we link lab and field experiments to predict the effects of climate change?

8.2 Lifecycle Approach

From most papers on insect physiology, a newcomer to the field would not deduce that insects have complex lifecycles—three stages in hemimetabolous species and four in holometabolous insects. That most of us work on intra-stage physiology is reasonable, given the usual constraints on time and money. Nonetheless, the response of a population to climate change will emerge from interactions of all three or four stages with their respective environments, and the environments for each stage can be vastly different (Kingsolver et al., 2011). We see two major unanswered questions here.

How do different stages respond to environmental variation?

How strongly linked are physiological processes and systems across stages?

The latter question can be viewed both as a problem for individual insects (how much does experience in one stage affect physiology in the next) and as an evolutionary problem for populations (how much does evolution of physiology in one stage affect physiology in other stages (see Gefen et al. (2006) for an example)).

8.3 Interactions of Behavior, Plasticity, and Evolution

> "Behavioral avoidance, not physiological adaptations, is an organism's primary response to an environmental challenge. . . . This point is elementary, but by no means trivial"
>
> (Bartholomew, 1987)

This issue has been pondered for years, and while many experiments are ongoing, in our opinion, many fundamental questions still remain concerning these interactions. Here we list a few general unanswered questions that could be applied to almost any specific system in insect ecological physiology.

What behavioral mechanisms allow insects to buffer physiological stress, and how and why do they vary among insects?

When and how does physiological plasticity allow for different behaviors among individuals, populations, and species?

How do behavioral and physiological plasticity together influence the evolution of physiological traits?

8.4 Environmental Physiology of Insects from the Field

Many insect physiologists—ourselves included—derive experimental subjects from laboratory colonies. Although there are many advantages to such an approach (convenience, control over extraneous factors, developmental synchronization, etc.), laboratory physiology is fundamentally ill-suited to understanding insect physiology in nature, where, after all, physiology acquires its ecological and evolutionary significance. The consequence for physiologists is that often we use insects that have experienced developmental conditions, or sometimes histories of laboratory selection, that dramatically alter their physiology. We suspect that many lab-reared insects are healthier, or at least have easier access to food, than insects in nature. Understanding the physiology of such domesticated animals may not be particularly relevant to the physiology of wild insects. Unfortunately, we have little idea of the average decrement in fitness imposed by insect diseases and experimental extremes, although it probably is large. Developing a true ecological physiology of insects will require doing experiments on field-collected insects that are compromised in various ways. Below we list four examples of ways that standard lab experiments may not reflect natural systems.

8.4.1 Effects of Insect Pathologies on Physiology and Performance

Although lab-reared insects can suffer from a variety of problems stemming from inadequate nutrition, crowding, poor hygiene, and so on, they generally do *not* experience two major problems that compromise the health of insects in nature: exposure to the full range of possible diseases (including parasites and parasitoids) and environmental extremes. Fortunately, insect pathology and immunology are beginning to be recognized for the fascinating and important fields they are by ecological and environmental physiologists (Diamond and Kingsolver; Siva-Jothy et al., 2005; Smilanich et al., 2009), and we see this as an important area for future research. A related area that has received less attention is the potential for developmental exposure to environmental extremes to have major, long-term effects on insect morphology, behavior, and physiology. For example, a few days of hypoxia at the egg stage can reduce insect adult body size (Heinrich et al., 2011), and exposure to brief periods of environmentally-relevant warm temperatures can change mushroom body size and learning (Wang et al., 2007).

These findings suggest that brief instances of environmental challenge at key developmental stages can have major, life-long effects (for a counterexample, see (Potter et al., 2010). These findings echo current research areas of great interest in humans.

How does disease affect the environmental physiology of insects?

What are the consequences of developmental exposure to environmental extremes for later performance and fitness?

8.4.2 Effects of Rich Diets

More common than inadequate nutrition in the lab is nutritional overabundance: Western eating habits obviously have led to widespread obesity in the past 30 years, and the laboratory eating habits of our insects may be having similar effects. The magnitude of possible effects is illustrated by Ojeda-Avila et al.'s (2003) study on fat content in *Manduca sexta*. Larvae were given artificial diets containing different amounts of sugar and protein, and the dietary treatments led to some differences in larval, pupal, and adult fat content. However, all larvae reared on artificial diets had dramatically higher levels of body fat (up to 35%) than did those reared on tobacco leaves (maximum levels ~10% fat). We know how strongly obesity in humans affects physiology.

To what extent are lab-reared insects physiologically impaired by overly rich diets?

How do the physiologies of insects in "field condition" differ from those of lab animals fed ad lib? *Should we stop using model insects in lab settings for studies of evolutionary and ecological physiology?*

8.4.3 Effects of Sensory Environment During Development on Physiology and Behavior

Insects reared in the lab experience unnatural sensory environments, as lab rearing regimes typically lack the physical, chemical, and social heterogeneity of natural environments. In *Drosophila*, the development of brain structures including the mushroom bodies, central complex, and optic lobes is repressed when newly-eclosed adults are reared under olfactory, visual, and social deprivation (Barth and Heisenberg, 1997; Techanau, 1984). Similarly, adult crickets reared in enriched sensory environments have more mushroom body neuroblasts and better learning ability compared to crickets reared in less complex conditions (Cayre et al., 2007). In mantids, visual experience during development affects the ability to gauge distance for striking (Walcher and Kral, 1994). Together these studies suggest that morphological and functional responses to the sensory environment may be very important in ecological physiology.

> *How does the sensory environment affect sensors, integrators and responses to the environment?*
>
> *How much phenotypic variation in nature is due to environmentally-dependent changes in sensory and neural integration?*

8.4.4 Effects of Natural Genetic Variation on Responses

In the past, genetic variation has been viewed by many physiologists as a bogeyman: we all know that he's important but no one wants the complications that accompany inviting him into their experiments. However, much of the future of physiology will consist of trying to understand how insect populations respond to, for example, global climate change. Such responses are inextricably linked to the physiological *diversity* within populations.

> *How much intraspecific variation exists in environmental responses of insects, and how important is this variation to population and evolutionary responses to environmental variation?*

What is the poor insect physiologist to do? Many of us use sensitive, expensive instruments that are impossible (or scary) to take into the field. We suggest three routes forward. First, insect physiologists should try to work more often on insects collected from the field and used right away in experiments. Second, rather than ignoring or minimizing intra-individual variation, we should acknowledge its importance and its primary role in determining how populations respond to environmental change. Third, we could bypass many of the problems associated with doing physiology in the lab by explicitly attempting to do more experiments in the wild. Although such an approach would require giving up some control over extraneous factors, it would force us to use natural, variable insect systems. Moreover, this approach has the appeal of forcing us to return to our roots as builders and tinkerers, if we are to develop robust, field-ready devices.

8.5 Paleophysiology

In their long evolutionary history, insects have been exposed to astonishing environmental variability, including major changes in temperatures, atmospheric oxygen levels between 15 and 32%, and more than 10-fold changes in atmospheric levels of carbon dioxide (Berner, 2006). This variation has been accompanied by major changes in the insect species present, by the average and maximum body sizes of insects, and by major extinctions and radiations. Geologists, paleontologists, and evolutionary biologists are very interested in how such changes may have affected the flora and fauna, and collaborations between paleontologists and environmental physiologists have the potential to provide new insights into the evolution of insects.

Did atmospheric hyperoxia allow the evolution of gigantic insects?

Did atmospheric hypoxia contribute to extinction of insects in the Triassic?

Did elevated atmospheric oxygen levels and atmospheric pressures allow the evolution of flight?

How did the large elevations in atmospheric carbon dioxide levels affect insect–plant interactions?

How did insects respond to Paleozoic glacial cycles?

8.6 Fundamental Principles of Scaling

A controversial area in physiology today concerns the mechanisms by which animal metabolic rates scale. Some consider this problem solved, while others consider this the best-documented, least-understood pattern in biology. Insects offer outstanding systems for studying the mechanisms responsible for metabolic scaling due to their small size, great diversity, fractal-like tracheal system, and the many genetic tools available to manipulate size, metabolism, and respiratory anatomy.

How does manipulation of the oxygen delivery system affect metabolism and metabolic scaling?

What genetic and developmental pathways dictate the lower mass-specific metabolic rates of larger insects?

8.7 Remote Monitoring of Insect Physiology, Behavior, and Environment

Technological advances in computing, communication, and miniaturization have conspired to give increasingly useful and data-rich means of monitoring animals in their environments. Older techniques of radio-telemetry have given way to sophisticated electronics packages that can both keep track of the animals onto which they are placed and monitor aspects of the animal's physiology and behavior. The best examples have come from large vertebrates—fish, mammalian herbivores, marine mammals, and birds—for which miniaturization is less of an issue.

Because of their small size, insects have largely not been studied by these techniques. But their day is coming. Initial work on remote monitoring of insects has focused on using radar of various kinds. Radar-based research is enabling studies of insect migration and foraging behavior, and has revealed previously unknown capacities of insects to orient and compensate for wind and weather (Chapman et al., 2011). It is also now possible to remotely control free-flying insects, and we will soon be receiving wireless information back from free-flying insects (Sato et al., 2009). As the electronics become even smaller we surely will see the addition of other kinds of data collection (e.g. physiological state of insects, GPS-tracking of foraging and feeding).

Over the long term, we envision the development of even more useful sets of electronics that will monitor individual insects over their entire lifetimes—what we call automated field ethography (AFE). AFE units will consist of tiny robots that are not attached directly to insects, but rather follow them around closely, without disturbing them. These systems will record what they are doing—when they rest, eat, mate, fly, and interact with predators and parasites. These units will simultaneously collect information on local microenvironmental conditions. AFE will also provide rich information on causes of death. An enormous hurdle in insect ecology and evolution is the difficulty of attributing death to particular causes. Usually insects *just disappear*, and we are left to guess about actual causes of mortality. Lifetime monitoring of insects could revolutionize our view of how much time insects spend performing various actions, the kinds of physiological stresses they experience, and their fates.

8.8 High-end Imaging

New imaging tools have yielded spectacular advances in the environmental physiology of insects. Phase-contrast x-ray imaging has allowed visualization of insect tracheal, digestive, and cardiovascular systems of insects in action in ways completely impossible before (Westneat et al., 2008). Magnetic resonance has the potential to reveal pulsatile hemolymph and air flow inside living insects, and to provide exciting information on such functions in relatively undisturbed animals (Hallock, 2008). Electron magnetic resonance imaging, combined with injected chips or solutions of paramagnetic materials, can be used to measure internal gas levels (Timmins et al., 1999). Computerized x-ray tomography will increasingly be used to investigate insect morphology and trade-offs in animal design (Socha and DeCarlo, 2008). Most of these techniques have barely been used by insect environmental physiologists. A few specific questions that may be pushed forward by such new technologies:

How do respiratory and circulatory systems interact during different behaviors and metabolic rates?

Why do insects compress their tracheae if diffusion is sufficient to provide adequate oxygen?

How does dehydration and over-hydration affect insect cardiovascular function?

Where does ice form, and how do insects control ice formation during freezing?

8.9 Internationalization of Environmental and Ecological Insect Physiology

Traditionally, the field has been dominated by European and North American scientists. This has limited insect ecological physiology in many ways, partly due to human cultural biases, and partly due to a focus on insects of temperate regions. Patterns of diversity can be quite varied across different hemispheres and continents, partly due to biogeographic factors and partly due to evolutionary history (Chown et al., 2004a; Chown et al., 2004b).

It is exciting to see the field continue to blossom on all continents. Strong development of the field in all countries will bring many benefits, among them increased information on broad patterns of species diversity and new insights from under-represented cultural perspectives. We predict that such studies will transform agriculture, improve healthcare, and provide new answers about the evolution of our interactive biosphere.

Bibliography

Abisgold, J. D. & Simpson, S. J. (1987). The physiology of compensation by locusts for changes in dietary protein. *Journal of Experimental Biology,* 129, 329–46.

Abisgold, J. D. & Simpson, S. J. (1988). The effect of dietary protein levels and haemolymph composition on the sensitivity of the maxillary palp chemoreceptors of locusts. *Journal of Experimental Biology,* 135, 215–29.

Acker, T. & Acker, H. (2004). Cellular oxygen sensing need in Cns function: physiological and pathological implications. *Journal of Experimental Biology,* 207, 3171–88.

Adamo, S. (2008). Norepinephrine and octopamine: linking stress and immune function across phyla. *Invertebrate Survival Journal,* 5, 12–19.

Addo-Bediako, A., Chown, S. L., & Gaston, K. J. (2001). Revisiting water loss in insects: A large scale view. *Journal of Insect Physiology,* 47, 1377–88.

Affolter, M. & Caussinus, E. (2008). Tracheal branching morphogenesis in *Drosophila*: new insights into cell behaviour and organ architecture. *Development,* 135, 2055–64.

Agarwal, S. & Sohal, R. S. (1994). DNA oxidative damage and life expectancy in houseflies. *Proceedings of the National Academy of Sciences of the United States of America,* 91, 12332–35.

Aguilar, R., Maestro, J. L., & Bellés, X. (2006). Effects of myoinhibitory peptides on food intake in the German cockroach. *Physiological Entomology,* 31, 257–61.

Ai, H. & Kuwasawa, K. (1995). Neural pathways for cardiac reflexes triggered by external mechanical stimuli in larvae of *Bombyx mori. Journal of Insect Physiology,* 41, 1119–31.

Akasaka, T., Klinedinst, S., Ocorr, K., Bustamante, E. L., Kim, S. K., & Bodmer, R. (2006). The ATP-sensitive potassium (KATP) channel-encoded dSUR gene is required for *Drosophila* heart function and is regulated by tinman. *Proceedings of the National Academy of Science, USA,* 103, 11999–12004.

Akman, L., Yamashita, A., Watanabe, H., Oshima, K., Shiba, T., Hattori, M., & Aksoy, S. (2002). Genome sequence of the endocellular obligate symbiont of tsetse flies, Wigglesworthia glossinidia. *Nature Genetics,* 32, 402–407.

Albers, M. A. & Bradley, T. J. (2004). The evolution of saline-tolerance in mosquito larvae. *Integrative and Comparative Biology,* 44, 514–514.

Ali, I. & Steele, J. E. (1997). Hypertrehalosemic hormones increase the concentration of free fatty acids in trophocytes of the cockroach (Periplaneta americana) fat body. *Comparative Biochemistry and Physiology A,* 118, 1225–31.

Allen, C. R., Garmestani, A. F., Havlicek, T. D., Marquet, P. A., Peterson, P. D., Restrepo, C., Stow, C. A., & Weeks, B. E. (2006). Patterns in body size distributions: sifting among alternative hypotheses. *Ecology Letters,* 9, 630–48.

Allsopp, M. H., De Lange, W. J., & Veldtman, R. (2008). Valuing insect pollination services with cost of replacement. *PLOS One,* 3, e3128.

Altner, H. & Loftus, R. (1985). Ultrastructure and function of insect thermoreceptors and hygroreceptors. *Annual Review of Entomology,* 30, 273–95.

Altstein, M. & Nassel, D. R. (2010). *Neuropeptide Signaling in Insects. Neuropeptide Systems as Targets for Parasite and Pest Control.* Berlin: Springer-Verlag Berlin.

Andersen, S. O. (2010). Insect cuticular sclerotization: a review. *Insect Biochemistry and Molecular Biology*, 40, 166–78.

Anderson, R. V., Tracy, C. R., & Abramsky, Z. (1979). Habitat selection in two species of short-horned grasshoppers – role of thermal and hydric stresses. *Oecologia,* 38, 359–74.

Andorfer, C. A. & Duman, J. G. (2000). Isolation and characterization of cDNA clones encoding antifreeze proteins of the pyrochroid beetle (IDendroides canadensis). *Journal of Insect Physiology,* 46, 365–72.

Angilletta, M. J. J. (2009). *Thermal Adaptation: A Theoretical and Empirical Synthesis,* New York, Oxford University Press.

Apodaca, C. K. & Chapman, L. J. (2004). Larval damselflies in extreme environments: behavioral and physiological response to hypoxic stress. *Journal of Insect Physiology,* 50, 767–75.

Apple, J. L., Wink, M., Wills, S. E., & Bishop, J. G. (2009). Successional change in phosphorus stoichiometry explains the inverse relationship between herbivory and lupin density on Mount St. Helens. *PLOS One,* 4, e7807.

Applebaum, S. W. (1985). Biochemistry of digestion. *In:* Kerkut, G. A. & Gilbert, L. I. (eds.) *Comparative Insect Physiological Biochemistry and Pharmocology.* Pergamon Press.

Aprille, J. R., Lagace, C. J., Modica-Napolitano, J., & Trimmer, B. A. (2004). Role of nitric oxide and mitochondria in control of firefly flash. *Integrative and Comparative Biology,* 44, 213–219.

Araujo, R. N., Soares, A. C., Paim, R. M. M., Gontijo, N. F., Gontijo, A. F., Lehane, M. J., & Pereira, M. H. (2009). The role of salivary nitrophorins in the ingestion of blood by the triatomine bug *Rhodnius prolixus* (Reduviidae: Triatominae). *Insect Biochemistry and Molecular Biology,* 39, 83–89.

Archer, M. A., Bradley, T. J., Mueller, L. D., & Rose, M. D. (2007). Using experimental evolution to study the physiological mechanisms of desiccation resistance in *Drosophila melanogater. Physiological & Biochemical Zoology,* 80, 386–98.

Arendt, J. (2007). Ecological correlates of body size in relation to cell size and cell number: patterns in flies, fish, fruits and foliage. *Biological Reviews,* 82, 241–56.

Arieli, R. & Lehrer, C. (1988). Recording of locust breathing frequency by barometric method exemplified by hypoxic exposure. *Journal of Insect Physiology,* 34, 325–28.

Arquier, N., Geminard, C., Bourouis, M., Jarretou, G., Honegger, B., Paix, A., & Leopold, P. (2008). Drosophila ALS regulates cell growth and metabolism through functional interactions with insulin-like peptides. *Cell Metabolism,* 7, 333–38.

Arquier, N., Vigne, P., Duplan, E., Hsu, T., Therond, P. P., Frelin, C., & D'angelo, G. (2006). Analysis of the hypoxia-sensing pathway in *Drosophila melanogaster. Biochemical Journal,* 393, 471–80.

Arrese, E. L., Mirza, S., Rivera, L., Howard, A. D., Chetty, P. S., & Soulages, J. L. (2008). Expression of lipid storage droplet protein-1 may define the role of AKH as a lipid mobilizing hormone in *Manduca sexta. Insect Biochemistry and Molecular Biology,* 38, 993–1000.

Arrese, E. L., Patel, R. T., & Soulages, J. L. (2006). The main triglyceride-lipase from the insect fat body is an active phospholipase A1: identification and characterization. *Journal of Lipid Research,* 47, 2656–67.

Arrese, E. L. & Soulages, J. L. (2010). Insect fat body: energy, metabolism, and regulation. *Annual Review of Entomology,* 55, 207–25.

Atkinson, D. (1994). Temperature and organism size-A biological law for ectotherms? *Advances in Ecological Research,* 25, 1–58.

Atkinson, D. (1996). Ectotherm life-history responses to developmental temperature. *In:* Johnston, I. A. & Bennett, A. F. (eds.) *Animals and Temperature. Phenotypic and Evolutionary Adaptation.* Cambridge: Cambridge University Press.

Attardo, G. M., Hansen, I. A., & Raikhel, A. S. (2005). Nutritional regulation of vitellogenesis in mosquitoes: Implications for anautogeny. *Insect Biochemistry and Molecular Biology,* 35, 661–75.

Audsley, N., Matthews, J., & Weaver, R. J. (2005). Neuropeptides associated with the frontal ganglion of larval Lepidoptera. *Peptides,* 26, 11–21.

Audsley, N., Mcintosh, C., & Phillips, J. E. (1992). Actions of ion-transport peptide from locust corpus cardiacum on several hindgut transport processes. *Journal of Experimental Biology,* 173, 275–88.

Audsley, N., Meredith, J., & Phillips, J. E. (2006). Haemolymph levels of *Schistocerca gregaria* ion transport peptide and ion transport-like peptide. *Physiological Entomology,* 31, 154–63.

Audsley, N. & Weaver, R. J. (2009). Neuropeptides associated with the regulation of feeding in insects. *General and Comparative Endocrinology,* 162, 93–104.

Auerswald, L. & Gäde, G. (2002). Physiological and biochemical aspects of flight metabolism in cocoon-enclosed adults of the fruit beetle, Pachnoda sinuata. *Journal of Insect Physiology,* 48, 239–48.

Azevedo, R. B. R., French, V., & Partridge, L. (2002). Temperature modulates epidermal cell size in Drosophila melanogaster. *Journal of Insect Physiology,* 48, 231–37.

Azpiazu, N. & Frasch, M. (1993). Tinman and bagpipe – two homeo box genes that determine cell fates in the dorsal mesoderm of Drosophila. *Genes & Development,* 7, 1325–40.

Azuma, M., Harvey, W. R., & Wieczorek, H. (1995). Stoichiometry of K^+/H^+ antiport helps to explain extracellular pH 11 in a model epithelium. *Federation of European Biochemical Societies Letters,* 361, 153–56.

Bader, R., Colomb, J., Pankratz, B., Schröck, A., Stocker, R. F., & Pankratz, M. J. (2007). Genetic dissection of neural circuit anatomy underlying feeding behavior in Drosophila: Distinct classes of hugin-expressing neurons. *The Journal of Comparative Neurology,* 502, 848–56.

Badisco, L., Claeys, I., Van Loy, T., Van Hiel, M., Franssens, V., Simonet, G., & Broeck, J. V. (2007). Neuroparsins, a family of conserved arthropod neuropeptides. *General and Comparative Endocrinology,* 153, 64–71.

Bailey, L. (1954). The respiratory currents of the tracheal system of the adult honey-bee. *Journal of Experimental Biology,* 31, 589–93.

Baker, H. G. & Baker, I. (1983). Chemical constituents of nectar in relation to pollination mechanisms and phylogeny. *Handbook of Experimental Pollination Biology,* 131–71.

Bakken, G. S., Santee, W. R., & Erskine, D. J. (1985). Operative and standard operative temperature-tools for thermal energetics studies. *American Zoologist,* 25, 933–43.

Balashov, Y. S. (2006). The origin and evolution of parasitism on vertebrates in insects, mites and ticks. *Parazitologiia,* 40, 409–24.

Bale, J. S. & Hayward, S. A. L. (2010). Insect overwintering in a changing climate. *Journal of Experimental Biology,* 213, 980–94.

Bantignies, F. & Cavalli, G. (2006). Cellular memory and dynamic regulation of polycomb group proteins. *Current Opinion in Cell Biology,* 18, 275–83.

Barbehenn, R. V. (2001). Roles of peritrophic membranes in protecting herbivorous insects from ingested plant allelochemicals. *Archives of Insect Biochemistry and Physiology,* 47, 86–99.

Barbero, F., Bonelli, S., Thomas, J. A., Balletto, E., & Schonrogge, K. (2009a). Acoustical mimicry in a predatory social parasite of ants. *Journal of Experimental Biology,* 212, 4084–90.

Barbero, F., Thomas, J. A., Bonelli, S., Balletto, E., & Schonrogge, K. (2009b). Queen ants make distinctive sounds that are mimicked by a butterfly social parasite. *Science,* 323, 782–85.

Barnes, A. I., Boone, J. M., Jacobson, J., Partridge, L., & Chapman, T. (2006). No extension of lifespan by ablation of germ line in *Drosophila. Proceedings of the Royal Society B-Biological Sciences,* 273, 939–47.

Barrieu, F., Marty-Mazars, D., Thomas, D., Chaumont, F., Charbonnier, M., & Marty, F. (1999). Desiccation and osmotic stress increase the abundance of mRNA of the tonoplast aquaporin BobTIP26-1 in cauliflower cells. *Planta,* 209, 77–86.

Barth, M. & Heisenberg, M. (1997). Vision affects mushroom bodies and central complex in *Drosophila melanogaster. Learning and Memory,* 4, 219–29.

Bartholomew, G. A. (1987). Interspecific comparison as a tool for ecological physiologists. *In:* Feder, M. E., Bennett, A. F., Burggren, W., & Huey, R. B. (eds.) *New Directions in Ecological Physiology.* Cambridge: Cambridge University Press.

Bartholomew, G. A. & Casey, T. M. (1977) Body temperature and oxygen consumption during rest and activity in relation to body size in some tropical beetles. *Journal of Thermal Biology,* 2, 173–76.

Bartholomew, G. A. & Casey, T. M. (1978). Oxygen consumption of moths during rest, pre-flight warm-up, and flight in relation to body size and wing morphology. *Journal of Experimental Biology,* 76, 11–25.

Baudelot, M. E. (1864). Sur la respiration des insectes. *Annales des Sciences Naturelles Series 5,* 11, 45–48.

Beament, J. W. L. (1945). The cuticular lipoids of insects. *Journal of Experimental Biology,* 21, 115–31.

Beattie, A. J. (1985). *The Evolutionary Ecology of Ant-Plant Mutualisms,* Cambridge, Cambridge University Press.

Becher, P. G., Bengtsson, M., Hansson, B. S., & Witzgall, P. (2010). Flying the fly: Long-range flight behavior of *Drosophila melanogaster* to attractive odors. *Journal of Chemical Ecology,* 36, 599–607.

Bechtold, D. A. & Luckman, S. M. (2007). The role of RFamide peptides in feeding. *Journal of Endocrinology,* 192, 3–15.

Beck, J., Mühlenberg, E. & Fiedler, K. (1999). Mud-puddling behavior in tropical butterflies: in search of proteins or minerals. *Oecologia,* 119, 140–48.

Becker, A., Schlöder, P., Steele, J. E., & Wegener, G. (1996). The regulation of trehalose metabolism in insects. *Experientia,* 52, 433–39.

Behmer, S. & Douglas, A. E. (2011). Plant sterols and host plant suitability for a phloem-feeding insect. *Functional Ecology,* 25, 484–91.

Behmer, S. T. (2009). Insect herbivore nutrient regulation. *Annual Review of Entomology,* 54, 165–87.

Behmer, S. T. & Elias, D. O. (1999). The nutritional significance of sterol metabolic constraints on the generalist grasshopper *Schistocerca americana. Journal of Insect Physiology,* 45, 339–48.

Behmer, S. T. & Elias, D. O. (2000). Sterol metabolic constraints as a factor contributing to the maintenance of diet mixing in grasshoppers (Orthoptera: Arrididae). *Physiological and Biochemical Zoology,* 73, 219–30.

Behmer, S. T. & Joern, A. (2008). Coexisting generalist herbivores occupy unique nutritional feeding niches. *Proceedings of the National Academy of Sciences,* 105, 1977–82.
Behmer, S. T. & Nes, W. D. (2003). Insect sterol nutrition and physiology: a global overview. *Advances in Insect Physiology,* 31, 1–72.
Behmer, S. T., Simpson, S. J., & Raubenheimer, D. (2002). Herbivore foraging in chemically heterogenous environments: nutrients and secondary metabolites. *Ecology,* 83, 2489–2501.
Bellah, K. L. (1984). A central action of octopamine on ventilation frequency in corydalus cornutus. *The Journal of Experimental Zoology,* 231, 289–92.
Belovsky, G. E. & Slade, J. B. (2000). Insect herbivory accelerates nutrient cycling and increases plant production. *Proceedings of the National Academy of Sciences USA,* 97, 14412–14417.
Bennet-Clark, H. (2007). The first description of resilin. *Journal of Experimental Biology,* 210, 3879–81.
Bennett, V. A., Pruitt, N. L., & Lee, R. E. (1997). Seasonal changes in fatty acid composition associated with cold-hardening in third instar larvae of Eurosta solidaginis. *Journal of Comparative Physiology B-Biochemical Systemic and Environmental Physiology,* 167, 249–55.
Benoit, J. B. & Denlinger, D. L. (2010). Meeting the challenges of on-host and off-host water balance in blood-feeding arthropods. *Journal of Insect Physiology,* 56, 1366–76.
Benton, R. (2008). Chemical sensing in *Drosophila*. *Current Opinion in Neurobiology,* 18, 357–63.
Berenbaum, M. (1983). Coumarins and caterpillars: A case for coevolution. *Evolution,* 37, 163–79.
Berenbaum, M. R. (2002). Postgenomic chemical ecology: from genetic code to ecological interactions. *Journal of Chemical Ecology,* 28, 873–96.
Bernays, E. A. (1981). Plant tannins and insect herbivores: an appraisal. *Ecological Entomology,* 6, 353–60.
Bernays, E. A. (1986). Diet-induced head allometry among foliage-chewing insects and its importance for graminivores. *Science,* 231, 495–97.
Bernays, E. A. (1990). Water regulation. *In:* Chapman, R. F. & Joem, A. (eds.) *The Biology of Grasshoppers.* New York: John Wiley & Sons.
Bernays, E. A. (1997). Feeding by lepidopteran larvae is dangerous. *Ecological Entomology,* 22, 121–23.
Bernays, E. A., Blaney, W. M., & Chapman, R. F. (1972). Changes in chemoreceptor sensilla on the maxillary palps of *Locusta migratoria* in relation to feeding. *Journal of Experimental Biology,* 57, 745–53.
Bernays, E. A. & Chapman, R. F. (1973). The regulation of feeding in *Locusta migratoria*. Internal inhibitory mechanisms. *Entomological Experimental Applications,* 16, 329–42.
Bernays, E. A. & Chapman, R. F. (1998). Phenotypic plasticity in numbers of antennal chemoreceptors in a grasshopper: effects of food. *Journal of Comparative Physiology A,* 183, 69–76.
Bernays, E. A. & Chapman, R. F. (2000). Plant secondary compounds and grasshoppers: Beyond plant defenses. *Journal of Chemical Ecology,* 26, 1773–94.
Bernays, E. A., Jarzembowski, E. A., & Malcolm, S. B. (1991). Evolution of insect morphology in relation to plants [and discussion]. *Philosophical Transactions: Biological Sciences,* 333, 257–64.

Bernays, E. A. & Lewis, A. C. (1986). The effect of wilting on the palatability of plants to *Schistocerca gregaria*, the desert locust. *Oecologia*, 70, 132–35.

Bernays, E. A. & Simpson, S. J. (1982). Control of food intake. *Advances in Insect Physiology*, 16, 59–118.

Bernays, E. A. & Woodhead, S. (1984). The need for high levels of phenylalanine in the diet of *Schistocerca gregaria* nymphs. *Journal of Insect Physiology*, 30, 489–93.

Berner, D., Blanckenhorn, W. U., & Körner, C. (2005). Grasshoppers cope with low host plant quality by compensatory feeding and food selection: N limitation challenged. *Oikos*, 111, 525–33.

Berner, R. A. (2006a). Carbon, sulfur and O_2 across the Permian-Triassic boundary. *Journal of Geochemical Exploration*, 88, 416–18.

Berner, R. A. (2006b). GEOCARBSULF: A combined model for Phanerozoic atmospheric O_2 and CO_2. *Geochimica et Cosmochimica Acta*, 70, 5653–64.

Bertram, S. M. & Johnson, L. (1998). An electronic technique for monitoring the temporal aspects of acoustic signals of captive organisms. *Bioacoustics*, 9, 107–18.

Bertram, S. M., Bowen, M., Kyle, M., & Schade, J. D. (2008). Extensive natural intraspecific variation in stoichiometric (C:N:P) composition in two terrestrial insect species. *Journal of Insect Science*, 8, 26.

Bertsch, A. (1984). Foraging in male bumblebees (*Bombus-lucorum* L) – maximizing energy or minimizing water load. *Oecologia*, 62, 325–36.

Beuron, F., LE Cahérec, F., Guillam, M.-T., Cavalier, A., Garret, A., Tassan, J.-P., Delamarche, C., Schultz, P., Mallouh, V., Rolland, J.-P., Hubert, J.-F., Gouranton, J., & Thomas, D. (1995). Structural analysis of a MIP family protein from the digestive tract of *Cicadella viridis*. *Journal of Biological Chemistry*, 270, 17414–22.

Beyenbach, K. W. (2003a). Regulation of tight junction permeability with switch-like speed. *Current Opinion in Nephrology and Hypertension*, 12, 543–50.

Beyenbach, K. W. (2003b). Transport mechanisms of diuresis in Malpighian tubules of insects. *J Exp Biol*, 206, 3845–56.

Beyenbach, K. W., Pannabecker, T. L., & Nagel, W. (2000). Central role of the apical membrane H^+-ATPase in electrogenesis and epithelial transport in Malpighian tubules. *Journal of Experimental Biology*, 203, 1459–68.

Beyenbach, K. W. & Piermarini, P. M. (2009). Osmotic and ionic regulation in insects. *In:* Evans, D. H. (ed.) *Osmotic and Ionic Regulation: Cells and Animals*. Boca Raton: CRC Press.

Beyenbach, K. W., Skaer, H., & Dow, J. A. T. (2010). The developmental, molecular, and transport biology of Malpighian tubules. *Annual Review of Entomology*, 55, 351–74.

Bhalerao, S., Sen, A., Stocker, R., & Rodrigues, V. (2003). Olfactory neurons expressing identified receptor genes project to subsets of glomeruli within the antennal lobe of *Drosophila melanogaster*. *Journal of Neurobiology*, 54, 577–92.

Bidla, G., Lindgren, M., Theopold, U., & Dushay, M. S. (2005). Hemolymph coagulation and phenoloxidase in *Drosophila* larvae. *Developmental and Comparative Immunology*, 29, 669–79.

Bier, E. & Bodmer, R. (2004). *Drosophila*, an emerging model for cardiac disease. *Gene*, 342, 1–11.

Bifano, T. D., Alegria, T. G. P., & Terra, W. R. (2010). Transporters involved in glucose and water absorption in the *Dysdercus peruvianus* (Hemiptera: Pyrrhocoridae) anterior midgut. *Comparative Biochemistry and Physiology B-Biochemistry & Molecular Biology*, 157, 1–9.

Bignell, D. E. & Anderson, J. M. (1980). Determination of pH and oxygen status in the guts of lower and higher termites. *Journal of Insect Physiology,* 26, 183–88.

Billingsley, P. F. & Lehane, M. J. (1996). Structure and ultrastructure of the insect midgut. *In:* Lehane, M. J. & Billingsley, P. F. (eds.) *Biology of the Insect Midgut.* London: Chapman and Hall.

Bird, R. B., Stewart, W. E., & Lightfood, E. N. (2002). *Transport Phenomena,* New York, Wiley.

Blanckenhorn, W. U. (2000). The evolution of body size: What keeps organisms small? *Quarterly Review of Biology,* 75, 385–407.

Blanckenhorn, W. U. & Demont, M. (2004). Bergmann and converse Bergmann latitudinal clines in arthropods: Two ends of a continuum? *Integrative and Comparative Biology,* 44, 413–24.

Blaney, W. M. & Simmonds, M. S. J. (1990). The Chemoreceptors. *In:* Chapman, R. F. & Joern, A. (eds.) *Biology of Grasshoppers.* New York: John Wiley and Sons.

Blatch, S. A., Meyer, K. W., & Harrison, J. F. (2010). Effects of dietary folic acid level and symbiotic folate production on fitness and development in the fruit fly *Drosophila melanogaster*. *Fly,* 4, 1–8.

Blatt, J. & Roces, F. (2001). Haemolymph sugar levels in foraging honeybees (*Apis mellifera carnica*): dependence on metabolic rate and *in vivo* measurement of maximal rates of trehalose synthesis. *Journal of Experimental Biology,* 204, 2709–16.

Blumenthal, E. M. (2003). Regulation of chloride permeability by endogenously produced tyramine in the *Drosophila* Malpighian tubule. *American Journal of Physiology – Cellular Physiology,* 284, C718–28.

Blumenthal, E. M. (2005). Modulation of tyramine signaling by osmolality in an insect secretory epithelium. *American Journal of Physiology – Cell Physiology,* 289, C1261–1267.

Bodmer, R. (1993). The gene tinman is required for specification of the heart and visceral muscles in *Drosophila. Development,* 118, 719–29.

Bodmer, R. (1995). Heart development in *Drosophila* and its relationship to vertebrates. *Trends in Cardiovascular Medicine,* 5, 21–28.

Bodmer, R., Wessells, R. J., Johnson, E. C., & Dowse, H. B. (2005). Heart development and function. *In:* Gilbert L.I., I. K. G. S. (ed.) *Comprehensive Molecular Insect Science.* Oxford: Elsevier.

Bohni, R., Riesgo-Escovar, J., Oldham, S., Brogiolo, W., Stocker, H., Andruss, B. F., Beckingham, K., & Hafen, E. (1999). Autonomous control of cell and organ size by CHICO, a *Drosophila* homolog of vertebrate IRS1-4. *Cell,* 97, 865–75.

Bonduriansky, R. & Day, T. (2009). Nongenetic inheritance and its evolutionary implications. *Annual Review of Ecology, Evolution and Systematics,* 40, 103–25.

Borgnia, M., Nielsen, S., Engel, A., & Agre, P. (1999). Cellular and molecular biology of the aquaporin water channels. *Annual Review of Biochemistry,* 68, 425–58.

Borrell, B. J. & Medeiros, M. J. (2004). Thermal stability and muscle efficiency in hovering orchid bees (Apidae : Euglossini). *Journal of Experimental Biology,* 207, 2925–33.

Bourtzis, K. & T.A., M. (eds.) (2003). *Insect Symbiosis,* Boca Raton, FL: CRC Press.

Bouligand, Y. (1965). On a twisted fibrillar arrangement common to several biologic structures. *C R Acad. Sci. Hebd Seances Acad. Sci. D* 261, 4864–67.

Bradley, T. J. (1987). Physiology of osmoregulation in mosquitos. *Annual Review of Entomology,* 32, 439–62.

Bradley, T. J., Brethorst, L., Robinson, S., & Hetz, S. K. (2003). Changes in the rate of CO_2 release following feeding in the insect *Rhodnius prolixus*. *Physiological and Biochemical Zoology,* 76, 302–09.

Bradley, T. J., Briscoe, A. D., Brady, S. G., Contreras, H. L., Danforth, B. N., Dudley, R., Grimaldi, D., Harrison, J. F., Kaiser, A., Merlin, C., Reppert, S. M., Vandenbrooks, J. M., & Yanoviak, S. P. (2009). Episodes in insect evolution. *Integrative and Comparative Biology,* 49, 590–606.

Bradley, T. J. & Phillips, J. E. (1977). The location and mechanism of hyperosmotic fluid secretion in the rectum of the saline-water mosquito larvae *Aedes Taeniorhynchus. Journal of Experimental Biology,* 66, 111–26.

Bradley, T. J. & Satir, P. (1979). Evidence of microfilament-associated mitochondrial movement. *Journal of Supramolecular Structure,* 12, 165–75.

Bradley, T. J., Williams, A. E., & Rose, M. R. (1999). Physiological responses to selection for desiccation resistance in *Drosophila melanogaster. American Zoologist,* 39, 337–45.

Bradshaw, W. E. & Holzapfel, C. M. (2006). Climate change – Evolutionary response to rapid climate change. *Science,* 312, 1477–78.

Brès M. & Hatzfeld C. (1977). Three-gas diffusion—experimental and theoretical study. *Plügers Archiv,* 371, 227–33.

Brodbeck, B. & Strong, D. (1987). Amino acid nutrition of herbivorous insects and stress to host plants. *In:* Barbosa, P. & Schultz, J. C. (eds.) *Insect Outbreaks.* San Diego: Academic Press.

Brogiolo, W., Stocker, H., Ikeya, T., Rintelen, S., Fernandez, R., & Hafen, E. (2001). An evolutionarily conserved function of the *Drosophila* insulin receptor and insulin-like peptides in growth control. *Current Biology,* 11, 213–21.

Bronstein, J. L. (2009). The evolution of facilitation and mutualism. *Journal of Ecology,* 97, 1160–70.

Bronstein, J. L. & Holland, J. N. (2008). Mutualism. *In:* Jorgensen, S. E. & Fath, B. D. (eds.) *Encyclopedia of Ecology.* Oxford: Elsevier.

Bronstein, J. L., Huxman, T., Horvath, B., Farabee, M., & Davidowitz, G. (2009). Reproductive biology of *Datura wrightii*: the benefits of a herbivorous pollinator. *Annals of Botany,* 103, 1435–43.

Brown, J. H. & Mauer, B. A. (1986). Body size, ecological dominance and Cope's rule. *Nature Australia,* 324, 248–50.

Brown, J. H., Sibly, R. M., & Kodric-Brown, A. (2011). *Metabolic Ecology: A Scaling Approach,* Wiley Blackwell.

Brugge, V. T., Ianowski, J. P., & Orchard, I. (2009). Biological activity of diuretic factors on the anterior midgut of the blood-feeding bug, *Rhodnius prolixus. General and Comparative Endocrinology,* 162, 105–12.

Brune, A. & Kuhl, M. (1996). pH profiles of the extremely alkaline hindguts of soil-feeding termites (Isoptera: Termitidae) determined with microelectrodes. *Journal of Insect Physiology,* 42, 1121–27.

Buck, J. (1962). Some physical aspects of insect respiration. *Annual Review of Entomology,* 7, 27–56.

Buck, J. & Keister, M. (1955). Cyclic CO_2 release in diapausing *Agapema* pupae. *Biological Bulletin,* 109, 144–63.

Buckner, J. S. & Caldwell, J. M. (1980). Uric-acid levels during the last larval instar of *Manduca sexta*, an abrupt transition from excretion to storage in the fat body. *Journal of Insect Physiology,* 26, 27–32.

Buckner, J. S., Henderson, T. A., Ehresmann, D. D., & Graf, G. (1990). Structure and composition of urate storage granules from the fat body of *Manduca sexta. Insect Biochemistry,* 20, 203–14.

Buckner, J. S. & Newman, S. M. (1990). Uric acid storage in the epidermal cells of *Manduca-sexta* – Localization and movement during the larval pupal transformation. *Journal of Insect Physiology,* 36, 219.

Bullard, B., Garcia, T., Benes, V., Leake, M. C., Linke, W. A., & Oberhauser, A. F. (2006). The molecular elasticity of the insect flight muscle proteins projectin and kettin. *Proceedings of the National Academy of Sciences of the United States of America,* 103, 4451–56.

Bullock, J. M. R., Drechsler, P., & Federle, W. (2008). Comparison of smooth and hairy attachment pads in insects: friction, adhesion and mechanisms for direction-dependence. *Journal of Experimental Biology,* 211, 3333–43.

Bullock, J. M. R. & Federle, W. (2011). The effect of surface roughness on claw and adhesive hair performance in the dock beetle *Gastrophysa viridula. Insect Science,* 18, 298–304.

Burkett, B. & Schneiderman, H. A. (1974a). Roles of oxygen and carbon dioxide in the control of spiracular function in *Cecropia* pupae. *Biological Bulletin,* 147, 274–93.

Burkett, B. N. & Schneiderman, H. A. (1974b). Roles of oxygen and carbon dioxide in the control of spiracular function in cecropia pupae. *Biological Bulletin,* 147, 274–93.

Burmester, T. & Hankeln, T. (2007). The respiratory proteins of insects. *Journal of Insect Physiology,* 53, 285–94.

Burmester, T., Massey, H. C., Zakharkin, S. O., & Benes, H. (1998). The evolution of hexamerins and the phylogeny of insects. *Journal of Molecular Evolution,* 47, 93–108.

Burrows, M. (1996). *The Neurobiology of an Insect Brain,* New York, Oxford University Press.

Burrows, M., Shaw, S. R., & Sutton, G. P. (2008). Resilin and chitinous cuticle form a composite structure for energy storage in jumping by froghopper insects. *BMC Biology,* 6.

Bursell, E. (1981). The role of proline in energy metabolism. In: *Energy Metabolism in Insects,* Plenum, New York, 135–54.

Buscarlet, L. A., Proux, J., & Gerster, R. (1978). Use of double labeled H_2O in a study of metabolic balance in *Locusta migratoria migratorioides. Journal of Insect Physiology,* 24, 225–32.

Bush, G. L. (1994). Sympatric speciation in animals: new wine in old bottles. *Trends in Ecology & Evolution,* 9, 285–88.

Bustami, H. P., Harrison, J. F., & Hustert, R. (2002). Evidence for oxygen and carbon dioxide receptors in insect CNS influencing ventilation. *Comparative Biochemistry and Physiology A,* 133, 595–604.

Bustami, H. P. & Hustert, R. (2000). Typical ventilatory pattern of the intact locust is produced by the isolated CNS. *Journal of Insect Physiology,* 46, 1285–93.

Butler, P. J. (2004). Metabolic regulation in diving birds and mammals. *Respiratory Physiology & Neurobiology,* 141, 297–315.

Buxton, P. A. (1926). The colonization of the sea by insects: with an account of the habits of Pontomyia the only known submarine insect. *Proceedings of the Zoological Society of London,* (1926), 807–14.

Caccia, S., Casartelli, M., Grimaldi, A., Losa, E., De Eguileor, M., Pennacchio, F., & Giordana, B. (2007). Unexpected similarity of intestinal sugar absorption by SGLT1 and apical GLUT2 in an insect (Aphidius ervi, Hymenoptera) and mammals. *American Journal of Physiology – Regulatory, Integrative and Comparative Physiology,* 292, R2284–91.

Cameron, P., Hiroi, M., Ngai, J. & Scott, K. (2010)The molecular basis for water taste in *Drosophila. Nature,* 465, 91–96.

Campbell, E. M., Ball, A., Hoppler, S., & Bowman, A. S. (2008). Invertebrate aquaporins: a review. *Journal of Comparative Physiology B-Biochemical Systemic and Environmental Physiology,* 178, 935–55.

Campbell, G. S. & Norman, R. M. (1997). *An Introduction to Environmental Biophysics,* New York, Springer.
Carson, W. P., Cronin, J. P., & Long, Z. T. (2004). A general rule for predicting when insects will have strong top-down effects. *In:* Weisser, W. W. & Siemann, E. (eds.) *Insects and Ecosystem Function.* Berlin: Springer-Verlag.
Cartar, R. V. (1991a). Colony energy requirements affect response to predation risk in foraging bumble bees. *Ethology,* 87, 90–96.
Cartar, R. V. (1991b). A test of risk-sensitive foraging in wild bumble bees. *Ecology,* 72, 888–95.
Cartar, R. V. (1992). Adjustment of foraging effort and task switching in energy-manipulated wild bumblebee colonies. *Animal Behaviour,* 44, 75–87.
Cayre, M., Scotto-Lomassese, S., Malaterre, J., Strambi, C., & Strambi, A. (2007). Understanding the regulation and function of adult neurogenesis: Contribution from an insect model, the house cricket. *Chemical Senses,* 32, 385–95.
Centanin, L., Dekanty, A., Romero, N., Irisarri, M., Gorr, T. A., & Wappner, P. (2008). Cell autonomy of HIF effects in *Drosophila*: Tracheal cells sense hypoxia and induce terminal branch sprouting. *Developmental Cell,* 14, 547–58.
Centanin, L., Gorr, T. A., & Wappner, P. (2010). Tracheal remodelling in response to hypoxia. *Journal of Insect Physiology,* 56, 447–54.
Chacon-Almeida, V. M. L., Soares, A. F. E., & Malheiros, E. B. (1999). Induction of the split sting trait in Africanized *Apis mellifera* (Hymenoptera : Apidae) by cold treatment of pupae. *Annals of the Entomological Society of America,* 92, 549–55.
Chamberlin, M. & Phillips, J. E. (1982a). Metabolic support of chloride-dependent short-circuit current across locust rectum. *Journal of Experimental Biology,* 99, 349–61.
Chamberlin, M. E. & Phillips, J. E. (1982b). Regulation of hemolymph amino acid levels and active secretion of proline by Malpighian tubules of locusts. *Canadian Journal of Zoology,* 60, 2745–52.
Chambers, P., Sword, G., Angel, J., Behmer, S., & Bernays, E. A. (1996). Foraging by generalist grasshoppers: dietary mixing and the role of crypsis. *Animal Behaviour,* 52, 155–65.
Champagne, D. E., Nussenzveig, R. H., & Ribeiro, J. M. C. (1995). Purification, partial characterization, and cloning of nitric oxide-carrying heme proteins (nitrophorins) from salivary glands of the blood-sucking insect *Rhodnius prolixus*. *Journal of Biological Chemistry,* 270, 8691–95.
Chandler, S. M., Wilkinson, T. L., & Douglas, A. E. (2008). Impact of plant nutrients on the relationship between a herbivorous insect and its symbiotic bacteria. *Proceedings of the Royal Society B-Biological Sciences,* 275, 565–70.
Chapman, J. W., Drake, V. A., & Reynolds, D. R. (2011). Recent insights from radar studies of insect flight. *Annual Review of Entomology,* 56, 337–56.
Chapman, R. F. (1982). Insect chemoreceptors. *Advances in Insect Physiology,* 16, 247–356.
Chapman, R. F. (1990). Food selection. *In:* Chapman, R. F. & Joern, A. (eds.) *Biology of Grasshoppers.* New York: John Wiley and Sons.
Chapman, R. F. (1998). *The Insects: Structure and Function,* Cambridge, Cambridge University Press.
Chapman, R. F. (2003). Contact chemoreception in feeding by phytophagous insects. *Annual Review of Entomology,* 48, 455–84.
Chappell, M. A. (1982). Temperature regulation of carpenter bees (*Xylocopa californica*) foraging in the Colorado desert of southern California. *Physiological Zoology,* 55, 267–80.

Chappell, M. A. & Rogowitz, G. L. (2000). Mass, temperature and metabolic effects on discontinuous gas exchange cycles in eucalyptus-boring beetles (Coleoptera: Cerambycidae). *Journal of Experimental Biology,* 203, 3809–20.

Chappell, M. A. & Whitman, D. W. (1990). Grasshopper thermoregulation. *In:* Chapman, R. F. & Joern, A. (eds.) *Biology of Grasshoppers.* New York: John Wiley & Sons.

Charles, J. P. (2010). The regulation of expression of insect cuticle protein genes. *Insect Biochemistry and Molecular Biology,* 40, 205–213.

Chen, A. & Wagner, R. (1992). Hemolymph constituents of the stable fly *Stomoxys calcitrans. Comparative Biochemistry and Physiology,* 102 A, 133–37.

Chen, C. P. & Denlinger, D. L. (1990). Activation of phosphorylase in response to cold and heat-stress in the flesh fly, *Sarcophaga crassipalpis. Journal of Insect Physiology,* 36, 549–53.

Chen, H., Chen, Z., & Zhou, Y. (2005). Rice water weevil (Coleoptera: Curculionidae) in mainland China: invasion, spread and control. *Crop Protection,* 24, 695–702.

Chen, Q. F., Ma, E., Behar, K. L., Xu, T., & Haddad, G. G. (2002). Role of trehalose phosphate synthase in anoxia tolerance and development in *Drosophila melanogaster. Journal of Biological Chemistry,* 277, 3274–79.

Cheng, K., Simpson, S. J., & Raubeheimer, D. (2008). A geometry of regulatory scaling. *The American Naturalist,* 172, 681–83.

Cheng, L. (1985). Biology of halobates (Heteroptera: Gerridae). *Annual Review of Entomology,* 30, 111–35.

Cheung, W. W. K. & Marshall, A. T. (1973). Water and ion regulation in cicadas in relation to xylem feeding. *Journal of Insect Physiology,* 19, 1801–16.

Chino, H. (1981). Lipid transport by haemolymph lipoprotein. A possible mutiple role of diacylglycerol-carrying protein. In: *Energy Metabolism in Insects,* Plenum, New York,155–68.

Chino, H., Downer, R. G. H., Wyatt, G. R., & Gilbert, L. I. (1981). Lipophorins, a major class of lipoproteins of insect haemolymph. *Insect Biochemistry,* 11, 491–91.

Chippindale, A. K., Gibbs, A. G., Sheik, M., Yee, K. J., Djawdan, M., Bradley, T. J., & Rose, M. R. (1998). Resource acquisition and the evolution of stress resistance in *Drosophila melanogaster. Evolution,* 52, 1342–52.

Chittka, A. & Chittka, L. (2010). Epigenetics of royalty. *PLoS Biol,* 8, e1000532.

Chown, S. L. (2002). Respiratory water loss in insects. *Comparative Biochemistry and Physiology A,* 133, 791–804.

Chown, S. L. & Davis, A. L. (2003). Discontinuous gas exchange and the significance of respiratory water loss in scarabaeine beetles. *Journal of Experimental Biology,* 206, 3547–56.

Chown, S. L. & Gaston, K. J. (1999). Exploring links between physiology and ecology at macro scales: the role of respiratory metabolism in insects. *Biological Reviews,* 74, 87–120.

Chown, S. L., Gaston, K. J., & Robinson, D. (2004a). Macrophysiology: large-scale patterns in physiological traits and their ecological implications. *Functional Ecology,* 18, 159–67.

Chown, S. L., Gibbs, A. G., Hetz, S. K., Klok, C. J., Lighton, J. R. B., & Marais, E. (2006). Discontinuous gas exchange in insects: A clarification of hypotheses and approaches. *Physiological and Biochemical Zoology,* 79, 333–43.

Chown, S. L., Marais, E., Terblanche, J. S., Klok, C. J., Lighton, J. R. B., & Blackburn, T. M. (2007). Scaling of insect metabolic rate is inconsistent with the nutrient supply network model. *Functional Ecology,* 21, 282–90.

Chown, S. L. & Nicolson, S. W. (2004). *Insect Physiological Ecology: Mechanisms and Patterns,* Oxford University Press.

Chown, S. L., Sinclair, B. J., Leinaas, H. P., & Gaston, K. J. (2004b). Hemispheric asymmetries in biodiversity – A serious matter for ecology. *PLoS Biology,* 2, 1701–07.

Chown, S. L. & Terblanche, J. S. (2007). Physiological diversity in insects: Ecological and evolutionary contexts. *Advances in Insect Physiology, Vol 33.* London: Academic Press Ltd.

Christensen, K., Gallacher, A., Martin, L., Tong, D., & Elgar, M. (2010). Nutrient compensatory foraging in a free-living social insect. *Naturwissenschaften,* 97, 941–44.

Clancy, K. M. & King, R. M. (1993). Defining the western spruce budworm's nutritional niche with response surface methodology. *Ecology,* 74, 442–54.

Clark, M. G., Bloxham, D. P., Holland, P. C., & Lardy, H. A. (1973). Estimation of the fructose diphosphatase-phosphofructokinase substrate cycle in the flight muscle of *Bombus affinis. Biochemical Journal,* 134, 589–97.

Clark, M. S. & Worland, M. R. (2008). How insects survive the cold: molecular mechanisms – a review. *Journal of Comparative Physiology B-Biochemical Systemic and Environmental Physiology,* 178, 917–33.

Clark, T. M. (1999). Evolution and adaptive significance of larval midgut alkalinization in the insect superorder Mecopterida. *Journal of Chemical Ecology,* 25, 1945–60.

Clark, T. M., Vieira, M. A. L., Huegel, K. L., Flury, D., & Carper, M. (2007). Strategies for regulation of hemolymph pH in acidic and alkaline water by the larval mosquito *Aedes aegypti* (L.) (Diptera; Culicidae). *Journal of Experimental Biology,* 210, 4359–67.

Clausen, C. P. (1940). *Entomphagous Insects,* New York, McGraw-Hill.

Clayton, R. B. (1964). The utilization of sterols by insects. *Journal of Lipid Research,* 5, 3–19.

Clements, A. N. (1992). *The Biology of Mosquitoes,* Springer.

Clench, H. K. (1966). Behavioral thermoregulation in butterflies. *Ecology,* 47, 1021–34.

Clissold, F. J., Sanson, G. D., & Read, J. (2006). The paradoxical effects of nutrient ratios and supply rates on an outbreaking insect herbivore, the Australian plague locust. *Journal of Animal Ecology,* 75, 1000–1013.

Clissold, F. J., Sanson, G. D., Read, J., & Simpson, S. J. (2009). Gross vs. net income: How plant toughness affects performance of an insect herbivore. *Ecology,* 90, 3393–3405.

Clyne, P. J., Warr, C. G., & Carlson, J. R. (2000). Candidate taste receptors in *Drosophila. Science,* 287, 1830–34.

Coast, G. (2007). The endocrine control of salt balance in insects. *General and Comparative Endocrinology,* 152, 332–38.

Coast, G. M. (1995). Synergism between diuretic peptides controlling ion and fluid transport in insect malpighian tubules. *Regulatory Peptides,* 57, 283–96.

Coast, G. M. (2001). The neuroendocrine regulation of salt and water balance in insects. *Zoology,* 103, 179–88.

Coast, G. M. (2004). Continuous recording of excretory water loss from *Musca domestica* using a flow-through humidity meter: hormonal control of diuresis. *Journal of Insect Physiology,* 50, 455–68.

Coast, G. M. (2009). Neuroendocrine control of ionic homeostasis in blood-sucking insects. *Journal of Experimental Biology,* 212, 378–86.

Coast, G. M., Garside, C. S., Webster, S. G., Schegg, K. M., & Schooley, D. A. (2005). Mosquito natriuretic peptide identified as a calcitonin-like diuretic hormone in *Anopheles gambiae* (Giles). *Journal of Experimental Biology,* 208, 3281–91.

Coast, G. M., Meredith, J., & Phillips, J. E. (1999). Target organ specificity of major neuropeptide stimulants in locust excretory systems. *Journal of Experimental Biology,* 202, 3195–3203.

Coast, G. M., Orchard, I., Phillips, J. E., & Schooley, D. A. (2002). Insect diuretic and antidiuretic hormones. *Advances in Insect Physiology*, 29, 279–409.
Cochran, D. G. (1985). Nitrogen excretion in cockroaches. *Annual Review of Entomology*, 30, 29–49.
Cocroft, R. B. & Rodríguez, R. L. (2005) The behavioral ecology of insect vibrational communication. *BioScience*, 55, 323–34.
Cohen, R. W. (2001). Diet Balancing in the Cockroach *Rhyparobia madera*: Does serotonin regulate this behavior? *Journal of Insect Behavior*, 14, 99–111.
Coley, P. D. (1983). Herbivory and defensive characteristics of tree species in a lowland tropical forest. *Ecological Monographs*, 53, 209–33.
Colinet, H., Lee, S. F., & Hoffmann, A. (2010). Functional characterization of the Frost gene in *Drosophila melanogaster*: Importance for recovery from chill coma. *PLOS One*, 5.
Colombani, J., Bianchini, L., Layalle, S., Pondeville, E., Dauphin-Villemant, C., Antoniewski, C., Carré, C., Noselli, S., & Léopold, P. (2005). Antagonistic actions of ecdysone and insulins determine final size in *Drosophila*. *Science*, 310, 667–70.
Consortium, T. H. G. S. (2006). Insights into social insects from the genome of the honeybee Apis mellifera. *Nature*, 443, 931–49.
Contreras, H. L. & Bradley, T. J. (2009). Metabolic rate controls respiratory pattern in insects. *Journal of Experimental Biology*, 212, 424–28.
Cook, J. M., Bean, D., Power, S. A., & Dixon, D. J. (2004). Evolution of a complex coevolved trait: active pollination in a genus of fig wasps. *Journal of Evolutionary Biology*, 17, 238–46.
Cook, S. C. & Behmer, S. T. (2010). Macronutrient regulation in the tropical terrestrial ant *Ectatomma ruidum* (Formicidae): a field study in Costa Rica. *Biotropica*, 42, 135–39.
Cook, S. C., Eubanks, M. D., Gold, R. E., & Behmer, S. T. (2010). Colony-level macronutrient regulation in ants: mechanisms, hoarding and associated costs. *Animal Behaviour*, 79, 429–37.
Coope, G. R. (1979). Late Cenozoic fossil Coleoptera: Evolution, biogeography and ecology. *Annual Review of Ecology and Systematics*, 10, 247–67.
Cooper-Driver, G., Finch, S., & Swain, T. (1977). Seasonal variation in secondary plant compounds in relation to palatablity of *Pteridium aquilinum*. *Biochemical and Systematic Evolution*, 5, 177–83.
Cooper, P. D. (1983). Validation of the doubly labeled water method for measuring water flux and energy metabolism in tenebrionid beetles. *Physiological & Biochemical Zoology*, 56, 41–46.
Cooper, P. D. & Vulcano, R. (1997). Regulation of pH in the digestive system of the cricket, *Teleogryllus commodus* Walker. *Journal of Insect Physiology*, 43, 495–99.
Cossins, A. R. & Raynard, R. S. (1987). Adaptive responses of animal cell membranes to temperature. *Symposium for the Society of Experimental Biology*, 41, 95–111.
Coulson, S. C. & Bale, J. S. (1992). Effect of rapid cold hardening on reproduction and survival of offspring in the housefly *Musca domestica*. *Journal of Insect Physiology*, 38, 421–24.
Crailsheim, K. (1988). Intestinal transport of glucose solution during honeybee flight. *In:* Nachtigall, W. (ed.) *The Flying Honeybee; Aspects of Energetics.* New York: Gustav Fischer, Stuttgart.
Crespi, B. J. & Sandoval, C. P. (2000). Phylogenetic evidence for the evolution of ecological specialization in *Timema* walking sticks. *Journal of Evolutionary Biology*, 13, 249–62.

Crill, W. D., Huey, R. B., & Gilchrist, G. W. (1996). Within- and between-generation effects of temperature on the morphology and physiology of *Drosophila melanogaster*. *Evolution,* 50, 1205–18.

Crosthwaite, J. C., Sobek, S., Lyons, D. B., Bernards, M. A., & Sinclair, B. J. (2011). The overwintering physiology of the emerald ash borer, Agrilus planipennis Fairmaire (Coleoptera: Buprestidae). *Journal of Insect Physiology,* 57, 166–73.

Culliton, B. (2001). One gene, many proteins. *Genome News Network,* Feb. 12.

Curtis, N. J., Ringo, J. M., & Dowse, H. B. (1999). Morphology of the pupal heart, adult heart, and associated tissues in the fruit fly, *Drosophila melanogaster*. *Journal of Morphology,* 240, 225–35.

Dacks, A. M., Dacks, J. B., Christensen, T. A., & Nighorn, A. J. (2006). The cloning of one putative octopamine receptor and two putative serotonin receptors from the tobacco hawkmoth, *Manduca sexta*. *Insect Biochemistry and Molecular Biology,* 36, 741–47.

Dadd, R. H. (1961). The nutritional requirements of locusts-IV. Requirements for vitamins of the B complex. *Journal of Insect Physiology,* 6, 1–12.

Dadd, R. H. (1970a). Arthropod nutrition. *In:* Florkin, M. & Scheer, B. T. (eds.) *Chemical Zoology.* Academic Press.

Dadd, R. H. (1970b). Digestion in insects. *In:* Florkin, M. & Scheer, B. T. (eds.) *Chemical Zoology.* Academic Press.

Dahanukar, A., Foster, K., Van Naters, W., & Carlson, J. R. (2001). A Gr receptor is required for response to the sugar trehalose in taste neurons of *Drosophila*. *Nature Neuroscience,* 4, 1182–86.

Danks, H. V. (2000). Dehydration in dormant insects. *Journal of Insect Physiology,* 46, 837–52.

Davidowitz, G., D'Amico, L. J., & Nijhout, H. F. (2003). Critical weight in the development of insect body size. *Evolution & Development,* 5, 188–97.

Davidowitz, G., D'Amico, L. J., & Nijhout, H. F. (2004). The effects of environmental variation on a mechanism that controls insect body size. *Evolutionary Ecology Research,* 6, 49–62.

Davidowitz, G. & Nijhout, H. F. (2004). The physiological basis of reaction norms: The interaction among growth rate, the duration of growth and body size. *Integrative and Comparative Biology,* 44, 443–49.

Davies, S. A., Huesmann, G. R., Maddrell, S. H. P., Odonnell, M. J., Skaer, N. J. V., Dow, J. A. T., & Tublitz, N. J. (1995). CAP(2b), a cardioacceleratory peptide, is present in *Drosophila* and stimulates tubule fluid secretion via cGMP. *American Journal of Physiology-Regulatory Integrative and Comparative Physiology,* 269, R1321–26.

Davis, A. L. V., Chown, S. L., Mcgeoch, M. A., & Scholtz, C. H. (2000). A comparative analysis of metabolic rate in six *Scarabaeus* species (Coleptera: Scarabaeidae) from southern Africa: further caveats when inferring adaptation. *Journal of Insect Physiology,* 46, 553–62.

De Brito Sanchez, M., Chen, C., LI, J., Liu, F., Gauthier, M., & Giurfa, M. (2008). Behavioral studies on tarsal gustation in honeybees: sucrose responsiveness and sucrose-mediated olfactory conditioning. *Journal of Comparative Physiology A: Neuroethology, Sensory, Neural, and Behavioral Physiology,* 194, 861–69.

De Bruyne, M., Foster, K., & Carlson, J. R. (2001). Odor coding in the *Drosophila* antenna. *Neuron,* 30, 537–52.

De Groot, B. L. & Grubmuller, H. (2005). The dynamics and energetics of water permeation and proton exclusion in aquaporins. *Current Opinion in Structural Biology,* 15, 176–83.

Deere, J. A. & Chown, S. L. (2006). Testing the beneficial acclimation hypothesis and its alternatives for locomotor performance. *The American Naturalist,* 168, 630.

Dejours, P. (1975). *Principles of Comparative Respiratory Physiology,* Amsterdam, North-Holland.

Dekker, T., Ibba, I., Siju, K. P., Stensmyr, M. C., & Hansson, B. S. (2006). Olfactory shifts parallel superspecialism for toxic fruit in *Drosophila melanogaster* Sibling, D. Sechellia. *Current Biology,*16, 101–109.

Delfinado-Baker, M. (1988). The tracheal mite of honeybees: a crisis in beekeeping. *In:* Needham, G. R., Page, R. E. J., Delfinado-Baker, M., & Bowman, C. E. (eds.) *Africanized Honey Bees and Bee Mites* Chichester, West Sussex, UK: E. Horwood.

Denlinger, D. L. (2002). Regulation of diapause. *Annual Review of Entomology,* 47, 93–122.

Denlinger, D. L., Willis, J. H., & Fraenkel, G. (1972). Rates and cycles of oxygen consumption during pupal diapause in *Sarcophaga* flesh flies. *Journal of Insect Physiology,* 18, 871–82.

Denlinger, D. L., Yocum, G. D., & Rinehart, J. P. (2005). Hormonal control of diapause. *In:* Lawrence, I. G., Kostas, I., & Sarjeet, S. G. (eds.) *Comprehensive Molecular Insect Science.* Amsterdam: Elsevier.

Denlinger, D. L. A. R. E. L. (2010). *Low Temperature Biology of Insects,* Cambridge, Cambridge University Press.

Denny, M. W. (1993). *Air and Water: The Biology and Physics of Life's Media,* Princeton, Princeton University Press.

Dethier, V. (1976). *The Hungry Fly,* Cambridge, Cambridge University Press.

Devries, A. L. (1971). Glycoproteins as biological antifreeze agents in antarctic fishes. *Science,* 172, 1152–55.

Diamond, S. E. & Kingsolver, J. G. (2011). Host plant quality, selection history and trade-offs shape the immune responses of *Manduca sexta. Proceedings of the Royal Society B: Biological Sciences,* 278, 289–97.

Diegisser, T., Johannesen, J., & Seitz, A. (2008). Performance of host-races of the fruit fly, Tephritis conura on a derived host plant, the cabbage thistle Cirsium oleraceum: Implications for the original host shift. *Journal of Insect Science,* 8, 1–6.

Diegisser, T., Seitz, A., & Johannesen, J. (2007). Morphological adaptation in host races of *Tephritis conura. Entomologia Experimentalis Et Applicata,* 122, 155–64.

Dillon, M. E. & Frazier, M. R. (2006). *Drosophila melanogaster* locomotion in cold thin air. *Journal of Experimental Biology,* 209, 364–71.

Dillon, M. E., Frazier, M. R., & Dudley, R. (2006). Into thin air: Physiology and evolution of alpine insects. *Integrative and Comparative Biology,* 46, 49–61.

Dillon, R. J. & Dillon, V. M. (2004). The gut bacteria of insects: Nonpathogenic interactions. *Annual Review of Entomology,* 49, 71–92.

Dircksen, H. (2009). Insect ion transport peptides are derived from alternatively spliced genes and differentially expressed in the central and peripheral nervous system. *Journal of Experimental Biology,* 212, 401–12.

Dobson, H. E. M. & Bergstrom, G. (2000). The ecology and evolution of pollen odors. *Plant Systematics and Evolution,* 222, 63–87.

Dobzhansky, T. (1948). Genetics of natural populations. XVI. Altitudinal and seasonal changes produced by natural selection in certain populations of *Drosophila pseudoobscura* and *Drosophila persimilis. Genetics,* 33, 158–76.

Donini, A., Gaidhu, M. P., Strasberg, D. R., & O'donnell, M. J. (2007). Changing salinity induces alterations in hemolymph ion concentrations and Na^+ and Cl^- transport kinetics of the anal papillae in the larval mosquito, *Aedes aegypti*. *Journal of Experimental Biology,* 210, 983–92.

Doong, H., Vrailas, A., & Kohn, E. C. (2002). What's in the "BAG"? – a functional domain analysis of the BAG-family proteins. *Cancer Letters,* 188, 25–32.

Doucet, D., Tyshenko, M. G., Davies, P. L., & Walker, V. K. (2002). A family of expressed antifreeze protein genes from the moth, *Choristoneura fumiferana*. *European Journal of Biochemistry,* 269, 38–46.

Doucet, D., Walker, V. K., & Qin, W. (2009). The bugs that came in from the cold: molecular adaptations to low temperatures in insects. *Cellular and Molecular Life Sciences,* 66, 1404–18.

Douglas, A. E. (1989). Mycetocyte symbiosis in insects. *Biological Reviews,* 69, 409–34.

Douglas, A. E. (2007). Symbiotic microorganisms: untapped resources for insect pest control. *Trends in Biotechnology,* 25, 338–42.

Douglas, A. E. (2009). The microbial dimension in insect nutritional ecology. *Functional Ecology,* 23, 38–47.

Douglas, A. E. (2010). *The Symbiotic Habit,* Princeton, Princeton University Press.

Dow, J. A. T. (1986). Insect midgut function. *In:* Evans, P. D. & Wigglesworth, V. B. (eds.) *Advances in Insect Physiology.* London: Academic Press.

Dow, J. A. T. & Harvey, W. R. (1988). Role of midgut electrogenic K^+ pump potential difference in regulating lumen K^+ and pH in larval lepidoptera. *Journal of Experimental Biology,* 140, 455–63.

Dow, J. A. T., Kelly, D. C., Davies, S. A., Maddrell, S. H. P., & Brown, D. (1995). A novel member of the major intrinsic protein family in *Drosophila* – aquaporins involved in insect Malpighian (Renal) tubule fluid secretion. *Journal of Physiology-London,* 489P, P110–P111.

Downer, K. E., Haselton, A. T., Nachman, R. J., & Stoffolano, J. J. G. (2007). Insect satiety: Sulfakinin localization and the effect of drosulfakinin on protein and carbohydrate ingestion in the blow fly, *Phormia regina* (Diptera: Calliphoridae). *Journal of Insect Physiology,* 53, 106–12.

Downer, R. G. H. & Chino, H. (1985). Turnover of protein and diacylglycerol components of lipophorin in insect haemolymph. *Insect Biochemistry,* 15, 627–30.

Downes, C. J., Carpenter, A., Hansen, L. D., & Lill, R. E. (2003). Microcalorimetric and mass spectrometric methods for determining the effects of controlled atmospheres on insect metabolism. *Thermochimica Acta,* 397, 19–29.

Drake, L. L., Boudko, D. Y., Marinotti, O., Carpenter, V. K., Dawe, A. L., & Hansen, I. A. (2010). The aquaporin gene family of the yellow fever mosquito, *Aedes aegypti*. *PLOS One,* 5, e15578.

Drès, M. & Mallet, J. (2002). Host races in plant-feeding insects and their importance in sympatric speciation. *Philosophical Transactions: Biological Sciences,* 357, 471–92.

Drosopoulos, S. & Claridge, M. F. (eds.) (2005). *Insect Sounds and Communication: Physiology, Behaviour, Ecology, and Evolution,* Boca Raton: CRC Press.

Duchesne, L., Hubert, J. F., Verbavatz, J. M., Thomas, D., & Pietrantonio, P. V. (2003). Mosquito (*Aedes aegypti*) aquaporin, present in tracheolar cells, transports water, not glycerol, and forms orthogonal arrays in *Xenopus oocyte* membranes. *European Journal of Biochemistry,* 270, 422–29.

Dudley, R. (1992). Aerodynamics of flight. *In:* Biewener, A. A. (ed.) *Biomechanics (Structures and Systems): A Practical Approach.* Oxford: Oxford University Press.
Dudley, R. (1995). Extraordinary flight performance of orchid bees (Apidae: Euglossini) hovering in heliox (80% He/20% O_2). *Journal of Experimental Biology,* 198, 1065–70.
Dudley, R. (1998). Atmospheric oxygen, giant Paleozoic insects and the evolution of aerial locomotor performance. *Journal of Experimental Biology,* 201, 1043–50.
Dudley, R. (2000). *The Biomechanics of Insect Flight. Form, Function, Evolution,* Princeton, Princeton University Press.
Dudley, R. & Chai, P. (1996). Animal flight mechanics in physically variable gas mixtures. *Journal of Experimental Biology,* 199, 1881–85.
Dulcis, D., Davis, N. T., & Hildebrand, J. G. (2001). Neuronal control of heart reversal in the hawkmoth *Manduca sexta. Journal of Comparative Physiology A,* 187, 837–49.
Dulcis, D. & Levine, R. B. (2005). Glutamatergic innervation of the heart initiates retrograde contractions in adult *Drosophila melanogaster. Journal of Neuroscience,* 25, 271–80.
Duncan, F. D. & Byrne, M. J. (2000). Discontinuous gas exchange in dung beetles: Patterns and ecological implications. *Oecologia,* 122, 452–58.
Duncan, F. D. & Byrne, M. J. (2002). Respiratory airflow in a wingless dung beetle. *Journal of Experimental Biology,* 205, 2489–97.
Duncan, F. D. & Byrne, M. J. (2005). The role of the mesothoracic spiracles in respiration in flighted and flightless dung beetles. *Journal of Experimental Biology,* 208, 907–14.
Dussourd, D. E., Ubik, K., Harvis, C., Resch, J., Meinwald, J., & Eisner, T. (1988). Biparental defensive endowment of eggs with acquired plant alkaloid in the moth *Utetheisia ornatrix. Proceedings of the National Academy of Science, USA,* 85, 5992–96.
Dussutour, A. & Simpson, S. J. (2008). Carbohydrate regulation in relation to colony growth in ants. *Journal of Experimental Biology,* 211, 2224–32.
Dussutour, A. & Simpson, S. J. (2009). Communal nutrition in ants. *Current Biology,* 19, 740–44.
Dzialowski, E. M. (2005). Use of operative temperature and standard operative temperature models in thermal biology. *Journal of Thermal Biology,* 30, 317–34.
Ebbesen, P., Eckardt, K. U., Ciampor, F., & Pettersen, E. O. (2004). Linking measured intercellular oxygen concentration to human cell functions. *Acta Oncologica,* 43, 598–600.
Echevarria, M., Ramirez-Lorca, R., Hernandez, C. S., Gutierriez, A., Mendez-Ferrer, S., Gonzalez, E., Toledo-Aral, J. J., Ilundain, A. A., & Whittembury, G. (2001). Identification of a new water channel (Rg-MIP) in the Malpighian tubules of the insect *Rhodnius prolixus. Pflugers Archiv-European Journal of Physiology,* 442, 27–34.
Eckstrand, I. A. & Richardson, R. H. (1980). Comparison of some water-balance characteristics in several *Drosophila* species which differ in habitat. *Environmental Entomology,* 9, 716–20.
Eckstrand, I. A. & Richardson, R. H. (1981). Relationships between water-balance properties and habitat characteristics in the sibling Hawaiian Drosophilids, *Drosophiaa-mimica* and *Drosophila-kambysellisi. Oecologia,* 50, 337–41.
Edney, E. B. (1977). *Water Balance in Land Arthropods,* Berlin, Springer.
Edwards, H. A. (1983). Electrophysiology of mosquito anal papillae. *Journal of Experimental Biology,* 102, 343–46.
Edwards, H. A. & Harrison, J. B. (1983). An osmoregulatory syncytium and associated cells in a fresh-water mosquito. *Tissue & Cell,* 15, 271–80.

Ehresmann, D. D., Buckner, J. S., & Graf, G. (1990). Uric acid translocation from the fat body of Manduca sexta during the pupal-adult transformation: effects of 20-hydroxyecdysone. *Journal of Insect Physiology,* 36, 173–80.

Eichler, S. & Schaub, G. A. (1998). The effects of aposymbiosis and of an infection with *Blastocrithidia triatomae* (Trypanosomatidae) on the tracheal system of the reduviid bugs *Rhodnius prolixus* and *Triatoma infestans*. *Journal of Insect Physiology,* 44, 131–40.

Eigenheer, R. A., Nicolson, S. W., Schegg, K. M., Hull, J. J., & Schooley, D. A. (2002). Identification of a potent antidiuretic factor acting on beetle Malpighian tubules. *Proceedings of the National Academy of Science, USA,* 99, 84–9.

Eigenheer, R. A., Wiehart, U. M., Nicolson, S. W., Schoofs, L., Schegg, K. M., Hull, J. J., & Schooley, D. A. (2003). Isolation, identification and localization of a second beetle antidiuretic peptide. *Peptides,* 24, 27–34.

Elser, J. J., Dobberfuhl, D. R., Mackay, N. A., & Schampel, J. H. (1996). Organism size, life history, and N:P stoichiometry Toward a unified view of cellular and ecosystem processes. *Bioscience,* 46, 674–84.

Elser, J. J., Fagan, W. F., Denno, R. F., Dobberfuhl, D. R., Folarin, A., Huberty, A., Interlandi, S., Kilham, S. S., Mccauley, E., Schulz, K. L., Siemann, E. H., & Sterner, R. W. (2000). Nutritional constraints in terrestrial and freshwater food webs. *Nature,* 408, 578–80.

Emlen, D. J., Szafran, Q., Corley, L. S., & Dworkin, I. (2006). Insulin signaling and limb-patterning: candidate pathways for the origin and evolutionary diversification of beetle "horns". *Heredity,* 97, 179–91.

Engelhard, E. K., Kam-Morgan, L. N. W., Washburn, J. O., & Volkman, L. E. (1994). The insect tracheal system: A conduit for the systemic spread of *Autographa californica* M nuclear polyhedrosis virus. *Proceedings of the National Academy of Sciences, USA,* 91, 3224–27.

Erlich, P. R. & Raven, P. H. (1964). Butterflies and plants: A study in coevolution. *Evolution,* 18, 586–608.

Esch, H., Goller, F., & Heinrich, B. (1991). How do bees shiver. *Naturwissenschaften,* 78, 325–28.

Etges, W. J. & Klassen, C. S. (1989). Influences of atmospheric ethanol on adult Drosophila mojavensis: altered metabolic rates and increases in fitness among poplulations. *Physiological Zoology,* 62, 170–93.

Evgen'ev, M. B., Garbuz, D. G., Shilova, V. Y., & Zatsepina, O. G. (2007). Molecular mechanisms underlying thermal adaptation of xeric animals. *Journal of Biosciences,* 32, 489–99.

Faeth, S. H. & Shochat, E. (2010). Inherited microbial symbionts increase herbivore abundances and alter arthropod diversity on a native grass. *Ecology,* 91, 1329–43.

Fagan, W. F., Siemann, E., Mitter, C., Denno, R. F., Huberty, A. F., Woods, H. A., & Elser, J. J. (2002). Nitrogen in insects: implications for trophic complexity and species diversification. *American Naturalist,* 160, 784–802.

Falconer, D. S. (1989). *Introduction to Quantitative Genetics,* New York, John Wiley and Sons.

Falk, R., Bleiseravivi, N., & Atidia, J. (1976). Labellar taste organs of *Drosophila melanogaster. Journal of Morphology,* 150, 327–41.

Farooqi, T. (2007). Octopamine-mediated modulation of insect senses. *Neurochemical. Research,* 32, 1511–29.

Farrell, B. D. (1998). "Inordinate fondness" explained: Why are there so many beetles? *Science,* 281, 555–59.

Faruki, S. I., Das, D. R., Khan, A. R., & Khatun, M. (2007). Effects of ultraviolet (254nm) irradiation on egg hatching and adult emergence of the flour beetles, *Tribolium castaneum, T-confusum* and the almond moth, *Cadra cautella. Journal of Insect Science,* 7.

Fatt, I. (1982). *Polarographic Oxygen Sensor: Its Theory Of Operation and its Application in Biology, Medicine And Technology.* Malabar, Krieger Publishing Company.

Feder, M. E. (1999). Organismal, ecological, and evolutionary aspects of heat-shock proteins and the stress response: Established conclusions and unresolved issues. *American Zoologist,* 39, 857–64.

Feder, M. E., Blair, N., & Figueras, H. (1997). Natural thermal stress and heat-shock protein expression in *Drosophila* larvae and pupae. *Functional Ecology,* 11, 90–100.

Feder, M. E. & Hofmann, G. E. (1999). Heat-shock proteins, molecular chaperones, and the stress response: evolutionary and ecological physiology. *Annual Review of Physiology,* 61, 243–82.

Feeny, P. P. (1976). Plant apparency and chemical defense. *In:* Wallace, J. W. & Mansell, J. L. (eds.) *Biochemical Interaction Between Plants and Insects.* New York: Plenum.

Fernandez-Winckler, F. & Da Cruz-Landim, C. (2008). A morphological view of the relationship between indirect flight muscle maturation and the flying needs of two species of advanced eusocial bees. *Micron,* 39, 1235–42.

Fewell, J. H., Harrison, J. F., Lighton, J. R. B., & Breed, M. D. (1996). Foraging energetics of the ant, *Paraponera clavata. Oecologia,* 105, 419–27.

Fewell, J. H., Harrison, J. F., Stiller, T. M., & Breed, M. D. (1992). Distance effects on resource profitability and recruitment in the giant tropical ant, *Paraponera clavata. Oecologia,* 92, 542–47.

Fewell, J. H., Ydenberg, R. C., & Winston, M. L. (1991). Individual foraging effort as a function of colony population in the honey bee, *Apis mellifera L. Animal Behaviour,* 42, 153–55.

Feyereisen, R. (2005). Insect cytochrome P450. *In:* Gilbert, K. L., and S. S. Gil (ed.) *Comprehensive Molecular Insect Science.* Oxford: Elsevier.

Fielding, D. J. & Defoliart, L. S. (2005). Density and temperature-dependent melanization of fifth-instar *Melanoplus sanguinipes*: interpopulation comparisons. *Journal of Orthoptera Research,* 14, 107–13.

Fielding, D. J. & Defoliart, L. S. (2008). Discriminating tastes: self-selection of macronutrients in two populations of grasshoppers. *Physiological Entomology,* 33, 264–73.

Fields, P. G., Fleurat-Lessard, F., Lavenseau, L., Febvay, G., Peypelut, L., & Bonnot, G. (1998). The effect of cold acclimation and deacclimation on cold tolerance, trehalose and freee amino acid levels in *Sitophilus granarius* and *Cryptolestes ferrugineus* (Coleoptera). *Journal of Insect Physiology,* 44, 955–65.

Filippova, M., Ross, L. S., & Gill, S. S. (1998). Cloning of the V-ATPase B subunit cDNA from Culex quinquefasciatus and expression of the B and C subunits in mosquitoes. *Insect Molecular Biology,* 7, 223–32.

Fine, P. V. A., Mesones, I., & Coley, P. D. (2004). Herbivores promote habitat specialization by trees in Amazonian forests. *Science,* 305, 663–65.

Finley, J., Camazine, S., & Frazier, M. (1996). The epidemic of honeybee colony losses during the 1995-1996 season. *American Bee Journal,* 136, 805–08.

Fischer, K. & Fiedler, K. (2001). Dimorphic growth patterns and sex-specific reaction norms in the butterfly *Lycaena hippothoe sumadiensis. Journal of Evolutionary Biology,* 14, 210–18.

Fischer, K., O'brien, D. M., & Boggs, C. L. (2004). Allocation of larval and adult resources to reproduction in a fruit-feeding butterfly. *Functional Ecology,* 18, 656–63.

Fisher, C. W. & Brady, U. E. (1983). Activation, properties and collection of haemolymph phenoloxidase of the american cockroach, *Periplaneta americana. Comparative Biochemistry and Physiology Part C: Comparative Pharmacology,* 75, 111–14.

Florkin, M. & Jeuniax, C. (1974). Hemolymph: Composition. *In:* Rockstein, M. (ed.) *The Physiology of Insecta.* New York: Academic Press.

Folk, D. G. & Bradley, T. J. (2003). Evolved patterns and rates of water loss and ion regulation in laboratory-selected populations of *Drosophila melanogaster. Journal of Experimental Biology,* 206, 2779–86.

Folk, D. G. & Bradley, T. J. (2004). The evolution of recovery from desiccation stress in laboratory-selected populations of *Drosophila melanogaster. Journal of Experimental Biology,* 207, 2671–78.

Folk, D. G. & Bradley, T. J. (2005). Adaptive evolution in the lab: Unique phenotypes in fruit flies comprise a fertile field of study. *Integrative and Comparative Biology,* 45, 492–99.

Folk, D. G., Han, C., & Bradley, T. J. (2001). Water acquisition and partitioning in *Drosophila melanogaster*: effects of selection for desiccation-resistance. *Journal of Experimental Biology,* 204, 3323–31.

Forbes, A. A. & Feder, J. L. (2006). Divergent preferences of Rhagoletis pomonella host races for olfactory and visual fruit cues. *Entomologia Experimentalis Et Applicata,* 119, 121–27.

Fore, T. (1996). Winter colony loss reported by state apiary inspectors surveyed by American Beekeeping Federation. *Speedy Bee,* 25, 16.

Forkner, R. E., Marquis, R. J., & Lill, J. T. (2004). Feeney revisited: condensed tannins as anti-herbivore defences in leaf-chewing herbivore communities of *Quercus. Ecological Entomology,* 29, 174–87.

Förster, T. D. & Hetz, S. K. (2010). Spiracle activity in moth pupae-The role of oxygen and carbon dioxide revisited. *Journal of Insect Physiology,* 5, 492–501.

Fraenkel, G. & Blewett, M. (1943). The vitamin B-complex requirements of several insects. *Biochemical Journal,* 37, 686–92.

Fraenkel, G. & Blewett, M. (1947). The importance of folic acid and unidentified members of the vitamin B complex in the nutrition of certain insects. *Biochemical Journal,* 41, 469–75.

Frazier, M. R., Harrison, J. F., Kirkton, S. D., & Roberts, S. P. (2008). Cold rearing improves cold-flight performance in *Drosophila* via changes in wing morphology. *Journal of Experimental Biology,* 211, 2116–22.

Frazier, M. R., Huey, R. B,. & Berrigan, D. (2006). Thermodynamics constrains the evolution of insect population growth rates: "Warmer is better". *American Naturalist,* 168, 512–20.

Frazier, M. R., Woods, H. A., & Harrison, J. F. (2001). Interactive effects of rearing temperature and oxygen on the development of *Drosophila melanogaster. Physiological and Biochemical Zoology,* 74, 641–50.

Froger, A., Tallur, B., Thomas, D., & Delamarche, C. (1998). Prediction of functional residues in water channels and related proteins. *Protein Science: A Publication of the Protein Society,* 7, 1458.

Fu, D. & Lu, M. (2007). The structural basis of water permeation and proton exclusion in aquaporins. *Molecular Membrane Biology,* 24, 366–74.

Fujimoto, M. & Nakai, A. (2010). The heat shock factor family and adaptation to proteotoxic stress. *FEBS Journal,* 277, 4112–25.

Fujiwara, Y. & Denlinger, D. L. (2007a). High temperature and hexane break pupal diapause in the flesh fly, *Sarcophaga crassipalpis,* by activating ERK/MAPK. *Journal of Insect Physiology,* 53, 1276–82.

Fujiwara, Y. & Denlinger, D. L. (2007b). p38 MAPK is a likely component of the signal transduction pathway triggering rapid cold hardening in the flesh fly *Sarcophaga crassipalpis. Journal of Experimental Biology,* 210, 3295–300.

Fujiwara, Y., Shindome, C., Takeda, M., & Shiomi, K. (2006a). The roles of ERK and P38 MAPK signaling cascades on embryonic diapause initiation and termination of the silkworm, *Bombyx mori. Insect Biochemistry and Molecular Biology,* 36, 47–53.

Fujiwara, Y., Tanaka, Y., Iwata, K. I., Rubio, R. O., Yaginuma, T., Yamashita, O., & Shiomi, K. (2006b). ERK/MAPK regulates ecdysteroid and sorbitol metabolism for embryonic diapause termination in the silkworm, *Bombyx mori. Journal of Insect Physiology,* 52, 569–75.

Full, R. J. (1997). Invertebrate locomotor systems. *In:* Dantzler, W. H. (ed.) *Handbook of Physiology Section 13: Comparative Physiology.* New York: Oxford University Press.

Furuya, K., Milchak, R. J., Schegg, K. M., Zhang, J., Tobe, S. S., Coast, G. M., & Schooley, D. A. (2000). Cockroach diuretic hormones: characterization of a calcitonin-like peptide in insects. *Proceedings of the National Academy of Sciences of the United States of America,* 97, 6469–74.

Gäde, G. (1985). Anaerobic energy metabolism. *In:* Hoffmann, K. H. (ed.) *Environmental Physiology and Biochemistry of Insects.* Berlin: Springer-Verlag.

Gäde, G. (2004). Regulation of intermediary metabolism and water balance of insects by neuropeptides. *Annual Review of Entomology,* 49, 93–113.

Gäde, G. & Auerswald, L. (2002). Beetles' choice – proline for energy output: control by AKHs. *Comparative Biochemistry and Physiology B-Biochemistry & Molecular Biology,* 132, 117–29.

Gäde, G. & Auerswald, L. (2003). Mode of action of neuropeptides from the adipokinetic hormone family. *General and Comparative Endocrinology,* 132, 10–20.

Gade, G., Auerswald, L., & Marco, H. G. (2006). Flight fuel and neuropeptidergic control of fuel mobilisation in the twig wilter, *Holopterna alata* (Hemiptera, Coreidae). *Journal of Insect Physiology,* 52, 1171–81.

Galen, C. (1996). Rates of floral evolution: adaptation to bumblebee pollination in an alpine wildflower, *Polemonium viscosum. Evolution,* 55, 1963–71.

Galindo, K. & Smith, D. P. (2001). A large family of divergent *Drosophila* odorant-binding proteins expressed in gustatory and olfactory sensilla. *Genetics,* 159, 1059–72.

Galizia, C. G. & Rössler, W. (2010). Parallel olfactory systems in insects: Anatomy and function. *Annual Review of Entomology,* 55, 399–420.

Galizia, C. G. & Szyszka, P. (2008). Olfactory coding in the insect brain: molecular receptive ranges, spatial and temporal coding. *Entomologia Experimentalis Et Applicata,* 128, 81–92.

Gardner, K. E., Forster, R. L., & O'donnell S. (2007) Experimental analysis of worker division of labor in bumblebee nest thermoregulation (*Bombus huntii,* Hymenoptera : Apidae). *Behavioral Ecology and Sociobiology,* 61(5), 783–92.

Garland, T., Bennett, A. F., & Rezende, E. L. (2005). Phylogenetic approaches in comparative physiology. *Journal of Experimental Biology,* 208, 3015–35.

Garlick, K. M. & Robertson, R. M. (2007). Cytoskeletal stability and heat shock-mediated thermoprotection of central pattern generation in *Locusta migratoria. Comparative Biochemistry and Physiology a-Molecular & Integrative Physiology,* 147, 344–48.

Garrett, M. & Bradley, T. J. (1984). The pattern of osmotic regulation in larvae of the mosquito *Culiseta-Inornata. Journal of Experimental Biology,* 113, 133–41.

Garrett, M. A. & Bradley, T. J. (1987). Extracellular accumulation of proline, serine, and trehalose in the haemolymph of asmoconforming brackish-water mosquitoes. *Journal of Experimental Biology,* 129, 231–38.

Gaston, K. J. & Blackburn, T. M. (1996). Range size-body size relationships: evidence of scale depedence. *Oikos,* 75, 479–85.

Ge, X. & Wang, X. (2009). Calculations of freezing point depression, boiling point elevation, vapor pressure and enthalpies of vaporization of electrolyte solutions by a modified three-characteristic parameter correlation model. *Journal of Solution Chemistry,* 38, 1097–117.

Gefen, E., Marlon, A. J., & Gibbs, A. G. (2006). Selection for desiccation resistance in adult *Drosophila melanogaster* affects larval development and metabolite accumulation. *Journal of Experimental Biology,* 209, 3293–300.

Gehring, W. J. & Wehner, R. (1995). Heat shock protein synthesis and thermotolerance in *Cataglyphis,* an ant from the Sahara desert. *Proceedings of the National Academy of Sciences of the USA,* 92, 2994–98.

Geister, T. L., Lorenz, M. W., Hoffmann, K. H., & Fischer, K. (2009). Energetics of embryonic development: effects of temperature on egg and hatchling composition in a butterfly. *Journal of Comparative Physiology B-Biochemical Systemic and Environmental Physiology,* 179, 87–98.

Géminard, C., Rulifson, E. J., & Léopold, P. (2009). Remote control of insulin secretion by fat cells in *Drosophila. Cell Metabolism,* 10, 199–207.

Ghabrial, A., Luschnig, S., Metzstein, M. M., & Krasnow, M. A. (2003). Branching morphogenesis of the *Drosophila* tracheal system. *Annual Review of Cell and Developmental Biology,* 19, 623–47.

Gibbs, A. G. (1999). Laboratory selection for the comparative physiologist. *Journal of Experimental Biology,* 202, 2709–18.

Gibbs, A. G. (2002a). Lipid melting and cuticular permeability: New insights into an old problem. *Journal of Insect Physiology,* 48, 391–400.

Gibbs, A. G. (2002b). Water balance in desert *Drosophila*: Lessons from non-charismatic microfauna. *Comparative Biochemistry and Physiology A,* 133, 781–89.

Gibbs, A. G., Chippindale, A. K., & Rose, M. R. (1997). Physiological mechanisms of evolved desiccation resistance in *Drosophila melanogaster. Journal of Experimental Biology,* 200, 1821–32.

Gibbs, A. G., Fukuzato, F., & Matzkin, L. M. (2003). Evolution of water conservation mechanisms in *Drosophila. Journal of Experimental Biology,* 206, 1183–92.

Gibbs, A. G. & Johnson, R. A. (2004). The role of discontinuous gas exchange in insects: the chthonic hypothesis does not hold water. *Journal of Experimental Biology,* 207, 3477–82.

Gibbs, A. G. & Matzkin, L. M. (2001). Evolution of water balance in the genus *Drosophila. Journal of Experimental Biology,* 204, 2331–38.

Gijzen, H. J., Vanderdrift, C., Barugahare, M., & Opdencamp, H. J. M. (1994). Effect of host diet and hindgut microbial composition on cellulytic activity in the hindgut of the American cockroach, *Periplaneta americana. Applied and Environmental Microbiology,* 60, 1822–26.

Gilbert, L. E. (1972). Pollen feeding and reproductive biology of *Heliconius* butterflies. *Proceedings of the National Academy of Science, USA,* 69, 1403–07.

Gilchrist, G. W. & Huey, R. B. (1999). The direct response of *Drosophila melanogaster* to selection on knockdown temperature. *Heredity,* 83, 15–29.

Gilchrist, G. W., Huey, R. B., Balanya, J., Pascual, M., & Serra, L. (2004). A time series of evolution in action: A latitudinal cline in wing size in South American Drosophila subobscura. *Evolution,* 58, 768–80.

Gilchrist, G. W., Huey, R. B., & Serra, L. (2001). Rapid evolution of wing size clines in *Drosophila subobscura. Genetica,* 112–113, 273–86.

Gillooly, J. F. & Ophir, A. G. (2010). The energetic basis of acoustic communication. *Proceedings of the Royal Society B-Biological Sciences,* 277, 1325–31.

Gleixner, E., Abriss, D., Adryan, B., Kraemer, M., Gerlach, F., Schuh, R., Burmester, T., & Hankeln, T. (2008). Oxygen-induced changes in hemoglobin expression in *Drosophila*. *FEBS Journal,* 275, 5108–16.

Glenn, J. D., King, J. G., & Hillyer, J. F. (2010). Structural mechanics of the mosquito heart and its function in bidirectional hemolymph transport. *Journal of Experimental Biology,* 213, 541–50.

Glenner, H., Thomsen, P. F., Hebsgaard, M. B., Serensen, M. V., & Willerslev, E. (2006a). The origin of insects. *Science,* 314, 1883–83.

Glenner, H., Thomsen, P. F., Hebsgaard, M. B., Sorensen, M. V., & Willerslev, E. (2006b). The origin of insects. *Science,* 314, 1883–84.

Gmeinbauer, R. & Crailsheim, K. (1993). Glucose utilization during flight of honeybee (*Apis mellifera*) workers, drones and queens. *Journal of Insect Physiology,* 39, 959–67.

Goh, S. & Phillips, J. E. (1978). Dependence of prolonged water absorption by *in vitro* locust rectum on ion transport. *Journal of Experimental Biology,* 72, 25–41.

Goldschmidt, R. B. (1935). Gen und Augeneigenschaft (Untersuchungen und Drosophila), *I. Z Indukt Abstamm Vererbungsl.*, 69, 38–69.

Goldschmidt, R. B. (1938). *Physiological Genetics,* McGraw-Hill, New York.

Goldsworthy, G. J., Chung, J. S., Simmonds, M. S. J., Tatari, M., Varouni, S., & Poulos, C. P. (2003). The synthesis of an analogue of the locust CRF-like diuretic peptide, and the biological activities of this and some C-terminal fragments. *Peptides,* 24, 1607–13.

Golob, P. (1997). Current status and future perspectives for inert dusts for control of stored product insects. *Journal of Stored Products Research,* 33, 69–79.

Gomez, N. N., Venette, R. C., Gould, J. R., & Winograd, D. F. (2009). A unified degree day model describes survivorship of *Copitarsia corruda* Pogue & Simmons (Lepidoptera: Noctuidae) at different constant temperatures. *Bulletin of Entomological Research,* 99, 65–72.

Gorr, T. A., Gassmann, M., & Wappner, P. (2006). Sensing and responding to hypoxia via HIF in model invertebrates. *Journal of Insect Physiology,* 52, 349–64.

Goto, S. G. (2001). A novel gene that is up-regulated during recovery from cold shock in *Drosophila melanogaster. Gene,* 270, 259–64.

Goto, S. G., Philip, B. N., Teets, N. M., Kawarasaki, Y., Lee, R. E., & Denlinger, D. L. (2011). Functional characterization of an aquaporin in the Antarctic midge *Belgica antarctica. Journal of Insect Physiology,* 57 (8), 1106–14.

Gotthard, K., Berger, D., & Walters, R. (2007). What keeps insects small? Time limitation during oviposition reduces the fecundity benefit of female size in a butterfly. *The American Naturalist,* 169, 768–79.

Gould, J., Venette, R., & Winograd, D. (2005). Effect of temperature on development and population parameters of *Copitarsia decolora* (Lepidoptera : Noctuidae). *Environmental Entomology,* 34, 548–56.

Gouranton, J. (1968). Ultrastructures en rapport avec un transit d'eau. *Etude de la chambre filtrante de Cicadella viridis L.(Homoptera, Jassidae). J. Microscopie,* 7, 559–74.

Gouveia, S. M., Simpson, S. J., Raubeheimer, D., & Zanotto, F. P. (2000). Patterns of respiration in *Locusta migratoria* nymphs when feeding. *Physiological Entomology,* 25, 88–93.

Graham, J. B., Dudley, R., Aguilar, N. M., & Gans, C. (1995). Implications of the later Palaeozoic oxygen pulse for physiology and evolution. *Nature,* 375, 117–20.

Graham, L. A. & Davies, P. L. (2005). Glycine-rich antifreeze proteins from snow fleas. *Science,* 310, 461–61.

Gray, J. R. & Robertson, R. M. (1998). Effects of heat stress on axonal conduction in the locust flight system. *Comparative Biochemistry and Physiology a Molecular & Integrative Physiology*, 120, 181–86.

Greenberg, S. & Ar, A. (1996). Effects of chronic hypoxia, normoxia and hyperoxia on larval development in the beetle *Tenebrio molitor*. *Journal of Insect Physiology*, 42, 991–96.

Greenfield, M. D. (2002). *Signalers & Receivers: Mechanisms and evolution of arthropod communication*, Oxford, Oxford University Press.

Greenlee, K. J. & Harrison, J. F. (1998). Acid-base and respiratory responses to hypoxia in the grasshopper *Schistocerca americana*. *Journal of Experimental Biology*, 201, 2843–55.

Greenlee, K. J. & Harrison, J. F. (2004a). Development of respiratory function in the American locust *Schistocerca americana* I. Across-instar effects. *Journal of Experimental Biology*, 207, 497–508.

Greenlee, K. J. & Harrison, J. F. (2004b). Development of respiratory function in the American locust *Schistocerca americana* II. Within-instar effects. *Journal of Experimental Biology*, 207, 509–17.

Greenlee, K. J. & Harrison, J. F. (2005). Respiratory changes throughout ontogeny in the tobacco hornworm caterpillar, *Manduca sexta*. *Journal of Experimental Biology*, 208, 1385–92.

Greenlee, K. J., Henry, J. R., Kirkton, S. D., Westneat, M. W., Fezzaa, K., Lee, W. K., & Harrison, J. F. (2009). Synchrotron imaging of the grasshopper tracheal system: morphological components of tracheal hypermetry and the effect of age and stage on abdominal air sac volumes and convection. *American Journal of Physiology: Comparative, Regulatory and Integrative Physiology*, 297, 1343–50.

Greive, H. & Surholt, B. (1990). Dependence of fructose-bis-phosphatase from flight muscles of the bumblebee (*Bombus terrestris* L.) on calcium ions. *Comparative Biochemistry and Physiology*, 97B, 197–200.

Grimaldi, D. & Engel, M. S. (2005). *Evolution of the Insects*, New York, Cambridge University Press.

Grüter, C., Moore, H., Firmin, N., Helanterä, H., & Ratnieks, F. L. W. (2011). Flower constancy in honey bee workers (Apis mellifera) depends on ecologically realistic rewards. *Journal of Experimental Biology*, 214, 1397–1402.

Gulinson, S. L. & Harrison, J. F. (1996). Control of resting ventilation rate in grasshoppers. *Journal of Experimental Biology*, 199, 379–89.

Gullan, P. J. & Cranston, P. S. (2005). *The Insects: An Outline of Entomology*, Malden, MA, Blackwell Publishing.

Gündüz, A. E. & Douglas, A. E. (2009). Symbiotic bacteria enable insect to use a nutritionally inadequate diet. *Proceedings of the Royal Society B: Biological Sciences*, 276, 987–91.

Gunn, D. L. (1933). The temperature and humidity relations of the cockroach (*Blatta orientalis*): I. Desiccation. *Journal of Experimental Biology*, 10, 274–85.

Haack, R. A. & Slansky, F., Jr. (1987). Nutritional ecology of wood-feeding Coleoptera, Lepidoptera and Hymenoptera. *In:* Slansky, F., Jr. & Rodriguez, J. G. (eds.) *Nutritional Ecology of Insects, Mites, Spiders and Related Invertebrates*. London: John Wiley and Sons.

Hacohen, N., Kramer, S., Sutherland, D., Hiromi, Y., & Krasnow, M. A. (1998). *sprouty* encodes a novel antagonist of FGF signaling that patterns apical branching of the *Drosophila* airways. *Cell*, 92, 253–63.

Hadley, N. F. (1994). *Water Relations of Terrestrial Arthopods*, San Diego, CA, Acadamic Press.

Hadley, N. F. & Draper, L. E. (1969). A method for determining anaerobic pH values of semi-micro samples. *Journal of the Arizona Academy of Sciences,* 5, 248–50.
Hadley, N. F. & Quinlan, M. C. (1993). Discontinuous CO_2 release in the eastern lubber grasshopper *Romalea guttata* and its effect on respiratory transpiration. *Journal of Experimental Biology,* 177, 169–80.
Hadley, N. F., Toolson, E. C., & M.C., Q. (1989). Regional differences in cuticular permeability in the desert cicada *Diceroprocta apache*: implications for evaporative cooling. *Journal of Experimental Biology,* 141, 219–30.
Hagner-Holler, S., Schoen, A., Erker, W., Marden, J. H., Rupprecht, R., Decker, H., & Burmester, T. (2004). A respiratory hemocyanin from an insect. *Proceedings of the National Academy of Sciences of the United States of America,* 101, 871–74.
Hahn, D. A. & Denlinger, D. L. (2007). Meeting the energetic demands of insect diapause: nutrient storage and utilization. *Journal of Insect Physiology,* 53, 760–63.
Hahn, D. A. & Denlinger, D. L. (2011). Energetics of insect diapause. *Annual Review of Entomology,* 56, 103–21.
Hahn, D. A. & Wheeler, D. E. (2003). Presence of a single abundant storage hexamerin in both larvae and adults of the grasshopper, *Schistocerca americana. Journal of Insect Physiology,* 49, 1189–97.
Haine, E. R., Rolff, J., & Siva-Jothy, M. T. (2007). Functional consequences of blood clotting in insects. *Developmental and Comparative Immunology,* 31, 456–64.
Hallem, E. A., Dahanukar, A., & Carlson, J. R. (2006). Insect odor and taste receptors *Annual Review of Entomology,* 51, 113–35.
Hallock, K. (2008). Magnetic resonance microscopy of flows and compressions of the circulatory, respiratory, and digestive systems in pupae of the tobacco hornworm, *Manduca sexta. Journal of Insect Science,* 8, 10–17.
Hamada, F. N., Rosenzweig, M., Kang, K., Pulver, S. R., Ghezzi, A., Jegla, T. J., & Garrity, P. A. (2008a). An internal thermal sensor controlling temperature preference in *Drosophila. Nature,* 454, 217–220.
Hamada, F. N., Rosenzweig, M., Kang, K., Pulver, S. R., Ghezzi, A., Jegla, T. J., & Garrity, P. J. (2008b). An internal thermal sensor controlling temperature preference in *Drosophila. Nature,* 454, 217–20.
Hamdoun, A. & Epel, D. (2007). Embryo stability and vulnerability in an always changing world. *Proceedings of the National Academy of Sciences,* 104, 1745–50.
Hamilton, W. J. & Seely, M. K. (1976). Fog basking by the Namib desert beetle, *Onymacris unguicularis. Nature,* 262, 284–87.
Hanegan, J. L. (1973). Control of heart-rate in Cecropia Moths – response to thermal stimulation. *Journal of Experimental Biology,* 59, 67–76.
Hanegan, J. L. & Heath, J. E. (1970). Temperature dependence of neural control of moth flight system. *Journal of Experimental Biology,* 53, 629–39.
Hanrahan, J. W., Meredith, J., Phillips, J. E., & Brandys, D. (1984). Methods for the study of transport and control in insect hindgut. *In:* Bradley, T. J. & Miller, T. A. (eds.) *Measurement of Ion Transport and Metabolic Rate in Insects.* New York: Springer-Verlag.
Hansen, I. A., Attardo, G. M., Roy, S. G., & Raikhel, A. S. (2005). Target of rapamycin-dependent activation of S6 kinase Is a central step in the transduction of nutritional signals during egg development in a mosquito. *Journal of Biological Chemistry,* 280, 20565–72.
Harak, M., Lamprecht, I., Kuusik, A., Hiiesaar, K., Metspalu, L., & Tartes, U. (1999). Calorimetric investigations of insect metabolism and development under the influence of a toxic plant extract. *Thermochimica Acta,* 333, 39–48.

Harrison, J. (1986). Caste-specific changes in honey bee flight capacity. *Physiological Zoology,* 59, 175–87.

Harrison, J. (2009). Tracheal systems. *In:* Resh, V. H. & Carde, R. T. (eds.) *Encyclopedia of Insects.* 2nd ed. New York: Academic Press.

Harrison, J., Frazier, M. R., Henry, J. R., Kaiser, A., Klok, C. J., & Rascon, B. (2006). Responses of terrestrial insects to hypoxia or hyperoxia. *Respiratory Physiology & Neurobiology,* 154, 4–17.

Harrison, J. F. (1988a). Temperature effects on haemolymph acid-base status *in vivo* and *in vitro* in the two-striped grasshopper *Melanoplus bivittatus. Journal of Experimental Biology,* 140, 421–35.

Harrison, J. F. (1995). Nitrogen metabolism and excretion in locusts. *In:* Walsh, P. J. & Wright, R. (eds.) *Nitrogen Metabolism and Excretion.* Boca Raton, FL: CRC Press.

Harrison, J. F., Camazine, S., Marden, J. H., Kirkton, S. D., Rozo, A., & Yang, X. L. (2001). Mite not make it home: Tracheal mites reduce the safety margin for oxygen delivery of flying honeybees. *Journal of Experimental Biology,* 204, 805–14.

Harrison, J. F. & Fewell, J. H. (1995). Thermal effects on feeding behavior and net energy intake in a grasshopper experienceing large diurnal fluctuations in body temperature. *Physiological Zoology,* 68, 453–73.

Harrison, J. F. & Fewell, J. H. (2002). Environmental and genetic influences on flight metabolic rate in the honey bee, *Apis mellifera. Comparative Biochemistry and Physiology A,* 133, 323–33.

Harrison, J. F., Fewell, J. H., Roberts, S. P., & Hall, H. G. (1996). Achievement of thermal stability by varying metabolic heat production in flying honeybees. *Science,* 274, 88–90.

Harrison, J. F. & Haddad, G. G. (2011). Effects of oxygen on growth and size: synthesis of molecular, organismal and evolutionary studies with *Drosophila melanogaster. Annual Review of Physiology,* 73, 95–113.

Harrison, J. F., Kaiser, A., & Vandenbrooks, J. M. (2009). Mysteries of oxygen and insect size. *In:* Morris, S. & Vosloo, A. (eds.) *4th CPB Meeting in Africa: Mara (2008). Molecules to Migration: The Pressures of Life.* Bologna, Italy: Medimond Publishing Co.

Harrison, J. F., Kaiser, A., & Vandenbrooks, J. M. (2010). Atmospheric oxygen level and the evolution of insect body size. *Proceedings of the Royal Society B-Biological Sciences,* 277, 1937–46.

Harrison, J. F. & Lighton, J. R. B. (1998). Oxygen-sensitive flight metabolism in the dragonfly *Erythemis simplicicollis. Journal of Experimental Biology,* 201, 1739–44.

Harrison, J. F. & Phillips, J. E. (1992). Recovery from acute haemolymph acidosis in unfed locusts II. Role of ammonium and titratible acid excretion. *Journal of Experimental Biology,* 165, 97–110.

Harrison, J. F. & Roberts, S. P. (2000). Flight respiration and energetics. *Annual Review of Physiology,* 62, 179–205.

Harrison, J. F., Wong, C. J. H., & Phillips, J. E. (1990). Hemolymph Buffering in the Locust *Schistocerca gregaria. Journal of Experimental Biology,* 154, 573–79.

Harrison, J. M. (1988b). Temperature effects on intra- and extracellular acid-base status in the American locust, *Schistocerca nitens. Journal of Comparative Physiology B,* 158, 763–70.

Hartung, D. K., Kirkton, S. D., & Harrison, J. F. (2004). Ontogeny of tracheal system structure: A light and electron-microscopy study of the metathoracic femur of the American locust, *Schistocerca americana. Journal of Morphology,* 262, 800–12.

Haruhito, K. & Haruo, C. (1982). Transport of hydrocarbons by the lipophorin of insect hemolymph. *Biochimica et Biophysica Acta (BBA) – Lipids and Lipid Metabolism,* 710, 341–48.

Harvey, W. R. (2009). Voltage coupling of primary H^+ V-ATPases to secondary Na^+ or K^+-dependent transporters. *Journal of Experimental Biology,* 212, 1620–29.

Haselton, A., Downer, K., Zylstra, J., & Stoffolano, J. (2009). Serotonin Inhibits Protein Feeding in the Blow Fly, Phormia regina; (Meigen). *Journal of Insect Behavior,* 22, 452–63.

Hawkins, B. A. (1995). Latitudinal body-size gradients for the bees of the eastern United States. *Ecological Entomology,* 20, 195–98.

Hayes, T. K. et al. (1986). Insect hypertrehalosemic hormone: Isolation and primary structure from *Blaberus discoidalis. Biochemical and Biophysical Research Communications,* 140, 674–78.

Hayes, T. K., Pannabecker, T. L., Hinckley, D. J., Holman, G. M., Nachman, R. J., Petzel, D. H., & Beyenbach, K. W. (1989). Leucokinins, a new family of ion transport stimulators and inhibitors in insect malpighian tubules. *Life Sciences,* 44, 1259–66.

Hayward, S. A. L., Pavlides, S. C., Tammariello, S. P., Rinehart, J. P., & Denlinger, D. L. (2005). Temporal expression patterns of diapause-associated genes in flesh fly pupae from the onset of diapause through post-diapause quiescence. *Journal of Insect Physiology,* 51, 631–40.

Hazel, J. R. (1995). Thermal adaptation in biological membranes: is homeoviscous adaptation the explanation? *Annual Review of Physiology,* 57, 19–42.

Hazel, W. N. (2002). The environmental and genetic control of seasonal polyphenism in larval color and its adaptive significance in a swallowtail butterfly. *Evolution,* 56, 342–48.

Hedin, J. & Ranius, T. (2002). Using radio telemetry to study dispersal of the beetle Osmoderma eremita, an inhabitant of tree hollows. *Computers and Electronics in Agriculture,* 35, 171–80.

Hegedus, D., Erlandson, M., Gillott, C., & Toprak, U. (2009). New Insights into peritrophic matrix synthesis, architecture, and function. *Annual Review of Entomology,* 54, 285–302.

Heinrich, B. (1970a). Nervous control of the heart during thoracic temperature regulation in a sphinx moth. *Science,* 169, 606–07.

Heinrich, B. (1970b). Thoracic temperature stabilized by blood circulation in a free flying moth. *Science,* 168, 580–82.

Heinrich, B. (1971a). Temperature regulation of sphinx moth, *Manduca sexta.* II. Regulation of heat loss by control of blood circulation. *Journal of Experimental Biology,* 54, 153–66.

Heinrich, B. (1971b). Temperature regulation of sphinx moth, *Manduca sexta.* 1. Flight energetics and body temperature during free and tethered flight. *Journal of Experimental Biology,* 54, 141–52.

Heinrich, B. (1972). Physiology of brood incubation in the bumblebee queen *Bombus vosnesenskii. Nature,* 239, 223–25.

Heinrich, B. (1973). The energetics of the bumblebee. *Scientific American,* 278(4), 96–102.

Heinrich, B. (1975a). Energetics of pollination. *Ann. Rev. Ecol. Syst.,* 6, 139–70.

Heinrich, B. (1975b). Thermoregulation in bumblebees. *Journal of Comparative Physiology,* 96, 155–66.

Heinrich, B. (1976). Heat exchange in relation to blood flow between thorax and abdomen in bumblebees. *Journal of Experimental Biology,* 64, 561–85.

Heinrich, B. (1980a). Mechanisms of body temperature regulation in honeybees, *Apis mellifera* I. Regulation of head temperature. *Journal of Experimental Biology,* 85, 61–72.
Heinrich, B. (1980b). Mechanisms of body temperature regulation in honeybees, *Apis mellifera* II. Regulation of thoracic temperature at high air temperatures. *Journal of Experimental Biology,* 85, 73–87.
Heinrich, B. (1993). *The Hot-blooded Insects: Strategies and Mechanisms of Thermoregulation,* Cambridge, MA, Harvard University Press.
Heinrich, B. & Bartholomew, G. A. (1971). An analysis of pre-flight warm-up in the shpinx moth, *Mandura sexta. Journal of Experimental Biology,* 55, 223–39.
Heinrich, E. C., Farzin, M., Klok, C. J., & Harrison, J. F. (2011). The effect of developmental stage on the sensitivity of cell and body size to hypoxia in *Drosophila melanogaster. The Journal of Experimental Biology,* 214, 1419–27.
Heinrich, R. & Ganter, G. K. (2007). Nitric oxide/cyclic GMP signaling and insect behavior. *Advances in Experimental Biology,* 1, 107–27.
Heitler, W. J., Goodman, C. S., & Fraserrowell, C. H. (1977). Effects of temperature on threshold of identified neurons in locust. *Journal of Comparative Physiology,* 117, 163–82.
Helms, K. R. & Vinson, S. B. (2008). Plant resources and colony growth in an invasive ant: the importance of honeydew producing Hemiptera in carbohydrate transfer across trophic levels. *Environmental Entomology,* 37, 487–93.
Hennig, K., Columbani, J., & Neufeld, T. P. (2006). TOR coordinates bulk and targeted endocytosis in the *Drosophila melanogaster* fat body to regulate cell growth. *Journal of Cell Biology,* 173, 963–74.
Henricksson, J., Haukioja, E., Ossipov, V., Ossipova, S., Sillanpaa, S., & Kapari, L. (2003). Effects of host shading on consumption and growth of the geometrid Eirrita autumnata: interactive roles of water, primary and secondary compounds. *Oikos,* 103, 3–16.
Henry, J. R. & Harrison, J. F. (2004). Plastic and evolved responses of larval tracheae and mass to varying atmospheric oxygen content in *Drosophila melanogaster. Journal of Experimental Biology,* 207, 3559–67.
Henwood, K. (1975). A field-tested thermoregulation model for two diurnal Namib desert tenebrionid beetles. *Ecology,* 56, 1329–42.
Hepburn, H. R., Youthed, C., Illgner, P., Radloff, S. E., & Brown, R. E. (1998). Production of aerodynamic power in mountain honeybees (apis mellifera). *Naturwissenschaften,* 85, 389–90.
Herold, R. C. & Borei, H. (1963). Cytochrome changes during honey bee flight muscle development. *Developmental Biology,* 8, 67–79.
Hetz, S. K. & Bradley, T. J. (2005). Insects breathe discontinuously to avoid oxygen toxicity. *Nature,* 433, 516–19.
Hetz, S. K., Wasserthal, L. T., Hermann, S., Kaden, H., & Oelssner, W. (1994). Direct oxygen measurements in the tracheal system of lepidopterous pupae using miniaturized amperometric sensors. *Bioelectrochemistry and Bioenergetics,* 33, 165–70.
Heymann, N. & Lehmann, F.-O. (2006). The significance of spiracle conductance and spatial arrangement for flight muscle function and aerodynamic performance in flying *Drosophila. Journal of Experimental Biology,* 209, 1662–77.
Higashi, M., Abe, T., & Burns, T. P. (1992). Carbon-nitrogen balance and termite ecology. *Proceedings: Biological Sciences,* 249, 303–08.

Hildebrand, J. G. & Shepherd, G. M. (1997). Mechanisms of olfactory discrimination: Converging evidence for common principles across phyla. *Annual Review of Neuroscience*, 20, 595–631.

Hillerton, J. E. & Vincent, J. F. V. (1979). The stabilization of insect cuticles. *Journal of Insect Physiology*, 25, 957–65.

Hinton, H. E. (1960). The structure and function of the respiratory horns of the eggs of some flies. *Philosophical Transactions of the Royal Society of London Series B*, 243, 45–73.

Hinton, H. E. (1981). *Biology of Insect Eggs*, Oxford, Pergamon Press.

Hirayama, C., Konno, K. & Shinbo, H. (1996). Utilization of ammonia as a nitrogen source in the silkworm, *Bombyx mori*. *Journal of Insect Physiology*, 42, 983–88.

Hoback, W. W. & Stanley, D. W. (2001). Insects in hypoxia. *Journal of Insect Physiology*, 47, 533–42.

Hodkinson, I. (2005). Terrestrial insects along elevation gradients: species and community responses to altitude. *Biol. Rev.*, 80, 489–513.

Hoffmann, A. A., Hallas, R., Sinclair, C., & Mitrovski, P. (2001). Levels of variation in stress resistance in *Drosophila* among strains, local populations, and geographic regions: Patterns for desiccation, starvation, cold resistance, and associated traits. *Evolution*, 55, 1621–30.

Hoffmann, A. A. & Parsons, P. A. (1993). Selection for adult desiccation resistance in *Drosophila melanogaster*: fitness components, larval resistance and stress correlations. *Biological Journal of the Linnean Society*, 48, 43–54.

Hoffmann, A. A., Shirriffs, J., & Scott, M. (2005). Relative importance of plastic vs genetic factors in adaptive differentiation: geographical variation for stress resistance in *Drosophila melanogaster* from eastern Australia. *Functional Ecology*, 19, 222–27.

Holbrook, G. & Schal, C. (2004). Maternal investment affects offspring phenotypic plasticity in a viviparous cockroach. *Proceedings of the National Academy of Science, USA*, 101, 5595–97.

Holldobler, B. & Wilson, E. O. (1990). *The Ants*, Cambridge, Harvard University Press.

Holldobler, B. & Wilson, E. O. (2010). *The Leafcutter Ants: Civilization by Instinct*, New York, W.W. Norton and Company.

Holman, G. M., Cook, B. J., & Nachman, R. J. (1986). Isolation, primary structure and synthesis of 2 neuropeptides from *Leucophaeae maderae* – members of a new family of cephalomyotropins. *Comparative Biochemistry and Physiology C: Pharmacology, Toxicology and Endocrinology*, 84, 205–11.

Holman, G. M., Nachman, R. J., & Coast, G. M. (1999). Isolation, characterization and biological activity of a diuretic myokinin neuropeptide from the housefly, *Musca domestica*. *Peptides*, 20, 1–10.

Honegger, B., Galic, M., Kohler, K., Wittwer, F., Brogiolo, W., Hafen, E., & Stocker, H. (2008). Imp-L2, a putative homolog of vertebrate IGF-binding protein 7, counteracts insulin signaling in Drosophila and is essential for starvation resistance. *Journal of Biology*, 7, 10.

Honek, A. (1993). Intraspecific variation in body size and fecundity in insects: a general relationship. *Oikos*, 66, 483–92.

Hong, S. T., Bang, S., Paik, D., Kang, J. K., Hwang, S., Jeon, K., Chun, B., Hyun, S., Lee, Y., & Kim, J. (2006). Histamine and its receptors modulate temperature-preference behaviors in Drosophila. *Journal of Neuroscience*, 26, 7245–56.

Hosler, J. S., Burns, J. E., & Esch, H. E. (2000). Flight muscle resting potential and species-specific differences in chill-coma. *Journal of Insect Physiology*, 46, 621–27.

Hosokawa, T., Kikuchi, Y., Nikoh, N., Shimada, M., & Fukatsu, T. (2006). Strict host-symbiont cospeciation and reductive genome evolution in insect gut bacteria. *PLoS Biol,* 4, e337.

Houchmandzadeh, B., Wieschaus, E., & Leibler, S. (2002). Establishment of developmental precision and proportions in the early *Drosophila* embryo. *Nature,* 415, 798–802.

Houlihan, D. F. (1969). Respiratory physiology of the larvae of *Donacia simplex,* a root-piercing beetle. *Journal of Insect Physiology,* 15, 1517–36.

Houlihan, D. F. (1970). Respiration in low oxygen partial pressures: the adults of *Donacia simplex* that respire from the roots of aquatic plants. *Journal of Insect Physiology,* 16, 1607–22.

Huang, H. & Haddad, G. G. (2007). *Drosophila* dMRP4 regulates responsiveness to O_2 deprivation and development under hypoxia. *Physiol. Genomics,* 29, 260–66.

Hubert, J.-F., Thomas, D., Cavalier, A., & Gouranton, J. (1989). Structural and biochemical observations on specialized membranes of the "filter chamber", a water-shunting complex in sap-sucking homopteran insects. *Biology of the Cell,* 66, 155–63.

Huberty, A. F. & Denno, R. F. (2006a). Consequences of nitrogen and phosphorus limitation for the performance of two planthoppers with divergent life history strategies. *Oecologia,* 149, 444–59.

Huberty, A. F. & Denno, R. F. (2006b). Trade-off in investment between dispersal and ingestion capability in phytophagous insects and its ecological implications. *Oecologia,* 148, 226–34.

Huesmann, G., Cheung, C., Loi, P., Lee, T., Swiderek, K., & Tublitz, N. J. (1995). Amino acid sequence of CAP 2b, an insect cardioacceleratory peptide from the tobacco hawkmoth *Manduca sexta. FEBS Letters,* 371, 311–14.

Huey, R. B. & Berrigan, D. (1996). Testing evolutionary hypotheses of acclimation. *In:* Johnston, I. A. & Bennett, A. F. (eds.) *Animals and Temperature. Phenotypic and Evolutionary Adaptation.* Cambridge: Cambridge University Press.

Huey, R. B., Berrigan, D., Gilchrist, G. W., & Herron, J. C. (1999). Testing the adaptive significance of acclimation: a strong inference approach. *American Zoologist,* 39, 323–36.

Huey, R. B., Crill, W. D., Kingsolver, J. G., & Weber, K. E. (1992). A method for rapid measurement of heat or cold resistance of small insects. *Functional Ecology,* 6, 489–94.

Huey, R. B., Gilchrist, G. W., Carlson, M. L., Berrigan, D., & Serra, L. (2000). Rapid evolution of a geographic cline in size in an introduced fly. *Science,* 287, 308–09.

Huey, R. B. & Kingsolver, J. G. (1989). Evolution of thermal sensitivity of ectotherm performance. *Trends in Ecology and Evolution,* 4, 131–35.

Hulbert, A. & Else, P. (2004). Basal metabolic rate: history, composition, regulation, and usefulness. *Physiological and Biochemical Zoology,* 77, 869–76.

Hunt, J. H. & Nalepa, C. A. (1994). *Nourishment and Evolution in Insect Societies,* Boulder, Westview Press.

Husband, R. W. & Shina, R. N. (1970). A review of the genus *Locustacarus* with a key to the genera of the family Podapolipidae (Acarina). *Ann. Entomol. Soc. Am.,* 63, 1152–62.

Hustert, R. (1975). Neuromuscular coordination and proprioceptive control of rhythmical abdominal ventilation in intact *Locusta migratoria migratorioides. Journal of Comparative Physiology,* 97, 159–79.

Hustert, R. (1999). Accessory hemolymph pump in the mesothoracic legs of locusts, (*Schistocerca gregaria forskal*) (Orthoptera, Acrididae). *International Journal of Insect Morphology & Embryology,* 28, 91–96.

Hutchinson, G. E. (1981). Thoughts on aquatic insects. *BioScience,* 31, 495–500.

Ichiki, R. & Shima, H. (2003). Immature life of *Compsilura concinnata* (Meigen) (Diptera: Tachinidae). *Annals of the Entomological Society of America,* 96, 161–67.

Inoshita, T. & Tanimura, T. (2006). Cellular identification of water gustatory receptor neurons and their central projection pattern in *Drosophila. Proceedings of the National Academy of Sciences of the United States of America,* 103, 1094–99.

Irwin, J. T. & Lee, R. E. J. (2003). Cold winter microenvironments conserve energy and improve overwintering survival and potential fecundity of the goldenrod gall fly, *Eurosta solidaginis. Oikos,* 100, 71–78.

Itoh, T., Yokohari, F., & Tominaga, Y. (1984). 2 types of antennal hygroreceptive and thermoreceptive sensilla of the cricket, *Gryllus bimaculatus* (De Geer). *Zoological Science,* 1, 533–43.

Iwasaki, M., Itoh, T., Yokohari, F., & Tominaga, Y. (1995). Identification of antennal hygroreceptive sensillum and other sensilla of the firefly, *Luciola cruciata. Zoological Science,* 12, 725–32.

Jablonka, E. & Gaz, R. (2009). Transgenerational epigenetic inheritance: prevalance, mechanisms, and implications for the study of heredity and evolution. *The Quarterly Review of Biology,* 84, 131–76.

Jacobsen, D. & Brodersen, K. P. (2008). Are altitudinal limits of equatorial stream insects reflected in their respiratory performance? *Freshwater Biology,* 53, 2295–308.

Jagge, C. L. & Pietrantonio, P. V. (2008). Diuretic hormone 44 receptor in Malpighian tubules of the mosquito *Aedes aegypti*: evidence for transcriptional regulation paralleling urination. *Insect Molecular Biology,* 17, 413–26.

James, P. E., Goda, F., Grinberg, O. Y., Szybinski, K. G., & Swartz, H. M. (1997). Intrarenal pO(2) measured by EPR oximetry and the effects of bacterial endotoxin. *In:* Nemoto, E. M. & Lamanna, J. C. (eds.) *Oxygen Transport to Tissue Xviii.*

Janz, N., Nyblom, K., & Nylin, S. (2001). Evolutionary dynamics of host-plant specialization: a case study of the tribe Nymphalini. *Evolution,* 55, 783–96.

Jarecki, J., Johnson, E., & Krasnow, M. A. (1999). Oxygen regulation of airway branching in *Drosophila* is mediated by branchless FGF. *Cell,* 99, 211–20.

Jensen, L. T., Cockerell, F. E., Kristensen, T. N., Rako, L., Loeschcke, V., Mckechnie, S. W., & Hoffmann, A. A. (2010). Adult heat tolerance variation in *Drosophila melanogaster* is not related to Hsp70 expression. *Journal of Experimental Zoology Part A- Ecological Genetics and Physiology,* 313A, 35–44.

Jensen, L. T., Nielsen, M. M., & Loeschcke, V. (2008). New candidate genes for heat resistance in *Drosophila melanogaster* are regulated by HSF. *Cell Stress & Chaperones,* 13, 177–82.

Jiggins, C. & Bridle, J. (2004). Speciation in the apple maggot fly: a blend of vintages? *Trends in Ecology & Evolution,* 19, 111–14.

Joern, A. & Behmer, S. T. (1997). Importance od dietary nitrogen and carbohydrates to survival, growth, and reproduction in adults of the grasshopper *Ageneotettix deorum* (Orthoperera : Acrididae). *Oecologia,* 112, 201–08.

Joern, A. & Behmer, S. T. (1998). Impact of diet quality on demographic attributes in adult grasshoppers and nitrogen limitation hypothesis. *Ecological Entomology,* 23, 174–84.

Johnson, C. G. (1969). *Migration and Dispersal of Insects by Flight,* London, Methuen.

Johnson, E., Ringo, J. M., & Dowse, H. B. (1997). Modulation of *Drosophila* heartbeat by neurotransmitters. *Journal of Comparative Physiology B,* 167, 89–97.

Johnson, E. C., Bohn, L. M., & Taghert, P. H. (2004). Drosophila CG8422 encodes a functional diuretic hormone receptor. *Journal of Experimental Biology,* 207, 743–48.

Johnson, K. S. & Felton, G. W. (1996a). Physiological and dietary influences on midgut redox conditions in generalist lepidopteran larvae. *Journal of Insect Physiology,* 42, 191–98.

Johnson, K. S. & Felton, G. W. (1996b). Potential influence of midgut pH and redox potential on protein utilization in insect herbivores. *Archives of Insect Biochemistry and Physiology,* 32, 85–105.

Johnson, K. S. & Felton, G. W. (2000). Digestive proteinase activity in corn earworm (Helicoverpa zea) after molting and in response to lowered redox potential. *Archives of Insect Biochemistry and Physiology,* 44, 151–61.

Johnson, S. A. & Nicolson, S. W. (2001). Pollen digestion by flower-feeding Scarabaeidae: protea beetles (Cetoniini) and monkey beetles (Hopliini). *Journal of Insect Physiology,* 47, 725–33.

Jonas, J. L. & Joern, A. (2008). Host-plant quality alters grass/forb consumption by a mixed-feeding insect herbivore, *Melanoplus bivittatus* (Orthoptera: Acrididae). *Ecological Entomology,* 33, 546–54.

Jones, E. I. (2010). Optimal foraging when predation risk increases with patch resources: an analysis of pollinators and ambush predators. *Oikos,* 119, 835–40.

Jones, J. (1977). The circulatory system of insects. *Contraction of circulatory pumps.* Springfield, IL: Thomas.

Joos, B., Lighton, J. R. B., Harrison, J. F., Suarez, R. K., & Roberts, S. P. (1997). Effects of ambient oxygen tension on flight performance, metabolism, and water loss of the honeybee. *Physiological Zoology,* 70, 167–74.

Joplin, K. H., Yocum, G. D., & Denlinger, D. L. (1990). Diapause specific proteins expressed by the brain during the pupal diapause of the flesh fly, *Sarcophaga-crassipalpis. Journal of Insect Physiology,* 36, 775–83.

Junger, M., Rintelen, F., Stocker, H., Wasserman, J., Vegh, M., Radimerski, T., Greenberg, M., & Hafen, E. (2003). The *Drosophila* Forkhead transcription factor FOXO mediates the reduction in cell number associated with reduced insulin signaling. *Journal of Biology,* 2, 20.

Kaelin, W. G. J. & Ratcliffe, P. J. (2008). Oxygen sensing by Metazoans: the central role of the HIF hydroxylase pathway. *Molecular Cell,* 30, 393–402.

Kaiser, A., Klok, C. J., Socha, J. J., Lee, W.-K., Quinlan, M. C., & Harrison, J. F. (2007). Increase in tracheal investment with beetle size supports hypothesis of oxygen limitation on insect gigantism. *Proceedings of the National Academy of Sciences,* 104, 13198–203.

Kane, M. D. & Breznak, J. A. (1991). Effect of host diet on production of oranic-acids and methane by cockroach gut bacteria. *Applied and Environmental Microbiology,* 57, 2628–34.

Karan, D., Morin, J. P., Moreteau, B., & David, J. R. (1998). Body size and developmental temperature in *Drosophila melanogaster*: Analysis of body weight reaction norm. *Journal of Thermal Biology,* 23, 301–09.

Karasov, W. H. & Diamond, J. M. (1983). A simple method for measuring intestinal solute uptake *in vitro Journal of Comparative Physiology,* 152, 105–16.

Karasov, W. H. & Martinez Del Rio, C. (2007). *Physiological Ecology. How Animals Process Energy, Nutrients and Toxins,* Princeton, N.J., Princeton University Press.

Karban, R. (2011). The ecology and evolution of induced resistance against herbivores. *Functional Ecology,* 25, 339–47.

Karban, R., Shiojiri, K., Huntzinger, M., & Mccall, A. C. (2006). Damage-induced resistance in sagebrush: volatiles are key to intra- and interplant communication. *Ecology,* 87, 922–30.

Karl, I., Stoks, R., De Block, M., Janowitz, S. A., & Fischer, K. (2011). Temperature extremes and butterfly fitness: conflicting evidence from life history and immune function. *Global Change Biology,* 17, 676–87.

Karley, A. J., Ashford, D. A., Minto, L. B., Pritchard, J., & Douglas, A. E. (2005). The significance of gut sucrase activity for osmoregulation in the pea aphid, *Acyrthosiphon pisum. Journal of Insect Physiology,* 51, 1313–19.

Karowe, D. N. & Martin, M. M. (1989). The effects of quantity and quality of diet nitrogen on the growth, efficiency of food utilization, nitrogen budget, and metabolic rate of fifth instar *Spodoptera eridania* (Lepidoptera:Noctuidae). *Journal of Insect Physiology,* 35, 699–708.

Karunanithi, S., Barclay, J. W., Robertson, R. M., Brown, I. R., & Atwood, H. L. (1999). Neuroprotection at *Drosophila* synapses conferred by prior heat shock. *Journal of Neuroscience,* 19, 4360–69.

Kaspari, M., Yanoviak, S. P., & Dudley, R. (2008). On the biogeography of salt limitation: A study of ant communities. *Proceedings of the National Academy of Sciences,* 105, 17848–51.

Kathirithamby, J. (2009). Host-parasitoid associations in Strepsiptera. *Annual Review of Entomology,* 54, 227–49.

Katz, S. L. & Gosline, J. M. (1992). Ontogenentic scaling and mechanical behaviour of the tibiae of the African desert locust (*Schistocerca gregaria*). *Journal of Experimental Biology,* 168, 125–50.

Katz, S. L. & Gosline, J. M. (1994). Scaling modulus as a degree of freedom in the design of locust legs. *Journal of Experimental Biology,* 187, 207–23.

Kaufman, M. G. & Klug, M. J. (1991). The contribution of hindgut bacteria to dietary carbohydrate utilization by crickets (Orthoptera: Gryllidae). *Comparative Biochemistry and Physiology Part A: Physiology,* 98, 117–23.

Kaufmann, N., Mathai, J. C., Hill, W. G., Dow, J. A. T., Zeidel, M. L., & Brodsky, J. L. (2005). Developmental expression and biophysical characterization of a *Drosophila melanogaster* aquaporin. *American Journal of Physiology: Cellular Physiology,* 289, C397–407.

Keeley, L. L. (1985). Biochemistry and physiology of the insect fat body. *In:* Kerkut, G. A. & Gilbert, L. I. (eds.) *Comprehensive Insect Physiology, Biochemistry and Pharmacology.* New York: Pergamon.

Kelty, J. D. & Lee Jr, R. E. (2001). Rapid cold-hardening of *Drosophila melanogaster* (Diptera: Drosophilidae) during ecologically based thermoperiodic cycles. *Journal of Experimental Biology,* 204, 1659–66.

Kelty, J. D. & Lee, R. E. (1999). Induction of rapid cold hardening by cooling at ecologically relevant rates in *Drosophila melanogaster. Journal of Insect Physiology,* 45, 719–26.

Ken, T., Hepburn, H. R., Radloff, S. E., Yu, Y. S., Liu, Y. Q., Zhou, D. Y., & Neumann, P. (2005). Heat-balling wasps by honeybees. *Naturwissenschaften,* 92, 492–95.

Kerman, B. E., Cheshire, A. M., & Andrew, D. J. (2006). From fate to function: the *Drosophila* trachea and salivary gland as models for tubulogenesis. *Differentiation,* 74, 326–48.

Kerr, M., Davies, S. A., & Dow, J. A. T. (2004). Cell-specific manipulation of second messengers: a toolbox for integrative physiology in *Drosophila. Current Biology,* 14, 1468–74.

Kestler, P. (1985). Respiration and respiratory water loss. *In:* Hoffmann, K. H. (ed.) *Environmental Physiology and Biochemistry of Insects.* Berlin: Springer.

Khare, P. V., Burnabas, R. J., Kanojiya, M., Kulkarni, A. D., & Joshi, D. S. (2002). Temperature dependent eclosion rhythmicity in the high altitude Himalayan strains of *Drosophila ananassae. Chronobiol. Int.,* 19, 1041–52.

Kikawada, T., Saito, A., Kanamori, Y., Fujita, M., Snigorska, K., Watanabe, M., & Okuda, T. (2008). Dehydration-inducible changes in expression of two aquaporins in the sleeping chironomid, *Polypedilum vanderplanki*. *Biochimica Et Biophysica Acta-Biomembranes,* 1778, 514–20.

Kim, J., Chung, Y. D., Park, D. Y., Choi, S., Shin, D. W., Soh, H., Lee, H. W., Son, W., Yim, J., Park, C. S., Kernan, M. J., & Kim, C. (2003). A TRPV family ion channel required for hearing in *Drosophila. Nature,* 424, 81–84.

King-Jones, K. & Thummel, C. S. (2005). Less steroids make bigger flies. *Science,* 310, 630–31.

King, W. W. & Hadley, N. F. (1979). Water flux and metabolic rates of free-roaming scorpions using the doubly labeled water technique. *Physiological Zoology,* 52, 176–89.

Kingsolver, J. & Huey, R. B. (2008). Size, temperature and fitness. Three rules. *Evolutionary Ecology Research,* 10, 251–68.

Kingsolver, J. G. (1995a). Fitness consequences of seasonal polyphenism in western white butterflies. *Evolution,* 49, 942–54.

Kingsolver, J. G. (1995b). Viability selection on seasonally polyphenic traits: wing melanin pattern in western white butterflies. *Evolution,* 49, 932–41.

Kingsolver, J. G. (1996). Experimental manipulation of wing pigment pattern and survival in western white butterflies. *American Naturalist,* 147, 296–306.

Kingsolver, J. G. & Hedrick, T. L. (2008). Biomechanical acclimation: Flying cold. *Current Biology,* 18, R876–76.

Kingsolver, J. G. & Huey, R. B. (1998). Evolutionary analyses of morphological and physiological plasticity in thermally variable environments. *American Zoologist,* 38, 545–60.

Kingsolver, J. G. & Moffat, R. J. (1982). Thermoregulation and the determinants of heat-transfer in *Colias* butterflies. *Oecologia,* 53, 27–33.

Kingsolver, J. G. & Pfennig, D. W. (2004). Individual-level selection as a cause of Cope's rule of phyletic size increase. *Evolution,* 58, 1608–12.

Kingsolver, J. G. & Wiernasz, D. C. (1991). Development, function, and the quantitative genetics of wing melanin pattern in *Pieris* butterflies. *Evolution,* 45, 1480–92.

Kingsolver, J. G., Woods, H. A., Buckley, L. B., Potter, K. A., Maclean, H., & Higgins, J. K. 2011. Complex life cycles and the responses of insects to climate change. *Integrative and Comparative Biology*, 51, 719–732.

Kirk, W. D. J. (1987). How much pollen can thrips destroy? *Ecological Entomology,* 12, 31–40.

Kirkton, S. D. (2007). Effects of insect body size on tracheal structure and function. *Advances in Experimental Medicine and Biology,* 618, 221–28.

Kleinhenz, M., Bujok, B., Fuchs, S., & Tautz, H. (2003). Hot bees in empty broodnest cells: heating from within. *Journal of Experimental Biology,* 206, 4217–31.

Klok, C. J. & Chown, S. L. (2005). Temperature- and body mass-related variation in cyclic gas exchange characteristics and metabolic rate of seven weevil species: Broader implications. *Journal of Insect Physiology,* 51, 789–801.

Klok, C. J., Hubb, A. J., & Harrison, J. F. (2009). Single and multigenerational responses of body mass to atmospheric oxygen concentrations in *Drosophila melanogaster*: evidence for roles of plasticity and evolution. *Journal of Evolutionary Biology,* 22, 2496–504.

Klok, C. J., Sinclair, B. J., & Chown, S. L. (2004). Upper thermal tolerance and oxygen limitation in terrestrial arthropods. *Journal of Experimental Biology,* 207, 2361–70.

Klose, M. K., Armstrong, G., & Robertson, R. M. (2004). A role for the cytoskeleton in heat-shock-mediated thermoprotection of locust neuromuscular junctions. *Journal of Neurobiology,* 60, 453–62.

Klose, M. K., Atwood, H. L., & Robertson, R. M. (2008). Hyperthermic preconditioning of presynaptic calcium regulation in *Drosophila*. *Journal of Neurophysiology,* 99, 2420–30.

Klose, M. K., Boulianne, G. L., Robertson, R. M., & Atwood, H. L. (2009). Role of ATP-dependent calcium regulation in modulation of *Drosophila* synaptic thermotolerance. *Journal of Neurophysiology,* 102, 901–13.

Klowden, M. J. (2009). *Physiological Systems in Insects,* New York, Academic Press.

Kocmarek, A. L. & O'donnell, M. J. (2011). Potassium fluxes across the blood brain barrier of the cockroach, *Periplaneta americana*. *Journal of Insect Physiology,* 57, 127–35.

Kölsch, G. (2001). Anoxia tolerance and anaerobic metabolism in two tropical weevil species (Coleoptera, Curculionidae). *Journal of Comparative Physiology B,* 171, 595–602.

Kolsch, G., Jakobi, K., Wegener, G., & Braune, H. J. (2002). Energy metabolism and metabolic rate of the alder leaf beetle *Agelastica alni* (L.) (Coleoptera, Chrysomelidae) under aerobic and anaerobic conditions: a microcalorimetric study. *Journal of Insect Physiology,* 48, 143–51.

Kolsch, G., Krause, A., Goetz, N., & Plagmann, S. (2010). The salinity preference of members of the genus Macroplea (Coleoptera, Chrysomelidae, Donaciinae), fully aquatic leaf beetles that occur in brackish water. *Journal of Experimental Marine Biology and Ecology,* 390, 203–09.

Komai, Y. (1998). Augmented respiration in a flying insect. *Journal of Experimental Biology,* 201, 2359–66.

Komai, Y. (2001). Direct measurement of oxygen partial pressure in a flying bumblebee. *Journal of Experimental Biology,* 204, 2999–3007.

Konno, K., Hirayama, C., Shinbo, H., & Nakamura, M. (2009). Glycine addition improves feeding performance of non-specialist herbivores on the privet, *Ligustrum obtusifolium*: In vivo evidence for the physiological impacts of anti-nutritive plant defense with iridoid and insect adaptation with glycine. *Applied Entomology and Zoology,* 44, 595–601.

Konno, K., Hiyama, C., Yasui, H., Okada, D., Sugimura, M., Yukuhiro, F., Tamura, Y., Hattori, M., Shinbo, H., & Nakamura, M. (2010). GABA, β-Alanine and glycine in the digestive juice of privet-specialist insects: convergent adaptive traits against plant iridoids. *Journal of Chemical Ecology,* 9, 983–91.

Konno, K., Okada, S., & Hirayama, C. (2001). Selective secretion of free glycine, a neutralizer against a plant defense chemical, in the digestive juice of the privet moth larvae. *Journal of Insect Physiology,* 47, 1451–57.

Kopf, A., Rank, N. E., Roininen, H., Julkunen-Tiito, R., Pasteels, J. M., & Tahvanainen, J. (1998). The evolution of host-plant use and sequestration in the leaf beetle genus *Phratora* (Coleoptera:Chrysomelidae). *Evolution,* 52, 517–28.

Kopp, A., Barmina, O., Hamilton, A. M., Higgins, L., Mcintyre, L. M., & Jones, C. D. (2008). Evolution of gene expression in the *Drosophila* olfactory system. *Molecular Biology and Evolution,* 25, 1081–92.

Koricheva, J., Larsson, S., Haukioja, E., & Keinanen, M. (1998). Regulation of wood plant secondary metabolism by resource availability: hypothesis testing by means of meta-analysis. *Oikos,* 83, 212–26.

Korner, P. & Schmid-Hempel, P. (2005). Correlates of parasite load in bumblebees in an Alpine habitat. *Entomological Science,* 8, 151–60.

Kos, M., Broekgaarden, C., Kabouw, P., Oude Lenferink, K., Poelman, E. H., Vet, L. E. M., Dicke, M., & Van Loon, J. J. A. (2011). Relative importance of plant-mediated bottom-up and top-down forces on herbivore abundance on *Brassica oleracea. Functional Ecology,* 1365–2435.

Kosola, K. R., Dickmann, D. I., Paul, E. A., & Parry, D. (2001). Repeated insect defoliation effects on growth, nitrogen acquisition, carbohydrates, and root demograhy of poplars. *Oecologia,* 129, 65–74.

Kostal, V., Berkova, P., & Simek, P. (2003). Remodelling of membrane phospholipids during transition to diapause and cold-acclimation in the larvae of Chymomyza costata (Drosophilidae). *Comparative Biochemistry and Physiology B-Biochemistry & Molecular Biology,* 135, 407–19.

Kostal, V., Renault, D., Mehrabianova, A., & Bastl, J. (2007). Insect cold tolerance and repair of chill-injury at fluctuating thermal regimes: Role of ion homeostasis. *Comparative Biochemistry and Physiology a-Molecular & Integrative Physiology,* 147, 231–38.

Kostal, V. & Simek, P. (1998). Changes in fatty acid composition of phospholipids and triacylglycerols after cold-acclimation of an aestivating insect prepupa. *Journal of Comparative Physiology B-Biochemical Systemic and Environmental Physiology,* 168, 453–60.

Kostal, V. & Tollarova-Borovanska, M. (2009). The 70 kDa heat shock protein assists during the repair of chilling injury in the insect, Pyrrhocoris apterus. *PLOS One,* 4.

Kostal, V., Vambera, J., & Bastl, J. (2004). On the nature of pre-freeze mortality in insects: water balance, ion homeostasis and energy charge in the adults of *Pyrrhocoris apterus. Journal of Experimental Biology,* 207, 1509–21.

Krebs, R. A. & Feder, M. E. (1997a). Deleterious consequences of Hsp70 overexpression in *Drosophila melanogaster* larvae. *Cell Stress & Chaperones,* 2, 60–71.

Krebs, R. A. & Feder, M. E. (1997b). Natural variation in the expression of the heat-shock protein HSP70 in a population of *Drosophila melanogaster* and its correlation with tolerance of ecologically relevant thermal stress. *Evolution,* 51, 173–79.

Kreiss, E. J., Schmitz, A., & Schmitz, H. (2005). Morphology of the prothoracic discs and associated sensilla of Acanthocnemus nigricans (Coleoptera, Acanthocnemidae). *Arthropod Structure & Development,* 34, 419–28.

Krogh, A. (1919). The rate of diffusion of gases through animal tissues, with some remarks on the coefficient of invasion. *Journal of Physiology,* 52, 391–408.

Krogh, A. (1920a). Studien über die Tracheenrespiration. III. Die Kombination von mechanischer Ventilation mit Gasdiffusion nach Versuchen an Dytiscidenlarven. *Pflügers Archiv,* 179, 113–20.

Krogh, A. (1920b). Studien über Tracheenrespiration. II. über Gasdiffusion in den Tracheen. *Pflügers Archiv,* 179, 95–112.

Krolikowski, K. & Harrison, J. F. (1996). Haemolymph acid-base status, tracheal gas levels and the control of post-exercise ventilation rate in grasshoppers. *Journal of Experimental Biology,* 199, 391–99.

Kuwasawa, K., Ai, H. & Matsushita, T. (1999). Cardiac reflexes and their neural pathways in lepidopterous insects. *Comparative Biochemistry and Physiology A,* 124, 581–86.

Labandeira, C. C. (1997). Insect mouthparts: ascertaining the paleobiology of insect feeding strategies. *Annual Review of Ecology and Systematics,* 28, 153–93.
Labandeira, C. C. (1998). Early history of arthropod and vascular plant associations. *Annual Review of Earth Planetary Sciences,* 26, 329–77.
Lachenicht, M. W., Clusella-Trullas, S., Boardman, L., Le Roux, C., & Terblanche, J. S. (2010). Effects of acclimation temperature on thermal tolerance, locomotion performance and respiratory metabolism in *Acheta domesticus* L. (Orthoptera: Gryllidae). *Journal of Insect Physiology,* 56, 822–30.
Lang, F., Busch, G. L., Ritter, M., Volkl, H., Waldegger, S., Gulbins, E., & Haussinger, D. (1998). Functional significance of cell volume regulatory mechanisms. *Physiol. Rev.,* 78, 247–306.
Lange, A. B. & Cheung, I. L. (1999). The modulation of skeletal muscle contraction by FMRFamide-related peptides of the locust. *Peptides,* 20, 1411–18.
Langlais, K. K., Stewart, J. A., & Morton, D. B. (2004). Preliminary characterization of two atypical soluble guanylyl cyclases in the central and peripheral nervous system of *Drosophila melanogaster. Journal of Experimental Biology,* 207, 2323–38.
Larsen, T., Ventura, M., O'brien, D. M., Magid, J., Lomstein, B. A., & Larsen, J. (2011). Contrasting effects of nitrogen limitation and amino acid imbalance on carbon and nitrogen turnover in three species of Collembola. *Soil Biology and Biochemistry,* 43, 749–59.
Laurent, S., Masson, C., & Jakob, I. (2002). Whole-cell recording from honeybee olfactory receptor neurons: ionic currents, membrane excitability and odourant response in developing workerbee and drone. *European Journal of Neuroscience,* 15, 1139–52.
Law, J. H. & Wells, M. A. (1989). Insects as biochemical models. *Journal of Biological Chemistry,* 264, 16335–38.
Le Cahérec, F., Deschamps, S., Delamarche, C., Pellerin, I., Bonnec, G., Guillam, M.-T., Thomas, D., Gouranton, J., & Hubert, J.-F. (1996). Molecular cloning and characterization of an insect aquaporin. *European Journal of Biochemistry,* 241, 707–15.
Le Caherec, F., Guillam, M.-T., Beuron, F., Cavalier, A., Thomas, D., Gouranton, J., & Hubert, J.-F. (1997). Aquaporin-related proteins in the filter chamber of homopteran insects. *Cell and Tissue Research,* 290, 143–51.
Le Cahérec, F., Guillam, M.-T., Beuron, F., Cavalier, A., Thomas, D., Gouranton, J., & Hubert, J.-F. (1997). Aquaporin-related proteins in the filter chamber of homopteran insects. *Cell and Tissue Research,* 290, 143–51.
Lease, H. M. & Wolf, B. O. (2010). Exoskeletal chitin scales isometrically with body size in terrestrial insects. *Journal of Morphology,* 271, 759–68.
Lease, H. M., Wolf, B. O., & Harrison, J. F. (2006). Intraspecific variation in tracheal volume in the American locust, *Schistocerca americana,* measured by a new inert gas method. *Journal of Experimental Biology,* 209, 3476–83.
Lecaherec, F., Bron, P., Verbavatz, J. M., Garret, A., Morel, G., Cavalier, A., Bonnec, G., Thomas, D., Gouranton, J., & Hubert, J. F. (1996). Incorporation of proteins into (*Xenopus*) oocytes by proteoliposome microinjection: functional characterization of a novel aquaporin. *Journal of Cell Science,* 109, 1285–95.
Lee Jr, R. E. (1991). Principles of insect low temperature tolerance. *In:* Lee Jr, R. E. & Denlinger, D. L. (eds.) *Insects at Low Temperature.* New York and London: Chapman and Hall.

Lee, K. P., Behmer, S. T., Simpson, S. J., & Raubenheimer, D. (2002). A geometric analysis of nutrient regulation in the generalist caterpillar *Spodopteris littoralis* (Boisduval). *Journal of Insect Physiology,* 48, 655–65.

Lee, K. P. & Roh, C. (2010). Temperature-by-nutrient interactions affecting growth rate in an insect ectotherm. *Entomologia Experimentalis Et Applicata,* 136, 151–63.

Lee, K. S., Kim, S. R., Lee, S. M., Lee, K. R., Sohn, H. D., & Jin, B. R. (2001). Molecular cloning and expression of a cDNA encoding the aquaporin homologue from the firefly, *Pyrocoelia rufa. Korean Journal of Entomology,* 31, 269–80.

Lee, W. K. & Socha, J. J. (2009). Direct visualization of hemolymph flow in the heart of a grasshopper (*Schistocerca americana*). *BMC Physiology,* 9, 2.

Lee, Y., Lee, J., Bang, S., Hyun, S., Kang, J., Hong, S. T., Bae, E., Kaang, B. K., & Kim, J. (2005). Pyrexia is a new thermal transient receptor potential channel endowing tolerance to high temperatures in Drosophila melanogaster. *Nature Genetics,* 37, 305–10.

Legaspi, J. C. & Legaspi, B. C. (2005). Life table analysis for Podisus maculiventris immatures and female adults under four constant temperatures. *Environmental Entomology,* 34, 990–98.

Lehane, M. J. (2005). Managing the blood meal. *In:* Lehane, M. J. (ed.) *Biology of Blood-Sucking in Insects.* Cambridge: Cambridge University Press.

Lehane, M. J., Blakemore, D., Williams, S., & Moffatt, M. R. (1995). Regulation of digestive enzyme levels in insects. *Comparative Biochemistry and Physiology Part B: Biochemistry and Molecular Biology,* 110, 285–89.

Lehmann, F.-O. (2001). Matching spiracle opening to metabolic need during flight in *Drosophila. Science,* 294, 1926–29.

Lehmann, F.-O. (2002a). The constraints of body size on aerodynamics and energetics in flying fruit flies: an integrative view. *Zoology,* 105, 287–95.

Lehmann, F.-O. (2002b). The control of breathing at elevated locomotor activity in flying fruit flies. *Integrative and Comparative Biology,* 42, 1265–65.

Lehmann, F.-O. & Dickinson, M. H. (2001). The production of elevated flight force compromises manoeuvrability in the fruit fly *Drosophila melanogaster. Journal of Experimental Biology,* 204, 627–35.

Lehmann, F.-O., Dickinson, M. H., & Staunton, J. (2000). The scaling of carbon dioxide release and respiratory water loss in flying fruit flies (*Drosophila* spp.). *Journal of Experimental Biology,* 203, 1613–24.

Lehmann, F. O. & Heymann, N. (2005). Unconventional mechanisms control cyclic respiratory gas release in flying *Drosophila. Journal of Experimental Biology,* 208, 3645–54.

Lehmann, F. O. & Heymann, N. (2006). Dynamics of in vivo power output and efficiency of Nasonia asynchronous flight muscle. *Journal of Biotechnology,* 124, 93–107.

Leigh, A., Clse, J. D., Ball, M. C., Siebke, K., & Nicotra, A. B. (2006). Research note: Leaf cooling curves: measuring leaf temperature in sunlight. *Functional Plant Biology,* 33, 515–19.

Lelagadec, M. D., Chown, S. L., & Scholtz, C. H. (1998). Desiccation resistance and water balance in southern African keratin beetles (Coleoptera, Trogidae): The Influence of body size and habitat. *Journal of Comparative Physiology B,* 168, 112–22.

Lemos, F. J. A. & Terra, W. R. (1991). Digestion of bacteria and the role of midgut lysozyme in some insect larvae. *Comparative Biochemistry and Physiology Part B: Comparative Biochemistry,* 100, 265–68.

Leonhard, B. & Crailsheim, K. (1999). Amino acids and osmolarity in honeybee drone and haemolymph. *Amino Acids,* 17, 195–205.
Levenbrook, L. (1958). Intracellular water of larval tissues of the southern armyworm as determined by the use of C^{14}-carboxyl-inulin. *Journal of Cellular and Comparative Physiology,* 52, 329–39.
Levy, R. I. & Schneiderman, H. A. (1958). An experimental solution to the paradox of discontinuous respiration in insects. *Nature,* 23, 491–93.
Lewis, A. C. & Bernays, E. A. (1985). Feeding-behavior – selection of both wet and dry food for increased growth in *Schistocerca-gregaria* Nymphs. *Entomologia Experimentalis et Applicata,* 37, 105–112.
Li, A. Q., Popova-Butler, A., Dean, D. H., & Denlinger, D. L. (2007). Proteomics of the flesh fly brain reveals an abundance of upregulated heat shock proteins during pupal diapause. *Journal of Insect Physiology,* 53, 385–91.
Li, W., Schuler, M. A., & Berenbaum, M. R. (2003). Diversification of furanocoumarin-metabolizing cytochrome P450 monooxygenases in two papilionids: Specificity and substrate encounter rate. *Proceedings of the National Academy of Sciences of the United States of America,* 100, 14593–98.
Lighton, J. R. B. (1996). Discontinuous gas exchange in insects. *Annual Review of Entomology,* 41, 309–24.
Lighton, J. R. B. (2008). *Measuring Metabolic Rates: A Manual for Scientists,* New York, Oxford University Press.
Lighton, J. R. B. & Feener, D. H. (1989). A comparison of energetics and ventilation of desert ants during voluntary and forced locomotion. *Nature,* 342, 174–75.
Lighton, J. R. B., Schilman, P. E., & Holway, D. A. (2004). The hyperoxic switch: assessing respiratory water loss rates in tracheate arthropods with continuous gas exchange. *Journal of Experimental Biology,* 207, 4463–71.
Lighton, J. R. B. & Turner, R. J. (2008). The hygric hypothesis does not hold water: abolition of discontinuous gas exchange cycles does not affect water loss in the ant Camponotus vicinus. *Journal of Experimental Biology,* 211, 563–67.
Lipp, A., Wolf, H., & Lehmann, F. O. (2005). Walking on inclines: energetics of locomotion in the ant *Camponotus. Journal of Experimental Biology,* 208, 707–19.
Liu, K. J., Gast, P., Moussavi, M., Norby, S. W., Vahidi, N., Walczak, T., WU, M., & Swartz, H. M. (1993). Lithium phthalocyanine – a probe for electron paramagnetic resonanance oximetry in viable biological systems. *Proceedings of the National Academy of Sciences of the United States of America,* 90, 5438–42.
Liu, L., Leonard, A. S., Motto, D. G., Feller, M. A., Price, M. P., Johnson, W. A., & Welsh, M. J. (2003) Contribution of *Drosophila* DEG/ENaC genes to salt taste. *Neuron,* 39, 133–46.
Liu, L., Li, Y. H., Wang, R. O., Yin, C., Dong, Q., Hing, H., Kim, C., & Welsh, M. J. (2007). Drosophila hygrosensation requires the TRP channels water witch and nanchung. *Nature,* 450, 294–98.
Liu, Y., Zhou, S., Ma, L., Tian, L., Wang, S., Sheng, Z., Jiang, R.-J., Bendena, W. G., & Li, S. (2010). Transcriptional regulation of the insulin signaling pathway genes by starvation and 20-hydroxyecdysone in the Bombyx fat body. *Journal of Insect Physiology,* 56, 1436–44.
Llimargas, M. & Lawrence, P. A. (2001). even *Wnt* homologues in *Drosophila*: A case study of the developing tracheae. *Proceedings of the National Academy of Sciences. USA,* 98, 14487–92.

Loaiza, V., Jonas, J. L., & Joern, A. (2008). Does dietary P affect feeding and performance in the mixed-feeding grasshopper (Acrididae) *Melanoplus bivittatus*? *Environmental Entomology*, 37, 333–39.

Loaiza, V., Jonas, J. L., & Joern, A. (2010). Grasshoppers (Orthoptera: Acrididae) select vegetation patches in local-scale responses to foliar nitrogen but not phosphorus in native grassland. *Insect Science*, Article 63.

Locke, M. (1958). The co-ordination of growth in the tracheal system of insects. *Quarterly Journal of Microscopical Science*, 99, 373–91.

Locke, M. (1998a). Caterpillars have evolved lungs for hemocyte gas exchange. *Journal of Insect Physiology*, 44, 1–20.

Locke, M. (1998b). The fat body. In: *Microscopic Anatomy of Invertebrates*, F.W Harrison and M. Locke (eds). Wiley-Liss, inc.

Locke, M. (2001). The Wigglesworth Lecture: Insects for studying fundamental problems in biology. *Journal of Insect Physiology*, 47, 495–507.

Locke, M. & Nichol, H. (1992). Iron economy in insects: Transport, metabolism, and storage. *Annual Review of Entomology*, 37, 195–215.

Loeschcke, V. & Sorensen, J. G. (2005). Acclimation, heat shock and hardening – a response from evolutionary biology. *Journal of Thermal Biology*, 30, 255–57.

Loof, A. D. (2008). Ecdysteroids, juvenile hormone and insect neuropeptides: Recent successes and remaining major challenges. *General and Comparative Endocrinology*, 155, 3–13.

Lopez-Vaamonde, C., Godfray, C. J., & Cook, J. M. (2003). Evolutionary dynamics of host-plant use in a genus of leaf-mining moths. *Evolution*, 57, 1804–21.

Lorenz, M. W. (2007). Oogenesis-flight syndrome in crickets: Age-dependent egg production, flight performance, and biochemical composition of the flight muscles in adult female *Gryllus bimaculatus*. *Journal of Insect Physiology*, 53, 819–32.

Loudon, C. (1989). Tracheal hypertrophy in mealworms: design and plasticity in oxygen supply systems. *Journal of Experimental Biology*, 147, 217–35.

Louw, G. N. & Hadley, N. F. (1985). Water economy of the honeybee: a stoichiometric accounting. *Journal of Experimental Zoology*, 235, 147–50.

Louw, G. N. & Nicolson, S. W. (1983). Thermal, energetic and nutritional considerations in the foraging and reproduction of the carpenter bee *Xylocopa capitata*. *Journal of the Entomological Society of Southern Africa*, 46, 227–40.

Luschnig, S., Batz, T., Armbruster, K., & Krasnow, M. A. (2006). *Serpentine* and *vermiform* encode matrix proteins with chitin binding and deacetylation domains that limit tracheal tube length in *Drosophila*. *Current Biology*, 16, 186–94.

Maas, A. H. (1949). über die Auslösbarkeit von Temperatur-modifikationen wÄhrend der Embryonalentwicklung von *Drosophila melanogaster* Meigen. *Development Genes and Evolution*, 143, 515–72.

Macdonald, S. J., Thomas, G. H., & Douglas, A. E. (2011). Genetic and metabolic determinants of nutritional phenotype in an insect–bacterial symbiosis. *Molecular Ecology*, 20, 2073–84.

Machin, J., Kestler, P., & Lampert, G. J. (1992). The effect of brain homogenates on directly measured water fluxes through the pronotum of *Periplaneta americana*. *Journal of Experimental Biology*, 171, 395–408.

Machin, J., Odonnell, M. J., & Coutchie, P. A. (1982). Mechanisms of water-vapor absorption in insects. *Journal of Experimental Zoology*, 222, 309–20.

Macloskie, G. (1884). The structure of the tracheae of insects. *American Naturalist,* 18, 567–73.

Macmillan, H. A., Guglielmo, C. G., & Sinclair, B. J. (2009). Membrane remodeling and glucose in *Drosophila melanogaster*: A test of rapid cold-hardening and chilling tolerance hypotheses. *Journal of Insect Physiology,* 55, 243–49.

Macmillan, H. A. & Sinclair, B. J. (2011a). Mechanisms underlying insect chill-coma. *Journal of Insect Physiology,* 57, 12–20.

Macmillan, H. A. & Sinclair, B. J. (2011b). The role of the gut in insect chilling injury: cold-induced disruption of osmoregulation in the fall field cricket, *Gryllus pennsylvanicus*. *Journal of Experimental Biology,* 214, 726–34.

Maddrell, S. H. (1998). Why are there no insects in the open sea? *Journal of Experimental Biology,* 201, 2461–64.

Maddrell, S. H., Pilcher, D. E. M., & Gardiner, B. O. (1969). Stimulatory effect of 5-hydroxytrptamine (serotonin) on secretion by Malpighian tubules of insects. *Nature,* 222, 784–85.

Maddrell, S. H. P. (1964). Excretion in the blood-sucking bug, *Rhodnius Prolixus* Stal. II. The normal course of diuresis and the effect of temperature. *Journal of Experimental Biology,* 41, 163–76.

Maddrell, S. H. P., Herman, W. S., Farndale, R. W., & Riegel, J. A. (1993). Synergism of hormones controlling epithelial fluid transport in an insect. *Journal of Experimental Biology,* 174, 65–80.

Maddrell, S. H. P., Pilcher, D. E. M., & Gardiner, B. O. C. (1971). Pharmacology of the Malpighian tubules of *Rhodnius* and *Carausius*: the structure-activity relationship of tryptamine analogues and the role of cyclic amp. *Journal of Experimental Biology,* 54, 779–804.

Majerus, M. E. (1998). *Melanism. Evolution in Action,* Oxford, Oxford University Press.

Malamud, J. G., Miszin, A. P., & Josephson, R. K. (1988). The effects of octopamine on contraction kinetics and power output of the locust flight muscle. *Journal of Comparative Physiology,* 165, 827–35.

Mallatt, J. & Giribet, G. (2006). Further use of nearly complete 28S and 18S rRNA genes to classify Ecdysozoa: 37 more arthropods and a kinorhynch. *Molecular Phylogenetics and Evolution,* 40, 772–94.

Maloeuf, N. S. R. (1938). The basis of the rhythmic flashing of the firefly. *Annals of the Entomological Society of America,* 31, 374–80.

Manalo, D. J., Rowan, A., Lavoie, T., Natarajan, L., Kelly, B. D., Ye, S. Q., Garcia, J. G., & Semenza, G. L. (2005). Transcriptional regulation of vascular endothelial cell responses to hypoxia by Hif-1. *Blood,* 105, 659–69.

Manning, G. & Krasnow, M. A. (1993). Development of the *Drosophila* tracheal system. *In:* Bate, M. & Arias, A. M. (eds.) *The Development of* Drosophila melanogaster. Cold Spring Harbor: Cold Spring Harbor Press.

Manson, J. S., Otterstatter, M. C., & Thomson, J. D. (2010). Consumption of a nectar alkaloid reduces pathogen load in bumble bees. *Oecologia,* 162, 81–89.

Mao, W., Rupasinghe, S., Zangerl, A. R., Schuler, M. A., & Berenbaum, M. R. (2006). Remarkable substrate-specificity of CYP6AB3 in *Depressaria pastinacella*, a highly specialized caterpillar. *Insect Molecular Biology,* 15, 169–79.

Marais, E., Klok, C. J., Terblanche, J. S., & Chown, S. L. (2005). Insect gas exchange patterns: a phylogenetic perspective. *Journal of Experimental Biology,* 208, 4495–507.

Marden, J. H. (2000). Ontogenetic patterns of insect flight muscle: function and composition. *Annual Review of Physiology*, 62.

Marden, J. H., Fitzhugh, G. H., & Wolf, M. R. (1998). From molecules to mating success: Integrative biology of muscle maturation in a dragonfly. *American Zoologist,* 38, 528–44.

Marden, J. H., Wolf, M. R., & Weber, K. E. (1997). Aerial performance of *Drosophila melanogaster* from populations selected for upwind flight ability. *Journal of Experimental Biology,* 200, 2747–755.

Marrero, T. R. & Mason, E. A. (1972). Gaseous diffusion coefficients. *Journal of Physical and Chemical Reference Data,* 1, 3–118.

Marshall, K. E. & Sinclair, B. J. (2010). Repeated stress exposure results in a survival-reproduction tradeoff in *Drosophila melanogaster. Proceedings of the Royal Society B-Biological Sciences,* 277, 963–69.

Martin, I. & Grotewiel, M. S. (2006). Oxidative damage and age-related functional declines. *Mechanisms of Ageing and Development,* 127, 411–423.

Martin, M. M. (1977). Cellulose digestion in the midgut of the fungus-growing termite *Macrotermes natalensis*: the role of acquired digestive enzymes. *Science,* 199, 21453.

Martini, S. V., Goldenberg, R. C., Fortes, F. S. A., Campos-De-Carvalho, A. C., Falkenstein, D., & Morales, M. M. (2004). *Rhodnius prolixus* Malpighian tubule's aquaporin expression is modulated by 5-hydroxytryptamine. *Archives of Insect Biochemistry and Physiology,* 57, 133–41.

Massaro, R. C., Lee, L. W., Patel, A. B., Wu, D. S., Yu, M. J., Scott, B. N., Schooley, D. A., Schegg, K. M., & Beyenbach, K. W. (2004). The mechanism of action of the antidiuretic peptide Tenmo ADFa in Malpighian tubules of *Aedes aegypti. Journal of Experimental Biology,* 207, 2877–88.

Massey, F. P. & Hartley, S. E. (2009). Physical defences wear you down: progressive and irreversible impacts of silica on insect herbivores. *Journal of Animal Ecology,* 78, 281–91.

Matadha, D., Hamilton, G. C., & Lashomb, J. H. (2004). Effect of temperature on development, fecundity, and life table parameters of *Encarsia citrina* Craw (Hymenoptera : Aphelinidae), a parasitoid of Euonymus scale, *Unaspis euonymi* (Comstock), and *Quadraspidiotus perniciosus* (Comstock) (Homoptera : Diaspididae). *Environmental Entomology,* 33, 1185–91.

Matson, P. A., Parton, W. J., Power, A. G., & Swift, M. J. (1997). Agricultural intensification and ecosystem properties. *Science,* 277, 504–09.

Matsuura, H., Sokabe, T., Kohno, K., Tominaga, M., & Kadowaki, T. (2009). Evolutionary conservation and changes in insect TRP channels. *BMC Evolutionary Biology,* 9.

Matthews, H. J., Audsley, N., & Weaver, R. J. (2007). Interactions between allatostatins and allatotropin on spontaneous contractions of the foregut of larval *Lacanobia oleracea. Journal of Insect Physiology,* 53, 75–83.

Matthews, J. R. & Downer, R. G. H. (1974). Origin of trehalose in stress-induced hyperglycaemia in the American cockroach, *Periplaneta americana. Canadian Journal of Zoology,* 52, 1005–10.

Matthews, P. G. D. & Seymour, R. S. (2006). Diving insects boost their buoyancy bubbles. *Nature,* 441, 171–171.

Matthews, R. W., González, J. M., Matthews, J. R., & Deyrup, L. D. (2009). Biology of the parasitoid Melittobia (Hymenoptera: Eulophidae)*. *Annual Review of Entomology,* 54, 251–66.

Mattson Jr., W. J. (1980). Herbivory in relation to plant nitrogen content. *Annual Review of Ecology and Systematics,* 11, 119–61.

Maurer, P., Debieu, D., Leroux, P., Malosse, C., & Riba, G. (1992). Sterols and symbiosis in the leaf-cutting ant *Acromyrmex octospinosus* (Reich) (Hymenoptera, Formicidae: Attini). *Archives of Insect Biochemistry and Physiology,* 20, 13–21.

Mauricio, R. & Rausher, M. D. (1997). Experimental manipulation of putative selective agents provides evidence for the role of natural enemies in the evolution of plant defense. *Evolution,* 51, 1435–44.

May, M. L. (1979). Insect thermoregulation. *Annual Review of Entomology,* 24, 313–49.

May, M. L. (1995a). Dependence of flight behavior and heat production on air temperature in the green darner dragonfly *Anax junius* (Odonata: Aeshnidae). *Journal of Experimental Biology,* 198, 2385–92.

May, M. L. (1995b). Simultaneous control of head and thoracic temperature by the green darner dragonfly *Anax junius* (Odonata: Aeshnidae). *Journal of Experimental Biology,* 198, 2373–84.

Mazur, P. (1984). Freezing of living cells-mechanisms and implications. *American Journal of Physiology,* 247, C125–C142.

Mcbride, C. S. (2007). Rapid evolution of smell and taste receptor genes during host specialization in *Drosophila sechellia. Proceedings of the National Academy of Sciences,* 104, 4996–5001.

Mccabe, J. & Partridge, L. (1997). An interaction between environmental temperature and genetic variation for body size for the fitness of adult female *Drosophila melanogaster. Evolution,* 51(4), 1164–74.

Mcneill, S. & Southwood, T. R. E. (1978). The role of nitrogen in the development of insect plant relationships. *In:* Harborne, J. E. (ed.) *Biochemical Aspects of Plant and Animal Coevolution.* London: Academic Press.

Melcher, C. & Pankratz, M. J. (2005). Candidate gustatory interneurons modulating feeding behavior in the *Drosophila* brain. *PLoS Biol,* 3, e305.

Meng, X., Wahlström, G., Immonen, T., Kolmer, M., Tirronen, M., Predel, R., Kalkkinen, N., Heino, T. I., Sariola, H., & Roos, C. (2002). The *Drosophila* hugin gene codes for myo-stimulatory and ecdysis-modifying neuropeptides. *Mechanisms of Development,* 117, 5–13.

Mercier, J., Doucet, D., & Retnakaran, A. (2007). Molecular physiology of crustacean and insect neuropeptides. *Journal of Pesticide Science,* 32, 345–59.

Meredith, J. & Phillips, J. E. (1973). Ultrastructure of anal papillae from a seawater mosquito larva (*Aedes togoi* Theobald). *Canadian Journal of Zoology,* 51, 349–53.

Mevi-Schütz, J., Goverde, M., & Erhardt, A. (2003). Effects of fertilization and elevated CO on larval food and butterfly nectar amino acid preference in *Coenonympha pamphilus. Behavioral Ecology and Sociobiology,* 54, 36–43.

Michaud, M. R. & Denlinger, D. L. (2006). Oleic acid is elevated in cell membranes during rapid cold-hardening and pupal diapause in the flesh fly, *Sarcophaga crassipalpis. Journal of Insect Physiology,* 52, 1073–82.

Michaud, M. R. & Denlinger, D. L. (2007). Shifts in the carbohydrate, polyol, and amino acid pools during rapid cold-hardening and diapause-associated cold-hardening in flesh flies (Sarcophaga crassipalpis): a metabolomic comparison. *Journal of Comparative Physiology B-Biochemical Systemic and Environmental Physiology,* 177, 753–63.

Milkman, R. (1962). Temperature effects on day old drosophila pupae. *Journal of General Physiology,* 45, 777–99.

Mill, P. J. (1974). Respiration: aquatic insects. *In:* Rockstein, M. (ed.) *The Physiology of Insecta* New York: Academic.

Mill, P. J. (1998). Caterpillars have lungs. *Nature,* 391, 129–30.

Miller, G. A., Clissold, F. J., Mayntz, D., & Simpson, S. J. (2009). Speed over efficiency: locusts select body temperatures that favour growth rate over efficient nutrient utilization. *Proceedings of the Royal Society B: Biological Sciences*, 276, 3581–89.

Miller, M. S., Lekkas, P., Braddock, J. M., Farman, G. P., Ballif, B. A., Irving, T. C., Maughan, D. W., & Vigoreaux, J. O. (2008). Aging enhances indirect flight muscle fiber performance yet decreases flight ability in *Drosophila. Biophysical Journal,* 95, 2391–401.

Miller, P. L. (1960). Respiration in the desert locust I. The control of ventilation. *Journal of Experimental Biology,* 37, 224–36.

Miller, P. L. (1966a). The regulation of breathing in insects. *Advances in Insect Physiology,* 3, 279–354.

Miller, P. L. (1966b). The supply of oxygen to the active flight muscles of some large beetles. *Journal of Experimental Biology,* 45, 285–304.

Miller, P. L. (1974). Respiration – aerial gas transport. *In:* Rockstein, M. (ed.) *Physiology of Insecta, Volume IV.* 2nd ed. New York: Academic Press.

Miller, P. L. (1994). Submerged oviposition and responses to oxygen lack in *Enallagma cyathigerum* (Charpentier) (Zygoptera: Coenagrionidae). *Advances in Odonatology,* 6, 79–88.

Miller, T. (1985). Structure and physiology of the circulatory system. *In:* Gilbert, G. K. L. (ed.) *Comprehensive Insect Physiology, Biochemistry, and Pharmacology,* Pergamon Press.

Miller, T. A. (1997). Control of circulation in insects. *General Pharmacology,* 29, 23–38.

Mirth, C., Drexler, A., Truman, J. W., & Riddiford, L. M. (2007). Critical weight as a switch in the developmental response to starvation: the role of ecdysone in the maturation of *Drosophila* wing discs. *Comparative Biochemistry and Physiology A-Molecular & Integrative Physiology,* 148, S9–S9.

Mirth, C. K. & Riddiford, L. M. (2007). Size assessment and growth control: how adult size is determined in insects. *BioEssays,* 29, 344–355.

Mitchell, H. K. & Lipps, L. S. (1978). Heat shock and phenocopy induction in *Drosophila. Cell,* 15, 907–918.

Mitchell, H. K. & Petersen, N. S. (1982). Developmental abnormalities in *Drosophila* induced by heat-shock. *Developmental Genetics,* 3, 91–102.

Mitter, C., Farrell, B., & Wiegmann, B. (1988). The phylogenetic study of adaptive zones: Has phytophagy promoted insect diversification? *The American Naturalist,* 132, 107–28.

Moczek, A. P. & Rose, D. J. (2009). Differential recruitment of limb patterning genes during development and diversification of beetle horns. *Proceedings of the National Academy of Sciences,* 106, 8992–97.

Moe, S. J., Stelzer, R. S., Forman, M. R., Harpole, W. S., Daufresne, T., & Yoshida, T. (2005). Recent advances in ecological stoichiometry: insights for population and community ecology. *Oikos,* 109, 29–39.

Mole, S., Ross, J. A. M., & Waterman, P. G. (1988). Light-induced variation in phenolic levls of foliage of rain-forest plants. I. Chemical changes. *Journal of Chemical Ecology,* 14, 1–21.

Moran, N. A. (2007). Symbiosis as an adaptive process and source of phenotypic complexity. *Proceedings of the National Academy of Sciences,* 104, 8627–633.

Moreau, M., Arrufat, P., Latil, G., & Jeanson, R. (2011). Use of radio-tagging to map spatial organization and social interactions in insects. *Journal of Experimental Biology*, 214, 17–21.

Moreau, R. et al. (1984). Hemolymph trehalose and carbohydrates in starved male adult *Locusta migratoria* : possibility of endocrine modification. *Comparative Biochemical Physiology*, 78A(3), 481–85.

Morehouse, N. I. & Rutowski, R. L. (2010). Developmental responses to variable diet composition in a butterfly: the role of nitrogen, carbohydrates and genotype. *Oikos*, 119, 636–45.

Morgan, A. H. & O'Neil, H. D. (1931). The function of the tracheal gills in larvae of the caddis fly, *Macronema zebratum* Hagen. *Physiological Zoology*, 4, 361–379.

Morgan, E. D. & Mandava, N. B. (1990). *CRC Handbook of Natural Pesticides, Volume 6, Insect Attractants and Repellents*, Boca Raton, CRC Press.

Morgan, T. J. & Mackay, T. F. C. (2006). Quantitative trait loci for thermotolerance phenotypes in *Drosophila melanogaster*. *Heredity*, 96, 232–42.

Morin, P., Jr., Mcmullen, D. C., & Storey, K. B. (2005). HIF-1α involvement in low temperature and anoxia survival by a freeze tolerant insect. *Molecular and Cellular Biochemistry*, 280, 99–106.

Morita, H. & Shirais, A. (1968). Stimulation of labellar sugar receptor of fthe fleshfly by mono- and disaccarides. *Journal of General Physiology*, 52, 559–83.

Morris, C. E. & Sigurdson, W. J. (1989). Stretch-inactivated ion channels coexist with stretch-activated ion channels. *Science*, 243, 807–09.

Morrissey, R. E. & Baust, J. G. (1976). Ontogeny of cold tolerance in gall fly, *Eurosta-Solidagensis*. *Journal of Insect Physiology*, 22, 431–37.

Mortimer, N. T. & Moberg, K. H. (2009). Regulation of *Drosophila* embryonic tracheogenesis by dVHL and hypoxia. *Developmental Biology*, 329, 294–305.

Morton, D. B. (2004a). Atypical soluble guanylyl cyclases in *Drosophila* can function as molecular oxygen sensors. *Journal of Biological Chemistry*, 279, 50651–53.

Morton, D. B. (2004b). Invertebrates yield a plethora of atypical guanylyl cyclases. *Molecular Neurobiology*, 29, 97–115.

Morton, D. B., Stewart, J. A., Langlais, K. K., Clemens-Grisham, R. A., & Vermehren, A. (2008). Synaptic transmission in neurons that express the *Drosophila* atypical soluble guanylyl cyclases, Gyc-89Da and Gyc-89Db, is necessary for the successful completion of larval and adult ecdysis. *Journal of Experimental Biology*, 211, 1645–56.

Mousseau, T. A. (2000). Intra- and interpopulation genetic variation: explaining the past and predicting the future. *In:* Mousseau,T., A. B. Sinervo, and J.A. Endler (eds) *Adaptive Genetic Variation in the Wild*. New York: Oxford University Press.

Mousseau, T. A. & Dingle, H. (1991). Maternal effects in insect life histories. *Annual Review of Entomology*, 36, 511–34.

Moussian, B. (2010). Recent advances in understanding mechanisms of insect cuticle differentiation. *Insect Biochemistry and Molecular Biology*, 40, 363–75.

Mullins, D. E. & Cochran, D. G. (1974). Nitrogen metabolism in the american cockroach: an examination of whole body and fat body regulation of cations in response to nitrogen balance. *Journal of Experimental Biology*, 61, 557–70.

Mullins, D. E. & Cochran, D. G. (1975). Nitrogen metabolism in the American cockroach. II. An examination of negative nitrogen balance with respect to mobilization of uric acid stores. *Comparative Biochemistry and Physiology A*, 50, 501–10.

Musser, R. O., Cipollini, D. F., Hum-Musser, S. M., Williams, S. A., Brown, J. K., & Felton, G. W. (2005). Evidence that the caterpillar salivary enzyme glucose oxidase provides herbivore offense in solanaceous plants. *Archives of Insect Biochemistry and Physiology,* 58, 128–37.

Musser, R. O., Hum-Musser, S. M., Eichenseer, H., Peiffer, M., Ervin, G., Murphy, J. B., & Felton, G. W. (2002). Herbivory: Caterpillar saliva beats plant defences. *Nature,* 416, 599–600.

Mykles, D. L., Adams, M. E., Gade, G., Lange, A. B., Marco, H. G., & Orchard, I. (2010). Neuropeptide action in insects and crustaceans. *Physiological and Biochemical Zoology,* 83, 836–46.

Nachman, R., Pietrantonio, P., & Coast, G. (2009). Towards the development of novel pest management agents based upon insect kinin neuropeptide analogues. *Annals of the New York Academy of Sciences,* 1163, 251–61.

Nachman, R. J. & Coast, G. M. (2007). Structure-activity relationships for in vitro diuretic activity of CAP2b in the housefly. *Peptides,* 28, 57–61.

Naef-Daenzer, B., Fruh, D., Stalder, M., Wetli, P., Weise, E. (2005). Miniaturization (0.2 g) and evaluation of attachment techniques of telemetry transmitters. *Journal of Experimental Biology,* 208, 4063–8.

Nagy, K. A., Girard, I. A., & Brown, T. K. (1999). Energetics of free-ranging mammals, reptiles, and birds. *Annual Review of Nutrition,* 19, 247–77.

Nakano, R., Skals, N., Takanashi, T., Surlykke, A., Koike, T., Yoshida, K., Maruyama, H., Tatsuki, S., & Ishikawa, Y. (2008). Moths produce extremely quiet ultrasonic courtship songs by rubbing specialized scales. *Proceedings of the National Academy of Sciences of the United States of America,* 105, 11812–17.

Nardi, J. B., Mackie, R. I., & Dawson, J. O. (2002). Could microbial symbionts of arthropod guts contribute significantly to nitrogen fixation in terrestrial ecosystems? *Journal of Insect Physiology,* 48, 751–63.

Nasir, H. & Noda, H. (2003). Yeast-like symbiotes as a sterol source in anobiid beetles (Coleoptera, Anobiidae): Possible metabolic pathways from fungal sterols to 7-dehydrocholesterol. *Archives of Insect Biochemistry and Physiology,* 52, 175–82.

Nassel, D. R. (2002). Neuropeptides in the nervous system of Drosophila and other insects: multiple roles as neuromodulators and neurohormones. *Progress in Neurobiology,* 68, 1–84.

Nassel, D. R., Eckert, M., Muren, J. E. & Penzin, H. (1998). Species-specific action and distribution of tachykinin-related peptides in the foregut of the cockroaches *Leucophaea maderae* and *Periplaneta americana. Journal of Experimental Biology,* 201, 1615–26.

Nassel, D. R. & Homberg, U. (2006). Neuropeptides in interneurons of the insect brain. *Cell and Tissue Research,* 326, 1–24.

Nation, J. L. (2008). *Insect Physiology and Biochemistry,* Boca Raton, CRC Press.

Nault, L. R. & Styer, W. E. (1972). Effects of sinigrin on host selection by aphids. *Entomologia Experimentalis Et Applicata,* 15, 423–37.

Nelson, R. J., Denlinger, D. L., & Somers, D. E. (2010). *Photoperiodism: The Biological Clock,* Oxford, Oxford University Press.

Neukirch, A. (1982). Dependence of the life-span of the honeybee (*Apis mellifera*) upon flight performance and energy consumption. *Journal of Comparative Physiology B,* 146, 35–40.

Neven, L. G., Duman, J. G., Beals, J. M., & Castellino, F. J. (1986). Overwintering adaptations of the stag beetle, *Ceruchus piceus* – removal of ice nucleators in the winter to promote supercooling. *Journal of Comparative Physiology B-Biochemical Systemic and Environmental Physiology,* 156, 707–16.

Neven, L. G. & Hansen, L. D. (2010). Effects of temperature and controlled atmospheres on codling moth metabolism. *Annals of the Entomological Society of America,* 103, 418–23.

Newsholme, E. A., Crabtree, B., Higgins, S. J., Thornton, S. D., & Start, C. (1972). The activities of fructose diphosphatase in flight muscles from the bumble-bee and the role of this enzyme in heat generation. *Biochemical Journal,* 128, 89–97.

Nice, C. C. & Fordyce, J. A. (2006). How caterpillars avoid overheating: behavioral and phenotypic plasticity of pipevine swallowtail larvae. *Oecologia,* 146, 541–48.

Nichol, A. C. (2000). *Water load: a physiological limitation to bumblebee foraging behaviour?* PhD Thesis.

Nichols, R., Kaminski, S., Walling, E., & Zornik, E. (1999). Regulating the activity of a cardioacceleratory peptide. *Peptides,* 20, 1153–58.

Nicolson, S. W. (1991). Diuresis or clearance: is there a physiological role for the "diuretic hormone" of the desert beetle *Onymacris*? *Journal of Insect Physiology,* 37 (6), 447–52.

Nicolson, S. W. (2009). Water homeostasis in bees, with the emphasis on sociality. *Journal of Experimental Biology,* 212, 429–34.

Nicolson, S. W. & Louw, G. N. (1982). Simultaneous measurement of evaporative water-loss, oxygen-consumption, and thoracic temperature during flight in a carpenter bee. *Journal of Experimental Zoology,* 222, 287–96.

Nijhout, H. F. (2006). A quantitative analysis of the mechanism that controls body size in *Manduca sexta*. *Journal of Biology,* 5.

Nilsson, L. A. (1988). The evolution of flowers with deep corolla tubes. *Nature,* 334, 147–49.

Noble-Nesbitt, J. (1970). Water uptake from subsaturated atmospheres. Its site in insects. *Nature,* 225, 753–4

Noble-Nesbitt, J. (2010). Mitochondrial-driven sustained active water vapour absorption (WVA) in the firebrat, *Thermobia domestica* (Packard), during development and the moulting cycle. *Journal of Insect Physiology,* 56, 488–91.

Noble-Nesbitt, J. & Al-Shakur, M. (1988). Cephalic neuroendocrine regulation of integumentary water loss in the cockroach *Periplaneta americana*. *Journal of Experimental Biology,* 136, 451–59.

Noble-Nesbitt, J., Appel, A. G., & Croghan, P. C. (1995). Water and carbon dioxide loss from the cockroach *Periplaneta americana* (L.) measured using radioactive isotopes. *Journal of Experimental Biology,* 198, 235–240.

Norry, F. M., Bubliy, O. A., & Loeschke, V. (2001). Developmental time, body size and wing loading in *Drosophila buzzatii* from lowland and highland populations in Argentina. *Hereditas,* 135, 35–40.

Nozawa, M. & Nei, M. (2007). Evolutionary dynamics of olfactory receptor genes in Drosophila species. *Proceedings of the National Academy of Sciences,* 104, 7122–127.

Núñez-Farfán, J., Fornoni, J., & Valverde, P. L. (2007). The evolution of resistance and tolerance to herbivores. *Annual Review of Ecology, Evolution, and Systematics,* 38, 541–66.

O'brien, D. M., Boggs, C. L., & Fogel, M. L. (2005). The amino acids used in reproduction by butterflies: a comparative study of dietary sources using compound specific stable isotope analysis. *Physiological & Biochemical Zoology,* 78, 819–27.

O'brien, D. M., Min, K. J., Larsen, T., & Tatar, M. (2008). Use of stable isotopes to examine how dietary restriction extends *Drosophila* lifespan. *Current Biology,* 18, R155–R156.

O'donnell, M. J. (1977). Site of water vapor absorption in the desert cockroach, Arenivaga investigata. *Proceedings of the National Academy of Sciences of the United States of America,* 74, 1757–60.

O'donnell, M. J., Maddrell, S. H. P., & Gardiner, B. O. C. (1983). Transport of uric acid by the malpighian tubules of *Rhodnius Prolixus* and other insects. *Journal of Experimental Biology,* 103, 169–84.

O'meara, G. F. (1976). Saltmarsh mosquitoes (Diptera: Culicidae). *In:* Cheng, L. (ed.) *Marine Insects.* New York: North-Holland.

O'brien, D. M., Boggs, C. L., & Fogel, M. L. (2003). Pollen feeding in the butterfly *Heliconius charitonia.* Isotopic evidence for essential amino acid transfer from pollen to eggs. *Proceedings of the Royal Society B: Biological Sciences,* 270, 2631–36.

O'brien, D. M., Fogel, M. L., & Boggs, C. L. (2002). Renewable and non-renewable resources: amino acid turnover and allocation to reproduction in Lepidoptera. *Proceedings of the National Academy of Science, USA,* 99, 4413–18.

O'donnell, M. J. & Spring, J. H. (2000). Modes of control of insect Malpighian tubules: synergism, antagonism, cooperation and autonomous regulation. *Journal of Insect Physiology,* 46, 107–17.

O'neill, R. G. & Heller, S. (2005). The mechanosensitive nature of TRPV channels. *European Journal of Physiology,* 451, 193–203.

Ocorr, K., Reeves, N. L., Wessells, R. J., Fink, M., Chen, H. S. V., Akasaka, T., Yasuda, S., Metzger, J. M., Giles, W., Posakony, J. W., & Bodmer, R. (2007). KCNQ potassium channel mutations cause cardiac arrhythmias in *Drosophila* that mimic the effects of aging. *Proceedings of the National Academy of Sciences of the United States of America,* 104, 3943–48.

Ohashi, K., D'souza, D., & Thomson, J. D. (2010). An automated system for tracking and identifying individual nectar foragers at multiple feeders. *Behavioral Ecology and Sociobiology,* 64, 891–97.

Ohshima, I. (2008). Host race formation in the leaf-mining moth *Acrocercops transecta* (Lepidoptera: Gracillariidae). *Biological Journal of the Linnean Society,* 93, 135–45.

Ohtsu, T., Kimura, M. T., & Katagiri, C. (1998). How Drosophila species acquire cold tolerance – Qualitative changes of phospholipids. *European Journal of Biochemistry,* 252, 608–11.

Okech, B. A., Boudko, D. Y., Linser, P. J., & Harvey, W. R. (2008). Cationic pathway of pH regulation in larvae of *Anopheles gambiae. Journal of Experimental Biology,* 211, 957–68.

Oldham, S. & Hafen, E. (2003). Insulin/IGF and target of rapamycin signaling: a TOR de force in growth control. *Trends in Cell Biology,* 13, 79–85.

Oliver, B. & Leblanc, B. (2004). How many genes in a genome? *Genome biology,* 5, 1.

Ollerton, J. & Mccollin, D. (1998). Insect and angiosperm diversity in marine environments: a response to van der Hage. *Functional Ecology,* 12, 976–77.

Olsen, T. M., Sass, S. J., Li, N., & Duman, J. G. (1998). Factors contributing to seasonal increases in inoculative freezing resistance in overwintering fire-colored beetle larvae *Dendroides canadensis* (Pyrochroidae). *Journal of Experimental Biology,* 201, 1585–94.

Onken, H. & Moffett, D. F. (2009). Revisiting the cellular mechanisms of strong luminal alkalinization in the anterior midgut of larval mosquitoes. *Journal of Experimental Biology,* 212, 373–77.

Onken, H., Moffett, S. B., & Moffett, D. F. (2006). The isolated anterior stomach of larval mosquitoes (*Aedes aegypti*): voltage-clamp measurements with a tubular epithelium. *Comparative Biochemistry and Physiology,* 143A, 24–34.

Onken, H., Moffett, S. B., & Moffett, D. F. (2008). Alkalinization in the isolated and perfused anterior midgut of the larval mosquito, *Aedes aegypti*. *Journal of Insect Science,* 8, 46.

Ono, M., Igarashi, T., Ohno, E., & Sasaki, M. (1995). Unusual thermal defense by a honeybee against mass attack by hornets. *Nature,* 377, 334–36.

Ono, M., Okada, I., & Sasaki, M. (1987). Heat-production by balling in the Japanese honeybee, *Apis cerana* japonica as a defensive behavior against the hornet, Vespa simillima xanthoptera (Hymenoptera, Vespidae). *Experientia,* 43, 1031–32.

Opstad, R., Rogers, S. M., Behmer, S. T., & Simpson, S. J. (2004). Behavioural correlates of phenotypic plasticity in mouthpart chemoreceptor numbers in locusts. *Journal of Insect Physiology,* 50, 725–36.

Orchard, I. (2006). Serotonin: A coordinator of feeding-related physiological events in the blood-gorging bug, *Rhodnius prolixus*. *Comparative Biochemistry and Physiology – Part A: Molecular & Integrative Physiology,* 144, 316–24.

Otterstatter, M. C. & Whidden, T. L. (2004). Patterns of parasitism by tracheal mites (*Locustacarus buchneri*) in natural bumble bee populations. *Apidologie,* 35, 351–57.

Overgaard, J., Malmendal, A., Sorensen, J. G., Bundy, J. G., Loeschcke, V., Nielsen, N. C., & Holmstrup, M. (2007). Metabolomic profiling of rapid cold hardening and cold shock in *Drosophila melanogaster*. *Journal of Insect Physiology,* 53, 1218–32.

Overgaard, J., Sorensen, J. G., Petersen, S. O., Loeschcke, V., & Holmstrup, M. (2005). Changes in membrane lipid composition following rapid cold hardening in *Drosophila melanogaster*. *Journal of Insect Physiology,* 51, 1173–82.

Overgaard, J., Sorensen, J. G., Petersen, S. O., Loeschcke, V., & Holmstrup, M. (2006). Reorganization of membrane lipids during fast and slow cold hardening in *Drosophila melanogaster*. *Physiological Entomology,* 31, 328–35.

Overgaard, J., Tomcala, A., Sorensen, J. G., Holmstrup, M., Krogh, P. H., Simek, P., & Kostal, V. (2008). Effects of acclimation temperature on thermal tolerance and membrane phospholipid composition in the fruit fly *Drosophila melanogaster*. *Journal of Insect Physiology,* 54, 619–29.

Ozaki, M., Takahara, T., Kawahara, Y., Wada-Katsumata, A., Seno, K., Amakawa, T., Yamaoka, R., & Nakamura, T. (2003). Perception of noxious compounds by contact chemoreceptors of the blowfly, *Phormia regina*: Putative role of an odorant-binding protein. *Chemical Senses,* 28, 349–59.

Page, R. E. & Robinson, G. E. (1991). The genetics of division of labor in honey bee colonies. *Advances in Insect Physiology,* 23, 117–169.

Paiva-Silva, G. O., Cruz-Oliveira, C., Nakayasu, E. S., Maya-Monteiro, C. M., Dunkov, B. C., Masuda, H., Almeida, I. C. & Oliveira, P. L. (2006). A heme-degradation pathway in a blood-sucking insect. *Proceedings of the National Academy of Sciences,* 103, 8030–8035.

Paluzzi, J.-P. & Orchard, I. (2006). Distribution, activity and evidence for the release of an anti-diuretic peptide in the kissing bug *Rhodnius prolixus*. *Journal of Experimental Biology,* 209, 907–15.

Paluzzi, J. P., Russell, W. K., Nachman, R. J., & Orchard, I. (2008). Isolation, cloning, and expression mapping of a gene encoding an antidiuretic hormone and other CAPA-related peptides in the disease vector, *Rhodnius prolixus*. *Endocrinology,* 149, 4638–46.

Pan, M. L. & Telfer, W. H. (1996). Methionine-rich hexamerin and arylphorin as precursor reservoirs for reproduction and metamorphosis in female luna moths. *Archives of Insect Biochemistry and Physiology,* 33, 149–62.

Pan, M. L. & Telfer, W. H. (2001). Storage hexamer utilization in two lepidopterans: difference correlated with the timing of egg formation. *Journal of Insect Science,* 1.2, 9.

Park, J.-H., Attardo, G. M., Hansen, I. A., & Raikhel, A. S. (2006). GATA factor translation Is the final downstream step in the amino acid/target-of-rapamycin-mediated vitellogenin gene expression in the anautogenous mosquito *Aedes aegypti*. *Journal of Biological Chemistry,* 281, 11167–76.

Parker, A. R. & Lawrence, C. R. (2001). Water capture by a desert beetle. *Nature,* 414, 33–34.

Parker, R. J. & Auld, V. J. (2006). Roles of glia in the *Drosophila* nervous system. *Seminars in Cell & Developmental Biology,* 17, 66–77.

Parson, P. A. (1989). Acetaldehyde utilization in *Drosophila*: an example of hormesis. *Biological Journal of the Linnean Society,* 37, 183–89.

Parsons, P. A. (1970). Genetic heterogeneity in natural populations of *Drosophila melanogaster* for ability to withstand dessication. *Theoretical and Applied Genetics,* 40, 261–66.

Partridge, L., Barrie, B., Fowler, K., & French, V. (1994). Evolution and development of body size and cell size in *Drosophila melanogaster* in response to temperature. *Evolution,* 48, 1269–76.

Partridge, L. & Fowler, K. (1993). Responses and correlated responses to artificial selection on thorax length in *Drosophila melanogaster*. *Evolution,* 47, 213–26.

Pass, G. (2000). Accessory pulsatile organs: evolutionary innovations in insects. *Annual Review of Entomology,* 45, 495–518.

Patek, S. N., Baio, J. E., Fisher, B. L., & Suarez, A. V. (2006). Multifunctionality and mechanical origins: Ballistic jaw propulsion in trap-jaw ants *Proceedings of the National Academy of Science, USA,* 103, 12787–92.

Patel, M., Hayes, T. K., & Coast, G. M. (1995). Evidence for the hormonal function of a CRF-related diuretic peptide (*Locusta*-dp) in *Locusta migratoria*. *Journal of Experimental Biology,* 198, 793–804.

Patel, R. T., Soulages, J. L., Hariharasundaram, B., & Arrese, E. L. (2005). Activation of the lipid droplet controls the rate of lipolysis of triglycerides in the insect fat body. *Journal of Biological Chemistry,* 280, 22624–31.

Patrick, M. L., Aimanova, K., Sanders, H. R., & Gill, S. S. (2006). P-type Na^+/K^+-ATPase and V-type H^+-ATPase expression patterns in the osmoregulatory organs of larval and adult mosquito *Aedes aegypti*. *Journal of Experimental Biology,* 209, 4638–51.

Patrick, M. L. & Bradley, T. J. (2000a). The physiology of salinity tolerance in larvae of two species of *Culex* mosquitoes: The role of compatible solutes. *Journal of Experimental Biology,* 203, 821–30.

Patrick, M. L. & Bradley, T. J. (2000b). Regulation of compatible solute accumulation in larvae of the mosquito *Culex tarsalis*: osmolarity *versus* salinity. *Journal of Experimental Biology,* 203, 831–39.

Patrick, M. L., Gonzalez, R. J., & Bradley, T. J. (2001). Sodium and chloride regulation in freshwater and osmoconforming larvae of *Culex* mosquitoes. *Journal of Experimental Biology,* 204, 3345–54.

Pedersen, O., Vos, H., & Colmer, T. D. (2006). Oxygen dynamics during submergence in the halophytic stem succulent *Halosarcia pergranulata Plant, Cell and Environment,* 29, 1388–99.

Pellmyr, O. & Krenn, H. W. (2002). Origin of a complex key innovation in an obligate insect–plant mutualism. *Proceedings of the National Academy of Sciences,* 99, 5498–502.

Penick, C. A. & Tschinkel, W. R. (2008). Thermoregulatory brood transport in the fire ant, *Solenopsis invicta. Insectes Sociaux,* 55, 176–82.

Perkins, M. C., Woods, H. A., Harrison, J. F., & Elser, J. J. (2004). Dietary phosphorus affects the growth of larval *Manduca sexta. Archives of Insect Biochemistry and Physiology,* 55, 153–68.

Perz-Edwards, R. J., Irving, T. C., Baumann, B. A. J., Gore, D., Hutchinson, D. C., Krzic, U., Porter, R. L., Ward, A. B., & Reedy, M. K. (2011). X-ray diffraction evidence for myosin-troponin connections and tropomyosin movement during stretch activation of insect flight muscle. *Proceedings of the National Academy of Sciences of the United States of America,* 108, 120–25.

Petz, M., Stabentheiner, A., & Crailsheim, K. (2004). Respiration of individual honeybee larvae in relation to age and ambient temperature. *Journal of Comparative Physiology B,* 174, 511–18.

Petzel, D. H., Berg, M. M., & Beyenbach, K. W. (1987). Hormone-controlled cAMP-mediated fluid secretion in yellow-fever mosquito. *American Journal of Physiology,* 253, R701–R711.

Philip, B. N. & Lee, R. E. (2010). Changes in abundance of aquaporin-like proteins occurs concomitantly with seasonal acquisition of freeze tolerance in the goldenrod gall fly, Eurosta solidaginis. *Journal of Insect Physiology,* 56, 679–85.

Philip, B. N., Yi, S. X., Elnitsky, M. A., & Lee, R. E. (2008). Aquaporins play a role in desiccation and freeze tolerance in larvae of the goldenrod gall fly, *Eurosta solidaginis. Journal of Experimental Biology,* 211, 1114–19.

Phillips, J. (1981). Comparative physiology of insect renal function. *American Journal of Physiology,* 241, R241–R257.

Phillips, J. E. (1970). Apparent transport of water by insect excretory systems. *American Zoologist,* 10, 413–36.

Phillips, J. E. (1982). Hormonal control of renal functions in insects. *Federation Proceedings,* 41, 2348–54.

Phillips, J. E., Hanrahan, J., Chamberlin, M., & Thomson, B. (1986). Mechanisms and control of reabsorption in insect hindgut. *Advances in Insect Physiology,* 19, 329–422.

Phillips, J. E., Meredith, J., Audsley, N., Richardson, N., Macins, A., & Ring, M. (1998). Locust ion transport peptide (ITP): a putative hormone controlling water and ionic balance in terrestrial insects. *American Zoologist,* 38, 461–70.

Phillips, J. E., Wiens, C., Audsley, N., Jeffs, L., Bilgen, T., & Meredith, J. (1996). Nature and control of chloride transport in insect absorptive epithelia. *Journal of Experimental Zoology,* 275, 292–99.

Pick, C., Schneuer, M., & Burmester, T. (2009). The occurrence of hemocyanin in Hexapoda. *FEBS Journal,* 276, 1930–41.

Pick, C., Schneuer, M., & Burmester, T. (2010). Ontogeny of hemocyanin in the ovoviviparous cockroach Blaptica dubia suggests an embryo-specific role in oxygen supply. *Journal of Insect Physiology,* 56, 455–60.

Pickard, W. F. (1974). Transition regime diffusion and the structure of the insect tracheolar system. *Journal of Insect Physiology,* 20, 947–56.

Pietrantonio, P. V., Jagge, C., Keeley, L. L., & Ross, L. S. (2000). Cloning of an aquaporin-like cDNA and *in situ* hybridization in adults of the mosquito *Aedes aegypti* (Diptera: Culicidae). *Insect Molecular Biology,* 9, 407–18.

Pietrantonio, P. V., Jagge, C., & Mcdowell, C. (2001). Cloning and expression analysis of a 5HT7-like serotonin receptor cDNA from mosquito *Aedes aegypti* female excretory and respiratory systems. *Insect Molecular Biology,* 10, 357–69.

Pietrantonio, P. V., Jagge, C., Taneja-Bageshwar, S., Nachman, R. J., & Barhoumi, R. (2005). The mosquito *Aedes aegypti* (L.) leucokinin receptor is a multiligand receptor for the three *Aedes* kinins. *Insect Molecular Biology,* 14, 55–67.

Pincebourde, S. & Casas, J. (2006). Multitrophic biophysical budgets: Thermal ecology of an intimate herbivore insect-plant interaction. *Ecological Monographs,* 76, 175–94.

Pollock, V. P., Radford, J. C., Pyne, S., Hasan, G., Dow, J. A. T., & Davies, S. A. (2003). norpA and itpr mutants reveal roles for phospholipase C and inositol (1,4,5)-triphosphate receptor in *Drosophila melanogaster* renal function. *Journal of Experimental Biology,* (2006), 901–11.

Potter, K. A., Davidowitz, G., & Woods, H. A. (2010). Cross-stage consequences of egg temperature in the insect *Manduca sexta. Functional Ecology,* 25, 548–56.

Predel, R. & Wegener, C. (2006). Biology of the CAPA peptides in insects. *Cellular and Molecular Life Sciences,* 63, 2477–90.

Price, D. R. G., Karley, A. J., Ashford, D. A., Isaacs, H. V., Pownall, M. E., Wilkinson, H. S., Gatehouse, J. A., & Douglas, A. E. (2007). Molecular characterisation of a candidate gut sucrase in the pea aphid *Acyrthosiphon pisum. Insect Biochemistry and Molecular Biology,* 37, 307–17.

Price, P. W. (1997). *Insect Ecology,* New York, John Wiley & Sons.

Prier, K., Beckman, O., & Tublitz, N. (1994). Modulating a modulator: biogenic amines at subthreshold levels potentiate peptide-mediated cardioexcitation of the heart of the tobacco hawkmoth *Manduca sexta. Journal of Experimental Biology,* 197, 377–91.

Pringle, J. W. S. (1949). The excitation and contraction of the flight muscles of insects. *Journal of Physiology,* 108, 226–32.

Qin, W., Doucet, D., Tyshenko, M. G., & Walker, V. K. (2007). Transcription of antifreeze protein genes in *Choristoneura fumiferana. Insect Molecular Biology,* 16, 423–34.

Qin, W., Neal, S. J., Robertson, R. M., Westwood, J. T., & Walker, V. K. (2005). Cold hardening and transcriptional change in *Drosophila melanogaster. Insect Molecular Biology,* 14, 607–13.

Quicke, D. L. J., Wyeth, P., Fawke, J. D., Basibuyuk, H. H., & Vincent, J. F. V. (1998). Manganese and zinc in the ovipositors and mandibles of hymenopterous insects. *Zoological Journal of the Linnean Society,* 124, 387–96.

Quinlan, M. C. & Hadley, N. F. (1993). Gas exchange, ventilatory patterns, and water loss in two lubber grasshoppers: quantifying cuticular and respiratory transpiration. *Physiological Zoology,* 66, 628–42.

Quinlan, M. C., Tublitz, N. J., & O'donnell, M. J. (1997). Anti-diuresis in the blood-feeding insect Rhodnius prolixus Stal: The peptide Cap2b and cyclic Gmp inhibit Malpighian tubule fluid secretion. *Journal of Experimental Biology,* 200, 2363–67.

Radford, J. H. C., Terhzaz, S., Cabrero, P., Davies, S. A., & Dow, J. A. T. (2004). Functional characterisation of the Anopheles leucokinins and their cognate G-protein coupled receptor *Journal of Experimental Biology,* 207, 4573–86.

Ragland, G. J., Denlinger, D. L., & Hahn, D. A. (2010). Mechanisms of suspended animation are revealed by transcript profiling of diapause in the flesh fly. *Proceedings of the National Academy of Sciences of the United States of America,* 107, 14909–14.

Rahn, H. & Paganelli, C. V. (1968). Gas exchange in gas gills of diving insects. *Respiration Physiology,* 5, 145–64.

Rako, L. & Hoffmann, A. A. (2006). Complexity of the cold acclimation response in Drosophila melanogaster. *Journal of Insect Physiology,* 52, 94–104.

Ramirez, J. M. & Pearson, K. G. (1989). Distribution of intersegmental interneurones that can reset the respiratory rhythm of the locust. *Journal of Experimental Biology,* 141, 151–76.

Ramløv, H. & Lee Jr, R. E. (2000). Extreme resistance to desiccation in overwintering larvae of the gall fly *Eurosta solidaginis* (Diptera, Tephritidae). *Journal of Experimental Biology,* 203, 783–89.

Ramsay, J. A. (1935). The evaporation of water from the cockroach. *Journal of Experimental Biology,* 12, 373–83.

Ramsay, J. A. (1955a). The excretion of sodium, potassium and water by the Malpighian tubules of the stick insect, *Dixippus morosus* (Orthoptera, Phasmidae). *Journal of Experimental Biology,* 32, 200–16.

Ramsay, J. A. (1955b). The excretory system of the stick insect, *Dixippus Morosus* (Orthoptera, Phasmidae). *Journal of Experimental Biology,* 32, 183–99.

Ramsay, J. A. (1964). Rectal complex of mealworm *Tenebrio molitor* (Coleoptera: Tenebrionidae). *Philosophical Transactions of the Royal Society of London Series B-Biological Sciences,* 248, 279–314.

Ramsey, J. S., Macdonald, S. J., Jander, G., Nakabachi, A., Thomas, G. H., & Douglas, A. E. (2010). Genomic evidence for complementary purine metabolism in the pea aphid, Acyrthosiphon pisum, and its symbiotic bacterium *Buchnera aphidicola. Insect Molecular Biology,* 19, 241–48.

Ranger, C. M., Winter, R. E., Singh, A. P., Reding, M. E., Frantz, J. M., Locke, J. C., & Krause, C. R. (2011). Rare excitatory amino acid from flowers of zonal geranium responsible for paralyzing the Japanese beetle. *Proceedings of the National Academy of Sciences.*

Rascón, B. & Harrison, J. F. (2005). Oxygen partial pressure effects on metabolic rate and behavior of tethered flying locusts. *Journal of Insect Physiology,* 51, 1193–99.

Rashevsky, N. (1960). *Mathematical Biophysics: Physico-Mathematical Foundations of Biology,* New York, Dover.

Ratzka, A., Vogel, H., Kliebenstein, D. J., Mitchell-Olds, T., & Kroymann, J. (2002). Disarming the mustard oil bomb. *Proceedings of the National Academy of Sciences,* 99, 11223–28.

Raubenheimer, D. & Gade, G. (1994). Hunger-thirst Interactions in the locust, *Locusta migratoria. Journal of Insect Physiology,* 40, 631–39.

Raubenheimer, D. & Gäde, G. (1996). Separating food and water deprivation in locusts: effects on the patterns of consumption, locomotion and growth. *Physiological Entomology,* 21, 76–84.

Raubenheimer, D., Lee, K. P., & Simpson, S. J. (2005). Does Bertrand's rule apply to macronutrients? *Proceedings of the Royal Society B: Biological Sciences,* 272, 2429–34.

Raubenheimer, D. & Simpson, S. J. (1993). The geometry of compensatory feeding in the locust. *Animal Behavior,* 45, 953–64.

Raubenheimer, D., Simpson, S. J., & Mayntz, D. (2009). Nutrition, ecology and nutritional ecology: toward an integrated framework. *Functional Ecology,* 23, 4–16.

Reagan, J. D. (1996). Molecular cloning and function expression of a diuretic hormone receptor from the house cricket, *Acheta domesticus. Insect Biochemistry and Molecular Biology,* 26, 1.

Reeve, M. W., Fowler, K., & Partridge, L. (2000). Increased body size confers greater fitness at lower experimental temperature in male *Drosophila melanogaster. Journal Evolutionary Biology,* 13, 836–44.

Resh, V. H., Buchwalter, D. B., Lamberti, G. A., & Eriksen., C. H. (2008). Aquatic insect respiration. *In:* Merritt, R. W., Cummins, K. L., & Berg, M. B. (eds.) *An Introduction to the Aquatic Insects of North America.* 4th ed. Dubuque, Iowa: Kendall/ Hunt Pub Co.

Riabinina, O., Dai, M. J., Duke, T., & Albert, J. T. (2011). Active process mediates species-specific tuning of *Drosophila* ears. *Current Biology,* 21, 658–64.

Ribeiro, J. M. C. (1995). Blood-feeding arthropods – Live syringes or invertebrate pharmacologists. *Infectious Agents and Disease-Reviews Issues and Commentary,* 4, 143–52.

Ribeiro, J. M. C. & Francischetti, I. M. B. (2003). Role of arthropod saliva in blood feeding: Sialome and post-sialome perspectives. *Annual Review of Entomology,* 48, 73–88.

Richter, K., Haslbeck, M., & Buchner, J. (2010). The heat shock response: Life on the verge of death. *Molecular Cell,* 40, 253–66.

Riddiford, L. M. (2008). Juvenile hormone action: A (2007) perspective. *Journal of Insect Physiology,* 54, 895–901.

Rinehart, J. P., Li, A., Yocum, G. D., Robich, R. M., Hayward, S. A. L., & Denlinger, D. L. (2007). Up-regulation of heat shock proteins is essentail for cold survival during insect diapause. *Proceedings of the National Academy of Sciences of the United States of America,* 104, 11130–37.

Ring, R. A. & Danks, H. V. (1994). Desiccation and cryoprotection – overlapping adaptations. *Cryo-Letters,* 15, 181–90.

Ritchie, M. E. (2000). Nitrogen limitation and trophic vs. abiotic influences on insect herbivores in a temperate grassland. *Ecology,* 81, 1601–12.

Rizki, T. (1978). The circulatory system and associated cells and tissues. In: *In The Genetics and Biology of Drosophila,* Vol 2B (ed. M. Ashburner and T. R. F. Wright), pp. 397–452. London, New York: Academic Press.

Roberts, S. P. (2005). Effects of flight behaviour on body temperature and kinematics during inter-male mate competition in the solitary desert bee *Centris pallida. Physiological Entomology,* 30, 151–57.

Roberts, S. P. & Feder, M. E. (1999). Natural hyperthermia and expression of the heat shock protein Hsp70 affect developmental abnormalities in *Drosophila melanogaster. Oecologia,* 121, 323–29.

Roberts, S. P. & Feder, M. E. (2000). Changing fitness consequences of hsp70 copy number in transgenic Drosophila larvae undergoing natural thermal stress. *Functional Ecology,* 14, 353–57.

Roberts, S. P. & Harrison, J. F. (1999). Mechanisms of thermal stability during flight in the honeybee *Apis mellifera. Journal of Experimental Biology,* 202, 1523–33.

Roberts, S. P., Harrison, J. F., & Dudley, R. (2004). Allometry of kinematics and energetics in carpenter bees (*Xylocopa* varipuncta) hovering in variable-density gases. *Journal of Experimental Biology,* 207, 993–1004.

Roberts, S. P., Harrison, J. F., & Hadley, N. F. (1998). Mechanisms of thermal balance in flying *Centris pallida* (Hymenoptera : Anthophoridae). *Journal of Experimental Biology,* 201, 2321–31.

Roberts, S. P., Marden, J. H., & Feder, M. E. (2003). Dropping like flies: Environmentally induced impairment and protection of locomotor performance in adult *Drosophila melanogaster. Physiological and Biochemical Zoology,* 76, 615–21.

Roberts, S. P., Quinlan, M. C., & Hadley, N. F. (1994). Interactive effects of humidity and temperature on water loss in the lubber grasshopper *Romelea guttata*. *Comparative Biochemistry and Physiology,* 109A, 627–31.

Robertson, H. M., Warr, C. G., & Carlson, J. R. (2003). Molecular evolution of the insect chemoreceptor gene superfamily in *Drosophila melanogaster*. *Proceedings of the National Academy of Sciences of the United States of America,* 100, 14537–42.

Robertson, I. C. (1998). Flight muscle changes in male pine engraver beetles during reproduction: the effects of body size, mating status and breeding failure. *Physiological Entomology,* 23, 75–80.

Roces, F. & Lighton, J. R. B. (1995). Larger bites of leaf-cutting ants. *Nature,* 373, 392–93.

Rodgers, C. I., Armstrong, G. A. B., Shoemaker, K. L., Labrie, J. D., Moyes, C. D., & Robertson, R. M. (2007). Stress preconditioning of spreading depression in the locust CNS. *PLOS One,* 2.

Roeder, T. (2005). Tyramine and octopamine: ruling behavior and metabolism. *Annual Review of Entomology,* 50, 447–77.

Roff, D. A. (2002). *Life History Evolution,* Sunderland, MA, Sinauer.

Rogowitz, G. L. & Chappell, M. A. (2000). Energy metabolism of eucalyptus-boring beetles at rest and during locomotion: gender makes a difference. *Journal of Experimental Biology,* 203, 1131–39.

Romer, H., Lang, A., & Hartbauer, M. (2010). The signaller's dilemma: A cost-benefit analysis of public and private communication. *PLOS One,* 5.

Rosay, P., Davies, S. A., Yu, Y., Soezen, M. A., Kaiser, K., & Dow, J. A. T. (1997). Cell-type specific calcium signalling in a *Drosophila* epithelium. *Journal of Cell Science,* 110, 1683–92.

Rose, M. (1984). Laboratory evolution of postponed senescence in *Drosophila melanogaster*. *Evolution,* 38, 1004–10.

Rosenzweig, M., Brennan, K. M., Tayler, T. D., Phelps, P. O., Patapoutian, A., & Garrity, P. A. (2005). The *Drosophila* ortholog of vertebrate TRPA1 regulates thermotaxis. *Genes & Development,* 19, 419–24.

Rosenzweig, M., Kang, K. J., & Garrity, P. A. (2008). Distinct TRP channels are required for warm and cool avoidance in Drosophila melanogaster. *Proceedings of the National Academy of Sciences of the United States of America,* 105, 14668–73.

Rostgaard, S. & Jacobsen, D. (2005). Respiration rate of stream insects measured *in situ* along a large altitude range. *Hydrobiologia,* 549, 79–98.

Roulston, T. H. & Cane, J. H. (2000). Pollen nutritional content and digestibility for animals. *Plant Systematics and Evolution,* 222, 187–209.

Rourke, B. C. (2000). Geographic and altitudinal variation in water balance and metabolic rate in a California grasshopper, *Melanoplus sanguinipes*. *Journal of Experimental Biology,* 203, 2699–712.

Rowland, J. M. & Emlen, D. J. (2009). Two thresholds, three male forms result in facultative male trimorphism in beetles. *Science* 323, 773.

Royer, D. L., Berner, R. A., Montanez, I. P., & Tabor, N. J. (2004). CO_2 as a primary driver for Phanerozoic climate. *GSA Today,* 14, 4–10.

Ruxton, G. D. & Humphries, S. (2008). Can ecological and evolutionary arguments solve the riddle of the missing marine insects? *Marine Ecology – An Evolutionary Perspective,* 29, 72–75.

Saito, T., Hirai, K., & Way, M. O. (2005). The rice water weevil, *Lissorhoptrus oryzophilus* Kuschel (Coleoptera: Curculionidae). *Applied Entomology and Zoology,* 40, 31–39.

Salt, G. (1968). The resistance of insect parasitoids to the defense reactions of their hosts. *Biological Reviews,* 43, 200–32.

Samakovlis, C., Manning, G., Steneberg, P., Hacohen, N., Cantera, R., & Krasnow, M. A. (1996). Genetic control of epithelial tube fusion during *Drosophila* tracheal development. *Development,* 122, 3531—36.

Samarra, F. I. P., Klappert, K., Brumm, H., & Miller, P. J. O. (2009). Background noise constrains communication: acoustic masking of courtship song in the fruit fly *Drosophila montana. Behaviour,* 146, 1635–48.

Sanchez-Gracia, A., Vieira, F. G., & Rozas, J. (2009). Molecular evolution of the major chemosensory gene families in insects. *Heredity,* 103, 208–16.

Sand-Jensen, K., Pedersen, O., Binzer, T., & Borum, J. (2005). Contrasting oxygen dynamics in the freshwater isoetid *Lobelia dortmanna* and the marine seagrass *Zostera marina. Annals of Botany,* 96, 613–23.

Santo Domingo, J. W., Kaufman, M. G., Klug, M. J., Holben, W. E., Harris, D., & Tiedje, J. M. (1998). Influence of diet on the structure and function of the bacterial hindgut community of crickets. *Molecular Ecology,* 7, 761–67.

Sato, H., Berry, C. W., Peeri, Y., Baghoomian, E., Casey, B. E., Lavella, G., Vandenbrooks, J. M., Harrison, J. F., & Maharbiz, M. M. (2009). Remote radio control of insect flight. *Frontiers in Integrative Neuroscience.*

Saudou, F., Boschert, U., Amlaiky, N., Plassat, J. L., & Hen, R. (1992). A family of Drosophila serotonin receptors with distinct intracellular signalling properties and expression patterns. *The EMBO Journal,* 11, 7–17.

Savage, V. M., Gillooly, J. F., Brown, J. H., West, G. B., & Charnov, E. L. (2004). Effects of body size and temperature on population growth. *American Naturalist,* 163, 429–41.

Sayeed, O. & Benzer, S. (1996). Behavioral genetics of thermosensation and hygrosensation in *Drosophila. Proceedings of the National Academy of Sciences of the United States of America,* 93, 6079–84.

Schade, J. D., Kyle, M., Hobbie, S. E., Fagan, W. F., & Elser, J. J. (2003). Stoichiometric tracking of soil nutrients by a desert insect herbivore. *Ecology Letters,* 6, 96–101.

Scheiner, R. & Erber, J. (2009). Sensory thresholds, learning and the division of labor of foraging in the honey bee. *In:* Gadau, J. & Fewell, J. H. (eds.) *Organization of Insect Socities: From Genomes to Complexity.* Cambridge: Harvard University Press.

Scherfer, C., Karlsson, C., Loseva, O., Bidla, G., Goto, A., Havemann, J., Dushay, M. S., & Theopold, U. (2004). Isolation and characterization of hemolymph clotting factors in *Drosophila melanogaster* by a pullout method. *Current Biology,* 14, 625–29.

Schilman, P. E. & Roces, F. (2006). Foraging energetics of a nectar-feeding ant: metabolic expenditure as a function of food-source profitability. *Journal of Experimental Biology,* 209, 4091–101.

Schilman, P. E. & Roces, F. (2008). Haemolymph sugar levels in a nectar feeding ant: dependence on metabolic expenditure and carbohydrate deprivation. *Journal of Comparative Physiology B-Biochemical Systemic and Environmental Physiology,* 178.

Schimpf, N. G., Matthews, P. G. D., Wilson, R. S., & White, C. R. (2009). Cockroaches breathe discontinuously to reduce respiratory water loss. *Journal of Experimental Biology,* 212, 2773–80.

Schippers, M. P., Dukas, R., & Mcclelland, G. B. (2010). Lifetime- and caste-specific changes in flight metabolic rate and muscle biochemistry of honeybees, *Apis mellifera. Journal of Comparative Physiology B-Biochemical Systemic and Environmental Physiology,* 180, 45–55.

Schippers, M. P., Dukas, R., Smith, R. W., Wang, J., Smolen, K., & Mcclelland, G. B. (2006). Lifetime performance in foraging honeybees: behaviour and physiology. *Journal of Experimental Biology,* 209, 3828–36.

Schlenstedt, J., Balfanz, S., Baumann, A., & Blenau, W. (2006). Am5-HT7: molecular and pharmacological characterization of the first serotonin receptor of the honeybee *(Apis mellifera) Journal of Neurochemistry,* 98, 1985–98.

Schmid-Hempel, P., Kacelnik, A., & Houston, A. I. (1985). Honeybees maximize efficiency by not filling their crop. *Behavioral Ecology and Sociobiology,* 17, 61–66.

Schmid-Hempel, P. & Wolf, T. (1988). Foraging effort and lifespan of workers in a social insect. *Journal of Animal Ecology,* 57, 500–21.

Schmidt-Nielsen, K. (1997). *Animal Physiology. Adaptation and Environment.* 5th Edition, Cambridge, Cambridge University Press.

Schmitz, A., Gebhardt, M., & Schmitz, H. (2008). Microfluidic photomechanic infrared receptors in a pyrophilous flat bug. *Naturwissenschaften,* 95, 455–60.

Schmitz, A. & Harrison, J. F. (2004). Hypoxic tolerance in air-breathing invertebrates. *Respiratory Physiology & Neurobiology,* 141, 229–42.

Schmitz, A. & Perry, S. F. (1999). Stereological determination of tracheal volume and diffusing capacity of the tracheal walls in the stick insect Carausius morosus (Phasmatodea, Lonchodidae). *Physiological and Biochemical Zoology,* 72, 205–18.

Schmitz, A., Schatzel, H., & Schmitz, H. (2010). Distribution and functional morphology of photomechanic infrared sensilla in flat bugs of the genus *Aradus* (Heteroptera, Aradidae). *Arthropod Structure & Development,* 39, 17–25.

Schmitz, A. & Wasserthal, L. T. (1999). Comparative morphology of the spiracles of the Papilionidae, Sphingidae, and Saturniidae (Insecta : Lepidoptera). *International Journal of Insect Morphology & Embryology,* 28, 13–26.

Schmitz, H. & Bleckmann, H. (1997). Fine structure and physiology of the infrared receptor of beetles of the genus *Melanophila* (Coleoptera : Buprestidae). *International Journal of Insect Morphology & Embryology,* 26, 205–15.

Schmitz, H. & Bleckmann, H. (1998). The photomechanic infrared receptor for the detection of forest fires in the beetles *Melanophila acuminata* (Coleoptera: Buprestidae). *Journal of Comparative Physiology A,* 182, 647–57.

Schmitz, H., Schmitz, A. & Bleckmann, H. (2000). A new type of infrared organ in the Australian "fire-beetle" Merimna atrata (Coleoptera : Buprestidae). *Naturwissenschaften,* 87, 542–45.

Schmitz, H., Schmitz, A., & Bleckmann, H. (2001). Morphology of a thermosensitive multipolar neuron in the infrared organ of *Merimna atrata* (Coleoptera, Buprestidae). *Arthropod Structure & Development,* 30, 99–111.

Schmitz, H., Schmitz, A., Trenner, S., & Bleckmann, H. (2002). A new type of insect infrared organ of low thermal mass. *Naturwissenschaften,* 89, 226–29.

Schmitz, H. & Trenner, S. (2003). Electrophysiological characterization of the multipolar thermoreceptors in the "fire-beetle" *Merimna atrata* and comparison with the infrared sensilla of Melanophila acuminata (both Coleoptera, Buprestidae). *Journal of Comparative Physiology a-Neuroethology Sensory Neural and Behavioral Physiology,* 189, 715–22.

Schmitz, H. & Wasserthal, L. T. (1993). Antennal thermoreceptors and wing-thermosensitivity of heliotherm butterflies: their possible role in thermoregulatory behaviour. *Journal of Insect Physiology,* 39(12), 1007–19.

Schmolz, E., Bruders, N., Schricker, B., & Lamprecht, I. (1999). Direct calorimetric measurement of heat production rates in flying hornets (*Vespa crabro*; Hymenoptera). *Thermochimica Acta,* 328, 3–8.

Schofield, C. J. (2000). *Trypanosoma cruzi*–the vector-parasite paradox. *Memórias do Instituto Oswaldo Cruz,* 95, 535–44.

Schoofs, A. & Spieß, R. (2007). Anatomical and functional characterisation of the stomatogastric nervous system of blowfly (*Calliphora vicina*) larvae. *Journal of Insect Physiology,* 53, 349–60.

Schoonhoven, L. M., Van Loon, J. A., & Dicke, M. (2005). *Insect-Plant Biology,* Oxford, University Press.

Schwab, W. (2003). Metabolome diversity: too few genes, too many metabolites? *Phytochemistry,* 62, 837–49.

Schwarz, H. & Moussian, B. (2007). Electron-microscopic and genetic dissection of arthropod cuticle differentiation. *In:* Méndez-Vilas, A. & Díaz, J. (eds.) *Modern Research and Educational Topics in Microscopy.* Badajoz, Spain: Formatex.

Schweikl, H., Klein, U., Schindlbeck, M., & Wieczorek, H. (1989). A vacuolar-type ATPase, partially purified from potassium transporting plasma membranes of tobacco hornworm midgut. *Journal of Biological Chemistry,* 264, 11136–42.

Scott, H. G. & Stojanovich, C. J. (1963). Digestion of juniper pollen by Collembola. *The Florida Entomologist,* 46, 189–91.

Scott, K., Brady, R., Cravchik, A., Morozov, P., Rzhetsky, A., Zuker, C., & Axel, R. (2001). A chemosensory gene family encoding candidate gustatory and olfactory receptors in *Drosophila. Cell,* 104, 661–73.

Scriber, J. M. (1979). Effects of leaf-water supplementation upon post-ingestive nutritional indices of forb-, shrub-, vine-, and tree-feeding Lepidoptera. *Entomologia Experimentalis Et Applicata,* 25, 240–52.

Scriber, J. M. (1984). Host-plant suitability. *In:* Bell, W. J. & Carde, R. T. (eds.) *Chemical Ecology of Insects.* London: Chapman and Hall.

Seely, M. K. (1979). Irregular fog as a water source for desert dune beetles. *Oecologia,* 42, 213–27.

Seely, M. K. & Hamilton, W. J. (1976). Fog catchment sand trenches constructed by tenebrionid beetles, *Lepidochora*, from Namib Desert. *Science,* 193, 484–86.

Seely, M. K., Lewis, C. J., Obrien, K. A., & Suttle, A. E. (1983). Fog response of tenebrionid beetles in the Namib Desert. *Journal of Arid Environments,* 6, 135–43.

Seid, M. A., Castillo, A., & Wcislo, W. T. (2011). The allometry of brain miniaturization in ants. *Brain Behavior and Evolution,* 77, 5–13.

Self, S. K., Guthrie, F. E., & Hodgson, E. (1964). Metabolism of nicotine by tobacco-feeding insects. *Nature,* 204, 300–301.

Seymour, R. S., White, C. R., & Gibernau, M. (2009). Endothermy of dynastine scarab beetles (*Cyclocephala colasi*) associated with pollination biology of a thermogenic arum lily (*Philodendron solimoesense*). *Journal of Experimental Biology,* 212, 2960–68.

Shakesby, A. J., Wallace, I. S., Isaacs, H. V., Pritchard, J., Roberts, D. M., & Douglas, A. E. (2009a). A water-specific aquaporin involved in aphid osmoregulation. *Insect Biochemistry and Molecular Biology,* 39, 1–10.

Shakesby, A. J., Wallace, I. S., Isaacs, H. V., Pritchard, J., Roberts, D. M., & Douglas, A. E. (2009b). A water-specific aquaporin involved in aphid osmoregulation. *Insect Biochemistry and Molecular Biology,* 39, 1–10.

Shanbhag, S. & Tripathi, S. (2005). Electrogenic H^+ Transport and pH Gradients generated by a V-H^+-ATPase in the isolated perfused larval *Drosophila* midgut. *Journal of Membrane Biology,* 206, 61–72.

Shanbhag, S. R., Müller, B., & Steinbrecht, R. A. (2000). Atlas of olfactory organs of *Drosophila melanogaster*: 2. Internal organization and cellular architecture of olfactory sensilla. *Arthropod Structure & Development,* 29, 211–29.

Shanbhag, S. R. & Singh, R. N. (1992). Functional implications of the projections of neurons from individual labellar sensillum of *Drosophila melanogaster* as revealed by neuronal marker horseradish peroxidase. *Cell and Tissue Research,* 267, 273–82.

Shear, W. A. & Kukalova-Peck, J. (1990). The ecology of Paleozoic terrestrial arthropods: the fossil evidence. *Canadian Journal of Zoology,* 68, 1807–34.

Shi, W., Ding, S. Y., & Yuan, J. S. (2011). Comparison of insect gut cellulase and xylanase activity across different insect species with distinct food sources. *Bioenergy Research,* 4, 1–10.

Shields, V. D. C. & Hildebrand, J. G. (2000). Responses of a population of antennal olfactory receptor cells in the female moth Manduca sexta to plant-associated volatile organic compounds. *Journal of Comparative Physiology A – Neuroethology Sensory Neural and Behavioral Physiology,* 186, 1135–51.

Shik, J. (2010). The metabolic costs of building ant colonies from variably sized subunits. *Behavioral Ecology and Sociobiology,* 64, 12, 1981–1990.

Showler, A. T. & Moran, P. J. (2003). Effects of drought stressed cotton, *Gossypium hirsutum* L., on beet armyworm, *Spodoptera exigua* (Hübner) oviposition, and larval feeding preferences and growth. *Journal of Chemical Ecology,* 29, 1997–2011.

Shreve, S. M., Yi, S. X., & Lee, R. E. (2007). Increased dietary cholesterol enhances cold tolerance in Drosophila melanogaster. *Cryoletters,* 28, 33–37.

Silbermann, R. & Tatar, M. (2000). Reproductive costs of heat shock protein in transgenic Drosophila melanogaster. *Evolution,* 54, 2038–45.

Sim, C. & Denlinger, D. L. (2008). Insulin signaling and FOXO regulate the overwintering diapause of the mosquito *Culex pipiens*. *Proceedings of the National Academy of Sciences USA,* 105, 6777–81.

Simmons, A. M. (1993). Effects of constant and fluctuating temperatures and humidities on the survival of *Spodoptera frugiperda* pupae (Lepidoptera, Noctuidae). *Florida Entomologist,* 76, 333–40.

Simpson, S. J. (1990). The pattern of feeding. *In:* Chapman, R. F. & Joern, A. (eds.) *The Biology of Grasshoppers.* New York: John Wiley and Sons.

Simpson, S. J. & Abisgold, J. D. (1985). Compensation by locusts for changes in dietary nutrients: behavioural mechanisms. *Physiological Entomology,* 10, 443–52.

Simpson, S. J. & Bernays, E. A. (1983). The regulation of feeding-locusts and blowflies are not so different from mammals. *Appetite,* 4, 313–46.

Simpson, S. J. & Miller, G. A. (2007). Maternal effects on phase characteristics in the desert locust, *Schistocerca gregaria*: A review of current understanding. *Journal of Insect Physiology,* 53, 869–76.

Simpson, S. J. & Raubenheimer, D. (1993a). The central role of the hemolymph in the regulation of nutrient intake in insects. *Physiological Entomology,* 18, 395–403.

Simpson, S. J. & Raubenheimer, D. (1993b). A multi-level anyalysis of feeding behaviour: the geometry of nutritional decisions. *Philosophical Transactions of the Royal Society of London,* 342, 381–402.

Simpson, S. J. & Raubenheimer, D. (1995). The geometric analysis of feeding and nutrition: a user's guide. *Journal of Insect Physiology,* 41, 545–53.

Simpson, S. J. & Raubenheimer, D. (2001). The geometric analysis of nutrient-allelochemical interactions: a case study using locusts. *Ecology,* 82, 422–39.

Simpson, S. J. & Raubenheimer, D. (2009). Macronutrient balance and lifespan. *Aging,* 1, 875–80.

Simpson, S. J., Sibly, R. M., Lee, K. P., Behmer, S. T., & Raubenheimer, D. (2004). Optimal foraging when regulating intake of multiple nutrients. *Animal Behaviour,* 68, 1299–1311.

Sinclair, B. J., Addo-Bediako, A., & Chown, S. L. (2003). Climatic variability and the evolution of insect freeze tolerance. *Biological Reviews,* 78, 181–95.

Sinclair, B. J., Gibbs, A. G., Lee, W. K., Rajamohan, A., Roberts, S. P., & Socha, J. J. (2009). Synchrotron X-Ray visualisation of ice formation in insects during lethal and non-lethal freezing. *PLOS One,* 4.

Sinclair, B. J., Gibbs, A. G., & Roberts, S. P. (2007). Gene transcription during exposure to, and recovery from, cold and desiccation stress in *Drosophila melanogaster. Insect Molecular Biology,* 16, 435–43.

Sinclair, B. J. & Renault, D. (2010). Intracellular ice formation in insects: Unresolved after 50 years? *Comparative Biochemistry and Physiology Part A Molecular & Integrative Physiology,* 155, 14–18.

Singer, M. S., Mace, K. C., & Bernays, E. A. (2009). Self-medication as adaptive plasticity: increased ingestion of plant toxins by parasitized caterpillars. *PLOS One,* 4, e4796.

Sipes, S. D. & Tepedino, V. J. (2005). Pollen-host specificity and evolutionary patterns of host switching in a clade of specialist bees (Apoidea: Diadasia). *Biological Journal of the Linnean Society,* 86, 487–505.

Siva-Jothy, M. T., Moret, Y., & Rolff, J. (2005). Insect immunity: an evolutionary ecology perspective. *In:* Simpson, S. J. (ed.) *Advances in Insect Physiology.* Academic Press.

Slachta, M., Berkova, P., Vambera, J., & Kostal, V. (2002). Physiology of cold-acclimation in non-diapausing adults of *Pyrrhocoris apterus* (Heteroptera). *European Journal of Entomology,* 99, 181–87.

Slama, K. (1988). A new look at insect respiration. *Biological Bulletin,* 175, 289–300.

Slama, K. (1994). Regulation of respiratory acidemia by the autonomic nervous system (Coelopulse) in insects and ticks. *Physiological Zoology,* 67, 163–74.

Slama, K. (1999). Active regulation of insect respiration. *Annals of the Entomological Society of America,* 92, 916–29.

Slama, K. (2006). Heartbeat reversal after sectioning the dorsal vessel and removal of the brain of diapausing pupae of *Manduca sexta* (Lepidoptera: Sphingidae). *European Journal of Entomology,* 103, 17–26.

Slama, K. (2010). Physiology of heartbeat reversal in adult *Drosophila melanogaster* (Diptera: Drosophilidae). *European Journal of Entomology,* 107, 13–31.

Slama, K. & Lukas, J. (2011). Myogenic nature of insect heartbeat and intestinal peristalsis, revealed by neuromuscular paralysis caused by the sting of a braconid wasp. *Journal of Insect Physiology,* 57, 251–59.

Slama, L. & Farkas, R. (2005). Heartbeat patterns during the postembryonic development of *Drosophila melanogaster. Journal of Insect Physiology,* 51, 489–503.

Slansky, F. J. & Scriber, J. M. (1985). Food consumption and utilization. *In:* Kerkut, G. A. & Gilbert, L. I. (eds.) *Comprehensive Insect Physiology, Biochemistry and Pharmacology.* Oxford: Pergamon Press.

Smedley, S. R. & Eisner, T. (1996). Sodium: a male moth's gift to its offspring. *Proceedings of the National Academy of Science, USA,* 93, 809–13.

Smilanich, A. M., Dyer, L. A., Chambers, J. Q., & Bowers, M. D. (2009). Immunological cost of chemical defence and the evolution of herbivore diet breadth. *Ecology Letters,* 12, 612–21.

Smith-Espinoza, C. J., Richter, A., Salamini, F., & Bartels, D. (2003). Dissecting the response to dehydration and salt (NaCl) in the resurrection plant *Craterostigma plantagineum. Plant, Cell and Environment,* 26, 1307–15.

Smith, K. E., Vanekeris, L. A., & Linser, P. J. (2007). Cloning and characterization of AgCA9, a novel alpha-carbonic anhydrase from *Anopheles gambiae* Giles sensu stricto (Diptera: Culicidae) larvae. *Journal of Experimental Biology,* 210, 3919–30.

Smith, K. E., Vanekeris, L. A., Okech, B. A., Harvey, W. R., & Linser, P. J. (2008). Larval anopheline mosquito recta exhibit a dramatic change in localization patterns of ion transport proteins in response to shifting salinity: a comparison between anopheline and culicine larvae. *Journal of Experimental Biology,* 211, 3067–76.

Smits, A. W., Burggren, W. W., & Oliveras, D. (2000). Developmental changes in in vivo cardiac performance in the moth *Manduca sexta. Journal of Experimental Biology,* 203, 369–78.

Snodgrass, R. E. (1935). *Principles of Insect Morphology,* New York, McGraw Hill.

Soares-Costa, A., Dias, A. B., Dellamano, M., De Paula, F. F. P., Carmona, A. K., Terra, W. R., & Henrique-Silva, F. (2011). Digestive physiology and characterization of digestive cathepsin L-like proteinase from the sugarcane weevil *Sphenophorus levis. Journal of Insect Physiology,* 57, 462–68.

Socha, J. J. & Decarlo, F. (2008). Use of synchrotron tomography to image naturalistic anatomy in insects. *Developments in X-Ray Tomography VI: (2008), San Diego, CA, USA: SPIE,* (2008), 70780A–70787.

Socha, J. J., Forster, T. D., & Greenlee, K. J. (2010). Issues of convection in insect respiration: Insights from synchotron X-ray imaging and beyond. *Respiratory Physiology & Neurobiology,* 173S, S65–S73.

Socha, J. J., Lee, W.-K., Harrison, J. F., Waters, J. S., Fezzaa, K., & Westneat, M. W. (2008). Correlated patterns of tracheal compression and convective gas exchange in a carabid beetle. *Journal of Experimental Biology,* 211, 3409–20.

Socha, J. J., Westneat, M. W., Harrison, J. F., Waters, J. S., & Lee, W. K. (2007). Real-time phase-contrast x-ray imaging, a new technique for the study of animal form and function. *BMC Biology,* 5, 6.

Sohal, R. S., Allen, R., Farmer, K. J., & Newton, R. K. (1985). Iron induces oxidative stress and may alter the rate of aging in the housefly, *Musca domestica. Mechanisms of Aging and Development,* 32, 33–38.

Sokabe, T. & Tominaga, M. (2009). A temperature-sensitive TRP ion channel, painless, functions as a noxious heat sensor in fruit flies. *Communicative & Integrative Biology,* 2, 170–73.

Sokolova, Y. Y. & Lange, C. E. (2002). An ultrastructural study of *Nosema locustae* Canning (Microsporidia) from three species of Acrididae (Orthoptera). *Acta Protozool.,* 41, 229–37.

Sollars, V., Lu, X., Xiao, L., Wang, X., Garfinkel, M. D., & Ruden, D. M. (2002). Evidence for an epigenetic mechanism by which Hsp90 acts as a capacitor for morphological evolution. *Nature Genetics,* 33, 70–74.

Sombati, S. & Hoyle, G. (1984). Generation of specific behaviors in a locust by local release into neuropil of the natural neuromodulator octopamine. *Journal of Neurobiology,* 15, 481–506.

Sorensen, J. G., Kristensen, T. N., & Loeschcke, V. (2003). The evolutionary and ecological role of heat shock proteins. *Ecology Letters,* 6, 1025–37.

Sorensen, J. G. & Loeschcke, V. (2007). Studying stress responses in the post-genomic era: its ecological and evolutionary role. *Journal of Biosciences,* 32, 447–56.

Soulages, J. L., Van Antwerpen, R., & Wells, M. A. (1996). Role of diacylglycerol and apolipophorin-III in regulation of the physicochemical properties of the lipophorin surface: metabolic implications. *Biochemistry,* 35, 5191–98.

South, S. H., House, C. M., Moore, A. J., Simpson, S. J., & Hunt, J. (2011). Male cockroaches prefer a high carbohydrate diet that makes them more attractive to females: implications for the study of condition dependence *Evolution,* 65, 1594–1606.

Southwick, E. E., Roubik, D. W., & Williams, J. M. (1990). Comparative energy balance in groups of Africanized and European honey bees: ecological implications. *Comparative Biochemistry and Physiology,* 97A, 1–7.

Spangler, H. G. (1992). The influence of temperature on the wingbeat frequencies of free-flying honey bees, *Apis mellifera* L. (Hymenoptera: Apidae). *Bee Science,* 2, 181–86.

Spangler, H. G. & Buchmann, S. L. (1991). Effects of temperature on wingbeat frequency in the solitary bee *Centris caesalpiniae* (Anthophoridae:Hymenoptera). *Journal of the Kansas Entomological Society,* 64, 107–09.

Speakman, J. R. (1997). *Doubly Labeled Water: Theory and Practice,* New York, Springer Scientific.

Spence, J. R. & Andersen, N. M. (1994). Biology of water striders – interactions between systematics and ecology. *Annual Review of Entomology,* 39, 101–28.

Spiewok, S. & Schmolz, E. (2006). Changes in temperature and light alter the flight speed of hornets (*Vespa crabro* L.). *Physiological and Biochemical Zoology,* 79, 188–93.

Spring, J. H., Robichaux, S. R., & Hamlin, J. A. (2009). The role of aquaporins in excretion in insects. *Journal of Experimental Biology,* 212, 358–62.

Spring, J. H., Robichaux, S. R., Kaufmann, N., & Brodsky, J. L. (2007). Localization of a *Drosophila* DRIP-like aquaporin in the Malpighian tubules of the house cricket, *Acheta domesticus. Comparative Biochemistry and Physiology – Part A: Molecular & Integrative Physiology,* 148, 92–100.

Srygley, R. B. & Lorch, P. D. (2011). Weakness in the band: nutrient-mediated trade-offs between migration and immunity of Mormon crickets, *Anabrus simplex. Animal Behaviour,* 81, 395–400.

Stadler, B. & Dixon, A. F. G. (2008). *Mutualism: Ants and Their Insect Partners,* Cambridge, Cambridge University Press.

Stalker, H. D. & Carson, H. L. (1948). An altitudinal transect of *Drosophila robusta* Sturtevant. *Evolution,* 2, 295–305.

Stanley, R. G. & Linskens, H. F. (1974). *Pollen: Biology, Biochemistry, Management.* New York, Springer.

Staples, J. F., Koen, E. L., & Laverty, T. M. (2004). "Futile cycle" enzymes in the flight muscles of North American bumblebees. *Journal of Experimental Biology,* 207, 749–54.

Starck, J. M., Karasov, W. H., & Afik, D. (2000). Intestinal nutrient uptake measurements and tissue damage: Validating the everted sleeves method. *Physiological and Biochemical Zoology,* 73, 454–60.

Stearns, S. C. (1992). *The Evolution of Life Histories,* Oxford, Oxford University Press.

Steele, J. E. & Hall, S. (1985). Trehalose synthesis and glycogenolysis as sites of action for the corpus cardiacum in *Periplaneta americana. Insect Biochemistry,* 15(4), 529–36.

Steinbrecht, R. A. (1984). Chemo-, thermo-, and hygroreceptors. *Biology of the Integument.* Heidelberg, Springer-Verlag.

Stelzer, R. J., Stanewsky, R., & Chittka, L. (2010). Circadian foraging rhythms of bumblebees monitored by radio-frequency identification. *Journal of Biological Rhythms,* 25, 257–67.

Stensmyr, M. C., Dekker, T., & Hansson, B. S. (2003). Evolution of the olfactory code in the Drosophila melanogaster subgroup. *Proceedings of the Royal Society of London Series B -Biological Sciences,* 270, 2333–40.

Stephens, D. W. & Krebs, J. R. (1986). *Foraging Theory,* Princeton, Princeton University Press.

Sterner, R. W. & Elser, J. E. (2002). *Ecological Stoichiometry: The Biology of Elements from Molecules to the Biosphere,* Princeton, Princeton University Press.

Stillwell, R. C. (2010). Are latitudinal clines in body size adaptive? *Oikos,* 119, 1387–90.

Stireman, J. O., O'hara, J. E., & Wood, D. M. (2006). Tachinidae: evolution, behavior, and ecology. *Annual Review of Entomology,* 51, 525–55.

Stockhoff, B. A. (1993). Ontogenic change in diet selection for protein and lipid by gypsy moth larvae. *Journal of Insect Physiology,* 39, 677–86.

Stoehr, A. M. (2010). Responses of disparate phenotypically-plastic, melanin-based traits to common cues: limits to the benefits of adaptive plasticity? *Evolutionary Ecology,* 24, 287–98.

Stoehr, A. M. & Goux, H. (2008). Seasonal phenotypic plasticity of wing melanisation in the cabbage white butterfly, *Pieris rapae* L. (Lepidoptera : Pieridae). *Ecological Entomology,* 33, 137–43.

Stone, G. N. (1993). Endothermy in the solitary bee anthophora plumipes – independent measures of thermoregulatory ability, costs of warm-up and the role of body size. *Journal of Experimental Biology,* 174, 299–320.

Stone, G. N. & Willmer, P. G. (1989). Endothermy and temperature regulation in bees: a critique of "grab and stab" measurement of body temperature. *Journal of Experimental Biology,* 143, 211–23.

Storey, K. B. (1978). Purification and properties of fructose diphosphatase from bumblebee flight muscle Role of enzyme in control of substrate cycling. *Biochimica et Biophysica Acta,* 523, 443–53.

Storey, K. B. (1997). Organic solutes in freezing tolerance. *Comparative Biochemistry and Physiology Part A: Physiology,* 117, 319–26.

Storey, K. B. & Storey, J. M. (1988). Freeze tolerance in animals. *Physiological Reviews,* 68, 27–84.

Strachan, L. A., Tarnowski-Garner, H. E., Marshall, K. E., & Sinclair, B. J. (2011). The evolution of cold tolerance in Drosophila larvae. *Physiological and Biochemical Zoology,* 84, 43–53.

Strain, B. R. (1991). Available technologies for field experimentation with elevated CO_2 in global change research. In: *Ecosystem Experiments,* H.A. Mooney et al.(eds), 245–61.

Straw, A. D., Branson, K., Neumann, T. R., & Dickinson, M. H. (2011). Multi-camera real-time three-dimensional tracking of multiple flying animals. *Journal of the Royal Society Interface,* 8, 395–409.

Sumner, S., Lucas, E., Barker, J., & Isaac, N. (2007). Radio-tagging technology reveals extreme nest-drifting behavior in a eusocial insect. *Current Biology,* 17, 140–45.

Sutton, G. P. & Burrows, M. (2011). Biomechanics of jumping in the flea. *Journal of Experimental Biology,* 214, 836–47.

Sword, G. A., Lorch, P. D., & Gwynne, D. T. (2005). Migratory bands give crickets protection. *Nature,* 433, 703–703.

Tammero, L. F. & Dickinson, M. F. (2002). The influence of visual landscape on the free flight behavior of the fruit fly *Drosophila melanogaster. Journal of Experimental Biology,* 205, 327–43.

Taneja-Bageshwar, S., Strey, A., Zubrzak, P., Williams, H., Reyes-Rangel, G., Juaristi, E., & Nachman, R. J. (2008). Identification of selective and non-selective, biostable beta-amino acid agonists of recombinant insect kinin receptors from the southern cattle tick *Boophilus microplus* and mosquito *Aedes aegypti. Peptides,* 29, 302–09.

Tartes, U., Vanatoa, A., & Kuusik, A. (2002). The insect abdomen—a heartbeat manager in insects? *Comparative Biochemistry and Physiology – Part A: Molecular & Integrative Physiology,* 133, 611–23.

Tatar, M., Kopelman, A., Epstein, D., Tu, M. P., Yin, C. M., & Garofalo, R. S. (2001). A mutant Drosophila insulin receptor homolog that extends life-span and impairs neuroendocrine function. *Science,* 292, 107–110.

Tatar, M. & Yin, C. (2001). Slow aging during insect reproductive diapause: why butterflies, grasshoppers and flies are like worms. *Experimental Gerontology,* 36, 723–38.

Te Brugge, V. A., Lombardi, V. C., Schooley, D. A., & Orchard, I. (2005). Presence and activity of a Dippu-DH31-like peptide in the blood-feeding bug, *Rhodnius prolixus. Peptides,* 26, 29–42.

Te Brugge, V. A., Schooley, D. A., & Orchard, I. (2008). Amino acid sequence and biological activity of a calcitonin-like diuretic hormone (DH31) from *Rhodnius prolixus. Journal of Experimental Biology,* 211, 382–90.

Techanau, G. M. (1984). Fiber number in the mushroom bodies of adult Drosophila-melanogaster depends on age sex and experience. *Journal of Neurogenetics,* 1, 113–26.

Teets, N. M., Elnitsky, M. A., Benoit, J. B., Lopez-Martinez, G., Denlinger, D. L., & Lee, R. E. (2008). Rapid cold-hardening in larvae of the Antarctic midge Belgica antarctica: cellular cold-sensing and a role for calcium. *American Journal of Physiology-Regulatory Integrative and Comparative Physiology,* 294, R1938–R1946.

Teixeira, L., Rabouille, C., Rørth, P., Ephrussi, A., & Vanzo, N. F. (2003). Drosophila Perilipin/ADRP homologue Lsd2 regulates lipid metabolism. *Mechanisms of Development,* 120, 1071–81.

Telang, A., Booton, V., Chapman, R. F., & Wheeler, D. E. (2001). How female caterpillars accumulate their nutrient reserves. *Journal of Insect Physiology,* 47, 1055–64.

Telang, A., Buck, N. A., Chapman, R. E., & Wheeler, D. E. (2003). Sexual differences in postingestive processing of dietary protein and carbohydrate in caterpillars of two species. *Physiological & Biochemical Zoology,* 76, 247.

Telang, A., Buck, N. A., & Wheeler, D. E. (2002). Response of storage protein levels to variation in dietary protein levels. *Journal of Insect Physiology,* 48, 1021–29.

Terblanche, J. S., Marais, E., Hetz, S. K,. & Chown, S. L. (2008). Control of discontinuous gas exchange in *Samia cynthia*: effects of atmospheric oxygen, carbon dioxide and moisture. *Journal of Experimental Biology,* 211, 3272–80.

Terra, W. R. (1990). Evolution of digestive systems of insects. *Annual Review of Entomology,* 35, 181–200.

Terra, W. R. (2001). The origin and functions of the insect peritrophic membrane and peritrophic gel. *Archives of Insect Biochemistry and Physiology,* 47, 47–61.

Terra, W. R., Ferriera, C., Jordao, B. P., & Dillon, R. J. (1996). Digestive enzymes. *In:* Lehane, M. J., & Billingsley, P. F. (eds.) *Biology of the Insect Midgut.* London: Chapman and Hall.

Theopold, U., Schmidt, O., Söderhäll, K., & Dushay, M. S. (2004). Coagulation in arthropods: defence, wound closure and healing. *Trends in Immunology,* 25, 289–94.

Thompson, S. N. (2003). Trehalose – The Insect 'Blood' Sugar. *In:* Simpson, S. J. (ed.) *Advances in Insect Physiology.* Academic Press.

Thompson, S. N. & Redak, R. A. (2005). Feeding behavior and nutrient selection in an insect *Manduca sexta L.* and alterations induced by parasitism. *Journal of Comparative Physiology A,* 191, 909–23.

Thomson, R. B., Thomson, J. M., & Phillips, J. E. (1988). NH_4^+ transport in acid-secreting insect epithelium. *American Journal of Physiology,* 254, 348–56.

Thorne, N., Chromey, C., Bray, S., & Amrein, H. (2004). Taste perception and coding in *Drosophila. Current Biology,* 14, 1065–79.

Thorpe, W. H. & Crisp, D. J. (1947a). Studies on plastron respiration. I. The biology of *Aphelocheirus* (Hemiptera, Aphelocheiridae (Naucoridae)) and the mechanism of plastron respiration. *Journal of Experimental Biology,* 24, 227–69.

Thorpe, W. H. & Crisp, D. J. (1947b). Studies on plastron respiration. II. The respiratory efficiency of the plastron in *Aphelocheirus. Journal of Experimental Biology,* 24, 270–303.

Timmins, G. S., Bechara, E. J. H., & Swartz, H. M. (2000). Direct determination of the kinetics of oxygen diffusion to the photocytes of a bioluminescent elaterid larva, measurement of gas- and aqueous-phase diffusional barriers and modelling of oxygen supply. *Journal of Experimental Biology,* 203, 2479–84.

Timmins, G. S., Penatti, C. A., Bechara, E. J., & Swartz, H. M. (1999). Measurement of oxygen partial pressure, its control during hypoxia and hyperoxia, and its effect upon light emission in a bioluminescent elaterid larva. *Journal of Experimental Biology,* 202, 2631–38.

Timmins, G. S., Robb, F. J., Wilmot, C. M., Jackson, S. K., & Swartz, H. M. (2001). Firefly flashing is controlled by gating oxygen to light-emitting cells. *Journal of Experimental Biology,* 204, 2795–801.

Tomcala, A., Tollarova, M., Overgaard, J., Simek, P., & Kostal, V. (2006). Seasonal acquisition of chill tolerance and restructuring of membrane glycerophospholipids in an overwintering insect: triggering by low temperature, desiccation and diapause progression. *Journal of Experimental Biology,* 209, 4102–14.

Tonning, A., Hemphala, J., Tang, E., Nannmark, U., Samakovlis, C., & Uv, A. (2005). A transient luminal chitinous matrix is required to model epithelial tube diameter in the *Drosophila* trachea. *Developmental Cell,* 9, 423–30.

Toolson, E. C. (1987). Water profligacy as an adaptation to hot deserts: water loss rates and evaporative cooling in the sonoran desert cicada, *Diceroprocta apache* (Homoptera: cicadidae). *Physiological Zoology,* 60, 379–85.

Toolson, E. C. & Hadley, N. F. (1987). Energy-dependent facilitation of transcuticular water flux contributes to evaporative cooling in the Sonoran Desert cicada, *Diceroprocta apache* (Homoptera: Cicadidae). *Journal of Experimental Biology,* 131, 439–44.

Tracey, W. D., Wilson, R. I., Laurent, G., & Benzer, S. (2003). painless, a *Drosophila* gene essential for nociception. *Cell,* 113, 261–73.

Treherne, J. E. & Schofield, P. K. (1981). Mechanisms of ionic homeostasis in the central nervous-system of an insect. *Journal of Experimental Biology,* 95, 61–73.

Trougakos, I. P. & Margaritis, L. H. (2002). Novel morphological and physiological aspects of insect eggs. *In:* Hilker, M. & Meiners, T. (eds.) *Chemoecology of Insect Eggs and Egg Deposition.* Berlin: Blackwell Wissenschaftsverlag.

Truman, J. W., Hiruma, K., Allee, J. P., Macwhinnie, S. G. B., Champlin, D. T., & Riddiford, L. M. (2006). Juvenile hormone is required to couple imaginal disc formation with nutrition in insects. *Science,* 312, 1385–88.

Trumper, S. & Simpson, S. J. (1994). Mechanisms Regulating Salt Intake in 5th-Instar Nymphs of Locusta-Migratoria. *Physiological Entomology,* 19, 203–15.

Tsubaki, Y., Kato, C., & Shintani, S. (2006). On the respiratory mechanism during underwater oviposition in a damselfly *Calopteryx cornelia* Selys. *Journal of Insect Physiology,* 52, 499–505.

Tu, M.-P., Yin, C.-M., & Tatar, M. (2005). Mutations in insulin signaling pathway alter juvenile hormone synthesis in *Drosophila melanogaster. General and Comparative Endocrinology,* 142, 347–56.

Tublitz, N. (1989). Insect cardioactive peptides – neuro-hormonal regulation of cardiac activity by 2 cardioacceleratory peptides during flight in the tobacco hawkmoth, *Manduca sexta. Journal of Experimental Biology,* 142, 31–48.

Tublitz, N. J. & Truman, J. W. (1985). Insect cardioactive peptide. 2. Neuro-hormonal control of heart activity by 2 cardioacceleratory peptides in the tobacco hawkmoth, *Manduca sexta. Journal of Experimental Biology,* 114, 381–95.

Turlings, T. C. J., Loughrin, J. H., Mccall, P. J., Rose, U. S. R., Lewis, W. J., & Tumlinson, J. H. (1995). How caterpillar-damaged plants protect themselves by attracting parasitic wasps. *Proceedings of the National Academy of Science, USA,* 92, 4169–74.

Turunen, S. & Crailsheim, K. (1996). Lipid and sugar absorption. *In:* Lehane, M. J. & Billingsley, P. F. (eds.) *Biology of the Insect Midgut.* London: Chapman and Hall.

Tyshenko, M. G., Doucet, D., & Walker, V. K. (2005). Analysis of antifreeze proteins within spruce budworm sister species. *Insect Molecular Biology,* 14, 319–26.

Ullman, M. (2006). African desert locusts in Morocco in November (2004). *British Birds,* 99, 489–91.

Unwin, D. M. & Corbet, S. A. (1984). Wingbeat frequency, temperature and body size in bees and flies. *Physiological Entomology,* 9, 115–21.

Usherwood, P. N. R. (1975). *Insect Muscle,* London, Academic Press.

Usinger, R. L. (1957). Marine insects. *Geological Society of America Memoir,* 67, 1177–82.

Van't Land, J., Van Putten, P., Zwaan, B., Kamping, A., & Van Delden, W. (1999). Latitiudinal variation in wild populations of *Drosophila melanogaster*: heritabilities and reaction norms. *Journal of Evolutionary Biology,* 12, 222–32.

Van Den Broek, I. V. F. & Den Otter, C. J. (1999). Olfactory sensitivities of mosquitoes with different host preferences (*Anopheles gambiae* s.s., An. arabiensis, An. quadriannulatus, An. m. atroparvus) to synthetic host odours. *Journal of Insect Physiology,* 45, 1001–10.

Van Der Hage, J. C. H. (1996). Why are there no insects and so few higher plants, in the sea? New thoughts on an old problem. *Functional Ecology,* 10, 546–47.

Van Der Have, T. M. & De Jong, G. (1996). Adult size in ectotherms: temperature effects on growth and differentiation. *Journal of Theoretical Biology,* 183, 329–40.

Van Der Klook, W. G. (1963). The electrophysiology and the nervous control of the spiracular muscle of pupae of the giant silkmoths. *Comparative Biochemistry and Physiology,* 9, 317–33.

Van Loon, J. J. A. & Schoonhoven, L. M. (1999). Specialist deterrent chemoreceptors enable *Pieris* caterpillars to discriiminate between chemically different deterrents. *Entomologia Experimentalis Et Applicata,* 91, 29–35.

Van Voorhies, W. A. (1996). Bergmann size clines: a simple explanation for their occurrence in ectotherms. *Evolution,* 50, 1259–64.

Vance, J. T., Williams, J. B., Elekonich, M. M., & Roberts, S. P. (2009). The effects of age and behavioral development on honey bee (Apis mellifera) flight performance. *Journal of Experimental Biology,* 212, 2604–11.

Vannier, G. (1994). The thermobiological limits of some freezing tolerant insects: the supercooling and thermostupor points. *Acta Oecologica,* 15, 31–42.

Veenstra, J. A. (1988). Effects of 5-hydroxytrptamine on the Malpighian tubules of *Aedes aegypti. Journal of Insect Physiology,* 34, 299–304.

Veenstra, J. A. (1989). Isolation and structure of corazonin, a cardioactive peptide from the american cockroach. *FEBS Letters,* 250, 231–34.

Vermehren, A., Langlais, K. K., & Morton, D. B. (2006). Oxygen-sensitive guanylyl cyclases in insects and their potential roles in oxygen detection and in feeding behaviors. *Journal of Insect Physiology,* 52, 340–48.

Vermeij, G. J. & Dudley, R. (2000). Why are there so few evolutionary transitions between aquatic and terrestrial ecosystems? *Biological Journal of the Linnean Society,* 70, 541–54.

Vigoreaux, J. O. (ed.) (2006). *Nature's Versatile Engine: Insect Flight Muscles Inside and Out,* New York/Georgetown: Springer/Landes Bioscience.

Vinatier, F., Chailleux, A., Duyck, P. F., Salmon, F., Lescourret, F., & Tixier, P. (2010). Radiotelemetry unravels movements of a walking insect species in heterogeneous environments. *Animal Behaviour,* 80, 221–29.

Vincent, J. F. V. & Hillerton, J. E. (1979). The tanning of insect cuticle e a critical review and a revised mechanism. *Journal of Insect Physiology,* 25, 653–58.

Vincent, J. F. V. & Wegst, U. G. K. (2004). Design and mechanical properties of insect cuticle. *Arthropod Structure & Development,* 33, 187–99.

Viswanath, V., Story, G. M., Peier, A. M., Petrus, M. J., Hwang, S. W., Patapoutian, A., & Jegla, T. (2003). Ion channels – Opposite thermosensor in fruitfly and mouse. *Nature,* 423, 822–23.

Vogel, S. (1994). *Life in Moving Fluids: The Physical Biology of Flow,* Princeton, Princeton University Press.

Vogt, J. T. & Appel, A. G. (1999). Standard metabolic rate of the fire ant, *Solenopsis invicta* Buren: effects of temperature, mass, and caste. *Journal of Insect Physiology,* 45, 655–66.

Voigt, C. C., Michener, R., & Kunz, T. H. (2005). The energetics of trading nuptial gifts for copulations in katydids. *Physiological and Biochemical Zoology,* 78, 417–23.

Von Frisch, K. (1993). *The Dance Language and Orientation of Bees,* Cambridge, Harvard University Press.

Vondran, T., Apel, K. H., & Schmitz, H. (1995). The infrared receptor of *Melanophila acuminata* De Geer (Coleoptera: Buprestidae): Ultrastructural study of a unique insect thermoreceptor and its possible descent from a hair mechanoreceptor. *Tissue & Cell,* 27, 645–58.

Vosshall, L. B. & Stocker, R. F. (2007). Molecular architecture of smell and taste in *Drosophila. Annual Review of Neuroscience,* 30, 505–33.

Walcher, F. & Kral, K. (1994). Visual deprivation and distance estimation in the praying-mantis larva. *Physiological Entomology,* 19, 230–40.

Walczynska, A. (2009). Bioenergetic strategy of a xylem-feeder. *Journal of Insect Physiology,* 55, 1107–17.

Walters, K. R., Serianni, A. S., Sformo, T., Barnes, B. M., & Duman, J. G. (2009). A nonprotein thermal hysteresis-producing xylomannan antifreeze in the freeze-tolerant Alaskan beetle *Upis ceramboides. Proceedings of the National Academy of Sciences, USA,* 106, 20210–15.

Walters, K. R., Serianni, A. S., Voituron, Y., Sformo, T., Barnes, B. M., & Duman, J. G. (2011). A thermal hysteresis-producing xylomannan glycolipid antifreeze associated with cold tolerance is found in diverse taxa. *Journal of Comparative Physiology B-Biochemical Systemic and Environmental Physiology,* 181, 631–40.

Wang, G. R., Qiu, Y. T., Lu, T., Kwon, H. W., Pitts, R. J., Van Loon, J. J. A., Takken, W., & Zwiebel, L. J. (2009). Anopheles gambiae TRPA1 is a heat-activated channel expressed in thermosensitive sensilla of female antennae. *European Journal of Neuroscience,* 30, 967–74.

Wang, H., Ando, N., & Kanzaki, R. (2008). Active control of free flight manoeuvres in a hawkmoth, *Agrius convolvuli. Journal of Experimental Biology,* 211, 423–32.

Wang, J. J. & Tsai, J. H. (2001). Development, survival and reproduction of black citrus aphid, *Toxoptera aurantii* (Hemiptera : Aphididae), as a function of temperature. *Bulletin of Entomological Research,* 91, 477–87.

Wang, X., Green, D. S., Roberts, S. P., & De Belle, J. S. (2007). Thermal disruption of mushroom body development and odor learning in *Drosophila melanogaster. PLOS One,* 2, e1125.

Wang, X. & Proud, C. G. (2006). The mTOR pathway in the control of protein synthesis. *Physiology & Behavior,* 21, 362–369.

Wang, Z. R., Singhvi, A., Kong, P., & Scott, K. (2004). Taste representations in the *Drosophila* brain. *Cell,* 117, 981–91.

Waniek, P. J. (2009). The digestive system of human lice: current advances and potential applications. *Physiological Entomology,* 34, 203–10.

Warbrick-Smith, J., Raubenheimer, D., Simpson, S. J., & Behmer, S. (2009). Three hundred and fifty generations of extreme food specialisation: testing predictions of nutritional ecology. *Entomologia Experimentalis et Aplicata,* 132, 65–75.

Ward, P. D. (2006). *Out of Thin Air: Dinosaurs, Birds and Earth's Ancient Atmosphere,* Washington, D.C., Joseph Henry Press.

Warner, K. A. & Bierzychudek, P. (2009). Does marking with fluorescent powders affect the survival or development of larval *Vanessa cardui*? *Entomologia Experimentalis Et Applicata,* 131, 320–24.

Warren, M., Mcgeoch, M. A., Nicolson, S. W., & Chown, S. L. (2006). Body size patterns in *Drosophila* inhabiting a mesocosm: interactive effects of spatial variation in temperature and abundance. *Oecologia,* 149, 245–255.

Wartlick, O., Mumcu, P., Kicheva, A., Bittig, T., Seum, C., Julicher, F., & Gonzalez-Gaitan, M. (2011). Dynamics of Dpp signaling and proliferation control. *Science,* 331, 1154–1159.

Waser, N. M. & Ollerton, J. (2006). *Plant-Pollinator Interactions: From Specialization to Generalization,* Chicago, University of Chicago Press.

Wasserthal, L. T. (1975). The role of butterfly wings in regulation of body temperature. *Journal of Insect Physiology,* 21, 1921–30.

Wasserthal, L. T. (1980). Oscillating haemolymph circulation in the butterfly *Papilio machaon* L. Revealed by contact thermography and photocell measurements. *Journal of Comparative Physiology,* 139, 145–63.

Wasserthal, L. T. (1996). Interaction of circulation and tracheal ventilation in holometabolous insects. *Advances in Insect Physiology,* 26, 297–351.

Wasserthal, L. T. (1999). Functional morphology of the heart and of a new cephalic pulsatile organ in the blowfly *Calliphora vicina* (Diptera: Calliphoridae) and their roles in hemolymph transport and tracheal ventilation. *International Journal of Insect Morphology & Embryology,* 28, 111–129.

Wasserthal, L. T. (2001). Flight-motor-driven respiratory air flow in the hawkmoth *Manduca sexta. Journal of Experimental Biology,* 204, 2209–20.

Wasserthal, L. T. (2007). Drosophila flies combine periodic heartbeat reversal with a circulation in the anterior body mediated by a newly discovered anterior pair of ostial valves and "venous" channels. *Journal of Experimental Biology,* 210, 3707–19.

Watanabe, H. & Tokuda, G. (2001). Animal cellulases. *Cellular and Molecular Life Sciences,* 58, 1167–78.

Watanabe, M., Kikawada, T., Minagawa, N., Yukuhiro, F., & Okuda, T. (2002). Mechanism allowing an insect to survive complete dehydration and extreme temperatures. *Journal of Experimental Biology,* 205, 2799–802.

Waters, J. S., Holbrook, C. T., Fewell, J. H., & Harrison, J. F. (2010). Allometric scaling of metabolism, growth, and activity in whole colonies of the seed-harvester ant *Pogonomyrmex californicus. The American Naturalist,* 176, 501–10.

Weaver, M. & Krasnow, M. A. (2008). Dual origin of tissue-specific progenitor cells in *Drosophila* tracheal remodeling. *Science,* 321, 1496–99.

Weeks, A. R., Turelli, M., Harcombe, W. R., Reynolds, K. T., & Hoffmann, A. A. (2007). From parasite to mutualist: Rapid evolution of *Wolbachia* in natural populations of *Drosophila. PLoS Biol,* 5, e114.

Wehner, R., Fukushi, T., & Isler, K. (2007). On being small: Brain allometry in ants. *Brain Behavior and Evolution,* 69, 220–28.

Wei, Z., Baggerman, G., J. Nachman, R., Goldsworthy, G., Verhaert, P., De Loof, A., & Schoofs, L. (2000). Sulfakinins reduce food intake in the desert locust, *Schistocerca gregaria. Journal of Insect Physiology,* 46, 1259–65.

Weibel, E. R., Taylor, C. R., & Bolis, L. (1998). *Principles of Animal Design. The Optimization and Symmorphosis Debate,* Cambridge, Cambridge University Press.

Weibel, E. R., Taylor, C. R., & Hoppeler, H. (1991). The concept of symmorphosis: a testable hypothesis of structure-function relationship. *Proceedings of the National Academy of Science, U.S.A.,* 88, 10357–61.

Weis-Fogh, T. (1960). A rubber-like protein in insect cuticle. *Journal of Experimental Biology,* 37, 899–907.

Weis-Fogh, T. (1964a). Diffusion in insect wing muscle, the most active tissue known. *Journal of Experimental Biology,* 41, 229–56.

Weis-Fogh, T. (1964b). Functional design of the tracheal system of flying insects as compared with the avian lung. *Journal of Experimental Biology,* 41, 207–27.

Weis-Fogh, T. (1967). Respiration and tracheal ventilation in locusts and other flying insects. *Journal of Experimental Biology,* 47, 561–87.

Weisser, W. W. & Siemann, E. (2004). The various effects of insects on ecosystem functioning. *In:* Weisser, W. W. & Siemann, E. (eds.) *Insects and Ecosystem Function.* Berlin: Springer-Verlag.

Welte, M. A., Duncan, I., & Lindquist, S. (1995). The basis for a heat-induced developmental defect – defining crucial lesions. *Genes & Development,* 9, 2240–50.

Welte, M. A., Tetrault, J. M., Dellavalle, R. P., & Lindquist, S. L. (1993). A new method for manipulating transgenes: engineering heat tolerance in a complex, multicellular organism. *Current Biology,* 3, 842–53.

Wen, Z., Rupasinghe, S., Niu, G., Berenbaum, M. R., & Schuler, M. A. (2006). CYP6B1 and CYP6B3 of the black swallowtail (*Papilio polyxenes*): adaptive evolution through subfunctionalization. *Molecular Biology and Evolution,* 23, 2434–43.

West-Eberhard, M. J. (2003). *Developmental Plasticity and Evolution,* Oxford, Oxford University Press.

Westneat, M. W., Betz, O., Blob, R. W., Fezzaa, K., Cooper, W. J., & Lee, W. K. (2003). Tracheal respiration in insects visualized with synchrotron X-ray imaging. *Science,* 299, 558–60.

Westneat, M. W., Socha, J. J., & Lee, W. K. (2008). Advances in biological structure, function and physiology using synchotron x-ray imaging. *Annual Review of Physiology,* 70, 119–42.

Wharton, D. A., Pow, B., Kristensen, M., Ramlov, H., & Marshall, C. J. (2009). ICE-active proteins and cryoprotectants from the New Zealand alpine cockroach, *Celatoblatta quinquemaculata. Journal of Insect Physiology,* 55, 27–31.

Wharton, D. R. A., Wharton, M. L., & Lola, J. (1965). Blood volume and water content of the male American cockroach, *Periplaneta americana* L.—Methods and the influence of age and starvation. *Journal of Insect Physiology,* 11, 391–404.

Wheat, C. W., Vogel, H., Wittstock, U., Braby, M. F., Underwood, D., & Mitchell-Olds, T. (2007). The genetic basis of a plant–insect coevolutionary key innovation. *Proceedings of the National Academy of Sciences, USA,* 104, 20427–31.

Wheeler, D. E., Tuchinskaya, I., Buck, N. A., & Tabashnik, B. E. (2000). Hexameric storage proteins during metamorphosis and egg production in the diamondback moth: *Plutella xylostella* (Lepidoptera). *Journal of Insect Physiology,* 46, 951–58.

White, C. R., Blackburn, T. M., Terblanche, J. S., Marais, E., Gibernau, M., & Chown, S. L. (2007). Evolutionary responses of discontinuous gas exchange in insects. *Proceedings of the National Academy of Sciences, USA,* 104, 8357–61.

White, T. C. R. (1993). *The Inadequate Environment: Nitrogen and the Abundance of Animals,* Berlin, Springer-Verlag.

Whitman, D. W. (1986). Developmental thermal requirements for the grasshopper *Taeniopoda eques* (Orthoptera: Acrididae). *Annals of the Entomological Society of America,* 79, 711–14.

Whitman, D. W. (1987). Thermoregulation and daily activity patterns in a black grasshopper, *Taeniopoda eques. Animal Behaviour,* 35, 1814–26.

Whitman, D. W. (1988). Function and evolution of thermoregulation in the desert grasshopper *Taeniopoda eques. Journal of Animal Ecology,* 57, 369–83.

Wieczorek, H., Beyenbach, K. W., Huss, M., & Vitavska, O. (2009a). Vacuolar-type proton pumps in insect epithelia. *Journal of Experimental Biology,* 212, 1611–19.

Wieczorek, H., Beyenbach, K. W., Huss, M., & Vitavska, O. (2009b). Vacuolar-type proton pumps in insect epithelia. *Journal of Experimental Biology,* 212, 1611–19.

Wieczorek, H., Putzenlechner, M., Zeiske, W., & Klein, U. (1991). A vacuolar-type proton pump energizes K^+/H^+-antiport in an animal plasma membrane. *Journal of Biological Chemistry,* 266, 15340–47.

Wieczorek, H., Weerth, S., Schindlbeck, M., & Klein, U. (1989). A vacuolar-type proton pump in a vesicle fraction enriched with potassium transporting plasma membranes from tobacco hornworm midgut. *Journal of Biological Chemistry,* 264, 11143–48.

Wiegert, R. G. & Evans, F. C. (1967). Investigations of secondary productivity in grasslands. *In:* Petrusewicz, K. (ed.) *Secondary Productivity in Terrestrial Ecosystems.* Warsaw: Institute of Ecology, Polish Academy of Sciences.

Wiehart, U. I. M., Nicolson, S. W., Eigenheer, R. A., & Schooley, D. A. (2002). Antagonistic control of fluid secretion by the Malpighian tubules of *Tenebrio molitor*: effects of diuretic and antidiuretic peptides and their second messengers. *Journal of Experimental Biology,* 205, 493–501.

Wigglesworth, V. B. (1930). A theory of tracheal respiration in insects. *Royal Society of London Proceedings,* 106, 229–50.

Wigglesworth, V. B. (1931). The extent of air in the tracheoles of some terrestrial insects. *Proceedings of the Royal Society of London B.,* 109, 354–59.

Wigglesworth, V. B. (1933). The function of the anal gills of the mosquito larva. *Journal of Experimental Biology,* 10, 16–26.

Wigglesworth, V. B. (1935). The regulation of respiration in the flea, *Xenopsylla cheopsis*, Roths. (Pulicidae). *Proceedings of the Royal Society, London B,* 118, 397–419.

Wigglesworth, V. B. (1938). The regulation of osmotic pressure and chloride concentration in the haemolymph of mosquito larvae. *Journal of Experimental Biology,* 15, 235–47.

Wigglesworth, V. B. (1945). Transpiration through the cuticle of insects. *Journal of Experimental Biology,* 21, 97–114.

Wigglesworth, V. B. (1954). Growth and regeneration in the tracheal system of an insect, *Rhodnius prolixus* (Hemiptera). *Quarterly Journal of Microscopical Sciences,* 95, 115–37.

Wigglesworth, V. B. (1957). The physiology of insect cuticle. *Annual Review of Entomology,* 2, 37–54.

Wigglesworth, V. B. (1981). The natural history of insect tracheoles. *Physiological Entomology,* 6, 121–28.

Wigglesworth, V. B. (1983). The physiology of insect tracheoles. *Advances in Insect Physiology,* 17, 85–149.

Wikelski, M., Kays, R. W., Kasdin, N. J., Thorup, K., Smith, J. A., & Swenson, G. W. (2007). Going wild: what a global small-animal tracking system could do for experimental biologists. *Journal of Experimental Biology,* 210, 181–86.

Wikelski, M., Moxley, J., Eaton-Mordas, A., Lopez-Uribe, M. M., Holland, R., Moskowitz, D., Roubik, D. W., & Kays, R. (2010). Large-range movements of neotropical orchid bees observed via radio telemetry. *PLOS One,* 5.

Wilder, S. M. & Eubanks, M. D. (2010). Might nitrogen limitation promote omnivory among carnivorous arthropods? Comment. *Ecology,* 91, 3114–17.

Williams, A. E. & Bradley, T. J. (1998). The effect of respiratory pattern on water loss in desiccation-resistant *Drophila melanogaster. Experimential Biology,* 201, 2953–59.

Williams, C. M., Pelini, S. L., Hellmann, J. J., & Sinclair, B. J. (2010). Intra-individual variation allows an explicit test of the hygric hypothesis for discontinuous gas exchange in insects. *Biology Letters,* 6, 274–77.

Williams, J. B., Roberts, S. P., & Elekonich, M. M. (2008). Age and natural metabolically-intensive behavior affect oxidative stress and antioxidant mechanisms. *Experimental Gerontology,* 43, 538–49.

Williams, K. D., Busto, M., Suster, M. L., So, A. K. C., Ben-Shahar, Y., Leevers, S. J., & Sokolowski, M. B. (2006). Natural variation in *Drosophila melanogaster* diapause due to the insulin-regulated PI3-kinase. *Proceedings of the National Academy of Sciences, USA,* 103, 15911–15.

Willmer, P. & Stone, G. (1997). Temperature and water relations in desert bees. *Journal of Thermal Biology,* 22, 453–65.

Willmer, P. G. (1982). Microclimate and the environmental physiology of insects. *Advances in Insect Physiology,* 16, 1–57.

Willmer, P. G. (1988). The role of insect water balance in pollination ecology: *Xylocopa* and *Calotropis. Oecologia,* 76, 430–38.

Wilson, R. S. & Franklin, C. E. (2002a). The detrimental acclimation hypothesis. *Trends in Ecology & Evolution,* 17, 407–08.

Wilson, R. S. & Franklin, C. E. (2002b). Testing the beneficial acclimation hypothesis. *Trends in Ecology & Evolution,* 17, 66–70.

Wingrove, J. A. & O'farrell, P. H. (1999). Nitric oxide contributes to behavioral, cellular, and developmental responses to low oxygen in *Drosophila. Cell,* 98, 105–114.

Winkler, I. S., Mitter, C., & Scheffer, S. J. (2009). Repeated climate-linked host shifts have promoted diversification in a temperate clade of leaf-mining flies. *Proceedings of the National Academy of Sciences,* 106, 18103–08.

Winston, M. L. (1987). *The Biology of the Honey Bee,* Cambridge, Harvard University Press.

Winston, M. L. (1992). The biology and management of Africanized honey bees. *Annual Review of Entomology,* 37, 395–406.

Winston, P. W. & Bates, D. H. (1960). Saturated salt solutions for the control of humidity in biological research. *Ecology,* 41, 232–37.

Wittstock, U., Agerbirk, N., Stauber, E. J., Olsen, C. E., Hippler, M., Mitchell-Olds, T., Gershenzon, J., & Vogel, H. (2004). Successful herbivore attack due to metabolic diversion of a plant chemical defense. *Proceedings of the National Academy of Sciences, USA,* 101, 4859–64.

Witz, P., Amlaiky, N., Plassat, J. L., Maroteaux, L., Borrelli, E., & Hen, R. (1990). Cloning and characterization of a *Drosophila* serotonin receptor that activates adenylate cyclase. *Proceedings of the National Academy of Sciences, USA,* 87, 8940–44.

Wolf, T. J., Ellington, C. P., & Begley, I. S. (1999). Foraging castes in bumblebees: field conditions cause large individual differences. *Insectes Sociaux,* 46, 291–95.

Wolf, T. J., Ellington, C. P., Davis, S., & Feltham, M. J. (1996). Validation of the doubly labelled water technique for bumblebees *Bombus terrestris* (L.). *Journal of Experimental Biology,* 199, 959–72.

Wolfersberger, M. G. (2000). Amino acid transport in insects. *Annual Review of Entomology,* 45, 111.

Wolschin, F. & Gadau, J. (2009). Deciphering proteomic signatures of early diapause in Nasonia. *PLOS One,* 4.

Woodman, J. D., Cooper, P. D., & Haritos, V. S. (2008). Neural regulation of discontinuous gas exchange in *Periplaneta Americana. Journal of Insect Physiology,* 54, 472–80.

Woodring, J., Hoffmann, K. H., & Lorenz, M. W. (2007a). Activity, release and flow of digestive enzymes in the cricket, *Gryllus bimaculatus. Physiological Entomology,* 32, 56–63.

Woodring, J., Hoffmann, K. H., & Lorenz, M. W. (2007b). Feeding, nutrient flow, and digestive enzyme release in the giant milkweed bug, *Oncopeltus fasciatus*. *Physiological Entomology*, 32, 328–35.

Woods, A. & Chamberlin, M. E. (1999). Effects of dietary protein concentration on L-proline transport by *Manduca sexta* midgut. *Journal of Insect Physiology*, 45, 735–41.

Woods, H. A. (1999). Patterns and mechanisms of growth of fifth-instar *Manduca sexta* caterpillars following exposure to low- or high-protein food during early instars. *Physiological and Biochemical Zoology*, 72, 445–54.

Woods, H. A. (2010). Water loss and gas exchange by eggs of *Manduca sexta*: Trading off costs and benefits. *Journal of Insect Physiology*, 56, 480–87.

Woods, H. A. & Bernays, E. A. (2000). Water homeostasis by wild larvae of *Manduca sexta*. *Physiological Entomology*, 25, 82–87.

Woods, H. A., Bonnecaze, R. T., & Zrubek, B. (2005a). Oxygen and water flux across eggshells of *Manduca sexta*. *Journal of Experimental Biology*, 208, 1297–1308.

Woods, H. A., Fagan, W. F., Elser, J. J., & Harrison, J. F. (2004). Allometric and phylogenetic variation in insect phosphorus content. *Functional Ecology*, 18, 103–09.

Woods, H. A. & Harrison, J. F. (2002). Interpreting rejections of the beneficial acclimation hypothesis: when is physiological plasticity adaptive? *Evolution*, 56, 1863–66.

Woods, H. A. & Hill, R. I. (2004). Temperature-dependent oxygen limitation in insect eggs. *Journal of Experimental Biology*, 207, 2267–76.

Woods, H. A. & Kingsolver, J. G. (1999). Feeding rate and the structure of protein digestion and absorption in lepidopteran midguts. *Archives of Insect Biochemistry and Physiology*, 42, 74–87.

Woods, H. A., Moran, A. L., Arango, C. P., Mullen, L., & Shields, C. (2009). Oxygen hypothesis of polar gigantism not supported by performance of Antarctic pycnogonids in hypoxia. *Proceedings of the Royal Society B: Biological Sciences*, 276, 1069–75.

Woods, H. A. & Smith, J. N. (2010). Universal model for water costs of gas exchange by animals and plants. *Proceedings of the National Academy of Science, USA*, 107, 8469–74.

Woods, W. A., Heinrich, B., & Stevenson, R. D. (2005). Honeybee flight metabolic rate: does it depend upon air temperature? *Journal of Experimental Biology*, 208, 1161–73.

Wright, J. C. & Westh, P. (2006). Water vapour absorption in the penicillate millipede *Polyxenus lagurus* (Diplopoda : Penicillata : Polyxenida): microcalorimetric analysis of uptake kinetics. *Journal of Experimental Biology*, 209, 2486–94.

Wu, Q. & Brown, M. R. (2006). Signaling and function of insulin-like peptides in insects. *Annual Review of Entomology*, 51, 1–24.

Xu, H. & Robertson, R. M. (1994). Effects of temperature on properties of flight neurons in the locust. *Journal of Comparative Physiology a-Sensory Neural and Behavioral Physiology*, 175, 193–202.

Yamanaka, N., Zitnan, D., Kim, Y. J., Adams, M. E., Hua, Y. J., Suzuki, Y., Suzuki, M., Suzuki, A., Satake, H., Mizoguchi, A., Asaoka, K., Tanaka, Y., & Kataoka, H. (2006). Regulation of insect steroid biosynthesis by innervating peptidergic neurons. *Proceedings of the National Academy of Sciences*, 103, 8622–27.

Yan, L. J. & Sohal, R. S. (2000). Prevention of flight activity prolongs the life span of the housefly, Musca domestica, and attenuates the age-associated oxidative damage to specific mitochondrial proteins. *Free Radical Biology and Medicine*, 29, 1143–50.

Yao, C. A., Ignell, R., & Carlson, J. R. (2005). Chemosensory coding by neurons in the coeloconic sensilla of the *Drosophila* antenna. *Journal of Neuroscience*, 25, 8359–67.

Yi, S.-X. & Lee, R. E., Jr. (2011). Rapid cold-hardening blocks cold-induced apoptosis by inhibiting the activation of pro-caspases in the flesh fly *Sarcophaga crassipalpis*. *Apoptosis*, 16, 249–55.

Yi, S. X., Moore, C. W., & Lee, R. E. (2007). Rapid cold-hardening protects *Drosophila melanogaster* from cold-induced apoptosis. *Apoptosis*, 12, 1183–93.

Yocum, G. D., Zdarek, J., Joplin, K. H., Lee, R. E., Smith, D. C., Manter, K. D., & Denlinger, D. L. (1994). Alteration of the eclosion rhythm and eclosion behavior in the flesh fly, *Sarcophaga crassipalpis*, by low and high-temperature stress. *Journal of Insect Physiology*, 40, 13–21.

Yokohari, F. (1978). Hygroreceptor mechanism in antenna of cockroach *Periplaneta*. *Journal of Comparative Physiology*, 124, 53–60.

Yokohari, F. & Tateda, H. (1976). Moist and dry hygroreceptors for relative humidity of cockroach, Periplaneta-Americana L. *Journal of Comparative Physiology*, 106, 137–52.

Zachariassen, K. E., Andersen, J., Maloiy, G. M. O., & Kamau, J. M. Z. (1987). Transpiratory water-loss and metabolism of beetles from arid areas in east-Africa. *Comparative Biochemistry and Physiology A-Physiology*, 86, 403–08.

Zangerl, A. R. & Berenbaum, M. (1998). Damage inducibility of primary and secondary metabolites in *Pastinaca sativa*. *Chemoecology*, 8, 187–93.

Zanotto, F., Gouveia, S., Simpson, S. J., & Calder, D. (1997). Nutritional homeostasis in locusts: is there a mechanism for increased energy expenditure during carbohydrate overfeeding? *Journal of Experimental Biology*, 200, 2437–2448.

Zanotto, F. P., Simpson, S. J., & Raubeheimer, D. (1993). The regulation of growth by locusts through post-ingestive compensation for variation in the levels of dietary protein and carbohydrate. *Physiological Entomology*, 18, 425–34.

Zars, T. (2001). Two thermosensors in Drosophila have different behavioral functions. *Journal of Comparative Physiology a-Sensory Neural and Behavioral Physiology*, 187, 235–42.

Zera, A. J. (2006). Evolutionary genetics of juvenile hormone and ecdysteroid regulation in *Gryllus*: A case study in the microevolution of endocrine regulation. *Comparative Biochemistry and Physiology a-Molecular & Integrative Physiology*, 144, 365–79.

Zera, A. J. & Denno, R. F. (1997). Physiology and ecology of dispersal polymorphism in insects. *Annual Review of Entomology*, 42, 207–31.

Zera, A. J. & Harshman, L. G. (2001). The physiology of life history tradeoffs in animals. *Annual Review of Ecology and Systematics*, 32, 95–126.

Zera, A. J. & Zhao, Z. W. (2003a). Life-history evolution and the microevolution of intermediary metabolism: Activities of lipid-metabolizing enzymes in life-history morphs of a wing-dimorphic cricket. *Evolution*, 57, 586–96.

Zera, A. J. & Zhao, Z. W. (2003b). Morph-dependent fatty acid oxidation in a wing-polymorphic cricket: implications for the trade-off between dispersal and reproduction. *Journal of Insect Physiology*, 49, 933–43.

Zera, A. J. & Zhao, Z. W. (2006). Intermediary metabolism and life-history trade-offs: Differential metabolism of amino acids underlies the dispersal-reproduction trade-off in a wing-polymorphic cricket. *American Naturalist*, 167, 889–900.

Zhai, L., Berg, M. C., Cebeci, F. C., Kim, Y., Milwid, J. M., Rubner, M. F., & Cohen, R. E. (2006). Patterned superhydrophobic surfaces: toward a synthetic mimic of the Namib Desert beetle. *Nano Letters*, 6, 1213–17.

Zhang, Z. T., Stout, M. J., Shang, H. W., & Pousson, R. C. (2006). Adaptations of larvae and pupae of the rice water weevil, *Lissorhoptrus oryzophilus* Kuschel (Coleoptera : Curculionidae), to living in flooded soils *J. Kansas Entomological Society*, 79, 176–83.

Zhao, H. W., Zhou, D., Nizet, V., & Haddad, G. G. (2010). Experimental selection for *Drosophila* survival in extremely high O_2 environments. *PLOS One,* 5, e11701.

Zhao, Z. & Zera, A. J. (2006). Biochemical basis of specialization for dispersal vs. reproduction in a wing-polymorphic cricket: Morph-specific metabolism of amino acids. *Journal of Insect Physiology,* 52, 646–58.

Zheng, A., Edelman, S. W., Tharmarajah, G., Walker, D. W., Pletcher, S. D., & Seroude, L. (2005). Differential patterns of apoptosis in response to aging in Drosophila. *Proceedings of the National Academy of Sciences of the United States of America, USA,* 102, 12083–88.

Ziegler, C. (1994). Titin-related proteins in invertebrate muscles. *Comparative Biochemistry and Physiology a-Physiology,* 109, 823–33.

Zitnan, D., Kim, Y. J., Zitnanová, I., Roller, L., & Adams, M. E. (2007). Complex steroid-peptide-receptor cascade controls insect ecdysis. *General and Comparative Endocrinology,* 153, 88–96.

Zornik, E., Paisley, K., & Nichols, R. (1999). Neural transmitters and a peptide modulate Drosophila heart rate. *Peptides,* 20, 45–51.

Zrubek, B. & Woods, H. A. (2006). Insect eggs exert rapid control over an oxygen-water tradeoff. *Proceedings of the Royal Society B-Biological Sciences,* 273, 831–34.

Index